… *International*
REVIEW OF
Neurobiology

Volume 46

Neurosteroids
AND
Brain Function

International REVIEW OF Neurobiology

Volume 46

SERIES EDITORS

RONALD J. BRADLEY
Department of Psychiatry, School of Medicine
Louisiana State University Medical Center
Shreveport, Louisiana, USA

R. ADRON HARRIS
Waggoner Center for Alcohol and Drug Addiction Research
The University of Texas at Austin
Austin, Texas, USA

PETER JENNER
Division of Pharmacology and Therapeutics
GKT School of Biomedical Sciences
King's College, London, UK

EDITORIAL BOARD

PHILIPPE ASCHER
ROSS J. BALDESSARINI
TAMAS BARTFAI
COLIN BLAKEMORE
FLOYD E. BLOOM
DAVID A. BROWN
MATTHEW J. DURING
KJELL FUXE
PAUL GREENGARD
SUSAN D. IVERSEN

KINYA KURILYAMA
BRUCE S. MCWEWN
HERBERT Y. MELTZLER
NOBORU MIZZUNO
SALAVADOR MONCADO
TREVOR W. ROBBINS
SOLOMON H. SNYDER
STEPHEN G. WAXMAN
CHIEN-PING WU
RICHARD J. WYATT

Neurosteroids
AND
Brain Function

EDITED BY

GIOVANNI BIGGIO
Department of Experimental Biology
University of Cagliari
Cagliari, Italy

ROBERT H. PURDY
Department of Psychiatry
Veterans Affairs Medical Center Research Service
San Diego, California, USA
Department of Neuropharmacology
The Scripps Research Institute
La Jolla, California, USA

ACADEMIC PRESS
A Harcourt Science and Technology Company

San Diego   San Francisco   New York   Boston   London   Sydney   Tokyo

This book is printed on acid-free paper. ∞

Copyright © 2001 by ACADEMIC PRESS

All Rights Reserved.
No part of this publication may be reproduced or transmitted in any form or by any means, electronic or mechanical, including photocopy, recording, or any information storage and retrieval system, without permission in writing from the Publisher.

The appearance of the code at the bottom of the first page of a chapter in this book indicates the Publisher's consent that copies of the chapter may be made for personal or internal use of specific clients. This consent is given on the condition, however, that the copier pay the stated per copy fee through the Copyright Clearance Center, Inc. (222 Rosewood Drive, Danvers, Massachusetts 01923), for copying beyond that permitted by Sections 107 or 108 of the U.S. Copyright Law. This consent does not extend to other kinds of copying, such as copying for general distribution, for advertising or promotional purposes, for creating new collective works, or for resale. Copy fees for pre-2001 chapters are as shown on the title pages. If no fee code appears on the title page, the copy fee is the same as for current chapters.
0074-7742/01 $35.00

Explicit permission from Academic Press is not required to reproduce a maximum of two figures or tables from an Academic Press chapter in another scientific or research publication provided that the material has not been credited to another source and that full credit to the Academic Press chapter is given.

Academic Press
A Harcourt Science and Technology Company
525 B Street, Suite 1900, San Diego, California 92101-4495, USA
http://www.academicpress.com

Academic Press
Harcourt Place, 32 Jamestown Road, London NW1 7BY, UK
http://www.academicpress.com

International Standard Book Number: 0-12-366846-8

PRINTED IN THE UNITED STATES OF AMERICA
01 02 03 04 05 06 SB 9 8 7 6 5 4 3 2 1

CONTENTS

CONTRIBUTORS... xiii
PREFACE... xvii

Neurosteroids: Beginning of the Story
ETIENNE E. BAULIEU, P. ROBEL, AND M. SCHUMACHER

I.	Introduction..	2
II.	Biosynthesis and Metabolism of Neurosteroids: Related Behavioral Effects	7
III.	Receptors and Related Activities of Neurosteroids	13
IV.	Conclusions	25
	References..	26

Biosynthesis of Neurosteroids and Regulation of Their Synthesis
SYNTHIA H. MELLON AND HUBERT VAUDRY

I.	What Is a Neurosteroid?.....................................	34
II.	Enzymes Involved in Neurosteroidogenesis: Biochemistry and Molecular Biology...	35
III.	Enzymes Involved in Neurosteroidogenesis: Distribution in the Brain and Developmental Regulation.................................	49
IV.	Regulation of Neurosteroidogenic Enzymes....................	61
V.	Conclusions	65
	References..	66

Neurosteroid 7-Hydroxylation Products in the Brain
ROBERT MORFIN AND LUBOSLAV STÁRKA

I.	Neurosteroid Metabolism in the Brain........................	79
II.	7α-Hydroxylation Studies in the Brain	81
III.	7β-Hydroxylation Studies in the Brain	83
IV.	7α-Hydroxy-DHEA and 7β-Hydroxy-DHEA as Native Anti-glucocorticoids	84
V.	The Brain and Other Organs	88

VI.	Sex Steroid Metabolism in the Brain	90
VII.	Conclusions	92
	References	92

Neurosteroid Analysis
Ahmed A. Alomary, Robert L. Fitzgerald, and Robert H. Purdy

I.	Introduction	98
II.	Analysis of Neurosteroids by Radioimmunoassay	99
III.	Analysis of Neurosteroids by High-Performance Liquid Chromatography Coupled with Gas Chromatography–Mass Spectrometry	99
IV.	Analysis of Unconjugated Neurosteroids by Gas Chromatography–Mass Spectrometry	100
V.	Derivatization of Neurosteroids	103
VI.	Gas Chromatography Columns for Separation of Neurosteroid Derivatives	106
VII.	Isotopic Dilution for Quantitative Gas Chromatography–Mass Spectrometry	107
VIII.	Data Acquisition Modes in Mass Spectrometry	107
IX.	Mass Spectrometry/Mass Spectrometry	108
X.	Analysis of Neurosteroid Sulfates by Mass Spectrometry	108
XI.	Analysis of Sulfated Steroids by Soft Ionization Mass Spectrometry	109
XII.	Analysis of Steroid Sulfates by Atmospheric Pressure Chemical Ionization/Mass Spectrometry	111
XIII.	Thermospray Liquid Chromatography/Mass Spectrometry	111
XIV.	Future Analysis of Neurosteroid Sulfates	111
XV.	Conclusions	112
	References	112

Role of the Peripheral-Type Benzodiazepine Receptor in Adrenal and Brain Steroidogenesis
Rachel C. Brown and Vassilios Papadopoulos

I.	Introduction	118
II.	Peripheral-Type Benzodiazepine Receptor	119
III.	Structure of the PBR Complex	120
IV.	Role of the PBR in Steroidogenesis	124
V.	The PBR in Adrenal Steroid Biosynthesis	128
VI.	Role of the PBR in Brain Neurosteroid Biosynthesis	131
VII.	Role of the PBR in Pathology	134
VIII.	Other Proteins Involved in the Acute Regulation of Steroidogenesis	135
IX.	Conclusions	137
	References	137

Formation and Effects of Neuroactive Steroids in the Central and Peripheral Nervous System

ROBERTO COSIMO MELCANGI, VALERIO MAGNAGHI, MARIARITA GALBIATI, AND LUCIANO MARTINI

I.	Introduction..	146
II.	5α-Reductase and 3α Hydroxysteroid Dehydrogenase System..............	146
III.	Effect of Glial–Neuronal Interactions on the Formation of Neuroactive Steroids...	153
IV.	Effects of Neuroactive Steroids on Glial Cells of the Central Nervous System	155
V.	Formation of Neuroactive Steroids in the Peripheral Nervous System......	161
VI.	Effects of Neuroactive Steroids on the Peripheral Nervous System	162
VII.	Conclusions	170
	References...	170

Neurosteroid Modulation of Recombinant and Synaptic GABA$_A$ Receptors

JEREMY J. LAMBERT, SARAH C. HARNEY, DELIA BELELLI, AND JOHN A. PETERS

I.	Introduction..	178
II.	Transmitter-Gated Ion Channels and Neurosteroid Selectivity..............	180
III.	Influence of GABA$_A$-Receptor Subunit Composition on Neurosteroid Action	185
IV.	Mechanism of Neurosteroid Modulation of GABA$_A$ Receptors	188
V.	Neurosteroid Modulation of Inhibitory Synaptic Transmission	190
VI.	Structure–Activity Relationships for Steroids at the GABA$_A$ Receptor......	194
VII.	Multiple Steroid Binding Sites on the GABA$_A$ Receptor.....................	196
VIII.	Conclusions	197
	References...	199

GABA$_A$-Receptor Plasticity during Long-Term Exposure to and Withdrawal from Progesterone

GIOVANNI BIGGIO, PAOLO FOLLESA, ENRICO SANNA, ROBERT H. PURDY, AND ALESSANDRA CONCAS

I.	Introduction..	208
II.	Effects of Long-Term Exposure to PROG and GABA$_A$-Receptor Gene Expression and Function *in Vitro*.......................................	209
III.	Effects of PROG Withdrawal on GABA$_A$-Receptor Gene Expression and Function *in Vitro*	216

IV.	Effects of Long-Term Exposure to and Subsequent Withdrawal of PROG in Pseudo-pregnancy	218
V.	Effects of Long-Term Exposure to and Subsequent Withdrawal of PROG in Pregnancy	220
VI.	Oral Contraceptives and $GABA_A$-Receptor Plasticity	231
VII.	Mechanism of the Effect of Long-Term PROG Exposure on $GABA_A$-Receptor Plasticity	234
VIII.	Conclusions	235
	References	235

Stress and Neuroactive Steroids

MARIA LUISA BARBACCIA, MARIANGELA SERRA,
ROBERT H. PURDY, AND GIOVANNI BIGGIO

I.	Stress and $GABA_A$ Receptors	244
II.	Effect of Stress on Brain Concentrations of Neuroactive Steroids	246
III.	Possible Mechanisms Underlying the Stress-Induced Changes in Brain Neurosteroid Concentrations	261
IV.	Conclusions	263
	References	264

Neurosteroids in Learning and Memory Processes

MONIQUE VALLÉE, WILLY MAYO, GEORGE F. KOOB,
AND MICHEL LE MOAL

I.	Introduction	274
II.	Learning and Memory Processes and Animal Models	275
III.	Pharmacological Effects of Neurosteroids	280
IV.	Mechanisms of Action	292
V.	Physiological Significance	301
VI.	Conclusions and Future Perspectives	309
	References	312

Neurosteroids and Behavior

SHARON R. ENGEL AND KATHLEEN A. GRANT

I.	Introduction	321
II.	Anxiety and Stress	322

III.	Cognition	328
IV.	Aggression	333
V.	Sleep, Feeding, and Reinforcement	335
VI.	Discriminative Stimulus Effects	337
	References	342

Ethanol and Neurosteroid Interactions in the Brain

A. Leslie Morrow, Margaret J. VanDoren,
Rebekah Fleming, and Shannon Penland

I.	Introduction	349
II.	Role of 3α,5α-TH PROG in Ethanol Action	351
III.	Role of Neurosteroids in Alcohol Reinforcement	359
IV.	Role of Neurosteroids in Ethanol Tolerance	361
V.	Role of Neurosteroids in Ethanol Dependence	364
VI.	Conclusions and Future Directions	369
	References	369

Preclinical Development of Neurosteroids as Neuroprotective Agents for the Treatment of Neurodegenerative Diseases

Paul A. Lapchak and Dalia M. Araujo

I.	Neurosteroids and the Brain	380
II.	Synthesis of Central Nervous System Neurosteroids	380
III.	Receptor Signaling Pathways	381
IV.	Neurosteroids and Central Nervous System Plasticity	384
V.	Neurosteroids and Neuroprotection	384
VI.	Conclusions	391
	References	391

Clinical Implications of Circulating Neurosteroids

Andrea R. Genazzani, Patrizia Monteleone, Massimo Stomati,
Francesca Bernardi, Luigi Cobellis, Elena Casarosa, Michele Luisi,
Stefano Luisi, and Felice Petraglia

I.	Introduction	399
II.	Changes and Possible Role of Neurosteroids in Humans	402
	References	414

Neuroactive Steroids and Central Nervous System Disorders

MINGDE WANG, TORBJÖRN BÄCKSTRÖM, INGER SUNDSTRÖM,
GÖRAN WAHLSTRÖM, TOMMY OLSSON, DI ZHU,
INGA-MAJ JOHANSSON, INGER BJÖRN, AND MARIE BIXO

I.	Neuroactive Steroids and the Central Nervous System	422
II.	Concentrations of Neuroactive Steroids in the Brain	425
III.	Sensory–Motor and Cognitive Function	428
IV.	Estrogen and Alzheimer's Disease	429
V.	Neuroactive Steroids and Menstrual-Cycle-Linked Mood Changes	430
VI.	Side Effects of Oral Contraceptives	435
VII.	Neuroactive Steroids and Menopause	436
VIII.	Side Effects of Hormone Replacement Therapy	439
IX.	Neuroactive Steroids and Epilepsy	441
	References	448

Neuroactive Steroids in Neuropsychopharmacology

RAINER RUPPRECHT AND FLORIAN HOLSBOER

I.	Introduction	462
II.	Sources and Biosynthesis of Neuroactive Steroids	463
III.	Steroid Modulation of $GABA_A$ Receptors	465
IV.	Steroid Modulation of Other Neurotransmitter Receptors	466
V.	A Putative Specific Steroid-Binding Site on Ligand-Gated Ion Channels	468
VI.	Genomic Effects of Neuroactive Steroids	469
VII.	Neuropsychopharmacological Properties of Neuroactive Steroids	470
VIII.	Modulation of Endogenous Neuroactive Steroids as a Pharmacological Principle	471
IX.	Outlook	472
	References	474

Current Perspectives on the Role of Neurosteroids in PMS and Depression

LISA D. GRIFFIN, SUSAN C. CONRAD, AND SYNTHIA H. MELLON

I.	Introduction	479
II.	Premenstrual Syndrome or Premenstrual (Late Luteal) Dysphoric Disorder	480
III.	Allopregnanolone and Premenstrual Dysphoric Disorder	481
IV.	Biosynthesis of Allopregnanolone	483
V.	SSRIs in the Treatment of Premenstrual Dysphoric Disorder: Modulation of Neurosteroid Levels	485

VI.	SSRIs, Neurosteroids, and Depression	487
VII.	Conclusions	489
	References	489

| INDEX | 493 |
| CONTENTS OF RECENT VOLUMES | 507 |

CONTRIBUTORS

Numbers in parentheses indicate the pages on which the authors' contributions begin.

Ahmed A. Alomary (97), Department of Neuropharmacology, The Scripps Research Institute, La Jolla, California 92037; and Department of Veterans Affairs Medical Center and Veterans Medical Research Foundation, San Diego, California 92093

Dalia M. Araujo (379), VASDHS, San Diego, California 92161

Torbjörn Bäckström (421), Department of Obstetrics and Gynecology, Department of Medicine, and Department of Pharmacology, University of Umeå, S-901 87, Umeå, Sweden

Maria Luisa Barbaccia (243), Department of Neuroscience, University of Rome "Tor Vergata," 00133 Rome, Italy

Etienne E. Baulieu (1), INSERM Units Communications Hormonales (U33) and Stéroïdes et Systéme Nerveux (U 488), and Collège de France, 94276 Le Kremlin–Bicêtre Cedex, Paris, France

Delia Belelli (177), Department of Pharmacology and Neuroscience, Ninewells Hospital and Medical School, Dundee University, Dundee DD1 9SY, Scotland

Francesca Bernardi (399), Department of Reproductive Medicine and Child Development, Division of Gynecology and Obstetrics, University of Pisa, 56100 Pisa, Italy

Giovanni Biggio (207, 243), Department of Experimental Biology "Bernardo Loddo," University of Cagliari, 09100 Cagliari, Italy

Marie Bixo (421), Department of Obstetrics and Gynecology, Department of Medicine, and Department of Pharmacology, University of Umeå, S-901 87 Umeå, Sweden

Inger Björn (421), Department of Obstetrics and Gynecology, Department of Medicine, and Department of Pharmacology, University of Umeå, S-901 87 Umeå, Sweden

Rachel C. Brown (117), Division of Hormone Research, Department of Cell Biology, and the Interdisciplinary Program in Neuroscience, Georgetown University Medical Center, Washington, DC 20007

Elena Casarosa (399), Department of Reproductive Medicine and Child Development, Division of Gynecology and Obstetrics, University of Pisa, 56100 Pisa, Italy

Luigi Cobellis (399), Obstetrics and Gynecology, University of Siena, Siena, Italy

Alessandra Concas (207), Department of Experimental Biology "Bernardo Loddo," University of Cagliari, 09100 Cagliari, Italy

Susan C. Conrad (479), Department of Pediatrics, University of California–San Francisco, California 94143

Sharon R. Engel (321), Department of Physiology and Pharmacology, Wake Forest University School of Medicine, Winston-Salem, North Carolina 27157-1083

Robert L. Fitzgerald (97), Department of Veterans Affairs Medical Center and Veterans Medical Research Foundation; and Department of Pathology University of California–San Diego, La Jolla, California 92093

Rebekah Fleming (349), Bowles Center for Alcohol Studies and Departments of Psychiatry and Pharmacology, University of North Carolina at Chapel Hill, Chapel Hill, North Carolina 27599-7178

Paolo Follesa (207), Department of Experimental Biology "Bernardo Loddo," University of Cagliari, 09100 Cagliari, Italy

Mariarita Galbiati (145), Department of Endocrinology, University of Milan, 20133 Milan, Italy

Andrea R. Genazzani (399), Department of Reproductive Medicine and Child Development, Division of Gynecology and Obstetrics, University of Pisa, 56100 Pisa, Italy

Kathleen A. Grant (321), Department of Physiology and Pharmacology, Wake Forest University School of Medicine, Winston-Salem, North Carolina 27157-1083

Lisa D. Griffin (479), Department of Neurology, University of California–San Francisco, California 94143

Sarah C. Harney (177), Department of Pharmacology and Neuroscience, Ninewells Hospital and Medical School, Dundee University, Dundee DD1 9SY, Scotland

Florian Holsboer (461), Max Planck Institute of Psychiatry, 80804 Munich, Germany

Inga-Maj Johansson (421), Department of Obstetrics and Gynecology, and Department of Medicine, and Department of Pharmacology, University of Umeå, S-90187 Umeå, Sweden

George F. Koob (273), Department of Neuropharmacology, The Scripps Research Institute, La Jolla, California 92037

Jeremy J. Lambert (177), Department of Pharmacology and Neuroscience, Ninewells Hospital and Medical School, Dundee University, Dundee DD1 9SY, Scotland

Paul A. Lapchak (379), Department of Neuroscience, University of California–San Diego, La Jolla, California 92093-0624; VASDHS, San

Diego, California 92161; and Veterans Medical Research Foundation, San Diego, California 92161

Michel Le Moal (273), INSERM U.259, Institut François Magendie, Domaine de Carreire, 33077 Bordeaux, France

Michele Luisi (399), Department of Reproductive Medicine and Child Development, Division of Gynecology and Obstetrics University of Pisa, 56100 Pisa, Italy

Stefano Luisi (399), Department of Reproductive Medicine and Child Development, Division of Gynecology and Obstetrics University of Pisa, 56100 Pisa, Italy

Valerio Magnaghi (145), Department of Endocrinology, University of Milan, 20133 Milan, Italy

Luciano Martini (145), Department of Endocrinology, University of Milan, 20133 Milan, Italy

Willy Mayo (273), INSERM U.259, Institut François Magendie, Domaine de Carreire, 33077 Bordeaux, France

Roberto Cosimo Melcangi (145), Department of Endocrinology, University of Milan, 20133 Milan, Italy

Synthia H. Mellon (33, 479), Department of Obstetrics, Gynecology, and Reproductive Sciences, The Center for Reproductive Sciences, and The Metabolic Research Unit, University of California–San Francisco, San Francisco, California 94143-0556

Patrizia Monteleone (399), Department of Reproductive Medicine and Child Development, Division of Gynecology and Obstetrics, University of Pisa, 56100 Pisa, Italy

Robert Morfin (79), Laboratoire de Biotechnologie, Conservatoire National des Arts et Métiers, 75003 Paris, France

A. Leslie Morrow (349), Bowles Center for Alcohol Studies and Departments of Psychiatry and Pharmacology, University of North Carolina at Chapel Hill, Chapel Hill, North Carolina 27599-7178

Tommy Olsson (42), Department of Obstetrics and Gynecology, Department of Medicine, and Department of Pharmacology, University of Umeå, S-901 87 Umeå, Sweden

Vassilios Papadopoulos (117), Division of Hormone Research, Departments of Cell Biology, Pharmacology, and Neuroscience, and the Interdisciplinary Program in Neuroscience, Georgetown University Medical Center, Washington, DC 20007

Shannon Penland (349), Bowles Center for Alcohol Studies and Departments of Psychiatry and Pharmacology, University of North Carolina at Chapel Hill, Chapel Hill, North Carolina 27599-7178

John A. Peters (177), Department of Pharmacology and Neuroscience, Ninewells Hospital and Medical School, Dundee University, Dundee DD1 9SY, Scotland

Felice Petraglia (399), Obstetrics and Gynecology, University of Siena, Siena, Italy

Robert H. Purdy (97, 207, 243), Department of Psychiatry, Veterans Affairs Medical Center Research Service, San Diego, California; and Department of Neuropharmacology, The Scripps Research Institute, La Jolla, California 92037

R. Robel (1), INSERM Units Communications Hormonales (U33) and Stéroïdes et Systême Nerveux (U 488) and Collège de France, 94276 Le Kremlin–Bicetre Cedex, Paris, France

Rainer Rupprecht (46), Department of Psychiatry, Ludwig-Maximilians-University of Munich, 80336 Munich, Germany

Enrico Sanna (207), Department of Experimental Biology "Bernardo Loddo," University of Cagliari, 09100 Cagliari, Italy

M. Schumacher (1), INSERM Units Communications Hormonales (U33) and Stéroïdes et Systême Nerveux (U 488), and Collège de France, 94276 Le Kremlin–Bicetre Cedex, Paris, France

Mariangela Serra (243), Department of Experimental Biology "Bernardo Loddo," University of Cagliari, 09100 Cagliari, Italy

Luboslav Stárka (79), Institute of Endocrinology, 116 94 Prague 1, Czech Republic

Massimo Stomati (399), Department of Reproductive Medicine and Child Development, Division of Gynecology and Obstetrics, University of Pisa, 56100 Pisa, Italy

Inger Sundström (421), Department of Obstetrics and Gynecology, Department of Medicine, and Department of Pharmacology, University of Umeå, S-901 87, Umeå, Sweden

Monique Vallée (273), INSERM U.259, Institut François Magendie, Domaine de Carreire, 33077 Bordeaux, France

Margaret J. VanDoren (349), Bowles Center for Alcohol Studies and Departments of Psychiatry and Pharmacology, University of North Carolina at Chapel Hill, Chapel Hill, North Carolina 27599-7178

Hubert Vaudry (33), European Institute for Peptide Research (IFRMP23), Laboratory of Cellular and Molecular Neuroendocrinology, INSERM U-413, UA CNRS, University of Rouen, 76821 Mont-Saint-Aignan, France

Göran Wahlström (421), Department of Obstetrics and Gynecology, Department of Medicine, and Department of Pharmacology, University of Umeå, S-901 87, Umeå, Sweden

Mingde Wang (421), Department of Obstetrics and Gynecology, Department of Medicine, and Department of Pharmacology, University of Umeå, S-901 87 Umeå, Sweden

Di Zhu (421), Department of Obstetrics and Gynecology, Department of Medicine, and Department of Pharmacology, University of Umeå, S-901 87 Umeå, Sweden

PREFACE

The number of publications focused on neurosteroids has increased exponentially since a multidisciplinary group at the National Institute of Mental Health in the United States published a study on γ-aminobutyric acid-related neurosteroids in 1986. Among the multiple reasons for this increasing interest in neurosteroids are their roles in stress, alcohol dependence and withdrawal, and other physiological and pharmacological actions, including potentiation or inhibition of neurotransmitter action.

This book begins with a chapter by two of the authors of the original work in the early 1980s that describes the basis for their seminal discovery that dehydroepiandrosterone and its sulfate are neurosteroids. The book continues with chapters about neurosteroid formation and metabolism, followed by a chapter on current methods for the analysis of neurosteroids in biological tissues. Subsequent chapters review important preclinical studies of neurosteroid actions as modulators of plasticity, neurochemical function, and behavior. The book concludes with four chapters on current clinical investigations of neurosteroids.

Since the introduction of glucocorticoids into clinical practice more than half a century ago, recognition of the importance of other classes of steroid hormones—estrogens, progestins, androgens, and mineralocorticoids—has been intimately linked with their therapeutic applications. We believe that universal recognition of the physiological and neuropharmacological importance of neurosteroids will follow a different course. Most neurosteroids investigated to date do not have significant hormonal activity (an important exception is 5α-dihydroprogesterone). Instead, evidence is suggesting that several categories of important neuropsychopharmacological agents that modulate neurotransmitter systems may involve alterations of the biosynthesis and/or metabolism of neurosteroids in the nervous system, both central and peripheral.

It has been our privilege and pleasure to have had the participation of numerous outstanding investigators in the preparation of this volume. Their enthusiastic cooperation has been most warmly appreciated. Finally, we are indebted to Noelle Gracy for her guidance and sustaining assistance in the publication of this work.

Giovanni Biggio
Robert H. Purdy

Names and Abbreviation of Steroids

Chemical name	Trivial name	Abbreviation
3α-hydroxy-5α-prenan-20-one	allopregnanolone	3α,5α-TH PROG
3β-hydroxy-5α-prenan-20-one	epiallopregnanolone	3β,5α-TH PROG
3α-hydroxy-5β-prenan-20-one	pregnalone	3α,5β-TH PROG
3β-hydroxy-5β-prenan-20-one	epipregnanolone	3β,5β-TH PROG
3α, 21-dihydroxy-5α-pregnan-20-one	tetrahydrodeoxycorticosterone	3α,5α-TH DOC
3β-hydroxyandrost-5-en-17-one	dehydroepiandrosterone	DHEA
3β-hydroxyandrost-5-en-17-one sulfate	dehydroepiandrosterone sulfate	DHEAS
Sulfate ester	sulfate	S
3β-hydroxypregn-5-en-20-one	pregnenolone	PREG
3β-hydroxypregn-5-en-20-one sulfate	pregnenolone sulfate	PREGS
pregn-4-en-3, 20-one	progesterone	PROG
1,3,5(10)-estratriene-3, 17β-diol	estradiol	17β-E
1,3,5(10)-estratriene-3, 17α-diol	17α-estradiol	17α-E
17β-hydroxyandrost-4-en-3-one	testosterone	T
11β,21-dihydroxypregn-4-ene-20-one	corticosterone	B
21-hydroxypregn-4-ene-3,20-dione	deoxycorticosterone	DOC
11β,21-dihydroxypregn-4-ene-3, 18,20-trione	aldosterone	
5α-androstane-3β, 17β-diol	3β-androstanediol	3β,5α-diol
5α-androstane-3α, 17β-diol	3α-androstanediol	
5α-pregnane-3,20-dione	5α-dihydroprogesterone	5α-DH PROG
5β-pregnane-3,20-dione	5β-dihydroprogesterone	5β-DH PROG
3α-hydroxy-5α-androstan-17-one	androsterone	
3α-hydroxypregn-4-en-20-one		3α-DH PROG
3α-hydroxy-5α-preg-9(11)-en-20-one		
3α-hydroxy-3β-trifluoromethyl-5α-pregnan-20-one		
5α-pregnane-3α,20α-diol		
5β-pregnane-3α,20α-diol		
5α-pregnane-3α,20β-diol		
5β-pregnane-3α,20β-diol		
11β,17,21-trihydroxypregn-4-ene-3,20-dione	cortisol	F
17,20α,21-trihydroxypregn-4-ene-3,11-dione	20α-dihydrocortisone	
3α-hydroxy-5α-pregnane-11,20-dione	alphaxalone	
3β-hydroxy-5α-pregnane-11,20-dione	betaxolone	
3α-hydroxy-3β-methyl-5α-pregnan-20-one	ganaxolone	
2β-ethoxy-3α-hydroxy-11α-dimethylamino-5α-pregnan-20-one	minaxolone	

ABBREVIATIONS OF PRINCIPAL ENZYMES INVOLVED IN STEROID METABOLISM

Current name	Abbreviation used	Abbreviation not used
Cholesterol side-chain cleavage enzyme	P450scc	P450 11A
17α-hydroxylase/17,20-lyase	P450c17	
21-hydroxylase	P450c21	
11β-hydroxylase	P450c11β	P45011B1
Aldosterone synthase	P450c11AS	P45011B2
Aromatase	P450 aro	P45019
7α-hydroxylase	P450c7A and P4507B	
3β-hydroxysteroid dehydrogenase	3βHSD	
3α-hydroxysteroid oxidoreductase	3αHSD	3αHOR
17β-hydroxysteroid oxidoreductase	17βHSD	17βHOR
11β-hydroxysteroid oxidoreductase	11βHSD	11-oxidoreductase, 11βHOR
sulfotransferase	HST	ST
sulfatase	STS	

NEUROSTEROIDS: BEGINNING OF THE STORY

E. E. Baulieu,* P. Robel, and M. Schumacher

INSERM Units Communications Hormonales (U33) and Stéroïdes et Systéme Nerveux (U 488) and *Collège de France, Paris, France

I. Introduction
II. Biosynthesis and Metabolism of Neurosteroids: Related Behavioral Effects
III. Receptors and Related Activities of Neurosteroids
   A. Intracellular Receptors and Trophic Action (Myelination)
   B. Membrane Receptors and Behavioral Effects
   C. Neurosteroids Bind to Microtubule-Associated Protein 2 and Stimulate Microtubule Assembly: Latest Discovery and Beginning of a New Story
IV. Conclusions
    References

Neurosteroids are synthetized in the central and the peripheral nervous system, in glial cells, and also in neurons, from cholesterol or steroidal precursors imported from peripheral sources. They include 3β-hydroxy-Δ^5-compounds, such as pregnenolone (PREG) and dehydroepiandrosterone, their sulfate esters, and compounds known as reduced metabolites of steroid hormones, such as the tetrahydroderivative of progesterone 3α-hydroxy-5α-pregnan-20-one. These neurosteroids can act as modulators of neurotransmitter receptors, such as $GABA_A$, NMDA, and sigma 1 receptors. Progesterone itself is also a neurosteroid, and a progesterone receptor has been detected in peripheral and central glial cells. At different sites in the brain, neurosteroid concentrations vary according to environmental and behavioral circumstances, such as stress, sex recognition, or aggressiveness. A physiological function of neurosteroids in the central nervous system is strongly suggested by the role of hippocampal PREGS with respect to memory performance, observed in aging rats. In the peripheral nervous system, a role for PROG synthesized in Schwann cells has been demonstrated in remyelination after cryolesion of the sciatic nerve *in vivo* and in cultures of dorsal root ganglia. A new mechanism of PREG action discovered in the brain involves specific steroid binding to microtubule associated protein and increased tubulin polymerization for assembling microtubules. It may be important to study the effects of abnormal neurosteroid concentration/metabolism in view of the possible treatment of functional and trophic disturbances of the nervous system. © 2001 Academic Press.

I. Introduction

The term *neurosteroids* designates steroids that are synthesized in the nervous system (Baulieu, 1981). The story of their discovery and function, gracefully requested by the editors of this volume, is not long; however, this "young" research domain is already the matter of several hundreds of publications. The beginnings were rather slow: It is difficult for "steroidologists" to become suddenly acquainted with the cellular and functional complexity of the nervous system, as investigated by molecular, electrophysiological, imaging, and behavioral techniques; and neurobiologists are seldom aware of the many steroidal metabolites, most of which were considered only as intermediates of steroid hormone detoxification and thus as compounds of little biological significance. Two of us (E.E.B. and P.R.) have worked for many years on the synthesis, metabolism, and mode of action of hormonal steroids. The unique structure of the cyclopentanophenanthrene cycle (Bloch, 1965) has inspired and stimulated the research efforts of thousands of scientists, and it will continue to do so for a long time, currently in conjunction with the progress of molecular biology and genetics, new recording and imaging techniques, behavioral and cognitive neurosciences, and reproductive biology and medicine. Gregory Pincus, known as the initiator of hormonal contraception, certainly did not predict that his behavioral studies on pregnenolone (PREG) administration to human beings in the late 1940s (Pincus and Hoagland, 1944) were a sort of prescience of a new field, neurosteroids, largely involving 3β-hydroxy-Δ^5-steroids, such as dehydroepiandrosterone (DHEA) (Corpéchot *et al.*, 1981) and its precursor PREG (Corpéchot *et al.*, 1983).

One of us (E.E.B.) was particularly interested in the description of the intracellular progesterone receptor (PROG-R) (Milgrom *et al.*, 1970) and undertook to answer a query for the "logical" detection of a PROG-R in *Xenopus laevis* oocytes by S. Schorderet-Slatkine. She was working in Geneva on PROG-induced meiotic maturation, but she could not find a PROG-R similar to the intracellular protein that we had described in the uterus. Indeed, a plasma membrane receptor was demonstrated, clearly different from the already classical PROG-R, which modulated adenylate cyclase activity in a steroidal ligand-dependent manner (Baulieu *et al.*, 1978; Finidori-Lepicard *et al.*, 1981). If there is a membrane receptor mechanism for a steroid in one given type of animal cells, it was reasoned, then it should also be found elsewhere, as nature often uses the same molecular functions, even in different physiological contexts. The nervous system was, at least at first glance, ideally suited to a mechanism able to give a rapid response to a hormone, as a membrane receptor "should," compared with a genomic mechanism.

Thus was started a search for brain steroids in the rat, with no preconceived ideas because it was known that free steroids can cross the blood–brain barrier (BBB) easily. Therefore, we would likely find circulating hormones imported into the brain, but there was no reason to expect a local steroid synthesis.

Then, with Colette Corpéchot—so efficient in the discovery of dehydroepiandrosterone sulfate (DHEAS) secretion 20 years earlier (Baulieu *et al.*, 1965)—there was a big surprise: DHEAS was detected in the rat brain (Corpéchot *et al.*, 1981). Because we knew, after many experiments, that there was a minute amount of this steroid in the blood of rodents, we were reluctant to admit the authentic presence of DHEAS and DHEA in the brain, despite all of the confirmary biochemical and immunological techniques used. Fortunately, J. Sjövall, in Stockholm, confirmed our results, qualitatively and quantitatively, using the coupled gas chromatography–mass spectrometry technique to measure DHEAS in our extracts. It then followed naturally that we should systematically look for neurosteroids (Baulieu, 1981)—that is, steroids formed in the nervous system and, for some, presumably active via receptor systems either identical or similar to intracellular receptors of the peripheral steroids or those found at the membrane level. Interestingly, at approximately the same time, S. Lieberman (1981, personal communication) was interested in discovering steroid-metabolizing enzymes in the brain, with—independently, of course—the same goal: demonstrating steroid synthesis in nervous structures.

The next work to be done involved describing the synthesis and metabolic pathways of neurosteroids and establishing their physiological and pathological functions and mechanism(s) of action. Many difficulties awaited us, including, for many years, C. Corpéchot, I. Jung-Testas, B. Eychenne, M. El-Etr, C. Le Goascogne, and, more recently, Y. Akwa and K. Rajkowski, as well as students and scientists from other French and foreign laboratories who contributed much.

There were and still are some major difficulties:

1. We met many analytical problems, qualitative and quantitative, because of the low concentration and the lipoidal nature of neurosteroids, which must be separated from the abundant lipidic constituents of neural tissues. Strictly controlled conditions had to be established because neurosteroid concentrations vary according to the time of the day, the lighting schedule, the food, the presence of other animals, the habituation to handling, etc.
2. The overall dynamics of the synthesis of neurosteroids is unknown because their turnover cannot be determined and no appropriate techniques for describing their compartmentation are available.

3. Quantitative aspects were initially studied using radioimmunoassays. However, for better specificity and more sensitivity, we have recently shifted to mass spectrometric analysis (Lière et al., 2000) coupled with appropriate extraction and chromatography techniques.
4. Another quantitative problem has been particularly difficult to master, as many neurosteroids are also secreted by peripheral glands, may cross the BBB, and may easily attain peripheral nerves, eventually mixing with neurosteroids. In fact, the respective distribution and contribution of steroids imported into the nervous system and of those synthesized *in situ* remain difficult to assess. Consequently, we do not necessarily know what the respective targets are. Binding studies and thus the search for receptors are especially difficult in a lipid-rich milieu, as often encountered in work with nervous tissue and its membranes, again because of the liposolubility of steroids.

We essentially followed two strategic lines. The first was establishing the synthetic and metabolic pathways of steroids in the nervous system (Fig. 1), including the characterization of the corresponding enzymes and receptors. The second was determining changes and, if possible, functions of neurosteroids under various physiological or pathological conditions. This was frequently done in collaboration with specialized laboratories expert in the neurosciences (e.g., electrophysiology, behavior). Independent of our work, also in the 1980s, a pharmacological approach to the activities of steroid metabolites in the brain was successfully undertaken in many laboratories, initiated by Simmonds and colleagues (Harrison and Simmonds, 1984), who observed the interaction of alphaxolone, a synthetic anesthetic steroid, with the $GABA_A$ receptor. This field has become very active, rejuvenating the pioneering discoveries of H. Selye (1941), who had shown the neuroactivities of a number of steroidal compounds. The descriptions of "neuroactive" compounds reviewed in Widdows and Chadwick, 1990; Paul and Purdy, 1992; Costa and Paul, 1991; Lambert *et al.*, 1996; Baulieu *et al.*, 1999) remarkably complemented the neurosteroid concept.

In initial experiments, we measured steroids remaining in the brain after removal of potential glandular sources of steroids (i.e., after adrenalectomy and gonadectomy) (Table I). In rats, we essentially noted the persistence of DHEA and its conjugates after $\sim$2 weeks and the only partial decrease of PREG and its conjugates, as if the ablation of endocrine glands had led to the suppression of only the imported steroid. (PREG is a circulating steroid in the rat and is also synthesized in the brain.) We observed the persistence of a low but sizable level of PROG in male rats after operation, while, in contrast,

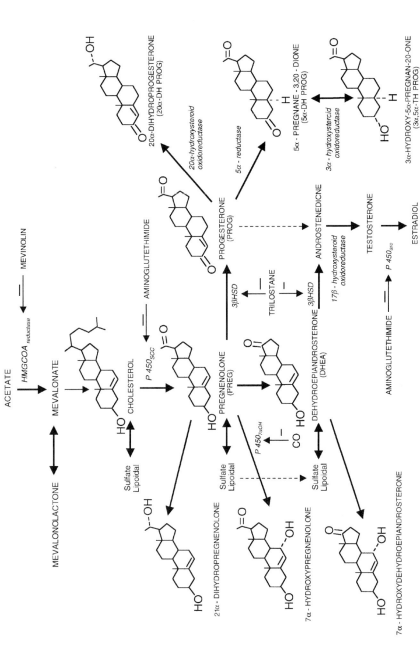

FIG. 1. Metabolism of steroids. Most biosynthetic and metabolic reactions cited in the text are indicated with the corresponding enzymes and some of their pharmacological blockers (marked —|).

TABLE I
NEUROSTEROIDS IN THE MALE RAT BRAIN

	PREG	PREGS	PREGL	DHEA	DHEAS	DHEAL	PROG
Brain ng/g							
Intact	8.9	14.2	9.4	0.24	1.70	0.45	2.2
	(±2.4)	(±2.5)	(±2.9)	(±0.33)	(±0.32)	(±0.13)	(±1.1)
Orx/adx	2.6	16.9	4.9	0.14	1.64	0.29	3.2
	(±0.8)	(±4.6)	(±1.3)	(±0.13)	(±0.43)	(±0.12)	(±1.6)
Plasma ng/ml							
Intact	1.2	2.1	2.4	0.06	0.20	0.18	1.9
	(±0.6)		(±0.9)	(±0.06)	(±0.08)	(±0.05)	(±0.7)
Orx/adx	0.3	nd	1.3	nm	nm	nm	0.1
	(±0.1)		(±0.3)				(±0.1)

Notes. (1) Mean ± sd (2) See text for steroid abbreviations. Orx, orchidectomized; Adx, adrenalectomized; nd, not determined; nm, not measurable.
From Corpéchot *et al.*, 1981, 1983, 1993, and unpublished results.

testosterone and corticosterone disappeared rapidly from their brain tissue (Corpéchot *et al.*, 1981, 1983, 1993).

Consistent with the hypothesis that the persistence of steroids after adrenalectomy–gonadectomy is not due to retention of compounds originally circulating in the blood, we observed the rapid release of radioactive DHEA and PREG taken up by the brain after their peripheral administration (Corpéchot *et al.*, 1983). A circadian rhythm of brain PREG and DHEA, out of phase with blood steroids, was also observed (Synguelakis *et al.*, 1985; Robel *et al.*, 1986), and brain PREG was found to be high for several days after birth in rats even when the adrenal steroid output was low (Robel and Baulieu, 1985). Interestingly, in adrenalectomized–orchiectomized males rats (and in sham-operated animals), 2 days after surgery, a temporary increase of DHEAS was found in the brain, possibly due to a local neural response to stress (Corpéchot *et al.*, 1981).

Essentially, the same results have been obtained with mice, although the basal values of DHEA, PREG, and their conjugates differ markedly according to strains. In intact guinea pigs and rabbits, also, these steroids were found in the brain at concentrations similar to those found in monkeys, where a limited study suggested that there is DHEA(S) of both adrenal and local origins in the brain (Robel *et al.*, 1987). Brain steroid concentrations measured in a few human cadavers were in the same $10^{-8 \pm 1}$ M range; the values, however, were widespread as a result of the heterogeneous sampling conditions (Lanthier and Patwardhan, 1986; Lacroix *et al.*, 1987). Globally, it may be observed that in primates, which have sizable concentrations of DHEA and DHEAS in the blood, brain DHEA(S) is more abundant than in rodents. Also noted is that DHEA, PREG, and their conjugates are found everywhere in the brain, even if there are some differences among certain regions (e.g., relatively more PREG in the olfactory bulb, more PREGS in the hippocampus, more DHEA(S) in the hypothalamus in rats). As a whole, the concentrations of several steroids, such as DHEA, PREG, their conjugates, and PROG and its 5α-metabolites, expressed in moles/volume equivalent of brain tissue weight, are relatively high in several instances and possibly even higher than appearances suggest because of compartmentation—the brain tissues being targets for paracrine/autocrine products.

II. Biosynthesis and Metabolism of Neurosteroids: Related Behavioral Effects

The 3β-hydroxy-Δ^5-steroids PREG and DHEA are, in steroidogenic glands, intermediary compounds between cholesterol and active 3-oxo-Δ^4-steroids, such as PROG and testosterone. Their synthesis in the nervous system could suggest a similar role, but it was found that these 3β-hydroxy-Δ^5-steroids have their own distinct activities.

Cholesterol itself can be synthesized locally within the brain and the peripheral nerves. For example, cholesterol, which is used for elaborating the fatty myelin sheaths, is derived from local synthesis, rather than being imported from the systemic circulation (Morrell and Jurewicks, 1996). It has also been shown that glial cells in culture synthesize cholesterol from low-molecular-weight precursors (Jung-Testas *et al.*, 1989; Hu *et al.*, 1989) and that cholesterol synthesis by myelinating glial cells is regulated by intracellular sterol levels, as it is in other tissues, such as the liver (Fu *et al.*, 1998). Proteins involved in the transport of cholesterol are also expressed by glial cells. Thus, receptors for low-density lipoproteins have been demonstrated on rat oligodendrocytes and astrocytes (Jung-Testas *et al.*, 1992), and functional peripheral benzodiazepine receptors (PBR) are present at the outer membrane of glial mitochondria (Papadopoulos, 1993). The PBR, a benzodiazepine-binding entity distinct from classical benzodiazepine sites on $GABA_A$ receptors (Yanagibachi *et al.*, 1988; Costa, 1991), is necessary for the transport of cholesterol to the inner mitochondrial membrane (Culty *et al.*, 1999), where the side-chain cleavage enzyme (SCC) is located (Oftebro *et al.*, 1979), transforming cholesterol (27 carbon atoms) into PREG (21 carbon atoms). The possibility of a link between pharmacological drug activity and neurosteroid synthesis is of striking interest (Lacor *et al.*, 1999).

After a number of unsuccessful experiments attempting to demonstrate enzymatic conversion of cholesterol to PREG in brain tissues and extracts, immunocytochemical evidence for the presence of the cytochrome P450scc, the specific hydroxylase involved in cholesterol side-chain cleavage, was

TABLE II
RAT OLIGODENDROCYTE MITOCHONDRIA: CONVERSION
OF [^{3}H]CHOLESTEROL TO [^{3}H]PREG AND [^{3}H]Δ5 PREGNENE-3β,
20α-DIOL

	[^{3}H]PREG	[^{3}H]Δ^{5}PREGNENE 3β, 20α-DIOL
	(pmol/mg prot/h)	
Oligodendrocytes	2.6 ± 0.5	1.9 ± 0.5
	(n = 5)	(n = 4)
Adrenals	14.8	
	(n = 2)	

obtained at the level of the white matter throughout the rat and the human brain (Le Goascogne et al., 1987, 1989; Iwahashi et al., 1990). The two associated enzymes, adrenodoxin and adrenodoxin reductase, were correlatively observed.

The presence and activity of P450scc in myelinating glial cells were verified in isolated oligodendrocytes and in their mitochondria (Table II) (Hu et al., 1987). These results have been confirmed by the detection of P450scc mRNA in oligodendrocytes (Mellon and Deschepper, 1993; Compagnone et al., 1995; Zwain and Yen, 1999). P450scc mRNA immunoreactivity and activity have also been detected in astrocytes (Mellon 1994; Zwain and Yen, 1999). In addition to glial cells, neurons express the P450scc and can synthesize PREG. The presence of cytochrome P450scc in selected neurons of the rat brain was first suggested by immunocytochemistry (Le Goascogne et al., 1987). P450scc-positive neurons have also been identified in the rat retina (Guarneri et al., 1994) and in dorsal root ganglia (DRG) of the mouse embryo (Compagnone et al., 1995). Rat cerebellar granule neurons in culture were found to express P450scc mRNA at levels comparable to those found in glial cells (Sanne and Krueger, 1995). The synthesis of PREGS in the hippocampus and its connection with N-methyl-D-aspartate (NMDA) receptors have been observed recently (T. Kimoto et al., 2000, personal communication). A recent study has shown that neurons isolated from the cerebral cortex of neonatal rat brains express P450scc mRNA and produce PREG (Zwain and Yen, 1999). Neurosteroid biosynthesis has been found in several vertebrate species (Mensah-Nyagan et al., 1999; Tsutsui et al., 1999).

Interestingly, the regulation of P450scc gene expression in the brain seems not to involve steroidogenic factor-1 (SF-1), a transcription factor required for high-level expression of steroidogenic enzymes in the gonads and adrenal glands. In contrast to P450scc mRNA, showing a widespread distribution in the brain, SF-1 mRNA is restricted to a few specific regions (Strömstedt and Waterman, 1995). Also, P450scc expression is not transcriptionally regulated by SF-1 in C6 glioma cells (Zhang *et al.*, 1995). The same study showed that regulation of the P450scc gene in glioma cells involves DNA sequences different from those playing a role in classic steroidogenic glands. The brain-specific transcription factors involved in the regulation of neurosteroidogenesis are still unknown.

The regulatory mechanisms governing P450scc function are unknown [even though some cAMP-induced increase in activity has been observed (Robel *et al.*, 1987), particularly in retina (Guarneri *et al.*, 1994)]. Although the protein hormones stimulating steroid synthesis in peripheral glands are probably not involved, the activities of a number of peptidic factors, such as IGF1, NGF, and BDNF, still must be investigated in this respect.

Paradoxically, the formation of DHEA and of DHEAS in the central nervous system (CNS) has not yet been clearly documented, although the isolation of these steroids in the brain was at the origin of the neurosteroid concept (Baulieu, 1981, Corpéchot *et al.*, 1981), and the recognized precursor in glandular cells, PREG, was rapidly identified at a higher concentration, as indeed a biosynthetic precursor should be (Corpéchot *et al.*, 1983). Currently, the possibility that DHEA (or conjugates) derives from cholesterol via unconventional pathways remains open (Prasad *et al.*, 1994), but alternatively technical problems (low activities, in particular) may be responsible for the lack of definitive evidence, which would be consistent with the recent demonstration of the specific 17α–hydroxylase–17, 20-desmolase enzyme (P450$_{17\alpha}$ mRNA) in the embryonic CNS (Compagnone *et al.*, 1995b) and the synthesis of DHEA from PREG in highly purified cultures of cortical astrocytes from neonatal rats (Zwain and Yen, 1999a).

Both PREG and DHEA are found in conjugated forms, sulfate esters and fatty acid esters ("lipoïdal"), the concentrations of which are often equal or superior to those of the corresponding free steroids (see Table I). The conversion in the brain of 3β-hydroxy-Δ^5-steroids to their sulfate esters is very likely and has not been documented, although preliminary evidence has been obtained for a low sulfotransferase activity (Rajkowski *et al.*, 1997); however, the possibility of formation of steroid sulfate-containing lipidic complexes ("sulfolipids") has not been excluded (Prasad *et al.*, 1994). Not all enzymes corresponding to the widely distributed steroid sulfatase activities of the brain have been cloned. Major conjugation forms of PREG and DHEA in the brain are their fatty acid esters (Jo *et al.*, 1989), designated "lipoidal" derivatives (Hochberg *et al.*, 1977). The acyltransferase responsible for their formation is enriched in the microsomal fraction of brain (Vourc'h *et al.*, 1992), and its activity is highest at the time of myelin formation. It is very likely that several isoforms exist because cholesterol- and corticosterone-esterifying activities, distinct from 3β-hydroxy-Δ^5-steroid acyltransferase, have been reported in the rat brain. The specific metabolism of these lipoidal 3β-hydroxy-Δ^5-steroids is unknown, as is the biological significance of the lipoidal derivatives.

Although PREG (and PROG) can be largely reduced to give 20α-hydroxy metabolites in glial cells and many neurons, no evidence for 17β-reduction of DHEA to give the weak estrogen Δ^5-androsten-3β-17β-diol has been documented. The 7α-hydroxylation of 3β-hydroxy-Δ^5-steroids can be performed by an enzyme, distinct from the classical cholesterol hydroxylase found in the liver (Akwa et al., 1992) and recently cloned (Rose et al., 1997). The 7α-hydroxy derivatives are of unknown biological significance.

As in steroidogenic gland cells and many peripheral tissues (Vande Wiele et al., 1963; Labrie, 1991), DHEA and PREG can be oxidized to 3-oxo-Δ^4 steroids (to Δ^4-androstenedione and PROG, respectively) in the nervous system by the 3β-hydroxysteroid dehydrogenase-isomerase enzyme (3βHSD) (Labrie et al., 1992), which can be inhibited by specific steroidal compounds, such as trilostane (Young et al., 1994). Several 3βHSD isoforms are present in most parts of the brain and in the peripheral nervous system (PNS), and they are found in glial cells and neurons (Guennoun et al., 1995, 1997; Sanne and Krueger, 1995). The metabolism of PREG and DHEA in astroglial cells is regulated by cell density: 3βHSD activity is strongly inhibited at high cell density (Akwa et al., 1993).

The formation and metabolism of PROG in the brain have been the subject of a number of studies since 5α-reduced PROG metabolites have attracted the attention of pharmacologists in the context of their effects on γ-aminobutyric acid (GABA)$_A$-R function. The two isoforms of the enzyme involved, 5α-reductase, have been cloned in several species. RNA encoding the type 2 isozyme is more abundant than type 1 mRNA in most male reproductive tissues. Type 1 isozyme predominates in other tissues, including the brain (Normington and Russell, 1992; Melcangi et al., 1994; Lauber and Lichtensteiger, 1996). The 5α-reduced (dihydrogenated) metabolite of PROG, 5α-DH PROG, is in turn converted to 3α- and 3β-hydroxy-5α-pregnan-20-ones ($3\alpha/\beta,5\alpha$-TH PROG). The corresponding 3α- and 3β-hydroxysteroid oxidoreductases in the brain have been detected.

Until now, there has been no demonstration of the synthesis of corticosteroids in sizable amounts in the nervous system, although reverse transcription–polymerase chain reaction (RT-PCR) has indicated the presence of cytochromes P450 involved in 11β-, 21-, and 18-hydroxylation (Ozaki et al., 1991; Iwahashi et al., 1993; Compagnone et al., 1995b; Gomez-Sanchez et al., 1996; Miyairi et al., 1988). For the synthesis of estrogens, the discovery of aromatase (P450aro)(Naftolin et al., 1975; MacLusky et al., 1994) in the brain may be viewed as the first evidence of a steroid metabolism of physiological significance in the nervous system, and therefore the formation of estradiol from testicular testosterone in hypothalamic structures may be considered

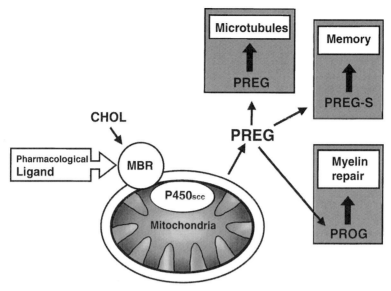

FIG. 2. Synthesis of neuropregnenolone and actions: simplified summary.

that of a neurosteroid in males of several species. However, there has been no systematic study of androst-4-ene-3,17-dione, which could be formed from neurosteroidal DHEA or PROG and which is also an aromatase substrate. Whether there may be formation of odorous Δ^{16}-androstene derivatives is unknown (Gower and Ruparelia, 1993).

In summary, even if many results are available, the global picture of neurosteroid metabolism is still incomplete and patchy, although some important results may be summarized schematically (Fig. 2). The formation of PREG in myelinating glial cells is well established qualitatively, but data are still not quantitatively satisfactory, considering the relatively high levels of this steroid in the CNS (Warner and Gustafsson, 1995). In addition, its regulatory aspects remain to be documented. The biosynthetic pathway of the neurosteroid DHEA is incompletely understood. Sulfates of 3β-hydroxy-Δ^5-steroids have been duly identified, but the related enzymology is obscure, whereas the metabolism and the significance of fatty acid esters are totally unknown. The 3βHSD and 5α-reductase enzymes are definitively active in the nervous system and in several cell types, and they are crucial for the formation of neurosteroids, such as PROG and $3\alpha,5\alpha$-TH PROG, in appropriate amounts to be neuroactive. A number of questions remain, including the possible transfers of steroids from one cell type to another with successive further metabolism during these passages because the relevant enzymes are differentially located in neurons and glial cells (Robel et al., 1987; Melcangi et al., 1993; Pelletier et al., 1994; Zwain and Yen, 1999b). Also important is the possibility that steroids entering the nervous system from the blood may not follow the same metabolic pathways as the same steroids synthesized in the nervous system. In that case, different effects on nervous functions may be discovered.

Physiologically, we can cite three results related to the metabolism of neurosteroids in the brain.

1. Early in our research, we observed an increase of brain DHEAS related to surgical (adrenalectomy and gonadectomy) stress conditions in rats (Corpéchot et al., 1981).
2. We also found that the exposure of male rats to females leads to a decrease of PREG in the rat olfactory bulb, an effect apparently due to a pheromonal stimulus, ovarian dependent in the females, and testosterone dependent in males (orchiectomy suppresses the response and testosterone reestablishes it) (Corpéchot et al., 1985).
3. A particular model of aggressiveness in castrated male mice, inhibited by testosterone or estrogen administration, has also been studied (Haug et al., 1989; Young et al., 1991). The stimulus is the introduction of a lactating female in a cage containing three resident orchiectomized males. The administration of DHEA (280 nmol for 15 days) decreases the males' aggressiveness. This does not seem to be due to a transformation of DHEA into brain testosterone (which has been measured), and, indeed, a DHEA derivative (3β-methylandrost-5-en-17-one), which is not converted to androgens or estrogens and is devoid of any hormonal activity, is at least as active as DHEA itself. Interestingly we found a progressive and significant decrease of PREGS in the brain, which may be related to the decrease of aggressiveness because PREGS is antagonist at $GABA_A$-R and stimulatory at NMDA-R. (Experiments have excluded that an increase of 3α-5α-TH PROG, activating the calming $GABA_A$-R, was responsible.) Other behavioral changes have been found to be correlated with changes of neurosteroid concentrations in the brain. In particular, in cases of increased production or administration of PROG or deoxycorticosterone, the elevation in the brain of the related 3α-hydroxy-5α-reduced tetrahydro-metabolites (3α-5α-TH PROG and $3\alpha,5\alpha$-tetrahydrodeoxycorticosterone), both agonists for the $GABA_A$-R, may therefore be responsible for corresponding behaviors in pregnancy or stress respectively (Purdy et al., 1991; Majewska, 1992).

III. Receptors and Related Activities of Neurosteroids

To the diversity of neurosteroids themselves should be added that of the receptor systems, which are of several kinds: nuclear, membrane, and, as we recently discovered, present at the cytoplasmic microtubules (Murakami *et al.*, 2000). Here, we report on the main findings relating receptors to activities of physiological and/or pharmacological importance.

A. Intracellular Receptors and Trophic Action (Myelination)

The distribution of intracellular steroid receptors in the brain has been described mainly on the basis of binding assays, autoradiography of tritiated steroids, and immunocytochemistry of receptor proteins (see McEwen, 1991; Schumacher *et al.*, 1999). Naturally, these techniques do not distinguish between receptors for peripheral steroids and those for neurosteroids, and, in any case, very little has been done in terms of cloning and sequencing to determine whether receptors in the nervous system are the same as in peripheral target tissues. Receptors for circulating neuroactive androgens, glucocorticosteroids, and mineralocorticosteroids have been described, as well as estrogen receptors, which can respond to both circulating estrogens and those synthesized in the hypothalamus. In addition, PROG receptors have been described that are estradiol inducible in hypothalamic neurons but not estradiol inducible in the cortex (Parsons *et al.*, 1982). It is interesting to note that neonatal hypothyroidism in rats is associated with a decrease of PROG-R and its mRNA in the cortex but not in the hypothalamus–preoptic area (results with estradiol receptor and its mRNA are not different) (Hirata *et al.*, 1994). We have biochemically and immunologically documented the presence of a PROG-R in myelinating glial cells (oligodendrocytes of rats of both sexes and mouse and rat Schwann cells) (Jung-Testas *et al.*, 1991, 1996).

Preliminary experiments on mixed glial cell cultures from rat brain (among which some astrocytes display weak PROG-R labeling by immunocytochemistry) indicated estrogen inducibility of the PROG-R in oligodendrocytes and effects of PROG on cell growth (inhibited) and morphological differentiation of both oligodendrocytes and astrocytes, with more cells expressing myelin basic protein and glial fibrillary acidic protein, respectively (Jung-Testas *et al.*, 1991). Whether the pharmacology of PROG analogs may differ when these analogs interact with the sexual (uterine, hypothalamic) or the glial (in CNS and PNS) PROG receptors is yet conjectural. In addition to the very presence of active PROG-R in glial cells, the capability of Schwann cells to synthesize PROG from PREG (which itself may derive from cholesterol) has been rigorously demonstrated (Koenig *et al.*, 1995), therefore forming a cellular system including a potential autocrine mechanism that could operate under controlling factors yet unknown.

Are there neurosteroids that act via a PROG "neuroreceptor"? We found PREG in the sciatic nerve of human cadavers at a mean concentration ≥ 100-fold the plasma level of the steroid (Morfin *et al.*, 1992), suggesting a possible biosynthesis, which we guessed to be in Schwann cells by analogy with what has previously been obtained in oligodendrocytes. The experimental simplicity of working with the PNS led to studying the regeneration of cryolesioned sciatic nerve in mice. We measured PREG and PROG in the sciatic nerve of normal animals and after adrenalectomy and gonadectomy. The levels of the two steroids were much higher in the nerve than in the plasma, whereas the corticosterone concentration was much lower in nerve than in plasma. After surgical endocrine ablation, PREG and PROG remained high in nerve, and corticosterone decreased in blood.

After cryolesion (Fig. 3), axons and their accompanying myelin sheaths degenerate quickly in the frozen zone and the distal segment (Wallerian

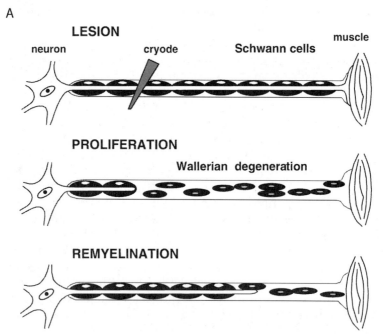

FIG. 3. Cryolesion of rat sciatic nerve. (**A**) Three steps. (**B**) Remyelination: control. (A) Effect of the blockade of 3β-HSD and thus of prosynthesis by trilostane. (B) An effect counteracted by PROG (C). (**C**) A similar result is obtained with RU486, an antagonist of PROG at PROG-R.

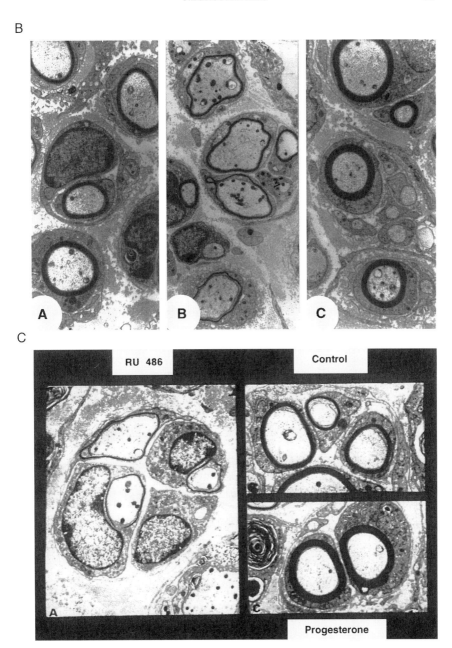

FIG. 3. (*Continued*)

degeneration). However, the intact basal lamina tubes provide an appropriate environment for regeneration. Schwann cells start to proliferate and myelinate the regenerating fibers after 1 week. Two weeks after surgery, myelin sheaths have reached approximately one-third of their final width. In the damaged portion of the nerve, PREG and PROG levels remain high 15 days after lesion.

The role of PROG in myelin repair, assessed after 2 weeks, was indicated by the decrease of thickness (number of lamellae) of myelin sheaths when trilostane, an inhibitor of the 3βHSD involved in the PREG-to-PROG transformation, was applied to the lesioned nerve or when, alternatively, RU 486 was locally delivered to competitively antagonize PROG action at the receptor level (as indicated previously, we had detected a PROG-R immunologically and by binding studies in Schwann cells). The inhibitory action of trilostane could not be attributed to toxicity because its effect was reversed by the simultaneous administration of PROG. We could even enhance remyelination with a high dose of either PREG or PROG. In this *in vivo* system, the myelin sheaths formed in response to neurosteroids, as determined by electron microscopy, were morphologically normal. However, the percentage of myelinated fibers in the regenerating nerves (75–85%) was not affected by the steroid environment, suggesting that PROG may not initiate myelination but rather stimulate ongoing myelination. In addition, axonal diameter 15 days after lesion was not affected by trilostane or RU 486 but was decreased by PROG. The data suggest that PROG promotes myelination independent of axonal growth. Recently, we have started to study the effect of PROG on myelin genes with the appropriate molecular biology techniques (Desarnaud *et al.*, 1998; Schumacher *et al.*, 1999).

The PROG effect on myelination was also observed in cultures of rat DRG (Fig. 4). After 4 weeks in culture, neurite elongation and Schwann cell proliferation had ceased and, in the appropriate medium, the presence of a physiological concentration of PROG (20 nM) for 2 weeks did not further increase the area occupied by the neurite network, the density of neurites, or the number of Schwann cells. It did, however, increase the number of myelin segments and the total length of myelinated axons about six fold. A relationship between neurosteroids and myelin formation has also been suggested by the low concentrations of PREG and PROG in the sciatic nerves of Trembler mice, severely hypomyelinated, in relation to a dominant mutation affecting the peripheral myelin protein PMP22 (Suter and Snipes, 1995). However, we certainly do not know whether the reduced concentrations of neurosteroids are a cause or a consequence of the defect in myelination. PBR expression is increased in astrocytes of mice with *Jimpy* and *Shiverer* syndromes, which may explain the elevated PROG level in these mutants (Le Goascogne *et al.*, 2000).

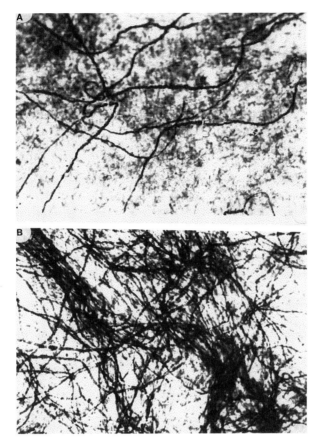

FIG. 4. PROG *in vitro* stimulates myelination of neurites (from mouse DRG in culture). (*Source:* Koenig *et al.,* 1995.)

Mechanistically, several hypotheses may be raised: PROG produced by Schwann cells may act on adjacent axons and activate the expression of neuronal signaling molecules required for myelination. Alternatively, PROG may function as an autocrine trophic factor and directly enhance the formation of new myelin sheaths. We are also analyzing the reciprocal influences of neurons and Schwann cells on each other in terms of Schwann cell proliferation and protein synthesis as well as metabolism of PREG and PROG in both cell types. The rather high concentrations of PROG in intact adult nerves suggest a role for this neurosteroid in the slow but continuous renewal of peripheral myelin.

It is quite interesting that PROG, a classical sex steroid, is also an active neurosteroid apparently synthesized and active independent of any sexual context. If similar results are obtained with CNS elements of human origin, then we may soon have to work on the possible therapeutic effect of PROG analogs for treating or preventing certain demyelinating pathological conditions.

B. MEMBRANE RECEPTORS AND BEHAVIORAL EFFECTS

As indicated, it was known that some steroid metabolites can produce a rapid depression of the CNS activity, and it was reported in 1984 (Harrison and Simmonds) that alphaxalone (3α-hydroxy-5α-pregnane-11,20-dione) specifically and potently enhances $GABA_A$-R-mediated hyperpolarization, while the 3β-hydroxy isomer is inactive.

A large number of biochemical and electrophysiological experiments have followed, demonstrating that natural (including neurosteroids) and synthetic steroids of very specific structure are in fact potent allosteric modulators of $GABA_A$-R function (Majewska et al., 1986) (Fig. 5). Now, in addition to the anesthetic and anticonvulsant activities of neurosteroids, hypnotic, anxiolytic, and analgesic effects of these compounds have been reported (Majewska, 1992; Gee et al., 1995; Bäckström, 1995; Lambert et al., 1996).

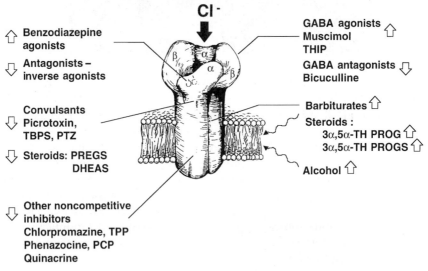

FIG. 5. A schematic diagram of the ligands stimulating (↑) or inhibiting (↓) $GABA_A$-R function. (review in MacDonald and Olsen, 1994).

Neurosteroids may also play an unexpected endocrine regulatory role via the GABA$_A$-R, for instance, on GnRH secretion (Brann *et al.*, 1995; El-Etr *et al.*, 1995; Genazzani *et al.*, 1996).

(3α,5α-TH PROG), its 5β-isomer—but not the corresponding 3β-hydroxysteroids—enhance GABA-evoked currents at concentrations as low as 1 nM. Interestingly, in the absence of GABA, the steroid provokes GABA-like activity at much higher concentrations. Studies using various patch–clamp configurations have suggested prolongation of the opening of the chloride channels of GABA$_A$-R in response to 3α,5α-TH PROG, and, more precisely, the increased probability that each channel will enter the naturally occurring open state, reminiscent of the effect of barbiturates, and increase the frequency of channel openings, similar to the effect of benzodiazepines. Further studies, including the use of recombinant GABA$_A$-R composed of different types of subunits after transfection into recipient cells, have indicated differences depending on the composition of the hetero-oligomeric receptor, suggesting distinct activities of neurosteroids in different parts of the brain (Puia *et al.*, 1990; Gee *et al.*, 1995; Lambert *et al.*, 1996). Neurosteroid action on the neurotransmitter receptors is probably not due to nonspecific disturbance of the membrane bilayer, but there is no absolute GABA$_A$-R subunit specificity yet demonstrated for neurosteroids as there is for benzodiazepine binding. The polymorphism of GABA$_A$-R in terms of subunit composition across the different cells of the nervous system, in neurons and glial cells, deserves more study to discover steroid derivatives with specific functions. Another type of complexity comes from data obtained when the interaction of PREGS with GABA$_A$-R is considered. At very low concentrations, in the nanomolar range, the steroid is a weak enhancer of GABA-evoked currents, but at a micromolar concentration it produces a noncompetitive voltage-independent inhibition (Majewska *et al.*, 1987) [due to reduced frequency of channel opening (Mienville and Vicini, 1989)]. These effects may be observed *in vivo* (Majewska *et al.*, 1989). DHEAS also is an inhibitor of the GABA$_A$-R (Majewska *et al.*, 1990; Spivak, 1994).

A global understanding of the potential effects of neurosteroids on neurotransmission should also take into account other modulatory activities displayed by the steroids. PREGS appears to allosterically potentiate the NMDA receptor (Wu *et al.*, 1991), and this effect may functionally reinforce the antagonistic effect of the same steroid on GABA$_A$-R and on the glycine receptor. PREGS also inhibits non-NMDA glutamate receptors. Other modulatory activities of neurosteroids have been described on glycine-activated chloride channels (Prince and Simmonds, 1992), on neural nicotinic acetylcholine receptors reconstituted in *Xenopus laevis* oocytes (Valera *et al.*, 1992), and on voltage-activated calcium channels (French-Müllen *et al.*, 1994). Sigma receptors, as pharmacologically defined by their effect on NMDA-R activity, have been studied in rat hippocampal preparations: Here, DHEAS acts as a sigma receptor agonist—differently from PREGS, which appears as a sigma inverse agonist and PROG, which behaves as a sigma antagonist (Monnet *et al.*, 1995). It is clear that all of these neurosteroid–neurotransmitter receptor interactions are themselves modulated by other hormones (i.e., during the estrous cycle), ligand kinetics (problems of desensitization), etc.

Are there other specific membrane receptors, the primary ligands of which are neurosteroids, in contrast with the preceeding examples indicating a modulatory function on neurotransmitter receptors and ion channels? Rapid effects of steroids may be explained by such novel receptors (i.e., Smith *et al.*, 1987; Schumacher *et al.*, 1990, 1993). A G-protein-coupled corticosteroid receptor has been identified in synaptic membranes from an amphibian brain (Orchinik *et al.*, 1991). Studies have indicated specific binding of PROG conjugated with a macromolecule (albumin in general) to different cellular or membrane preparations, but none has been able to determine its biological relevance (Tischkau and Ramirez, 1993). We also found selective binding of steroid sulfates to purified synaptosomal membranes of yet unknown biological significance (Robel and Baulieu, 1994).

Cognitive performance partly depends on neurosteroid activity at the membrane neurotransmitter receptor level. Our first experiments (Mayo *et al.*, 1993) dealt with the effect of PREGS and 3α-5α-TH PROG in the nucleus basalis magnocellularis in the rat, the steroids being pro- and antimnesic, respectively. Then, the study of cognitive performance in deficient aging rats was particularly rewarding (Vallée *et al.*, 1997) (Fig. 6). We found that PREGS is significantly lower in the hippocampus of aged (24 month old) than in young ($\sim$3-month-old) male animals. Interestingly the individual concentration of PREGS in the hippocampus of aged animals was widely distributed, between $\sim$2 and $\sim$28 ng/g of tissue. The animals had previously been classified according to their performance on two tasks for spatial memory, the Morris water maze and the two-trial test in a Y maze (Mayo *et al.*, 1993): Low levels of PREGS in hippocampus were correlated with poor performance in both tasks. When aged rats that had been classified as memory impaired received a single dose (by intraperitoneal or intrahippocampal injection) of PREGS, their performance was significantly improved, albeit transiently, while an increase of acetylcholine release in the hippocampus was measured. Both the physiological approach (measurement of endogenous neurosteroid levels) and the pharmacological approach (effect of administered PREGS) suggest strongly that hippocampal PREGS is involved in the maintenance of cognitive performance. These observations are consistent with the results from systemic or intracerebral administration of PREG or derivatives in rodents, enhancing their natural memory performances or

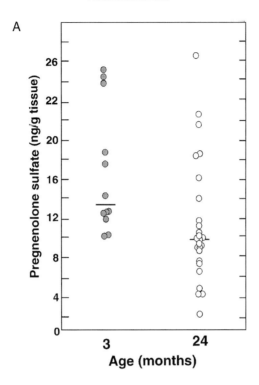
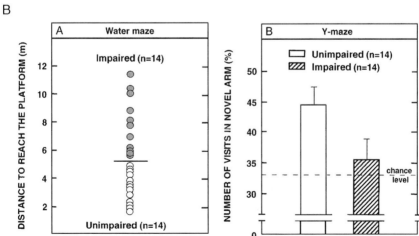

FIG. 6. (A) Hippocampal PREGS is lower in old rats than in young rats, but unevenly. (B) Memory tasks classify old rats as impaired and unimpaired animals. (C) Hippocampal PREGS and memory task impairment (water maze) are inversely correlated. (D) Administration of PREGS temporarily improves memory performance in impaired rats. (*Source:* Vallée *et al.*, 1997.)

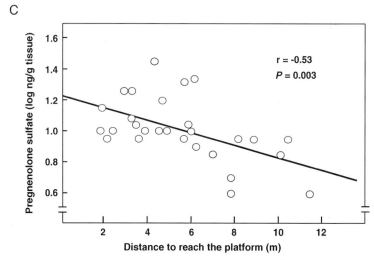

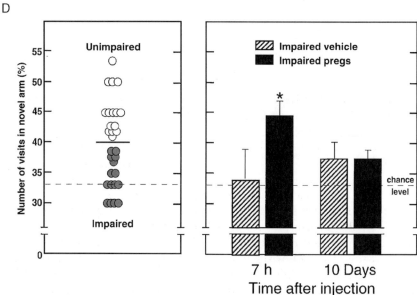

FIG. 6. (*Continued*)

antagonizing pharmacologically induced amnesia (Flood *et al.*, 1992; Mathis *et al.*, 1994). In most experiments, however, the procedure involves alteration of the motivational or emotional states of the animals, and the direct relationship between the administered dose and the endogenous steroid concentration has yet to be established. We have thus described, for the first time, changes of a neurosteroid with age, individual differences, and the correlation of these differences with behavioral performance, plus the improvement of performance in PREGS-deficient, impaired, aged animals by the steroid administration.

These results are determinant for attributing a physiological significance to a neurosteroid. Much remains to be done, including the study of prolonged administration (Ladurelle *et al.*, 2000), the different types of memory deficits observed in young animals, the mechanism of PREGS action (direct or via a metabolite?), the interactions with other factors controlling the memory potential, including other neurosteroids and glucocorticosteroids, clinically damaging (Sapolsky *et al.*, 1986). The molecular aspects of the involvement of PREGS in memory (i.e., the steroid's metabolism as well as receptors and intracellular signaling pathways involved) remain to be worked out. This is also the case for correlations between the pharmacological properties of neurosteroids at the neurotransmitter receptor level and behavior.

C. Neurosteroids Bind to Microtubule-Associated Protein 2 and Stimulate Microtubule Assembly: Latest Discovery and Beginning of a New Story

In 2000, Murakami *et al.* discovered a new modality of steroid action when studying PREG binding in the rat brain cytosol. High-affinity ($K_d \sim 30$ nM), saturable binding sites susceptible to proteolytic enzymes were demonstrated in the fetal and adult brain cytosol. Steroid specificity was demonstrated, with no or insignificant binding of testosterone, DHEA, DHEAS, estradiol, or corticosteroids. Interestingly, however, PROG and PREGS were efficient competitors to radioactive PREG. Attempts at purification and immunological detection suggested that the protein is a microtubule-associated protein, class 2 (MAP2). Highly purified MAP2 binds PREG and even more so in the presence of purified tubulin. PREG induced a large increase in both the rate and the extent of MAP2-induced tubulin polymerization into microtubules. The microtubules polymerized *in vitro* in the presence of PREG were of normal appearance at the electron microscopic level. Strikingly, PROG and PREGS, which were inactive for polymerizing tubulin, counteracted the stimulatory effect of PREG on microtubule assembly (Fig. 7). In cultured neurons from embryonic D17 fetuses, exposure to PREG increased the MAP2 immunostaining in cell bodies and led to its appearance in proximal neurites. These results suggest a novel mechanism of steroid action, neither at the gene nor at the membrane level, and may open a new field of research for the study of brain development, plasticity, aging, and degenerative alterations of the nervous system.

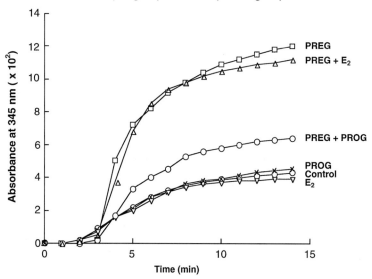

FIG. 7. **(A)** PREG stimulates MAP2-induced tubulin polymerization (A) and **(B)** increases MAP2 immunostaining in cultured fetal neurons (→proximal neurites). (Murakami et al., 2000).

IV. Conclusions

Neurosteroids are synthesized in the CNS and PNS, particularly in myelinating glial cells, as well as in astrocytes and many neurons. They act in the nervous system. Synthetic pathways may start from cholesterol or from steroidal precursor(s) imported from peripheral sources. That several cell types participate in the synthesis of a given neurosteroid by sequential chemical modifications remains a fair possibility. The most studied neurosteroids are PROG and its reduced metabolite $3\alpha,5\alpha$-TH PROG, the 3β-hydroxy-Δ^5-steroids PREG and DHEA, and their sulfate esters (PREGS and DHEAS). Selective changes in the concentration of neurosteroids in certain parts of the brain have been found in different behavioral or environmental situations, such as sex recognition or aggressiveness. Measured concentrations of neurosteroids are consistent with the affinities of receptor systems with which they interact in the nervous system. Both intracellular and membrane receptors responding to neurosteroids can be observed, and microtubule component(s) may participate in a new receptor system. Intracellular receptors are identical or similar to the "nuclear receptors" of steroid hormones found in peripheral target organs. A PROG-R, in particular, has been found in oligodendrocytes and Schwann cells, evoking the possibility of an autocrine system for PROG action. At the plasma membrane level, neurosteroids in the brain are, in fact, modulators of neurotransmitter receptors, and their concentrations are compatible with a physiological neuromodulatory role. This is the case not only with the $GABA_A$-R but also with the NMDA-R, sigma 1-R, and others. In addition to numerous experiments of pharmacological nature indicating well-defined properties of several neurosteroids, there are now results that demonstrate a physiological contribution of steroids formed endogenously and accumulated in the nervous system at least in part independently of the contribution from the steroidogenic glands. PROG synthesized in Schwann cells has "trophic" activity; it contributes to the synthesis of myelin in regenerating sciatic nerve in rats and mice, and the effect can be demonstrated *in vivo* and *in vitro*. PREGS has a behavioral effect in aged (2-year-old) rats. The concentration of PREGS in the hippocampus (and not in other parts of the brain) is inversely correlated with the quantified impairment of individual's performance in spatial memory tasks, while temporary improvement is obtained after bilateral intrahippocampal PREGS injections. Therefore, it is important to study the neuromodulatory role of neurosteroids in situations such as the estrous cycle, pregnancy, and stress as well as their influence on sexual behavior, memory, and the developmental and aging processes. Available data suggest that the neurosteroid concept may also be applicable to humans, and the robust and specific activities of neurosteroids may become useful for enlarging therapeutic approaches to functional and trophic alterations of the nervous system.

Acknowledgments

We thank researchers and students in our laboratories and colleagues from many others institutions who helped to collect the data reported in this report (see References). The work was supported mainly by INSERM. We gratefully acknowledge the support of Faculté de Mèdecine de Bicêtre, the Centre National pour la Recherche Scientifique (CNRS), the Collége de France, Roussel-Uclaf, l'Association pour la Recherche de la Sclérose en Plaque (ARSEP), l'Association Française contre les Myopathies (AFM), the F. Gould and Mathers Foundations, the Myelin Project, and Mme P. Schlumberger. We also thank very much Dr. K. Rajkowski for critical reading of the text, and F. Boussac, C. Legris, M. Ourtou, and J. C. Lambert for their help in preparing the manuscript.

References

Akwa, Y., et al. (1992). Neurosteroid metabolism. 7α-hydroxylation of dehydroepiandrosterone and pregnenolone by rat brain microsomes. *Biochem. J.* **228**, 959–964.
Akwa, Y., et al. (1993). Astrocytes and neurosteroids: Metabolism of pregnenolone and dehydroepiandrosterone: Regulation by cell density. *J. Cell. Biol.* **121**, 135–143.
Bäckström, T. (1995). Symptoms related to the menopause and sex steroid treatments. *In* "Non-Reproductive Actions of Sex Steroids" (G. R. Bock and J. A. Goode, eds.), pp. 171–180. Wiley, Chichester.
Baulieu, E. E., et al. (1965). An adrenal-secreted "Androgen": Dehydroisoandrosterone sulfate. Its metabolism and a tentative generalization on the metabolism of other steroid conjugates in man. *Rec. Progr. Hormone Res.* **21**, 411–500.
Baulieu, E. E., et al. (1978). Steroid induced meiotic division in *Xenopus laevis* oocytes: Surface and calcium. *Nature* **275**, 593–598.
Baulieu, E. E. (1981). Steroid hormones in the brain: Several mechanisms? *In* "Steroid Hormone Regulation of the Brain" (K. Fuxe, J. A. Gustafsson, and L. Wetterberg, eds.), pp. 3–14. Pergamon Press, Oxford.
Baulieu, E. E., Robel, P., and Schumacher, M. (Eds.) (1999). "Neurosteroids: A New Regulatory Function in the Nervous System." Humana Press, Totowa, NJ.
Bloch, K. (1965). The biological synthesis of cholesterol. *Science* **150**, 19–28.
Brann, D. W., et al. (1995). Emerging diversities in the mechanism of action of steroid hormones. *J. Steroid Biochem. Mol. Biol.* **52**, 113–133.
Compagnone, N. A., et al. (1995a). Expression of the steroidogenic enzyme P450scc in the central and peripheral nervous systems during rodent embryogenesis. *Endocrinology* **136**, 2689–2696.
Compagnone, N. A., et al. (1995b). Steroidogenic enzyme P450c17 is expressed in the embryonic central nervous system. *Endocrinology* **136**, 5212–5223.

Corpéchot, C., et al. (1981). Characterization and measurement of dehydroepiandrosterone sulfate in the rat brain. *Proc. Natl. Acad. Sci. USA* **78,** 4704–4707.

Corpéchot, C., et al. (1983). Pregnenolone and its sulfate ester in the rat brain. *Brain Res.* **270,** 119–125.

Corpéchot, C., et al. (1985). Neurosteroids: Regulatory mechanisms in male rat brain during heterosexual exposure. *Steroids* **45,** 229–234.

Corpéchot, C., et al. (1993). Neurosteroids: 3α-Hydroxy-5α-pregnan-20-one and its precursors in the brain, plasma, and steroidogenic glands of male and female rats. *Endocrinology* **13,** 1003–1009.

Costa, E., (1991). "Prologomena" to the biology of the diazepam binding inhibitor (DBI). *Neuropharmacology* **30,** 1357–1364.

Costa, E., and Paul, S. M. (Eds.) (1991). "Neurosteroids and Brain Function" Thieme Medical Publishers, New York.

Culty, M., et al. (1999). In vitro studies on the role of the peripheral-type benzodiazepine receptor in steroidogenesis. *J. Steroid Biochem Mol. Biol.* **69,** 123–130.

Desarnaud, F., et al., (1998). Progesterone stimulates the activity of the promoters of peripheral myelin protein-22 and protein zero genes in Schwann cells. *J. Neurochem.* **71,** 1765–1768.

El-Etr, M., et al. (1995). A progesterone metabolite stimulates the release of gonadotropin-releasing hormone from GT1-1 hypothalamic neurons via the γ-amicobutyric acid type A receptor. *Proc. Natl. Acad. Sci. USA* **92,** 3769–3773.

Finidori-Lepicard, J., et al. (1981). Steroid hormone as regulatory agent of adenylate cyclase: Inhibition by progesterone of the membrane bound enzyme in *Xenoppus laevis* oocytes. *Nature* **292,** 255–256.

Flood, J. F., et al. (1992). Memory-enhancing effects in male mice of pregnenolone and steroids metabolically derived from it. *Proc. Natl. Acad. Sci. USA* **89,** 1567–1571.

French-Müllen, J. M. H., et al. (1994). Neurosteroids modulate calcium currents in hippocampal CA1 neurons via a pertussis toxin-sensitive G-protein-coupled mechanism. *J. Neurosci.* **14,** 1963–1977.

Fu, Q., et al. (1998). Control of cholesterol biosynthesis in Schwann cells. *J. Neurochem.* **71,** 549–555.

Gee, K. W., et al. (1995). A putative receptor for neurosteroids on the $GABA_A$ receptor complex: The pharmacological properties and therapeutic potential of epalons. *Crit. Rev. Neurobiol.* **9,** 207–227.

Genazzani, A. R., et al. (1996). Neurosteroids and regulation of neuroendocrine function. In "The Brain: Sources and Target for Sex Steroid Hormones" (A. R. Genazzani, F. Petraglia, and R. H. Purdy, eds.), pp. 83–91. Parthenon, New York.

Gomez-Sanchez, C. E., et al. (1996). Corticosteroid synthesis in the central nervous system. *Endocr. Res.* **22,** 463–470.

Gower, D. B., and Ruparelia, B. A. (1993). Olfaction in humans with special reference to odorous 16-androstenes: Their occurrence, perception and possible social, psychological and sexual impact. *J. Endocrinol.* **137,** 167–187.

Guarneri, P., et al. (1994). Neurosteroidogenesis in rat retinas. *J. Neurochem.* **63,** 86–96.

Guennoun, R., et al. (1995). A key enzyme in the biosynthesis of neurosteroids, 3β-hydroxysteroid dehydrogenase/Δ^5-Δ^4-isomerase (3β-HSD), is expressed in rat brain. *Mol. Brain Res.* **30,** 287–300.

Guennoun, R., Schumacher, M., Robert, F., Delespierre, B., Gouezou, M., Eychenne, B., Akwa, Y., Robel, P., and Baulieu, E. E. (1997). Neurosteroids: Expression of functional 3β-hydroxysteroid dehydrogenase by rat sensory neurons and Schwann cells. *Euro. J. Neurosci.* **9,** 2236–2247.

Harrison, N. L., and Simmonds, M. A. (1984). Modulation of the GABA receptor complex by a steroid anaesthetic. *Brain Res.* **323**, 287–292.

Haug, M., et al. (1989). Suppressive effects of dehydroepiandrosterone and 3β-methyl androst-5-en-17-one on attack towards lactating female intruders by castrated male mice. *Physiol. Behav.* **46**, 955–959.

Hirata, S., et al. (1994). Effects of hypothyroidism on the gene expression of progesterone receptors in the neonatal rat brain. *J. Steroid Biochem. Mol. Biol.* **50**, 293–297.

Hochberg, R. B., et al. (1977). Detection in bovine adrenal cortex of a lipoidal substance that yields pregnenolone upon treatment with alkali. *Proc. Natl. Acad. Sci. USA* **74**, 941–945.

Hu, Z. Y., et al. (1987). Neurosteroids: Oligodendrocyte mitochondria convert cholesterol to pregnenolone. *Proc. Natl. Acad. Sci. USA* **84**, 8215–8219.

Hu, Z. Y., et al. (1989). Neurosteroids: Steroidogenesis in primary cultures of rat glial cells after release of aminoglutethimide blockade. *Biochem. Biophys. Res. Commun.* **161**, 917–922.

Iwahashi, K., et al. (1990). Studies of the immunohistochemical and biochemical localization of the cytochrome P-450scc-linked monooxygenase system in the adult rat brain. *Biochim. Biophys. Acta* **1035**, 182–189.

Iwahashi, K., et al. (1993). A localization study of the cytochrome P-450(21)-linked monooxygenase system in adult rat brain. *J. Steroid Biochem. Mol. Biol.* **44**, 163–169.

Jo, D. H., et al. (1989). Pregnenolone, dehydroepiandrosterone, and their sulfate and fatty acid esters in the rat brain. *Steroids* **54**, 287–297.

Jung-Testas, I., et al. (1989). Neurosteroids: Biosynthesis of pregnenolone and progesterone in primary cultures of rat glial cells. *Endocrinology* **125**, 2083–2091.

Jung-Testas, I., et al. (1991). Estrogen-inducible progesterone receptor in primary cultures of rat glial cells. *Exp. Cell Res.* **193**, 12–19.

Jung-Testas, I., et al. (1992). Low density lipoprotein-receptors in primary cultures of rat glial cells. *J. Steroid Biochem. Mol. Biol.* **42**, 597–605.

Jung-Testas, I., et al. (1996). Demonstration of progesterone receptors in rat Schwann cells. *J. Steroid Biochem. Mol. Biol.* **58**, 77–82.

Koenig, H., et al. (1995). Progesterone synthesis and myelin formation by Schwann cells. *Science* **268**, 1500–1503.

Labrie, F. (1991). Intracrinology. *Mol. Cell. Endocrinol.* **78**, C113–C118.

Labrie, F., et al. (1992). Structure and tissue-specific expression of 3 beta-hydroxysteroid dehydrogenase/5-ene-4-ene isomerase genes in human and rat classical and peripheral steroidogenic tissues. *J. Steroid Biochem. Mol. Biol.* **41**, 421–435.

Lacor, P., et al. (1999). Regulation of the expression of peripheral benzodiazepine receptors and their endogenous ligands during rat sciatic nerve degeneration and regeneration: A role for PBR in neurosteroidogenesis. *Brain Res.* **815**, 70–80.

Lacroix, C., et al. (1987). Simultaneous radioimmunoassay of progesterone, androst-4-enedione, pregnenolone, dehydroepiandrosterone and 17-hydroxy progesterone in specific regions of human brain. *J. Steroid Biochem.* **28**, 317–325.

Ladurelle, N., et al. (2000). Prolonged intracerebroventricular infusion of neurosteroids affects cognitive performance in the mouse. *Brain Res.* **858**, 371–379.

Lambert, J., et al. (1996). Neurosteroid modulation of native and recombinant $GABA_A$ receptors. *Cell. Mol. Neurobiol.* **16**, 155–174.

Lanthier, A., and Patwardhan, V. V. (1986). Sex steroids and 5-en-3β-hydroxysteroids in specific regions of the human brain and cranial nerves. *J. Steroid Biochem.* **25**, 445–449

Lauber, M. E., and Lichtensteiger, W. (1996). Ontogeny of 5 alpha-reductase (type 1) messenger ribonucleic acid expression in rat brain: Early presence in germinal zones. *Endocrinology* **137**, 2718–2730.

Le Goascogne, C., et al. (1987). Neurosteroids: Cytochrome P450 scc in rat brain. *Science* **237,** 1212–1215.
Le Goascogne, C., et al. (1989). The cholesterol side-chain cleavage complex in human brain white matter. *J. Neuroendocrinol.* **1,** 153–156.
Le Goascogne, C., Eychenne, B., Tonon, M.-C., Lachapelle, F., Baumann, N., and Robel, P. (2000). Neurosteroid progesterone is up-regulated in the brain of Jimpy and shiverer mice. *Glia* **29,** 14–24.
Liere, P., et al. (2000). Validation of an analytical procedure to measure trace amounts of neurosteroids in brain tissue by gas chromotography—mass spectrometry. *J. Chromatogr. B.* **739,** 301–312.
MacDonald, R. L., and Olsen, R. W. (1994). GABA$_A$ receptor channels. *Annu. Rev. Neurosci.* **17,** 569–602.
MacLusky, N. J., et al. (1994). Aromatase in the cerebral cortex, hippocampus and midbrain. Ontogeny and developmental implications. *Mol. Cell. Neurol. Sci.* **5,** 691–698.
Majewska, M. D. (1992). Neurosteroids: Endogenous bimodal modulators of the GABA$_A$ receptor: Mechanism of action and physiological significance. *Prog. Neurobiol.* **38,** 379–395.
Majewska, M. D., et al. (1986). Steroid hormone metabolites are barbiturate-like modulators of the GABA$_A$ receptor. *Science* **232,** 1004–1007.
Majewska, M. D., et al. (1987). Neurosteroid pregnenolone sulfate antagonizes electrophysiological responses to GABA in neurons. *Neurosci. Lett.* **90,** 279–284.
Majewska, M. D., et al. (1989). Pregnenolone sulfate antagonizes barbiturate-induced hypnosis. *Pharmacol. Biochem. Behavior* **33,** 701–703.
Majewska, M. D., et al. (1990). The neurosteroid dehydroepiandrosterone sulfate is an allosteric antagonist of the GABA$_A$ receptor. *Brain Res.* **526,** 143–146.
Mathis, C., et al. (1994). The neurosteroid pregnenolone sulfate blocks NMDA antagonist-induced deficits in a passive avoidance memory task. *Psychopharmacology* **116,** 201–206.
Mayo, W., et al. (1993). Infusion of neurosteroids into the *nucleus basalis magnocellularis* affects cognitive processes in the rat. *Brain Res.* **607,** 324–328.
McEwen, B. S. (1991). Non-genomic and genomic effects of steroids on neural activity. *Trends Pharmacol. Sci* **12,** 141–147.
Melcangi, R. C., et al. (1993). Differential localization of the 5 alpha-reductase and the 3 alpha-hydroxysteroid dehydrogenase in neuronal and glial cultures. *Endocrinology* **132,** 1252–1259.
Melcangi, R. C., et al. (1994). Progesterone 5-alpha-reduction in neuronal and in different types of glial cell cultures: type 1 and 2 astrocytes and oligodendrocytes. *Brain Res.* **639,** 202–206.
Mellon, S. H., and Deschepper, C. F. (1993). Neurosteroid biosynthesis: Genes for adrenal steroidogenic enzymes are expressed in the brain. *Brain Res.* **629,** 283–292.
Mensah-Nyagan, A. G., Do-Rego, J. L., Beaujan, D., Luu-The, V., Pelletier, G., and Vaudry, H. (1999). Neurosteroids: Expression of steroidogenic enzymes and regulation of steroid biosynthesis in the central nervous system. *Pharmacol Rev.* **51,** 63–81.
Mienville, J. L., and Vicini, S. (1989). Pregnenolone sulfate antagonizes GABA$_A$ receptor-mediated currents via a reduction of channel opening frequency. *Brain Res.* **489,** 190–194
Milgrom, E., et al. (1970). Progesterone in uterus and plasma. IV. Progesterone receptor(s) in guinea pig uterus cytosol. *Steroids* **16,** 741–754.
Miyairi, S., et al. (1988). Aromatization and 19-hydroxylation of androgens by rat brain cytochrome P-450. *Biochem. Biophys. Res. Commun.* **150,** 311–315.
Monnet, F. P., et al. (1995). Neurosteroids via σ receptors, modulate the [^{3}H]norepinephrine release evoked by *N*-methyl-D-aspartate in the rat hippocampus. *Proc. Natl. Acad. Sci. USA* **92,** 3774–3778.

Morell, P., and Jurewicks, H. (1996). Origin of cholesterol in myelin. *Neurochem. Res.* **21**, 463–470.
Morfin, R., et al. (1992). Neurosteroids: Pregnenolone in human sciatic nerves. *Proc. Natl. Acad. Sci. USA* **89**, 6790–6793.
Murakami, K., et al. (2000). Pregnenolone binds to microtubule-associated protein 2 and stimulates microtubule assembly. *Proc. Natl. Acad. Sci. USA* **97**, 3579–3584.
Naftolin, F., et al. (1975). The transformation of estrogens by central neuroendocrine tissues. *Rec. Progr. Horm. Res.* **31**, 295–319.
Normington, K., and Russell, D. W. (1992). Tissue distribution and kinetic characteristics of rat steroid 5α-reductase isozymes: Evidence for distinct physiological functions. *J. Biol. Chem.* **267**, 19548–19554.
Oftebro, H., et al. (1979). The presence of an adrenodoxin-like ferredoxin and cytochrome P-450 in brain mitochondria. *J. Biol. Chem.* **254**, 4331–4334.
Orchinik, M., et al. (1991). A corticosteroid receptor in neuronal membranes. *Science* **252**, 1848–1851.
Ozaki, H. S., et al. (1991). Cytochrome P-450 11-beta in rat brain. *J. Neurosci. Res.* **28**, 518–524.
Papadopoulos, V. (1993). Peripheral-type benzodiazepine/diazepam binding inhibitor receptor: Biological role in steroidogenic cell function. *Endocr. Rev.* **14**, 222–240.
Parsons, B., et al. (1982). Progestin receptor levels in rat hypothalamic and limbic nuclei. *J. Neurosci.* **2**, 1446–1452.
Paul, S. M., and Purdy, R. H. (1992). Neuroactive steroids. *FASEB J.* **6**, 2311–2322.
Pelletier, G., et al. (1994). Immunocytochemical localization of 5 alpha-reductase in rat brain. *Mol. Cell. Neurosci.* **5**, 394–399.
Pincus, G. P., and Hoagland, H. (1944). Effects of administered pregnenolone on fatiguing psychomotor performance. *Aviation Med.* **15**, 98–135.
Prasad, V. V. K., et al. (1994). Precursors of the neurosteroids. *Proc. Natl. Acad. Sci. USA* **91**, 3220–3223.
Prince, R. J., and Simmonds, M. A. (1992). Steroid modulation of the strychnine-sensitive glycine receptor. *Neuropharmacology* **31**, 201–205.
Puia, G., et al. (1990). Neurosteroids act on recombinant human $GABA_A$ receptors. *Neuron* **4**, 759–765.
Purdy, R. H., et al. (1991). Stress-induced elevations of gamma aminobutyric type. A receptor-active steroids in the rat brain. *Proc. Natl. Acad. Sci. USA* **88**, 4553–4557.
Rajkowski, K. M., et al. (1997). Hydroxysteroid sulfotransferase activity in the rat brain and liver as a function of age and sex. *Steroids* **62**, 427–436.
Robel, P., and Baulieu, E. E. (1985). Neuro-steroids: 3β-Hydroxy-Δ5-derivatives in the rodent brain. *Neurochem. Int.* **7**, 953–958.
Robel, P., and Baulieu, E. E. (1994). Neurosteroids: Biosynthesis and function. *Trends Endocrinol. Metab.* **5**, 1–18.
Robel, P., et al. (1986). Persistance d'un rythme circadien de la déhydroépiandrostérone dans le cerveau, mais non dans le plasma, de rats castrés et surrénalectomisés. *C.R. Acad. Sci. Paris* **303**, 235–238.
Robel, P., et al. (1987). Neurosteroids: 3β-Hydroxy-Δ5-derivatives in rat and monkey brain. *J. Steroid Biochem.* **27**, 649–655.
Rose K. A., et al. (1997). Cyp 7b, a novel brain cytochrome P450, catalyzethe synthesis of neurosteroids 7-hydroxy-dehydroepiandrosterone and 7-hydroxy-pregnenolone. *Proc. Natl. Acad. Sci. USA,* **94**, 4925–4930.
Sanne, J. L., and Krueger, K. E. (1995). Expression of cytochrome P450 side-chain cleavage enzyme and 3 beta-hydroxysteroid dehydrogenase in the rat central nervous system:

A study by polymerase chain reaction and in situ hybridization. *J. Neurochem.* **65,** 528–536.
Sapolsky, R. M., et al. (1986). The neuroendocrinology of stress and aging: The glucocorticoid cascade hypothesis. *Endocr. Rev.* **7,** 284–301.
Schumacher, M., et al. (1990). Behavioral effects of progesterone associated with rapid modulation of oxytocin receptors. *Science* **250,** 691–694.
Schumacher, M., et al. (1993). The oxytocin receptor: A target for steroid hormones. *Regul. Pept.* **45,** 115–119.
Schumacher, M., et al. (1999). Genomic and membrane actions of progesterone: Implications for reproductive physiology and behavior. *Behav. Brain Res.* **105,** 37–52.
Selye, H. (1941). The anesthetic effect of steroid hormones. *Proc. Soc. Exp. Biol. Med.* **46,** 116–121.
Smith, S. S., et al. (1987). Sex steroid effects on extrahypothalamic CNS. I. Estrogen augments neuronal responsiveness to iontophoretically applied glutamate in the cerebellum. *Brain Res.* **422,** 40–51.
Spivak, C. E. (1994). Desensitization and noncompetitive blockade of GABA$_A$ receptors in ventral midbrain neurons by a neurosteroid dehydroepiandrosterone sulfate. *Synapse* **16,** 113–122.
Strömstedt, M., and Waterman, M. R. (1995). Messenger RNAs encoding steroidogenic enzymes are expressed in rodent brain. *Brain Res. Mol. Brain Res.* **34,** 75–88.
Suter, U., and Snipes, G. J. (1995). Biology and genetics of hereditary motor and sensory neuropathies. *Annu. Rev. Neurosci.* **18,** 45–75.
Synguelakis, M., et al. (1985). Evolution circadienne de Δ5-3b-hydroxystéroïdes et de glucocorticostéroïdes dans le plasma et le cerveau de rat. *C.R. Acad. Sci. Paris* **301,** 823–826.
Tischkau, S. A., and Ramirez, V. D. (1993). A specific membrane binding protein for progesterone in rat brain: Sex differences and induction by estrogen. *Proc. Natl. Acad. Sci. USA* **90,** 1285–1289.
Tsutsui, K., et al. (1999). Neurosteroid biosynthesis in vertebrate brains. *Comp. Biochem. Physiol. C. Pharmacol. Toxicol. Endocrinol.* **124,** 121–129.
Valera, S., et al. (1992). Progesterone modulates a neuronal nicotinic acetylcholine receptor. *Proc. Natl. Acad. Sci. USA* **89,** 9949–9953.
Vallée, M., et al. (1997). Neurosteroids: Deficient cognitive performance in aged rats depends on low pregnenolone sulfate levels in hippocampus. *Proc. Natl. Acad. Sci USA* **94,** 14865–14870.
Vande Wiele, R. L., et al. (1963). Studies on the secretion and interconversion of the androgens. *Rec. Prog. Horm. Res.* **19,** 275–305.
Vourc'h, C., et al. (1992). Δ^5-3β-Hydroxysteroid acyl transferase activity in the rat brain. *Steroids* **57,** 210–215.
Warner, M., and Gustafsson, J. A. (1995). Cytochrome P450 in the brain: Neuroendocrine functions. *Front. Neuroendocrinol.* **16,** 224–236.
Widdows, K., and Chadwick, D. (Eds.) (1990). "Steroids and Neuronal Activity" Wiley, Wischester.
Wu, F. S., et al. (1991). Pregnenolone sulphate: A positive allosteric modulator at the *N*-methyl-D-aspartate receptor. *Mol. Pharmacol.* **40,** 333–336.
Yanagibachi, K., et al. (1988). The regulation of intracellular transport of cholesterol in bovine adrenal cells: Purification of a novel protein. *Endocrinology* **123,** 2075–2082.
Young, J., et al. (1991). Suppressive effects of dehydroepiandrosterone and 3β-methyl-androst-5-en-17-one on attack towards lactating female intruders by castrated male mice. II. Brain neurosteroids. *Biochem. Biophys. Res. Commun.* **174,** 892–897.

Young, J., et al. (1994). Neurosteroids: Pharmacological effects of a 3β-hydroxy-steroid dehydrogenase inhibitor. *Endocrine* **2**, 505–509.

Zhang, P., et al. (1995). Transcriptional regulation of P450scc gene expression in neural and steroidogenic cells: Implications for regulation of neurosteroidogenesis. *Mol. Endocrinol.* **9**, 1571–1582.

Zwain, I. H., and Yen, S. S. C. (1999a). Dehydroepiandrosterone: Biosynthesis and metabolism in the brain. *Endocrinology,* **140**, 880–887.

Zwain, I. H., and Yen, S. S. C. (1999b). Neurosteroidogenesis in astrocytes, oligodendrocytes, and neurons of cerebral cortex of rat brain. *Endocrinology,* **140**, 3843–3852.

BIOSYNTHESIS OF NEUROSTEROIDS AND REGULATION OF THEIR SYNTHESIS

Synthia H. Mellon[1]

Department of Obstetrics, Gynecology, and Reproductive Sciences
The Center for Reproductive Sciences, and The Metabolic Research Unit
University of California–San Francisco
San Francisco, California 94143-0556

Hubert Vaudry

European Institute for Peptide Research (IFRMP23)
Laboratory of Cellular and Molecular Neuroendocrinology
INSERM U-413, UA CNRS, University of Rouen
76821 Mont-Saint-Aignan, France

I. What Is a Neurosteroid?
II. Enzymes Involved in Neurosteroidogenesis: Biochemistry and Molecular Biology
   A. P450scc
   B. Adrenodoxin Reductase/Adrenodoxin
   C. StAR
   D. 3βHSD
   E. P450c17
   F. P450 Reductase
   G. Cytochrome b_5
   H. P450c21 (and Extra-adrenal 21-Hydroxylase Activity)
   I. P450c11
   J. 17-Ketosteroid Reductase/17β-Hydroxysteroid Dehydrogenase (17βHSD)
   K. P450 aro
   L. 11β-Hydroxysteroid Dehydrogenase (11βHSD)
   M. 5α-Reductase
   N. 3α-Hydroxysteroid Dehydrogenase/3α-Hydroxysteroid Oxidoreductase (3αHSD)
   O. Sulfotransferase (HST)
   P. Sulfatase (STS)
   Q. Other Neurosteroidogenic Enzymes
III. Enzymes Involved in Neurosteroidogenesis: Distribution in the Brain and Developmental Regulation
   A. Expression of P450scc in the Adult Rat Brain
   B. Developmental Regulation of P450scc Expression in the CNS and PNS
   C. Expression of Adrenodoxin and Adrenodoxin Reductase in the Brain
   D. Expression of StAR in the Brain
   E. Developmental Regulation of P450c17 Expression in the Nervous System
   F. Expression and Activity of P450 Reductase and b_5 in the Brain

[1]Author to whom correspondence should be sent.

G. 3βHSD Expression in the Adult Nervous System
   H. P450c11β and P450c11AS Expression in the Adult Nervous System
   I. Expression of 17βHSD in the Brain
   J. Expression of P450 aro in the Brain
   K. Expression of 11βHSD in the Brain
   L. Expression of 5α-Reductase in the Brain
   M. Expression of 3αHSD in the Brain
   N. Expression of HST in the Brain
   O. Activity and Expression of STS in the Brain
IV. Regulation of Neurosteroidogenic Enzymes
   A. P450scc
   B. P450c17
   C. 3βHSD
V. Conclusion
   References

The brain, like the gonads, adrenal glands, and placenta, is a steroidogenic organ. The steroids synthesized by the brain and by the nervous system, given the name neurosteroids, have a wide variety of diverse functions. In general, they mediate their actions not through classic steroid hormone nuclear receptors but through ion-gated neurotransmitter receptors. This chapter summarizes the biochemistry of the enzymes involved in the biosynthesis of neurosteroids, their localization during development and in adulthood, and the regulation of their expression, highlighting both similarities and differences between expression in the brain and in classic steroidogenic tissues. © 2001 Academic Press.

I. What Is a Neurosteroid?

The demonstration that steroids could be synthesized *de novo* in the brain and the simultaneous experiments describing novel functions for certain steroidal compounds at nonclassical γ-aminobutyric acid type A ($GABA_A$) and N-methyl-D-asparate, receptors brought a new field to light. The former demonstration, that steroids could be synthesized in the brain, came initially from observations made in the 1980s by Baulieu and colleagues, who found that steroids such as pregnenolone (PREG), dehydroepiandrosterone (DHEA), and their sulfate and lipoidal esters were present in higher concentrations in tissue from the nervous system (brain and peripheral nerve) than in the plasma. Although these compounds could be present as a result of peripheral synthesis and then sequestration in the brain, Baulieu and colleagues found that the steroids remained in the nervous system long after gonadectomy or adrenalectomy (Corpéchot *et al.*, 1981, 1983), suggesting that steroids might be synthesized *de novo* in the central and peripheral nervous systems (CNS and PNS, respectively) or might accumulate in those structures. Such steroids were named neurosteroids to refer to their unusual origin and to differentiate them from steroids derived from more classical steroidogenic organs, such as gonads, adrenal glands, and placentae. How were these steroids synthesized? To test whether steroids were actually made in the brain or were accumulated specifically in tissue from the nervous system, several laboratories, including ours, determined directly if enzymes known to be involved in steroidogenesis (i.e., adrenal glands, gonads, and placentae) could be responsible for neurosteroid synthesis. These studies have established unequivocally that the enzymes found in classic steroidogenic tissues are indeed found in the nervous system. Depending on the steroid synthesized, these steroids could affect gene expression through action at classic intracellular nuclear receptors, or they could affect neurotransmission through action at membrane ion-gated and other neurotransmitter receptors.

II. Enzymes Involved in Neurosteroidogenesis: Biochemistry and Molecular Biology

Enzymes involved in neurosteroidogenesis can be classified in two main groups: the cytochrome P450 group and the non-P450 group. Neurosteroidogenic enzymes, P450s and non-P450, are in general either mitochondrial or microsomal, although the exact localization of some enzymes is still controversial.

Cytochromes P450 are oxidases that function identically. They can be defined as heme-binding mono-oxygenases, able to catalyze the oxidative conversion of many steroids, lipids, and a variety of xenobiotics and environmental toxins (reviewed in Miller, 1988). Steroidogenic P450s are unusual in that they have limited and specific steroidal substrates. They reduce atmospheric oxygen with electrons from NADPH, and this reduction requires the action of specific co-factors, adrenodoxin reductase and adrenodoxin for mitochondrial P450s, and P450-reductase and b_5 for microsomal P450s.

Most of the enzymes present in the adrenal glands, gonads, and placenta have been found in the brain by measuring their enzymatic activity and/or their mRNA transcript level and/or their protein expression. The biosynthetic pathway of neurosteroids is presented in Fig. 1. The synthesis of specific steroid hormones in the adrenal glands, gonads, placenta, and brain is dependent on the tissue-, cell- and developmentally specific expression of these various enzymes. For example, 11β-hydroxylase (P450c11β)

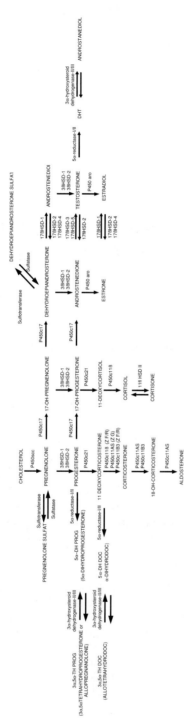

FIG. 1. The biosynthetic pathway of neurosteroids. The names for each enzyme are shown by each reaction. Mitochondrial cholesterol side-chain cleavage enzyme, P450scc, mediates 20α-hydroxylation, 22-hydroxylation, and scission of the C20–22 bond; 3βHSD mediates both 3β-hydroxysteroid dehydrogenase and Δ^5–Δ^4-isomerase activities; P450c11β, mitochondrial 11 hydroxylase, mediates 11 hydroxylation; P450c11AS, mitochondrial aldosterone synthase, mediates c11,18-hydroxylation and 18-oxidation; ZF/R and ZG refer to the adrenal zona fasciculata/reticularis, or zona glomerulosa, that express the particular P450c11 gene; 17βHSD (also called 17-ketosteroid reductase, or 17-KSR), mediates c17β reduction or c17 oxidation; 3αHSD mediates the conversion of dihydroprogesterone to allopregnanolone, of 5α-dihydroDOC to allotetrahydroDOC, or of dihydrotestosterone to androstanediol. P450 aro mediates the conversion of testosterone to estradiol or of androstenedione to estrone. Interconversion of free steroids to their sulfated derivatives, mediated by sulfotransferase (HST) and sulfatase (STS), are also shown.

and aldosterone synthase (P450c11AS) are expressed in the adrenal glands and not in the gonads or the placenta, resulting in glucocorticoid and mineralocorticoid production, while expression of 17α-hydroxylase/17, 20-lyase (P450c17) in the testis results in androgen production, and aromatase (P450 aro) expression in the gonads results in estrogen production.

Many of the steroid hydroxylases have multiple enzymatic activities, as demonstrated by the bioconversions shown in Fig. 1, while some bioconversions are mediated by more than one member of a particular gene family. Demonstration of a single protein mediating certain reactions has been rigorously demonstrated through purification of the proteins and cloning of the cDNAs encoding these proteins. This is true for P450c17, where cloning and expression of the single P450c17 cDNA demonstrated that this protein had both 17α-hydroxylase activity *and* 17,20-lyase activity, further demonstrating that these activities indeed reside within single proteins. In contrast, similar cDNA cloning has demonstrated the existence of multiple different genes encoding other steroidogenic enzymes (e.g., 17βHSDs and human 3αHSDs).

In the nervous system, there is not only region-specific expression of the steroidogenic enzymes but also cell-type-specific and developmental regulation of these enzymes, indicating a more complex scheme than that depicted in the Fig. 1. The synthesis of neurosteroids proceeds through some similar and some different pathways from those used in the adrenal glands, gonads, and placentae. The brain contains additional steroid-metabolizing enzymes, including sulfotransferases and sulfohydrolases, that convert classic steroid hormones into a variety of neuroactive compounds.

We have modified the original definition of neurosteroids to include both neuroactive compounds produced *de novo* and steroids metabolized to neuroactive compounds in the brain but derived from circulating precursors. Do steroids readily enter the brain? Most steroids are transported in the blood bound to protein, and entry into the brain depends on dissociation of the steroid from the protein. Albumin-bound steroids are freely cleared by the brain on a single pass (Pardridge and Mietus, 1979a,b, 1980a, 1980b). However, globulin-bound steroids, like testosterone or estradiol bound to the sex-hormone-binding globulin or corticosterone bound to the corticosteroid-binding globulin, are *not* transported into the brain. Progesterone (PROG), in contrast, bound to guinea pig PROG-binding globulin (PBG), *can* be transported into the brain, suggesting that the halftime of PROG dissociation from its binding globulin (1.8 sec) is similar to the brain capillary transit time (0.1–1.0 s) (Pardridge and Mietus, 1980a). Testosterone binds to the PROG-binding globulin even less tightly than does PROG. Thus, a substantial fraction of plasma PROG (10%) or testosterone (25%), bound to PBG, can still be transported into the brain. Gonadal steroids, such as PROG, 17-hydroxyprogesterone, testosterone, and estradiol, are sequestered in the brain, because their brain: plasma concentrations are greater than 1 (Marynick *et al.*, 1977; Pardridge *et al.*, 1980). Unlike gonadal steroids, corticosterone is *not* sequestered in the brain. There may also be a regional brain distribution of steroids derived from the plasma (Wang *et al.*, 1997).

Neuroactive steroids are considered to be inactive metabolites in the adrenal glands, gonads, and placentae and include 5α-DH PROG, allopregnanolone, DHEA, and its sulfated ester, DHEAS. Both 5α-dihydroprogesterone (5α-DH PROG) and allopregnanolone can be synthesized endogenously in the brain from cholesterol, or they can be synthesized in the brain from gonadal PROG precursor.

Other enzymatic activities giving rise to steroidal compounds have been described in the brain of several species. Hydroxylation and accumulation of 20α-reduced and 7α-hydroxylated metabolites of PROG (20α) and PREG (20α and 7α) have repeatedly been observed in the fetal and adult rodent brain (Akwa *et al.*, 1992; Rose *et al.*, 1997). Several steroidogenic enzymes expressed in nervous system display brain-specific isotypes not found in the liver or classical steroidogenic organs. In most cases, the expression of the steroidogenic enzymes in the CNS and PNS is developmentally, regionally, and cell-specifically regulated, ensuring the regulated synthesis of specific neurosteroids.

In this chapter, we review what is known about each of the steroidogenic enzymes that are crucial for the synthesis of the neuroactive steroids in different species and at different developmental stages.

A. P450scc

The first rate-limiting and hormonally regulated step in the synthesis of all steroid hormones is the conversion of cholesterol to PREG (Fig. 1). This reaction is catalyzed by the mitochondrial enzyme cholesterol side-chain cleavage (SCC), P450scc, in three successive chemical reactions: 20α-hydroxylation, 22-hydroxylation, and scission of the c20–c22 carbon bond of cholesterol. The products of this reaction are PREG and isocaproic acid. A single P450scc species is found in all steroidogenic tissue, including the brain (Mellon and Deschepper, 1993; Mellon, 1994).

P450scc is the rate-limiting step in steroidogenesis, and it is one of the slowest enzymes known, with a V_{max} of 1 mol cholesterol/mol enzyme/sec. The slowest part of this reaction may be the entry of cholesterol into the mitochondria and its binding to the active site of P450scc (see the section on StAR).

The human and rat genomes contain a single gene encoding P450scc (Chung *et al.*, 1986a; Matteson *et al.*, 1986a; Morohashi *et al.*, 1987; Oonk *et al.*, 1990), which is about 20 kb long, contains 9 exons, and in humans is located on chromosome 15. This gene encodes an mRNA of about 2.0 kb, which encodes a 521-amino acid protein. This protein is proteolytically cleaved, removing a 39-amino acid leader peptide that directs the protein to the mitochondria.

B. ADRENODOXIN REDUCTASE/ADRENODOXIN

P450scc functions as the terminal oxidase in a mitochondrial electron transport system. As described, electrons from NADPH are first accepted by a flavoprotein, adrenodoxin reductase, which is located in the mitochondrial matrix and is loosely associated with the inner membrane (Nakamura *et al.*, 1966; Omura *et al.*, 1966; Kimura and Suzuki, 1967). Adrenodoxin reductase transfers the electrons to an iron/sulfur protein, adrenodoxin, located in the mitochondrial matrix. Adrenodoxin first forms a complex with adrenodoxin reductase, dissociates after oxidation, and then binds to P450scc (or to the other mitochondrial P450s, P450c11β, P450c11AS, and P450c11B3). These proteins are also often called "ferredoxin oxidoreductase" and "ferredoxin." In humans, there is one gene encoding adrenodoxin reductase, found on chromosome 17, and multiple functional adrenodoxin genes on chromosome 11, encoding identical mRNAs and proteins, and two nonfunctional adrenodoxin pseudo-genes on chromosome 20. Adrenodoxin, but not adrenodoxin reductase, is transcriptionally regulated by tropic hormones, acting though cyclic adenosine monophosphate.

C. STAR

A novel protein, steroidogenic acute regulatory protein, or StAR, was identified as important for movement of cholesterol into the mitochondria in response to trophic hormone stimulation (Clark *et al.*, 1994; Stocco and Clark, 1996). In the adrenal glands and gonads, StAR expression and function are critical for steroidogenesis. StAR is a 30-kDa phosphoprotein that is synthesized as a 37-kDa precursor, containing an *N*-terminal mitochondrial targeting sequence. Individuals who are homozygous for mutations that inactivate StAR have a marked impairment in adrenal and gonadal steroidogenesis but not in placental steroidogenesis (Lin *et al.*, 1993; Bose *et al.*, 1996). Patients with congenital lipoid adrenal hyperplasia do not appear to have neurological defects, suggesting that StAR is not necessary for neurosteroidogenesis, as it is unnecessary for placental steroidogenesis (Saenger *et al.*, 1995). Nevertheless, StAR has been identified in the rodent brain, suggesting that it may play a role in regulating neurosteroidogenesis (Furukawa *et al.*, 1998).

D. 3βHSD

PREG produced from cholesterol can undergo one of two conversions: It may be 17α-hydroxylated to 17α-hydroxypregnenolone by P450c17 (see next section), or converted to PROG by the enzyme 3βHSD (Fig. 1). 3βHSD has two distinct enzymatic activities: 3β-dehydrogenation and isomerization of the double bond from C5,6 in the B-ring (Δ^5-steroids) to C4,5 in the A-ring (Δ^4-steroids) (Luu-The *et al.*, 1989b; Thomas *et al.*, 1989; Lorence *et al.*, 1990). This enzyme is encoded by multiple distinct genes, located on human chromosome 1, that are expressed in a tissue-specific manner. There are at least two forms of human 3βHSD and at least four forms of rodent 3βHSD. The human 3βHSD type I gene is expressed in the placenta, skin, mammary gland, and other tissues, including the brain, while a distinct human type II 3βHSD gene is expressed in the adrenal glands and gonads. In rats, it is unknown whether the type I 3βHSD isoform alone or additional isoforms are expressed in the brain. The enzymes can be classified in two groups: those that function as dehydrogenase/isomerases and those that function as 3-ketosteroid reductases.

E. P450c17

PREG and PROG may undergo 17α-hydroxylation to 17α-hydroxypregnenolone and 17α-hydroxyprogesterone. These steroids may then undergo scission of the c17,20-bond to form DHEA and androstenedione (Fig. 1). All of these four reactions are mediated by a single P450c17 enzyme. P450c17 is bound to the smooth endoplasmic reticulum and accepts electrons from P450 reductase. Because P450c17 has both 17α-hydroxylase and 17,20-lyase activities, it catalyzes a key branch point in steroidogenesis (Fig. 1). In the human adrenal glands, the regional expression of P450c17 directs steroidogenesis to the production of mineralocorticoids, glucocorticoids, or sex steroids. In the human zona glomerulosa, P450c17 is not expressed, and PREG is metabolized to mineralocorticoids; in the human zona fasciculata, P450c17 is expressed but the majority of the activity is 17α-hydroxylase activity, and hence pregnenolone is directed to glucocorticoids; and, in the zona reticularis, P450c17 has both 17α-hydroxylase and c17,20-lyase activities, and hence PREG is metabolized into sex steroids. Several factors are important in determining whether a steroid will undergo 17,20-bond scission after 17-hydroxylation, such as the presence of the co-factors and the potential competition for substrate between P450c17 and 3βHSD. In the human adrenal glands, 3βHSD mRNA and activity are low in the zona reticularis and high in the zona fasciculata (Voutilainen *et al.*, 1991; Coulter *et al.*, 1996; Endoh *et al.*, 1996). In addition, there appears to be a gradient of b_5 expression in the human adrenal glands, with the highest concentration in the zona reticularis, indicating that P450c17 expressed in this zone would have greater lyase activity (see the section on cytochrome b_5) (Yanase *et al.*, 1998). The 17α-hydroxylase reaction occurs more readily than the 17,20-lyase reaction. P450c17 prefers Δ^5-substrates, especially for 17,20-bond scission, accounting for the large concentrations of DHEA in the human adrenal glands.

The single human gene encoding P450c17 is located on chromosome 10 and contains eight exons (Matteson *et al.*, 1986b).

F. P450 REDUCTASE

Both P450c17 and 21-hydroxylase (P450c21) receive electrons from a mitochondrial membrane-bound flavoprotein, P450 reductase, which is distinct from adrenodoxin reductase. P450-reductase receives electrons from NADPH and transfers them one at a time to the microsomal P450. The second electron can also be provided by cytochrome b_5 (as discussed in the following section). The amount of microsomal P450 reductase is less than that of both P450c21 and P450c17, resulting in competition for this protein. Therefore, factors that influence the association of a specific P450 with the reductase will likely influence the pathway that will be followed (e.g., c17- vs c21-hydroxylation of PROG or c17,20-lyase vs c21-hydroxylation). Electron abundance also influences the activity of P450c17; increasing electron abundance favors both 17α-hydroxylase *and* 17,20-lyase activities, while limiting electron abundance favors only 17α-hydroxylase activity. Therefore, the ratio of P450-reductase to P450c17 seems to be critical in determining the pathway of steroidogenesis (Yanagibashi and Hall, 1986; Yamano *et al.*, 1989; Lin *et al.*, 1993).

G. CYTOCHROME B_5

Cytochrome b_5 is a small, heme-containing protein that supplies electrons for many cytochrome-P450-catalyzed reactions in the liver and for the reduction of methemoglobin in erythrocytes. Cytochrome b_5 is found both in a soluble form, necessary for its function in blood cells, and as a microsomal electron donor, playing the role of co-factor for microsomal P450s. Cytochrome b_5 has been shown *in vitro* to control the 17,20-lyase activity of P450c17 (Onoda and Hall, 1982; Kominami *et al.*, 1992; Auchus *et al.*, 1998). Cytochrome b_5 specifically augments the 17,20-lyase activity of P450c17 *in vitro*, and this augmentation is produced by an allosteric modulation of P450c17, not through an electron donor (Auchus *et al.*, 1998). These results suggest that expression of b_5 may be a mechanism by which c17,20-lyase activity, and hence DHEA production, is regulated in specific regions of the brain. Cytochrome b_5 expression in the human adrenal glands may be zone specific, as b_5 expression appears to be greater in the zona reticularis than in the zona fasciculata (Yanase *et al.*, 1998). This differential expression may account for zone-specific synthesis of glucocorticoids in the zona fasciculata vs c19 steroids in the zona reticularis.

H. P450c21 (AND EXTRA-ADRENAL 21-HYDROXYLASE ACTIVITY)

Both PROG and 17-hydroxyprogesterone can be hydroxylated at c21 to yield 11-deoxycorticosterone and 11-deoxycortisol by P450c21 (Fig. 1). This enzyme has been of great clinical interest because mutations in it result in congenital adrenal hyperplasia. P450c21 is found in the smooth endoplasmic reticulum. There are two P450c21 genes that lie in the middle of the human histocompatibility leukocyte antigen (HLA) locus on human chromosome 6. Only one gene is functional in both humans and mice, but both genes are expressed in cows.

21-Hydroxylase activity has been shown in a large number of extra-adrenal tissues, especially in the fetus and in pregnant women, resulting in the conversion of PROG to 11-deoxycorticosterone (Casey and MacDonald, 1982; Casey *et al.*, 1983). Tissues with this activity include kidney, testis, ovary, skin, urinary bladder, pancreas, thymus, spleen, aorta, and brain. Analysis of RNA from various human fetal tissues using RNase protection assays demonstrated that P450c21 mRNA is not found in these tissues (Mellon and Miller, 1989). Thus, this activity is not mediated by P450c21. The presence of an additional enzyme with 21-hydroxylase activity has been demonstrated by persistence of this activity even in humans who lack a functional P450c21 gene (Zhou *et al.*, 1997). The enzyme responsible for this activity has not been identified yet.

I. P450c11

The final steps in the synthesis of glucocorticoids and mineralocorticoids are mediated by two distinct adrenocortical enzymes, P450c11β and P450c11AS, encoded by two different genes located on human chromosome 8. The conversion of 11-deoxycorticosterone and 11-deoxycortisol to corticosterone and cortisol is mediated by the mitochondrial 11β-hydroxylase, P450c11β (Fig. 1). This enzyme is found specifically in the zona fasciculata/reticularis, not in the zona glomerulosa, and is regulated by adrenocorticotrophic hormone (ACTH). P450c11AS, or aldosterone synthase, is found exclusively in the zona glomerulosa and has three distinct activities: 11β-hydroxylase, 18-hydroxylase and 18-oxidase. It therefore converts 11-deoxycorticosterone to aldosterone (Fig. 1). This enzyme is mainly regulated by the renin/angiotensin system. A third P450c11 gene, called P450c11B3, has been isolated from rat adrenal glands. P450c11B3 mRNA is expressed only during the early neonatal period and is not expressed in the fetal or adult adrenal glands. Like P450c11β, P450c11B3 is expressed in the zona fasciculata/reticularis and is regulated by ACTH. However, P450c11B3 has enzymatic activity intermediate between P450c11β and P450c11AS. It has both 11β- and 18-hydroxylase activities but has no 18-oxidase activity and, therefore, can synthesize both corticosterone and 18-OH DOC (from DOC) and 18-OH corticosterone (from corticosterone). A gene corresponding to P450c11B3 has not been found in humans (Zhang and Miller, 1996) and has not been reported in other species.

J. 17-KETOSTEROID REDUCTASE/17β-HYDROXYSTEROID DEHYDROGENASE (17βHSD)

In the adrenal glands, DHEA is converted to Δ^5-androstenediol, and Δ^4-androstenedione is converted to testosterone by 17-ketosteroid reductase (17KSR). In the ovary, estrone is converted to estradiol by similar mechanisms. These reactions are reversible but, although the reverse reactions are mediated by the same enzyme, they are given the name 17βHSD (Fig. 1). Like the multiple 3βHSD enzymes, the three 17KSR/17βHSD reactions are catalyzed by more than one enzyme (reviewed in Labrie *et al.*, 1997). So far, seven types of 17βHSD have been cloned and have been given the names types I–VII. Type I mainly catalyzes the reductive conversion of estrone to estradiol but also catalyzes the conversion of DHEA to androstenediol (Peltoketo *et al.*, 1988; Luu-The *et al.*, 1989a; Tremblay *et al.*, 1989), while type IV mainly catalyzes the oxidative conversion of estradiol to estrone (Leenders *et al.*, 1994; Adamski *et al.*, 1996; Carstensen *et al.*, 1996). Type II catalyzes the conversion of testosterone to androstenedione, androstenediol to DHEA, and estradiol to estrone. Types III and V are "androgenic," as they mainly catalyze the conversion of androstenedione to testosterone (Fig. 1).

17βHSD is an NADPH-dependent, non-P450 enzyme that is bound to the endoplasmic reticulum. It is widely found in both steroidogenic and nonsteroidogenic tissues (reviewed in Labrie *et al.*, 1997).

K. P450 ARO

The aromatization of C18 estrogenic steroids from C19 androgenic steroids is mediated by the enzyme aromatase, P450 aro, found in the endoplasmic reticulum (Fig. 1). P450 aro converts androgens to estrogens by two hydroxylations at the C19 methyl and a third hydroxylation at C2. These three hydroxylations result in the loss of C19 and aromatization of the A-ring of the steroid. These reactions, all occurring on a single active site of P450 aro, use three pairs of electrons, donated by three molecules of NADPH and P450 reductase (Thompson and Siiteri, 1973; Thompson and Siiteri, 1974).

The gene for P450aro has been cloned, is more than 75 kb, and is located on human chromosome 15 (Chen *et al.*, 1988; Means *et al.*, 1989; Mahendroo *et al.*, 1991). This gene encodes two mRNAs that differ in the length of their 3' untranslated regions (Means *et al.*, 1989).

L. 11β-HYDROXYSTEROID DEHYDROGENASE (11βHSD)

The conversion of cortisol to cortisone is mediated by 11βHSD, and the reverse reaction is mediated by an 11-oxidoreductase activity (Fig. 1). Although enzymologic studies suggest that these reactions are mediated by two different proteins, cloning and expression of 11βHSD cDNA showed that one protein has both activities (reviewed in Bujalska *et al.*, 1997). Two isoforms of 11βHSD have been described. 11βHSD is a low-affinity NADH(H)-dependent enzyme (Moore *et al.*, 1993; Stewart *et al.*, 1994) and is found mainly in human liver, decidua, lung, gonad, pituitary, and cerebellum (Tannin *et al.*, 1991; Whorwood *et al.*, 1995). By contrast, 11βHSD-2 is an NAD-dependent enzyme and is localized to the placenta and to mineralocorticoid target tissues (kidney, colon, and salivary gland) (Brown *et al.*, 1993; Albiston *et al.*, 1994; Stewart *et al.*, 1994; Whorwood *et al.*, 1995; Brown *et al.*, 1996). As 11βHSD converts cortisol to cortisone, it is thought to protect mineralocorticoid receptors, which can be bound by both glucocorticoids and mineralocorticoids, from occupation by glucocorticoids. It is also thought to protect glucocorticoid receptors from occupation by glucocorticoids.

M. 5α-REDUCTASE

Testosterone is converted to the more potent androgen dihydrotestosterone (DHT), by the enzyme 5α-reductase (Fig. 1). In the brain, PROG is converted to 5α-DHP, and 11-deoxycorticosterone is likewise reduced at the

5α-position. This membrane-bound, non-P450, is found mainly in peripheral target tissues, such as genital skin and hair follicles.

Cloning and expression studies have demonstrated the existence of two 5α-reductase genes (Andersson and Russell, 1990; Andersson et al., 1991). The gene for type I is found on human chromosome 5, is about 35 kb, and encodes a 29-kDa protein but does not encode the protein that is required for sexual differentiation (Jenkins et al., 1992). The gene for type II is found on human chromosome 2 and has the same intron/exon structure as the gene for the type I enzyme (Thigpen et al., 1992a). Mutations in the gene for 5α-reductase type II cause classic 5α-reductase deficiency (Thigpen et al., 1992b). Human types I and II genes are differentially regulated during development (Thigpen et al., 1993). The type I isozyme is not detectable in the fetus, is only transiently expressed in the newborn skin and scalp, and is permanently expressed in the skin from the time of puberty. The type II isozyme is transiently expressed in the skin and scalp of newborns and is the predominant form in fetal genital skin, male accessory sex glands, and the normal prostate and prostatic hyperplasia and adenocarcinoma tissues. In rats, the type I mRNA is expressed in basal epithelial cells, while the type II mRNA is expressed in the stromal cells of regenerating ventral prostate. In the rodent brain, both types I and II mRNAs are expressed in a developmentally regulated fashion (Poletti et al., 1998a,b).

N. 3α-Hydroxysteroid Dehydrogenase/3α-Hydroxysteroid Oxidoreductase (3αHSD)

cDNA cloning experiments revealed that most HSDs belong to one of two families: the short chain dehydrogenase/reductase family (also known as short-chain alcohol dehydrogenases) and the aldo-keto reductase family [reviewed in (Penning et al., 1997)]. Mammalian 3αHSDs are members of the aldo-keto reductase family. The reactions catalyzed by 3αHSD are stereospecific and involve the interconversion of a carbonyl with a hydroxyl group. In the prostate, 3αHSD is involved in *inactivating* DHT, by conversion to the weak androgen 3α-androstanediol; whereas, in the nervous system, 3αHSD is involved in *activating* 5α-reduced steroids, such as 5α-DHP to the potent neurosteroid allopregnanolone. Thus, this enzyme is a key regulator of both steroid hormone receptor and ion-gated receptor occupancy and action.

Most biochemical studies have used the rat liver enzyme for purification, biochemical analyses, generation of antibodies, cDNA cloning, and enzyme structure determination (reviewed in Penning et al., 1997). A single cDNA species from rat liver was cloned (Cheng et al., 1991; Pawlowski et al., 1991;

Stolz et al., 1991; Usui et al., 1994), had an open reading frame of 966 nucleotides, and predicted a protein of 322 amino acids. The rat liver 3αHSD cDNA has high-sequence identity (>70%) with the clone for human liver type I 3αHSD (DD4 or chlordecone reductase), human liver DD1 (which is both a 3α- and 20αHSD), and human liver DD2 (human bile–acid binding protein) (Deyashiki et al., 1994; Hara et al., 1996). Human liver type II 3αHSD (Khanna et al., 1995), human type III prostatic 3αHSD (Dufort et al., 1996), and a human type II expressed in the brain (called "type II$_{brain}$") (Griffin and Mellon, 1999) cDNAs have also been cloned. Human types I and II 3αHSDs differ in their K_m values for 5α-DHT, with the type I enzyme having a lower K_m. The human type III and type II$_{brain}$ also differ in their enzyme activities and substrate specificities. The type II$_{brain}$ uses 5α-reduced androgens as substrates rather than 5α-reduced PROG (5α-DHP). The K_m of the type III for dihydroprogesterone is ∼7 nM; for 5α-DHT, the K_m is ∼2 μM, indicating that this enzyme has greater affinity (∼1000-fold) for 5α-DHP. The K_m for the type II$_{brain}$ (5α-DHT as substrate) is also ∼2 μM. In addition to having substrate-selective 3αHSD activity, the type II$_{brain}$ and type III 3αHSD have 20αHSD and 17βHSD activities as well. All 3αHSD cDNAs share high-sequence identity (over 80% at both amino acid and nucleic acid levels). The genes encoding the human types I and II enzymes span approximately 20 and 16 kb, respectively, and both contain nine exons of the same size and intron/exon boundaries (Khanna et al., 1995).

O. SULFOTRANSFERASE (HST)

Sulfation of free 3β-hydroxysteroids is a major enzymatic reaction of metabolism, excretion, and homeostasis of steroids and bile acids (reviewed in Hobkirk, 1985). Sulfation and sulfohydrolation activities have been reported in lung, kidney, adrenal gland, and testis and have been described as crucial for development. 16-Hydroxy-DHEA sulfate originating from the liver, DHEA-sulfate originating from the adrenal glands, and DHEA originating from the placenta serve as precursors for the production of estrone, estradiol, and estriol in the developing human placenta (Iwamori et al., 1976; Barker et al., 1994; Parker et al., 1994) (Fig. 1). Sulfotransferases (HSTs) are a family of cytosolic enzymes that conjugate steroid and phenolic substrates with inorganic sulfate derived from an active donor, adenosine 3′-phosphate 5′-phosphosulfate (PAPS); (De Meio, 1975; Mulder, 1981). The hydroxysteroid HST specifically uses Δ^5-steroid substrates that are hydroxylated at C3, C5, C17, or C21 (De Meio, 1975). The resulting steroid sulfate esters are hydrophilic; however, sulfation has a role greater than facilitating secretion because it can also change the pharmacological activity of steroids, such as changing the way in which PREG binds to the GABA$_A$ receptor (reviewed in Majewska, 1991, 1992). HST is mainly found in the adrenal glands in humans and in the liver in other mammals (Ogura *et al.,* 1989).

P. SULFATASE (STS)

The steroid sulfohydrolase is a sterol-sulfate sulfohydrolase also known as steroid sulfatase (STS) or steroid 3-sulfatase. It specifically hydrolyzes sulfate groups in the 3β-position of Δ^5-steroids, such as PREG, DHEA, and androstenediol (Fig. 1). STS is an important enzyme in steroid metabolism because its activity increases the pool of precursors that can be metabolized by other steroidogenic enzymes to produce biologically active sex steroids (De Meio, 1975). The human STS gene has been cloned and mapped to chromosome Xp22.3, proximal to the pseudo-autosomal region, and the genetic aspect of this enzyme has been widely documented (Ballabio and Shapiro, 1995). Deficiency of STS activity results in severe ichthyosis caused by accumulation of steroid sulfates in the stratum corneum of the skin (Shapiro, 1982; Mohandas *et al.,* 1987). Rat and mouse STS have recently been cloned and are fairly dissimilar. Mouse STS cDNA is 75% identical to rat STS cDNA and only 63% identical to human STS cDNA (Li *et al.,* 1996; Salido *et al.,* 1996).

Q. OTHER NEUROSTEROIDOGENIC ENZYMES

Other enzymatic activities that modify steroid hormones have been identified in the brain, but the activities of the resulting neurosteroid products have not yet been characterized. Those enzymes are the 7α-hydroxylase, a novel 7α-hydroxylase expressed primarily in the brain (Stapleton *et al.,* 1995; Rose *et al.,* 1997), 24-hydroxylase (Lund *et al.,* 1999), and 26α-hydroxylase (Ray *et al.,* 1997).

1. *24-Hydroxylase*

The turnover of cholesterol in the brain is thought to occur via conversion of excess cholesterol into 24S-hydroxycholesterol. This compound, an oxysterol, is readily secreted from the CNS into the plasma. Cloning of human and mouse cDNA sequences has shown that both proteins are localized to the endoplasmic reticulum, share 95% identity, and represent a new cytochrome P450 subfamily (CYP46) (Lund *et al.,* 1999). Transfection studies have revealed that the proteins encoded by the cDNAs convert cholesterol into 24S-hydroxycholesterol and, to a lesser extent, 25-hydroxycholesterol.

The cholesterol 24-hydroxylase gene contains 15 exons and is located on human chromosome 14q32.1.

Analysis of protein and RNA has demonstrated that cholesterol 24-hydroxylase is expressed predominantly in the brain. In the brain, 24-hydroxylase is expressed in neurons in several regions. The concentrations of 24S-hydroxycholesterol in serum are low in newborn mice, reach a peak between postnatal days (PD) 12 and 15, and thereafter decline to baseline levels; whereas the 24-hydroxylase protein is first detected in the brain of mice at birth and continues to accumulate with age.

2. 26α-Hydroxylase

Another novel member of the cytochrome P450 gene family has been cloned from retinoic-acid-induced neural differentiation of embryonic stem cells (Ray *et al.*, 1997) from zebrafish (White *et al.*, 1996) and from a human teratocarcinoma cell line (White *et al.*, 1997). This cDNA, called CYP26, encodes a 1.9-kb mRNA transcript that is regulated during embryonic stem cell neural differentiation, and it is also expressed in liver. CYP26 cDNA is 1701 nt long, containing an open reading frame of 1491 nt, that encodes a 56-kDa protein of 497 amino acids. Consistent with its induction by retinoic acid, CYP26 is expressed in the early mouse embryo from at least E8.5. However, there is only low, if any, expression in the body or head of E12.5 embryos. In the adult mouse, CYP26 mRNA is expressed mainly in the liver and the brain, with very low amounts in other tissues. By reverse transcription–polymerase chain reaction (RT–PCR) analysis, CYP26 mRNA was detected in a variety of human brain regions, including the olfactory bulb, temporal cortex, and hippocampus. This protein appears to be involved in metabolism of retinoic acid (White *et al.*, 1996), and other substrates have not been reported.

3. 7α-Hydroxylase

7α-Hydroxylase is a microsomal P450 that hydroxylates steroids at the C7α-position. Although its activity was originally described in liver protein extract, where 7α-hydroxylated bile acids are produced for detoxification 7α-hydroxylated derivatives of PREG and DHEA appear to modulate glucocorticoid action, functioning as antiglucocorticoids and modulating immune function (Morfin and Courchay, 1994). The stereospecific 7α-hydroxylation of steroids is mediated by at least two different enzymes, initially identified by competition experiments (Doostzadeh *et al.*, 1998). In addition, there appear to be at least two different classes of 7α-hydroxylases. 7-Hydroxylation of cholesterol and of Δ^4-3-oxosteroids is mediated by one specific P450 (Wood *et al.*, 1983; Swinney *et al.*, 1987; Arlotto *et al.*, 1989; Tzung *et al.*, 1994). This enzyme(s) is necessary for conversion of both cholesterol and 27-hydroxycholesterol into bile acids but does not mediate 7α-hydroxylation of PREG and DHEA. Another brain-enriched P450 7α-hydroxylase, called CYP7B, was identified and its cDNA cloned (Stapleton *et al.*, 1995; Rose *et al.*, 1997). This latter cDNA encodes a protein that catalyzes the 7α-hydroxylation of PREG and DHEA in the brain and also catalyzes 7α-hydroxylation of oxysterols. In addition to expression in the brain, CYP7B is found in rat liver and kidney but is barely detectable in testis, ovary, and adrenal gland. Expression in the liver is dimorphic, with less expression in female rats than in male rats. CYP7b is widely expressed in the mouse brain, regionally expressed in the hippocampus (Stapleton *et al.*, 1995), and the most highly concentrated in the corpus callosum. Among the P450s, CYP7B is most similar (39% at the amino acid sequence) to cholesterol 7α-hydroxylase (CYP7) and contains a postulated steroidogenic domain present in other steroid-metabolizing CYPs but clearly represents a novel type of CYP. The formation of 7α-hydroxylated metabolites of PREG and DHEA is low in prepubertal rats and increases five-fold in adults (Akwa *et al.*, 1992). It is unknown if brain expression of CYP7B parallels this enzymatic profile.

III. Enzymes Involved in Neurosteroidogenesis: Distribution in the Brain and Developmental Regulation

The presence of functional steroidogenic enzymes in the brain has been established using enzymatic activity measurements, mRNA expression (using RT-PCR, ribonuclease protection assays, or *in situ* hybridization), and protein expression (using Western blotting or immunocytochemistry). These studies, by numerous laboratories over the years, were originally done mainly in the adult rat brain but now include analyses of neurosteroidogenesis in the adult lungfish, frog, bird, and guinea pig. Only a limited number of reports document steroidogenic enzyme expression during embryogenesis in rodents (Lauber and Lichtensteiger, 1994, 1996; Compagnone *et al.*, 1995a,b, 1997). However, such studies are crucial because neurosteroids have diverse functions during development as well as in the adult.

A. EXPRESSION OF P450SCC IN THE ADULT RAT BRAIN

Initial support for the hypothesis that the brain is a steroidogenic organ derives from experiments demonstrating conversion of radioactive cholesterol to PREG (reviewed in Baulieu and Robel, 1990). Although these experiments suggested that the brain had steroidogenic capacity, the demonstration that CNS tissue and neuronal cells contained P450scc mRNA and protein was hampered by the extremely low amount of this mRNA. Using RT–PCR and Southern blotting, we showed that a very P450scc mRNA was regionally expressed in extremely low amounts (Mellon and Deschepper, 1993). In the adult rat, it was found most abundantly in the cortex, and to a lesser extent in the amygdala, hippocampus, and midbrain of both male and female rats. Purification of mixed primary glial cultures showed that type I astrocytes synthesized P450scc mRNA (Mellon and Deschepper, 1993). Western blotting and immunocytochemistry showed that P450scc protein was almost as abundant in neonatal cultures of forebrain astrocytes as in mouse adrenocortical Y-1 cells, whereas P450scc mRNA was present in orders of magnitude less abundant, suggesting that the protein was stable in the brain (Compagnone *et al.*, 1995a).

B. Developmental Regulation of P450scc Expression in the CNS and PNS

P450scc mRNA and protein are expressed very early in development. P450scc mRNA was present as early as embryonic day (ED) 7.5, and its amount in the whole embryo increased until ED 9.5. This increased expression was from placental expression, rather than from embryo expression (Durkee *et al.*, 1992; Compagnone *et al.*, 1995a). Although P450scc mRNA could be readily detected in the developing gonads and adrenal glands, it could not be easily detected in the developing nervous system. However, by analyzing P450scc protein with immunocytochemistry, we demonstrated P450scc expression as early as ED 9.5 in the mouse (ED 10.5 in the rat) in cells in the neural crest (Compagnone *et al.*, 1995a). Expression of P450scc continued mainly in structures derived from the neural crest during embryogenesis and was found in the neuroepithelium, the retina, the trigeminal, and the dorsal root ganglia (DRG) of the neuroectoderm and the thymus (Compagnone *et al.*, 1995a). Consistent with a neural crest origin, cells expressing P450scc belong to several different cell lineages. P450scc was found in neurons of the DRG, trigeminal ganglia, and in the neuroectoderm, and it was found in glial lineages in the CNS. P450scc was also found in cells not derived from the neural crest, such as motor neurons (Compagnone *et al.*, 1995a), Purkinje cells in the cerebellum (Ukena *et al.*, 1998), in oligodendrocytes (Hu *et al.*, 1987), and astrocytes (Mellon and Deschepper, 1993) in various regions of the brain from late embryogenesis to adulthood. In the PNS, P450scc was expressed from ED 10.5 in condensing DRG and in cranial ganglia (Compagnone *et al.*, 1995a). Other groups have also reported activity and expression of P450scc protein in peripheral nerves (Morfin *et al.*, 1992). Thus, P450scc is expressed in a variety of cell types both in the CNS and in the PNS. Its expression is initiated in the developing neural tube and in the neural crest, before organogenesis of the adrenal glands or of the gonads, suggesting a role for neurosteroids in neural development. Its expression in restricted areas of the brain does not seem to be developmentally regulated, as P450scc expression persists during adulthood.

C. Expression of Adrenodoxin and Adrenodoxin Reductase in the Brain

Adrenodoxin mRNA was found in virtually all rodent tissues examined (Mellon *et al.*, 1991), including the brain. This is not surprising, as this protein functions with all mitochondrial P450s, not just the steroidogenic P450s. Similarly, as expected, adrenodoxin protein has also been found in brain homogenates (Oftebro *et al.*, 1979). The exact site of expression of these proteins has not been systematically examined either in adult brain or during development. The expression of adrenodoxin reductase has not been documented yet in the CNS or in the PNS.

D. Expression of StAR in the Brain

StAR, a 30-kDa protein involved in the acute transport of cholesterol to the inner mitochondrial membrane, is stimulated by tropic hormones in the adrenal glands and gonads (Clark *et al.*, 1994). Its lack of expression in the placenta further suggested that StAR was expressed in steroidogenic tissues having an "acute" response, but not in tissues (placenta, and perhaps brain, also) in which there is not an acute response (Saenger *et al.*, 1995). In addition, patients lacking StAR activity had normal placental steroidogenesis and had no apparent neurological defects, suggesting that StAR may not be important for neurosteroidogenesis (Lin *et al.*, 1995; Saenger *et al.*, 1995). Similar phenotypes and lack of neurological impairment were seen in a mouse StAR knockout line (Caron *et al.*, 1997). As early studies failed to detect StAR mRNA in the brain (Lin *et al.*, 1995), its role in neurosteroidogenesis was questioned. However, more recent studies *have* demonstrated StAR transcripts in rat brains, at levels two to three orders of magnitude less than in the adrenal glands (Furukawa *et al.*, 1998). This is similar to the relative abundance of P450scc in the brain vs the adrenal glands. This study also demonstrated StAR's localization in the rat brain, and co-localization with P450scc and 3βHSD, using *in situ* hybridization. StAR transcripts were abundant in the cerebral cortex, hippocampus, dentate gyrus, olfactory bulb, cerebellar granule cell layer, and Purkinje cells. P450scc and 3βHSD co-localized with StAR in the hippocampus, dentate gyrus, cerebellar granule layer, and Purkinje cells. It is unclear whether

StAR plays the same role in the brain as it does in the adrenal glands and gonad, whether it is constitutively expressed in the brain, or whether its function is acutely regulated by some as yet undefined factor.

E. DEVELOPMENTAL REGULATION OF P450c17 EXPRESSION IN THE NERVOUS SYSTEM

Initial studies using highly sensitive RT-PCR analyses failed to detect expression of P450c17 mRNA in any region of the adult rat brain (Mellon and Deschepper, 1993). Those studies also indicated that the steroidogenic enzyme proteins may be more easily detected than their mRNAs because, using an antibody directed against human P450c17, we determined the pattern of expression of P450c17 in developing rodent embryos (Compagnone *et al.*, 1995b). P450c17 expression is restricted to neurons in specific regions of the developing brain. We first observed P450c17 in ED 10.5 cells migrating from the neural crest and condensing in DRGs. Immunopositive cells were also observed in the neural tube in the position of the lateral motor column. Neuronal cell bodies immunopositive for P450c17 were restricted to the hindbrain mesopontine system (from ED 14.5), the thalamus, and the neocortical subplate (from ED 16.5). Fibers extended in the areas of projections of these nuclei. We also found fibers coming from the trigeminal ganglion and the retina.

In all of the CNS regions, P450c17 expression was transient, and P450c17 immunostaining gradually disappeared in the first week of life. In the neonatal mouse and rat (Day 0 to Day 7), P450c17 was mainly detected in the brainstem and in the cerebellum, internal capsule, olfactory tract, hippocampus, stria terminalis, and thalamus. Few immunopositive cell bodies could be detected in any region of the adult rat brain, although fiber tracts could still readily be observed. In the PNS, P450c17 persisted in the adult, both in cell bodies and fibers. Thus, unlike P450scc, P450c17 is expressed in the CNS only during development but persists in the PNS throughout life.

DHEA has been one of the first neurosteroids identified in the adult rodent brain, although at low concentrations. However, P450c17 mRNA and protein are not found in cells of the CNS in adult animals. We propose that, in the rodent, peripheral expression of P450c17 provides the brain with DHEA found in adult rat brains. An alternative pathway has been proposed for DHEA generation in the brain involving a Fe^{2+} sensitive chemical reaction independent of P450 (Cascio *et al.*, 1998). However, it is unlikely that such a process occurs *in vivo* because the concentrations of Fe^{2+} necessary to induce production of DHEA (in the 10-mM range) in cells lacking P450c17 is far out of a physiological range. We, thus, believe that, in the adult, DHEA is delivered to the spinal cord and brain via the peripheral nerves and terminals that express P450c17 and synthesize DHEA.

F. EXPRESSION AND ACTIVITY OF P450 REDUCTASE AND b_5 IN THE BRAIN

P450-reductase activity, protein (Norris *et al.*, 1994), and mRNA (Simmons and Kasper, 1989; Norris *et al.*, 1994) have been located in the brain. Cytochrome-P450-reductase immunoreactivity was detected mainly in neurons, but also in some glial populations, and it appears to be expressed widely in the rat CNS (Norris *et al.*, 1994). Although no studies specifically addressed the co-localization of P450 reductase and P450c17 in the same cells, the regional distribution of these proteins suggests that P450c17 may have functional activities in the cortical thalamic and pontine regions, where we described its expression. The expression of b_5 has not been documented yet in the CNS or the PNS.

G. 3βHSD EXPRESSION IN THE ADULT NERVOUS SYSTEM

Demonstration of the conversion of PREG to PROG in cultured rat oligodendrocytes (Hu *et al.*, 1987; Jung-Testas *et al.*, 1989), astrocytes (Kabbadj *et al.*, 1993), cultured rodent glia and neurons (Bauer and Bauer, 1989), and in discrete regions of the rat and monkey brain (Weidenfeld *et al.*, 1980; Robel *et al.*, 1987) suggests that 3βHSD is expressed in those regions. Expression of 3βHSD protein and mRNA has been studied in brains from rats (Dupont *et al.*, 1992; Guennoun *et al.*, 1995; Ukena *et al.*, 1999) and in considerable detail in frogs (Mensah-Nyagan *et al.*, 1994) and lungfish (Mathieu *et al.*, 2000), demonstrating that this enzyme is expressed in the brains from both mammalian and nonmammalian vertebrates. There are conflicting reports as to whether 3βHSD type I alone (Guennoun *et al.*, 1995) or types I, II, and IV (Sanne and Krueger, 1995) are the isoforms that are expressed in the adult rat brain. While 3βHSD activity has been reported in both neurons and glia, 3βHSD mRNA and protein appear to be expressed only in neurons (Dupont *et al.*, 1992; Guennoun *et al.*, 1995; Ukena *et al.*, 1999). 3βHSD mRNA was found in the brain in much lower amounts than in the ovary, adrenal gland, and liver. Results from different laboratories are conflicting and relate to the specificity of the probe/antibody to a particular subtype of the enzyme and to the sensitivity of the technique used (Dupont *et al.*, 1992; Guennoun *et al.*, 1995; Ukena *et al.*, 1999). An extensive study using immunocytochemistry in the frog showed the restricted expression of 3βHSD in neurons (Mensah-Nyagan *et al.*, 1994) in regions consistent with the pattern of expression of 3βHSD type I mRNA in the rat (Guennoun *et al.*, 1995), except in the cerebellum and in the cortex, where only 3βHSD mRNA was reported (Guennoun *et al.*, 1995; Ukena *et al.*, 1999). However, reports of 3βHSD activity did not always correlate with expression of 3βHSD protein. The lack of significant activity in the rat cerebellum and cortex (Weidenfeld *et al.*, 1980; Robel *et al.*, 1987) correlates with the lack of 3βHSD protein in these regions in the frog (Mensah-Nyagan *et al.*, 1994), but not with the presence of 3βHSD mRNA in the rat (Guennoun *et al.*, 1995). A report described transient 3βHSD activity in cerebellum slices obtained from neonates (Ukena *et al.*, 1999). High levels of 3βHSD activity have been reported in the rat amygdala and the hippocampus (Weidenfeld *et al.*, 1980); and 3βHSD mRNA was detected in the rat hippocampus, but 3βHSD protein was not detected in this region in the frog (Mensah-Nyagan *et al.*, 1994).

H. P450c11β AND P450c11AS EXPRESSION IN THE ADULT NERVOUS SYSTEM

Analyses of RNA from different regions of the adult rat brain indicated that P450c11β, but not P450c11AS, was region-specifically expressed in the brain (Mellon and Deschepper, 1993). The abundance of P450c11β mRNA was greater than that of P450scc mRNA, as it could be detected readily by RNase protection assays. We showed that P450c11β mRNA was found in virtually all regions analyzed (cortex, hippocampus, hypothalamus, amygdala, cerebellum, and midbrain) with the greatest amount in the cortex. We also found that there may be differences between male and female rats in the expression of P450c11β mRNA in the hippocampus, as it appears to be found in greater amounts in the female.

Initial studies in our laboratory failed to detect P450c11AS mRNA in any region of the adult rat brain, analyzing RNA by both RNase protection assays or by RT-PCR (Mellon and Deschepper, 1993). However, more recently, others have shown that P450c11AS mRNA and protein are indeed expressed in adult rat brains (MacKenzie *et al.*, 2000). In the brain, P450c11β and P450c11AS proteins were detected in the cerebellum, especially in Purkinje cells, as well as in the hippocampus. Moreover, P450c11AS activity has also been demonstrated in the brain (Gomez-Sanchez *et al.*, 1997). Thus, the adult rat brain has the ability to synthesize aldosterone.

I. EXPRESSION OF 17βHSD IN THE BRAIN

Demonstration of 17βHSD activity in the brain of rats and monkeys (Reddy, 1979; Resko *et al.*, 1979) indicated that 17βHSD protein and mRNA would also be present. Analyses of both rat and frog brains have demonstrated that 17βHSD type I is expressed in the brain (Pelletier *et al.*, 1995; Mensah-Nyagan *et al.*, 1996a; Mensah-Nyagan *et al.*, 1996b). In adult rat brains, 17βHSD type I protein was detected in non-neuronal cells (GFAP-positive cells) (Pelletier *et al.*, 1995). In the frog brain, 17βHSD type I was also found in glia (Mensah-Nyagan *et al.*, 1996b). Regional distribution of 17βHSD protein and mRNA in the rodent and in the frog are consistent. In humans, 17βHSD activity was detected in the cortex, and greater activity was found in the subcortical white matter (Steckelbroeck *et al.*, 1999). Analysis of 17βHSD mRNAs in human brains demonstrated that 17βHSD-1, 17βHSD-3, and 17βHSD-4 were expressed in the temporal lobe of children and adults, whereas 17βHSD-2 and the pseudo-gene of 17βHSD-1 were not (Steckelbroeck *et al.*, 1999). In adults, 17βHSD-3 and 17βHSD-4 mRNA concentrations were significantly higher in the subcortical white matter than in the cortex, similar to what was found for enzyme activity, while 17βHSD-1 concentrations were similar in both tissues. In another study, in humans, 17βHSD-1, -3 and -4 mRNAs, but not 17βHSD-2 mRNA, were detected in the hippocampus (Beyenburg *et al.*, 2000). Similar amounts of these mRNAs were detected in hippocampi from men and women.

J. EXPRESSION OF P450 ARO IN THE BRAIN

Several lines of evidence demonstrate that testosterone is aromatized to estradiol in the brain of many species, from frogs and songbirds to humans (reviewed in Hutchison *et al.*, 1996; Lephart, 1996). Aromatase activity in the rat brain was shown to be limited to discrete regions (Roselli *et al.*, 1985). Aromatase activity was first detected on ED 15, increased up to ED 19, and then declined to adult levels (George and Ojeda, 1982; MacLusky *et al.*, 1985; Lephart *et al.*, 1992).

While three distinct P450 aro mRNA species of 2.7, 2.2, and 1.7 kb exist in many tissues, only the 2.7-kb P450 aro mRNA was found in the brain (Lephart *et al.*, 1992). In humans, the major transcript in the hypothalamic preoptic area and amygdala contained exon I-f, referred to as the "brain-specific exon I" (Honda *et al.*, 1994), while a minor transcript contained exon I-b (Honda *et al.*, 1994; Toda *et al.*, 1994). In rat cortex, an additional novel P450 aro transcript was identified that contained exons IV–X, but initiated transcription within intron III and did not contain exons I, II, or III (Kato *et al.*, 1997). Thus, the rat brain contains P450 aro transcripts that initiate at three different sites. It is not clear if the three different P450 aro transcripts correspond to different sites of transcription initiation or if they are variants that lack the heme-binding domain (Lephart, 1997).

The regulation of P450 aro mRNA expression during rodent development was studied by *in situ* hybridization (Lauber and Lichtensteiger, 1994). In neurons, P450 aro mRNA appeared to parallel aromatase activity, but P450 aro mRNA was also found in regions not previously associated with activity. The regional distribution of P450 aro mRNA was restricted to the preoptic/hypothalamic area on ED 16 and was more widely distributed by ED 18 to ED 20 but was still absent from the cortex, midbrain, and hindbrain structures. From PD 2 to adulthood, P450 aro mRNA abundance decreased in the preoptic area but remained constant in other areas. In the adult, P450 aro mRNA was still present with a similar pattern of expression. Female rats had the same distribution of P450 aro mRNA, but the number of cells expressing P450 aro mRNA in each region was less than in males rats (Wagner and Morrell, 1996).

The regions of the brain that contained P450 aro mRNA but did not contain activity or were not examined include the mediodorsal thalamus, subfornical organ and cingulate cortex, and the hippocampus. P450 aro immunoreactivity has been demonstrated in the same areas in which P450 aro mRNA has been detected (Sanghera *et al.*, 1991), indicating that the lack of aromatase activity may be due to the sensitivity of the assay.

K. Expression of 11βHSD in the Brain

Both 11βHSD-1 (Moisan *et al.*, 1990; Lakshmi *et al.*, 1991) and 11βHSD-2 (Roland *et al.*, 1995; Zhou *et al.*, 1995; Brown *et al.*, 1996) activities and mRNA have been detected in adult rat brains. During development, 11βHSD-2 mRNA was widely expressed in the CNS throughout the neuroepithelium, thalamus, and spinal cord until ED 12.5. It declined thereafter (Brown *et al.*, 1996). Expression in the thalamus and cerebellum persisted postnatally.

L. Expression of 5α Reductase in the Brain

The rat CNS is capable of converting PROG into 5α-reductase metabolites [reviewed in (Celotti *et al.*, 1992)], indicating that this enzyme is present in the CNS. 5α-reductase activity, protein, and mRNA have been detected in neurons, astrocytes, and glia (Celotti *et al.*, 1992; Melcangi *et al.*, 1993, 1994; Pelletier *et al.*, 1994; Lauber and Lichtensteiger, 1996), and the predominant isoform is type I (Normington and Russell, 1992). The occurrence of 5α-reductase activity has also been reported in the brain of teleost fish (Pasmanik and Callard, 1985; 1988) and the distribution of 5α-reductase-like immunoreactivity has been described in the brain of the lungfish *Protopterus annectens* (Mathieu *et al.*, 2000). Type I mRNA is more abundant during development than in the adult, but its overall level of expression during development remains relatively constant (Mahendroo *et al.*, 1997; Poletti *et al.*, 1998b). Expression of 5α-reductase type II mRNA in the brain is, however, developmentally regulated (Poletti *et al.*, 1998a; Poletti *et al.*, 1998b). RT-PCR analyses of embryonic and neonatal rat brain mRNAs demonstrated that, whereas type I mRNA was constitutively expressed in the rat brain throughout late gestation, neonatal period, and adulthood, the type II mRNA was transiently expressed only at the end of gestation (ED 18) through the early postnatal period, dramatically falling by PD 14. This pattern of expression of type II mRNA correlated with testosterone synthesis in the fetal testis and may also suggest that 5α-reductase type II is involved in regulating brain sex differentiation at a critical period. In the adult rat, immunocytochemical staining detected 5α-reductase type I protein throughout the brain, with no differences between male and female rats.

The pattern of expression of 5α-reductase during embryogenesis was greatly regulated during development (Lauber and Lichtensteiger, 1996; Compagnone and Mellon, 2000). First observed at ED 12 in the spinal cord and neuroepithelial walls of the brain ventricular system, its distribution widened to highly proliferative zones of the CNS from ED 14 to ED 16. At ED 16.5, 5α-reductase mRNA and protein were detected in the cortex, thalamus, cerebellum, medulla, spinal cord, and in peripheral ganglia. 5α-reductase mRNA was found in lesser amounts in the trigeminal ganglia and inferior ganglia of the nodose-petrosal nerves (IX and X) than in the spiral ganglia (Lauber and Lichtensteiger, 1996). However, 5α-reductase type I protein expression was found at similar level in all peripheral ganglia (trigeminal, DRGs, nodose-petrosal, or spiral ganglia) . In late embryogenesis (ED 18–ED 20), 5α-reductase type I mRNA expression and protein were similar in distribution, but decreased in all areas except in the trigeminal ganglia and the ventricular zones of striatum and amygdala, and the thalamus. In the neonate, expression of 5α-reductase was higher in the PNS (trigeminal ganglia, DRGs) than in the CNS and, by 2 weeks of age, 5α-reductase mRNA expression changed completely and was evenly distributed throughout the brain. At this time, white matter structures (e.g., optic chiasma, lateral olfactory tract, internal capsule, corpus callosum) were highly positive for 5α-reductase mRNA, but other regions (hippocampus, striatal neuroepithelium) were also positive (Lauber and Lichtensteiger, 1996). Our results demonstrate that 5α-reductase is expressed in the same location as its mRNA (Compagnone and Mellon, 2000). We and others further established a transient expression of 5α-reductase in the proliferative zones of the developing CNS, suggesting that 5α-reduced steroid may have a role in neuronal proliferation.

Like P450scc, 5α-reductase is expressed in several cell types, including neurons in the hindbrain, cerebellum, spinal cord, and PNS and in a majority of glial cells in the other regions of the CNS. 5α-Reductase and P450scc are colocalized in the same cells in the developing neocortex. In other locations, such as the neuroepithelium, the spinal cord, and the peripheral ganglia, where both P450scc and 5α-reductase type I were expressed, P450scc and 5α-reductase were not always found in the same cells. In the adult DRGs, 5α-reductase was also found in cells expressing P450c17 but not P450scc. The co-localization of P450c17 and 5α-reductase in the regions of the developing brain where they both are expressed remains to be tested. However, the expression of both enzymes in those regions suggests that steroids other than 5α-DH PROG or allopregnanolone may be synthesized in regions such as the hindbrain, the hippocampus, and the spinal cord. Further detection of 17βHSD and P450 aro in those same regions could argue for local androgen and/or estrogen production during embryogenesis in sexually dimorphic regions. In the lungfish brain, 5α-reductase- and 3βHSD-immunoreactive neurons are located in the same brain regions, including the lateral pallium, thalamic nuclei, periventricular preoptic nucleus, dorsal and ventral hypothalamic nuclei, mesencephalic tectum, periaqueductal gray, and nucleus cerebelli (Mathieu *et al.*, 2000).

Brain-expressed 5α-reductase type I may be crucial for regulation of estrogen synthesis during pregnancy. Studies created a transgenic mouse that had the 5α-reductase type I gene ablated (Mahendroo *et al.*, 1997). About half of homozygous mutant mice died in midgestation between ED 10.75 and ED 11, resulting from estrogen excess. Administration of exogenous estrogens to normal wild-type mice also caused fetal death, and antagonists of estrogen action could prevent fetal demise in the homozygous knockout animals, demonstrating that estrogens, rather than a lack of 5α-reduced steroids, was causative of fetal death. Thus, 5α-reductase type I plays a major role in the fetus in inactivating testosterone, thereby reducing available substrate for estrogen synthesis. The results from the 5α-reductase type I knockout mouse also indicate that it is the inactivation of testosterone mediated by 5α-reductase *type I*, and not by 5α-reductase *type II*, that is crucial for fetal development and survival. 5α-reductase type I mRNA is induced in the brain with pregnancy. Its increased expression in the brain may regulate estrogen levels in normal wild-type animals, perhaps through feedback mechanisms involving the hypothalamic-pituitary-gonadal axis.

M. Expression of 3αHSD in the Brain

Human 3αHSD type II, type II$_{brain}$, and type III mRNAs have been found in the brain, while human 3αHSD type I mRNA was found only in the liver

(Khanna et al., 1995; Griffin and Mellon, 1999). In humans, both type II$_{brain}$ and type III are found in various regions, but are found in different amounts. Most regions of the human brain express 3αHSD type II$_{brain}$ and type III, assessed by Northern blots of polyA$^+$ mRNA, including the cerebellum, cortex, medulla, spinal cord, occipital, frontal and temporal lobes, putamen, amygdala, caudate nucleus, corpus callosum, hippocampus, substantia nigra, thalamus, and subthalamic nucleus. Extensive biochemical studies demonstrated 3αHSD activity in various regions of the rodent brain, and its regulation by estrogens and lactation [reviewed in (Karavolas and Hodges, 1990)]. 3αHSD activity was found mainly in type I astrocytes, but was also detected in oligodendrocytes (Melcangi et al., 1994). Western blots of proteins extracted from different rodent brain regions demonstrated that the olfactory bulb contained the greatest abundance of 3αHSD protein (Khanna et al., 1995).

N. EXPRESSION OF HST IN THE BRAIN

The presence of abundant quantities of steroid ester sulfates in the brain of gonadectomized and adrenalectomized animals has been noticed since the early 1980s, when the concept of neurosteroid was proposed (Corpéchot et al., 1981, 1983). We and others have observed that several neurosteroids and their respective sulfate esters have different pharmacological properties [reviewed in (Majewska, 1991, 1992)] and different biological effects in the developing brain (Compagnone and Mellon, 1998), suggesting that both steroid HST and steroid sulfohydrolase activities may play crucial roles in the modulation of the physiological effects of neurosteroids. However, very little data are available on the expression and enzymatic activity of steroid HSTs in human or rodent fetal or adult brains. The distribution of HST-like immunoreactivity has been described in the brain of the frog *Rana ridibunda* using an antiserum against rat liver HST (Beaujean et al., 1999). Two populations of HST-immunoreactive cell bodies were found in the diencephalon, namely in the anterior preoptic area and in the dorsal part of the magnocellular preoptic nucleus. Several bundles of HST-positive nerve fibers were visualized in the telencephalon and diencephalon. The perikarya containing HST-like immunoreactivity were not labeled with antisera against glial fibrillary acidic protein or galactocerebrosides I and II, two selective markers of astrocytes and oligodendrocytes, respectively. In addition, the immunoreactive processes exhibited the varicose aspect of beaded nerve fibers. These observations clearly indicate that, in the frog brain, the HST-immunoreactive material is expressed exclusively in neurons (Beaujean et al., 1999).

Despite the presence of a higher concentration of steroid ester sulfate than of free steroids in the brain, and despite the ubiquitous presence of

PAPS, a co-substrate of the steroid HST, there is no conclusive evidence for expression of HST in the brain of mammals. Several studies failed to show HST activity in the human fetal brain (Iwamori *et al.*, 1976). Immunocytochemistry did not reveal any expression of HST in fetal or adult human brains nor in adult rat brains (Sharp *et al.*, 1993; Parker *et al.*, 1994). However, using human fetal brain slices, HST activity was detected (Knapstein *et al.*, 1968). Subsequently, others have characterized brain and liver HST activities from immature rats (Rajkowski *et al.*, 1997). These data demonstrated that brain HST activity (kinetics, pH, and specificity for substrate) was not comparable to the liver HST activity, suggesting that brain and liver HSTs may be different isozymes. These authors found that brain HST activity was relatively high in late embryogenesis, remained high until puberty, and then decreased in the adult, consistent with previous failures to detect HST activity in the adult rat brain. This ontogeny of HST activity in the brain is the opposite of the ontogeny in the liver, where HST activity is absent during embryogenesis and rises following birth (Rajkowski *et al.*, 1997). Detection of HST activity in the brain of the frog *Rana ridibunda* indicates that biosynthesis of sulfated neurosteroids in the CNS occurred early during evolution (Beaujean *et al.*, 1999).

O. ACTIVITY AND EXPRESSION OF STS IN THE BRAIN

Previous studies have demonstrated steroid STS activity in the brain as well as in a variety of tissues other than its major site of expression in the placenta (Iwamori *et al.*, 1976). We demonstrated the sites of expression of STS mRNA using a mouse STS riboprobe (Compagnone *et al.*, 1997). The sites of STS expression provide more precise indications of the target tissues in which active, nonsulfated steroid hormones resulting from STS activity may act during embryogenesis. In addition to the known roles of STS in the placenta and in the skin, our results emphasized other possible roles for STS in the function of other organs during embryogenesis. In the CNS and the PNS, the expression of mSTS mRNA may be related to the expression of steroidogenic enzymes involved in neurosteroid synthesis. We previously showed the expression of P450c17 in restricted areas of the developing nervous system (Compagnone *et al.*, 1995b), particularly in the neocortex, and demonstrated a role for both DHEA and DHEAS in neocortical neurons differentiation. Whereas DHEA promoted axonal growth, DHEAS promoted dendritic growth and cell clustering (Compagnone and Mellon, 1998). The expression of STS in the neocortex, hippocampus, and thalamus may regulate the DHEA:DHEAS ratio. Furthermore, as the thalamus has been shown to influence the survival of cortical subplate (Price and Lotto, 1996), where

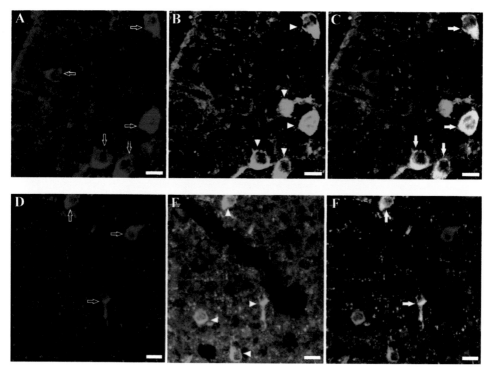

FIG. 2. Dual-channel confocal laser scanning microscope photomicrographs comparing the distribution of peripheral-type benzodiazepine receptor (PBR)- or γ-aminobutyric acid type-A-receptor β_2/β_3-subunit-like immunoreactivity and 3β-hydroxysteroid dehydrogenase (3βHSD)-like immunoreactivity in the frog hypothalamus. (**A–C**) Frontal section through the anterior preoptic area labeled with a rabbit antiserum against human placental Type 1 3βHSD revealed by a donkey anti-rabbit second antibody coupled to Texas Red (**A,** *open arrows*) or chicken anti-PBR immunoglobulins revealed by a mouse anti-chicken second antibody coupled to fluorescein isothiocyanate (**B,** *arrowheads*). A combination of the two images acquired in **A** and **B** showing that four neurons simultaneously contain 3βHSD- and PBR-like immunoreactivity (**C,** *solid arrows*). (**D–F**) Frontal section through the posterior tuberculum labeled with a rabbit antiserum against human placental Type 1 3βHSD revealed by a donkey anti-rabbit second antibody coupled to Texas Red (**D,** *open arrows*) or a monoclonal antibody against the β_2/β_3-subunits of the $GABA_A$ receptor revealed by a goat anti-mouse second antibody coupled to Alexa-488 (**E,** *arrowheads*). A combination of the two images acquired in **D** and **E** showing that two neurons simultaneously contain 3βHSD- and β_2/β_3-subunit-like immunoreactivity (**F,** *solid arrows*). Scale bars: 10 μm.

we detected P450c17 expressing neurons. The ratio of these two steroids may play crucial roles in the controlling axonal vs dendritic growth of corticothalamic fibers during neocortex organization. The afferent and efferent connections of the cerebral neocortex develop simultaneously toward the end of embryogenesis.

IV. Regulation of Neurosteroidogenic Enzymes

A. P450scc

Transcriptional regulation of P450scc gene expression differs among the tissues in which this gene is expressed, indicating that there are tissue-specific factors that regulate its expression. The extremely low abundance of P450scc mRNA in the CNS suggested that its transcriptional regulation was likely to be different in the brain than in the adrenal glands or gonads. We demonstrated that the P450scc gene was transcriptionally regulated in neural cells and that the brain used different DNA sequences and nuclear proteins from those used in traditional steroidogenic tissues (Zhang et al., 1995). In MA-10 and in Y-1 cells, but not in C6 cells, basal *and* cAMP-responsive elements were within 94 bp of the transcriptional start site. This site is bound by the orphan nuclear receptor SF-1. In both MA-10 and C6 cells, another element that increased both basal and cAMP-mediated transcription was between −130 and −94 bp. A third *cis*-acting region was between −230 and −130 in all three cell types. DNA sequences beyond 450 bp did not increase transcriptional activity further. The 2.5-kb P450scc promoter/regulatory region was transcriptionally inactive in rat GC somatotrope and mouse GT1-7 neurosecretory cells, indicating that neural expression of P450scc is most likely cell type and region specific.

Gel shift analysis of the −130/−94 region showed two protein:DNA complexes with nuclear extracts from C6 glioma cells, one protein:DNA complex with extract from MA-10 cells, and no complexes with extracts from four different adrenal cell lines, indicating that this region of the rat P450scc gene may be regulated in a tissue-specific fashion. Methylation interference assays indicated that proteins from C6 cells bind to a sequence GGGCGGG, resembling an Sp1 site. However, these C6 proteins did not bind to a consensus Sp1 sequence (Zhang *et al.*, 1995). Nevertheless, mutations of either group of triplet Gs resulted in loss of C6 nuclear protein binding. Functional assays indicated that the −130/−94 P450scc DNA mediated basal and cAMP-induced transcription in C6 cells, but only basal transcription in MA-10 cells. We have identified this protein as the autoimmune antigen Ku and have also demonstrated that members of the Sp transcription family also bind to the same region (Hammer *et al.*, 1999). Ku, as well as the Sps, are developmentally and regionally expressed in the rodent brain, and co-localize to varying extents with P450scc early in development. Thus, the P450scc gene is transcriptionally active in C6 glioma cells, and its transcription is regulated by cAMP. Furthermore, this transcriptional regulation uses nuclear factors distinct from those involved in its transcriptional regulation in adrenal glands and gonads and involves Ku and Sp family members.

B. P450c17

The transcriptional strategy for the expression of P450c17 in the brain is clearly different from that in the adrenal gland or in the gonad. We characterized a novel transcriptional regulator from Leydig MA-10 cells, termed StF-IT-1, that binds at bases −447/−399 of the rat P450c17 promoter, along with the known transcription factors COUP-TF, NGF-IB, and SF-1 (Zhang and Mellon, 1997). We have purified and sequenced this protein from immature porcine testes, identifying it as the nuclear phosphoprotein SET (Compagnone *et al.*, 2000). A role for SET in transcription had not been established. Binding bacterially expressed human and rat of SET to the DNA site at −418/−399 of the rat P450c17 gene transactivates P450c17 in neuronal and in testicular Leydig cells. We also found SET expressed in human NT2 neuronal precursor cells, implicating a role in neurosteroidogenesis. Immunocytochemistry and *in situ* hybridization in the mouse fetus showed that the ontogeny and distribution of SET in the developing nervous system are consistent with SET being crucial for initiating P450c17 transcription. Early in brain development, SET mRNA was expressed from ED 10.5 in the prosencephalon, the rhombencephalon from ED 11.5, the cortex, the rhombencephalon and the diencephalon at ED 13.5, and in the basal diencephalon, pituitary, and hindbrain, suggesting that SET is expressed in an anteroposterior gradient. At this same time in development, SET protein was co-expressed with P450c17 in the lateral motor column (from ED 11). In addition, SET mRNA expression was restricted to the dorsal and ventral segments of the neural tube. At ED 18.5, SET mRNA was widely expressed in the embryo. Sites of expression included the cochlea, paraspinal muscle, thymus, whiskers, skin, intestine, spinal cord, DRG, cartilage, and trigeminal ganglia. SET protein and mRNA expression were developmentally regulated and largely reduced around birth. Nevertheless, at P9, SET mRNA could still be detected in the hippocampus, cortex, thalamus, hypothalamus, septum, and in both the granular layer and Purkinje cells of the cerebellum, but its expression disappeared in the adult.

Both SET and P450c17 were expressed in structures derived from the migration of neural crest cells. Although P450c17 protein was found mainly in structures derived from the cranial neural crest (Compagnone *et al.*, 1995b), SET mRNA was found in structures derived from the cranial neural crest as well as from the trunk and cardiac neural crests.

SET mRNA and protein were found in the same neurons that express P450c17 (Compagnone *et al.*, 1995b)—for example, in the peri-locus cœruleus nucleus, trigeminal ganglia, and the pontine nucleus as well as in the cortical subplate. SET mRNA was also detected in regions of the brain where P450c17 was not expressed, suggesting that SET may regulate other genes in those structures. However, P450c17 was never expressed in regions that did not express SET.

Thus, our developmental analysis of SET expression shows (1) SET is expressed in cell tissues that express P450c17; (2) SET is also expressed in some tissues that do not express P450c17; (3) P450c17 is not expressed in tissues that do not express SET; (4) where SET and P450c17 are co-expressed, SET expression always precedes P450c17 expression; (5) SET expression in the developing CNS follows an anteroposterior gradient; and (6) sites of SET expression indicate that it may play a role in organogenesis of the neural tube, differentiation of blood cells, and development of the skeleton. These studies delineate an important new factor in the transcriptional regulation of P450c17 and, consequently, in the production of DHEA and sex steroids.

C. 3βHSD

The roles of classical neurotransmitters and neuropeptides in the regulation of 3βHSD activity have been extensively investigated in the frog brain (reviewed in Mensah-Nyagan *et al.*, 1999). Immunohistochemical labeling of hypothalamic slices with antisera against various subunits of the GABA$_A$ receptor revealed that most 3βHSD-containing neurons (labeled with an antiserum against human placental type I 3βHSD) also exhibited GABA$_A$ receptor α_3- and β_2/β_3-like immunoreactivities (Do-Rego *et al.*, 2000), indicating that GABA may regulate 3βHSD activity in the frog CNS. Indeed, incubation of frog hypothalamic explants with graded concentrations of GABA induced a dose-dependent inhibition of the conversion of [^{3}H]PREG into radioactive metabolites, including PROG and 17-hydroxyprogesterone. The effect of GABA on neurosteroid biosynthesis was mimicked by the GABA$_A$-receptor agonist muscimol and inhibited by the GABA$_A$-receptor antagonists bicuculline and SR95531. In contrast, the selective GABA$_B$ receptor agonist baclofen did not affect the formation of steroids in hypothalamic explants (Do-Rego *et al.*, 2000). These observations demonstrate that

GABA, acting through $GABA_A$ receptors, inhibits the activity of $3\beta HSD$ in the frog brain. Various neurosteroids, such as PROG, allopregnanolone, PREG, and tetrahydroxydeoxycorticosterone—all potent allosteric modulators of $GABA_A$ receptors—suggest the existence of an ultrashort regulatory loop by which neurosteroids may regulate their own biosynthesis through modulation of $GABA_A$-receptor activity at the level of $3\beta HSD$ neurons.

The $GABA_A$-receptor function is allosterically modulated by benzodiazepines (reviewed in Sieghart, 1995). The search for endogenous ligands of benzodiazepine receptors has led to the discovery of an 86-amino acid polypeptide termed diazepam-binding inhibitor (DBI) which, like β-carbolines, acts as an inverse agonist of central-type benzodiazepine receptors (reviewed in Lihrmann et al., 1994). Proteolytic cleavage of DBI generates several biologically active peptides, including the triakontatetraneuropeptide (TTN) (DBI_{17-50}) (Slobodyansky et al., 1989) and the octadecaneuropeptide (ODN)(DBI_{33-50})(Ferrero et al., 1986). The generic term endozepines is commonly used to designate DBI and its processing fragments (Tonon et al., 1994). We have investigated the effect of the endozepine ODN in the control of $3\beta HSD$ activity in the frog brain (Do-Rego et al., 2001). Using an antiserum against human ODN, we have observed that ODN-immunoreactive glial cells send thick processes in the close vicinity of $3\beta HSD$-expressing neurons in various hypothalamic nuclei. We have also found that ODN enhances in a dose-dependent manner the conversion of $[^3H]$PREG into various steroids, including PROG, 17-hydroxypregnenolone, 17-hydroxyprogesterone, DHEA, and 5α-DHT by frog hypothalamic explants. The β-carbolines, β-CCM and DMCM, two inverse agonists of central-type benzodiazepine receptors, mimicked the stimulatory effect of ODN on neurosteroid synthesis, while the central-type benzodiazepine receptor antagonist flumazenil markedly reduced the stimulatory response evoked by ODN, β-CCM, and DMCM on the production of neurosteroids. In addition, the ODN-induced stimulation of neurosteroidogenesis was significantly attenuated by GABA (Do-Rego et al., 2001). These results indicate that ODN, released by glial cell processes in the vicinity of $3\beta HSD$-containing neurons, stimulates the biosynthesis of neurosteroids through activation of the $GABA_A$/central-type benzodiazepine receptor complex.

The DBI gene is actively expressed in adrenal glands and gonads (Rheaume et al., 1990; Brown et al., 1992; Rouet-Smih et al., 1992). In adrenocortical cells (Krueger and Papadopoulos, 1990), Leydig cells (Papadopoulos et al., 1990), and granulosa cells (Amsterdam and Suh, 1991), DBI stimulates PREG formation by activating intracellularly peripheral-type benzodiazepine receptors (PBR) located on the outer mitochondrial membrane (reviewed in Papadopoulos, 1993). The DBI gene is also expressed in C6 glioma cells (Alho et al., 1994) and in rat astrocytes (Lamacz et al., 1996). In C6 cells, endozepines act as intracrine factors by interacting with mitochondrial PBR and stimulating PREG synthesis (Papadopoulos *et al.*, 1992). However, PBR-like immunoreactivity has also been observed on the plasma membrane (Oke *et al.*, 1992; Lesouhaitier *et al.*, 1996), and it has been shown that DBI stimulates testosterone secretion by intact rat Leydig cells (Garnier *et al.*, 1993). These data indicate that endozepines can stimulate steroid secretion from endocrine cells through activation of PBR located at the cell surface. Because TTN is a selective ligand for PBR, we have investigated the possible effect of TTN on neurosteroid secretion in the frog brain (Do-Rego *et al.*, 1998). Using an antiserum against the 18-kDa subunit of the human PBR, we have detected PBR-like immunoreactivity in numerous nuclei of the frog diencephalon and telencephalon (Fig. 2; see color insert). Double labeling of brain slices with the antisera against 3βHSD and the 18-kDa subunit of PBR has shown that most hypothalamic neurons that express 3βHSD also contain PBR-like immunoreactivity (Fig. 2; see color insert). Confocal laser scanning microscopic analysis revealed that the PBR-immunoreactive material was located both in the cytoplasm and at the cell membrane level (Do-Rego *et al.*, 1998). Exposure of hypothalamic explants to graded concentration of TTN (10^{-9} to 10^{-6} M) induced a dose-dependent increase in the conversion of [^{3}H]PREG into 17-hydroxypregnenolone and 17-hydroxyprogesterone with and ED$_{50}$ of approximately 10^{-8} M. The stimulatory effect of TTN on the formation of neurosteroids was mimicked by the PBR agonist Ro5-4864 and markedly reduced by the PBR antagonist PK11195. In contrast, the central-type benzodiazepine receptor antagonist flumazenil did not affect the TTN-induced stimulation of neurosteroid biosynthesis (Do-Rego *et al.*, 1998). These data indicate that TTN stimulates the synthesis of 3-keto-17α-hydroxysteroids in hypothalamic neurons through activation of PBR likely located at the plasma membrane level.

In the frog diencephalon, 3βHSD-containing neurons are particularly abundant in the anterior preoptic area and the dorsal and ventral hypothalamic nuclei (Mensah-Nyagan *et al.*, 1994). These nuclei are also richly innervated by processes containing various neuropeptides, including gonadotropin-releasing hormone, corticotropin-releasing factor, arginine vasotocin, and neuropeptide Y [reviewed in (Andersen *et al.*, 1992)] as well as tyrosine hydroxylase (Tuinhof *et al.*, 1994). Studies are progress to determine whether, besides endozepines, other neuropeptides and/or catecholamines may regulate neurosteroid biosynthesis.

V. Conclusion

The identification and localization of the enzymes required for neurosteroid synthesis has provided investigators with information about the regions of the nervous system that may be regulated in some fashion by neurosteroids. In addition, developmental studies have enabled investigators to create hypotheses about novel functions that particular neurosteroids may play during neural development. Identification and characterization of factors required for transcriptional regulation of these neurosteroidogenic enzymes has also demonstrated similarities and differences between neurosteroidogenesis and classic steroidogenesis. Future studies in which these transcription factors are modulated *in vivo,* combined with studies in which neurosteroidogenic enzymes are inhibited in a region- and developmentally regulated fashion, may provide important clues about the growing role of neurosteroids in neurodevelopment and behavior. Concurrently, identification of the various neuropeptides and neurotransmitters controlling the activity of neurosteroid-producing neural cells should provide crucial information concerning the possible involvement of neurosteroids in neurotransmission.

Acknowledgments

This work was supported by grants from the National Institutes of Health (HD27970), March of Dimes, Alzheimer's Association, and NARSAD Foundation (to S.H.M.), INSERM (U413), and the Counseil Regional de Haute-Normandie (to H.V.).

References

Adamski, J., Carstensen, J., Husen, B., Kaufmann, M., de Launoit, Y., Leenders, F., Markus, M., and Jungblut, P. W. (1996). New 17 beta-hydroxysteroid dehydrogenases. Molecular and cell biology of the type IV porcine and human enzymes. *Ann. N. Y. Acad. Sci.* **784,** 124–136.

Akwa, Y., Morfin, R. F., Robel, P., and Baulieu, E. E. (1992). Neurosteroid metabolism. 7 alpha-hydroxylation of dehydroepiandrosterone and pregnenolone by rat brain microsomes. *Biochem. J.* **288,** 959–964.

Albiston, A. L., Obeyesekere, V. R., Smith, R. E., and Krozowski, Z. S. (1994). Cloning and tissue distribution of the human 11 beta-hydroxysteroid dehydrogenase type 2 enzyme. *Mol. Cell. Endocrinol.* **105,** R11–7.

Alho, H., Varga, V., and Krueger, K. E. (1994). Expression of mitochondrial benzodiazepine receptor and its putative endogenous ligand diazepam binding inhibitor in cultured primary astrocytes and C-6 cells: Relation to cell growth. *Cell Growth Differ.* **5,** 1005–1014.

Amsterdam, A., and Suh, B. S. (1991). An inducible functional peripheral benzodiazepine receptor in mitochondria of steroidogenic granulosa cells. *Endocrinology* **129,** 503–510.

Andersen, A. C., Tonon, M. C., Pelletier, G., Conlon, J. M., Fasolo, A., and Vaudry, H. (1992). Neuropeptides in the amphibian brain. *Int. Rev. Cytol.* **138,** 89–210.

Andersson, S., Berman, D. M., Jenkins, E. P., and Russell, D. W. (1991). Deletion of steroid 5 alpha-reductase 2 gene in male pseudohermaphroditism. *Nature* **354,** 159–161.

Andersson, S., and Russell, D. W. (1990). Structural and biochemical properties of cloned and expressed human and rat steroid 5 alpha-reductases. *Proc. Natl. Acad. Sci. USA* **87,** 3640–3644.

Arlotto, M. P., Greenway, D. J., and Parkinson, A. (1989). Purification of two isozymes of rat liver microsomal cytochrome P450 with testosterone 7 alpha-hydroxylase activity. *Arch. Biochem. Biophys.* **270,** 441–457.

Auchus, R. J., Lee, T. C., and Miller, W. L. (1998). Cytochrome b_5 augments the 17,20-lyase activity of human P450c17 without direct electron transfer. *J. Biol. Chem.* **273,** 3158–3165.

Ballabio, A., and Shapiro, L. J. (1995). Steroid sulfatase deficiency and X-linked ichtyosis. In "The metabolic and molecular bases of inherited disease" (C. R. Scriver, A. L. Beaudet, W. S. Sly, and D. Valle, Eds.), pp. 2999–3022. McGraw-Hill, New York.

Barker, E. V., Hume, R., Hallas, A., and Coughtrie, W. H. (1994). Dehydroepiandrosterone sulfotransferase in the developing human fetus: Quantitative biochemical and immunological characterization of the hepatic, renal, and adrenal enzymes. *Endocrinology* **134,** 982–989.

Bauer, H. C., and Bauer, H. (1989). Micromethod for the determination of 3-beta-HSD activity in cultured cells. *J. Steroid Biochem.* **33,** 643–646.

Baulieu, E. E., and Robel, P. (1990). Neurosteroids: A new brain function? *J. Steroid Biochem. Mol. Biol.* **37,** 395–403.

Beaujean, D., Mensah-Nyagan, A. G., Do-Rego, J. L., Luu-The, V., Pelletier, G., and Vaudry, H. (1999). Immunocytochemical localization and biological activity of hydroxysteroid sulfotransferase in the frog brain. *J. Neurochem.* **72,** 848–857.

Beyenburg, S., Watzka, M., Blumcke, I., Schramm, J., Bidlingmaier, F., Elger, C. E., and Stoffel-Wagner, B. (2000). Expression of mRNAs encoding for 17β-hydroxisteroid dehydrogenase isozymes 1, 2, 3 and 4 in epileptic human hippocampus. *Epilepsy Res.* **41,** 83–91.

Bose, H. S., Sugawara, T., Strauss, J. F. R., and Miller, W. L. (1996). The pathophysiology and genetics of congenital lipoid adrenal hyperplasia. International Congenital Lipoid Adrenal Hyperplasia Consortium. *N. Engl. J. Med.* **335,** 1870–1878.

Brown, A. S., Hall, P. F., Shoyab, M., and Papadopoulos, V. (1992). Endozepine/diazepam binding inhibitor in adrenocortical and Leydig cell lines: Absence of hormonal regulation. *Mol. Cell Endocrinol.* **83,** 1–9.

Brown, R. W., Chapman, K. E., Edwards, C. R., and Seckl, J. R. (1993). Human placental 11 beta-hydroxysteroid dehydrogenase: Evidence for and partial purification of a distinct NAD-dependent isoform. *Endocrinology* **132,** 2614–2621.

Brown, R. W., Diaz, R., Robson, A. C., Kotelevtsev, Y. V., Mullins, J. J., Kaufman, M. H., and Seckl, J. R. (1996). The ontogeny of 11 beta-hydroxysteroid dehydrogenase type 2 and mineralocorticoid receptor gene expression reveal intricate control of glucocorticoid action in development. *Endocrinology* **137,** 794–797.

Bujalska, I., Shimojo, M., Howie, A., and Stewart, P. M. (1997). Human 11 beta-hydroxysteroid dehydrogenase: studies on the stably transfected isoforms and localization of the type 2 isozyme within renal tissue. *Steroids* **62,** 77–82.

Caron, K. M., Soo, S. C., Wetsel, W. C., Stocco, D. M., Clark, B. J., and Parker, K. L. (1997). Targeted disruption of the mouse gene encoding steroidogenic acute regulatory protein provides insights into congenital lipoid adrenal hyperplasia. *Proc. Natl. Acad. Sci. USA* **94,** 11540–11545.

Carstensen, J. F., Tesdorpf, J. G., Kaufmann, M., Markus, M. M., Husen, B., Leenders, F., Jakob, F., de Launoit, Y., and Adamski, J. (1996). Characterization of 17 beta-hydroxysteroid dehydrogenase IV. *J. Endocrinol.* **150,** S3–12.

Cascio, C., Prasad, V. V., Lin, Y. Y., Lieberman, S., and Papadopoulos, V. (1998). Detection of P450c17-independent pathways for dehydroepiandrosterone (DHEA) biosynthesis in brain glial tumor cells. *Proc. Natl. Acad. Sci.* **95,** 2862–2867.

Casey, M. L., and MacDonald, P. C. (1982). Extraadrenal formation of a mineralocorticosteroid: Deoxycorticosterone and deoxycorticosterone sulfate biosynthesis and metabolism. *Endocr. Rev.* **3,** 396–403.

Casey, M. L., Winkel, C. A., and MacDonald, P. C. (1983). Conversion of progesterone to deoxycorticosterone in the human fetus: Steroid 21-hydroxylase activity in fetal tissues. *J. Steroid Biochem.* **18,** 449–452.

Celotti, F., Melcangi, R. C., and Martini, L. (1992). The 5 alpha-reductase in the brain: Molecular aspects and relation to brain function. *Front. Neuroendocrinol.* **13,** 163–215.

Chen, S. A., Besman, M. J., Sparkes, R. S., Zollman, S., Klisak, I., Mohandas, T., Hall, P. F., and Shively, J. E. (1988). Human aromatase: cDNA cloning, Southern blot analysis, and assignment of the gene to chromosome 15. *DNA* **7,** 27–38.

Cheng, K. C., White, P. C., and Qin, K. N. (1991). Molecular cloning and expression of rat liver 3 alpha-hydroxysteroid dehydrogenase. *Mol. Endocrinol.* **5,** 823–828.

Chung, B. C., Matteson, K. J., Voutilainen, R., Mohandas, T. K., and Miller, W. L. (1986). Human cholesterol side-chain cleavage enzyme, P450scc: cDNA cloning, assignment of the gene to chromosome 15, and expression in the placenta. *Proc. Natl. Acad. Sci. USA* **83,** 8962–8966.

Clark, B. J., Wells, J., King, S. R., and Stocco, D. M. (1994). The purification, cloning, and expression of a novel luteinizing hormone-induced mitochondrial protein in MA-10 mouse Leydig tumor cells. Characterization of the steroidogenic acute regulatory protein (StAR). *J. Biol. Chem.* **269,** 28314–28322.

Compagnone, N. A., Bulfone, A., Rubenstein, J. L., and Mellon, S. H. (1995a). Expression of the steroidogenic enzyme P450scc in the central and peripheral nervous systems during rodent embryogenesis. *Endocrinology* **136,** 2689–2696.

Compagnone, N. A., Bulfone, A., Rubenstein, J. L., and Mellon, S. H. (1995b). Steroidogenic enzyme P450c17 is expressed in the embryonic central nervous system. *Endocrinology* **136,** 5212–5223.

Compagnone, N. A., and Mellon, S. H. (1998). Dehydroepiandrosterone: A potential signalling molecule for neocortical organization during development. *Proc. Natl. Acad. Sci. USA* **95,** 4678–4683.

Compagnone, N. A., and Mellon, S. H. (2000). Neurosteroids: Biosynthesis and function of these novel neuromodulators. *Front. Neuroendocrin.* **21,** 1–58.

Compagnone, N. A., Salido, E., Shapiro, L. J., and Mellon, S. H. (1997). Expression of steroid sulfatase during embryogenesis. *Endocrinology* **138,** 4768–4773.

Compagnone, N. A., Zhang, P., Vigne, J. L., and Mellon, S. H. (2000). Novel role for the nuclear phosphoprotein SET in transcriptional activation of P450c17 and initiation of neurosteroidogenesis. *Mol. Endocrinol.* **14,** 875–888.

Corpechot, C., Robel, P., Axelson, M., Sjovall, J., and Baulieu, E. E. (1981). Characterization and measurement of dehydroepiandrosterone sulfate in rat brain. *Proc. Natl. Acad. Sci. USA* **78,** 4704–4707.

Corpechot, C., Synguelakis, M., Talha, S., Axelson, M., Sjovall, J., Vihko, R., Baulieu, E. E., and Robel, P. (1983). Pregnenolone and its sulfate ester in the rat brain. *Brain Res.* **270,** 119–125.

Coulter, C. L., Goldsmith, P. C., Mesiano, S., Voytek, C. C., Martin, M. C., Mason, J. I., and Jaffe, R. B. (1996). Functional maturation of the primate fetal adrenal in vivo: II. Ontogeny of corticosteroid synthesis is dependent upon specific zonal expression of 3 beta-hydroxysteroid dehydrogenase/isomerase. *Endocrinology* **137,** 4953–4959.

De Meio, R. (1975). Sulfate activation and transfer. *In* "Metabolism of sulfur compounds," David M. Greenberg, ed., pp. 287–359. Academic Press, New York.

Deyashiki, Y., Ogasawara, A., Nakayama, T., Nakanishi, M., Miyabe, Y., Sato, K., and Hara, A. (1994). Molecular cloning of two human liver 3 alpha-hydroxysteroid/dihydrodiol dehydrogenase isoenzymes that are identical with chlordecone reductase and bile-acid binder. *Biochem. J.* **299,** 545–552.

Do-Rego, J. L., Mensah-Nyagan, A. G., Beaujean, D., Leprince, J., Tonon, M. C., Luu-The, V., Pelletier, G., and Vaudry, H. (2001). The octadecaneuropeptide ODN stimulates neurosteroid biosynthesis through activation of central-rype benzodiazepine receptors. *J. Neurochem* **76,** 128–138.

Do-Rego, J. L., Mensah-Nyagan, A. G., Beaujean, D., Vaudry, D., Sieghart, W., Luu-The, V., Pelletier, G., and Vaudry, H. (2000). GABA, acting through GABA$_A$ receptors, inhibits biosynthesis of neurosteroids in the frog hypothalamus. *Proc. Natl. Acad. Sci. USA* **97,** 13925–13930.

Do-Rego, J. L., Mensah-Nyagan, A. G., Feuilloley, M., Ferrara, P., Pelletier, G., and Vaudry, H. (1998). The endozepine triakontatetraneuropeptide diazepam-binding inhibitor [17–50] stimulates neurosteroid biosynthesis in the frog hypothalamus. *Neuroscience* **83,** 555–570.

Doostzadeh, J., Cotillon, A. C., and Morfin, R. (1998). Hydroxylation of pregnenolone at the 7 alpha- and 7 beta- positions by mouse liver microsomes. Effects of cytochrome p450 inhibitors and structure-specific inhibition by steroid hormones. *Steroids* **63,** 383–392.

Dufort, I., Soucy, P., Labrie, F., and Luu-The, V. (1996). Molecular cloning of human type 3 3 alpha-hydroxysteroid dehydrogenase that differs from 20 alpha-hydroxysteroid dehydrogenase by seven amino acids. *Biochem. Biophys. Res. Commun.* **228,** 474–479.

Dupont, E., Labrie, F., Luu-The, V., and Pelletier, G. (1992). Immunocytochemical localization of 3 beta-hydroxysteroid dehydrogenase/delta 5-delta 4-isomerase in human ovary. *J. Clin. Endocrinol. Metab.* **74,** 994–998.

Durkee, T. J., McLean, M. P., Hales, D. B., Payne, A. H., Waterman, M. R., Khan, I., and Gibori, G. (1992). P450(17 alpha) and P450scc gene expression and regulation in the rat placenta. *Endocrinology* **130,** 1309–1317.

Endoh, A., Kristiansen, S. B., Casson, P. R., Buster, J. E., and Hornsby, P. J. (1996). The zona reticularis is the site of biosynthesis of dehydroepiandrosterone and dehydroepiandrosterone sulfate in the adult human adrenal cortex resulting from its low expression of 3 beta-hydroxysteroid dehydrogenase. *J. Clin. Endocrinol. Metab.* **81,** 3558–3565.

Ferrero, P., Santi, M. R., Conti-Tronconi, B., Costa, E., and Guidotti, A. (1986). Study of an octadecaneuropeptide derived from diazepam binding inhibitor (DBI): Biological activity and presence in rat brain. *Proc. Natl. Acad. Sci. USA* **83,** 827–831.

Furukawa, A., Miyatake, A., Ohnishi, T., and Ichikawa, Y. (1998). Steroidogenic acute regulatory protein (StAR) transcripts constitutively expressed in the adult rat central nervous system: Colocalization of StAR, cytochrome P-450SCC (CYP XIA1), and 3beta-hydroxysteroid dehydrogenase in the rat brain. *J. Neurochem.* **71,** 2231–2238.

Garnier, M., Boujrad, N., Oke, B. O., Brown, A. S., Riond, J., Ferrara, P., Shoyab, M., Suarez-Quian, C. A., and Papadopoulos, V. (1993). Diazepam binding inhibitor is a paracrine/autocrine regulator of Leydig cell proliferation and steroidogenesis: Action via peripheral-type benzodiazepine receptor and independent mechanisms. *Endocrinology* **132,** 444–458.

George, F. W., and Ojeda, S. R. (1982). Changes in aromatase activity in the rat brain during embryonic, neonatal, and infantile development. *Endocrinology* **111,** 522–529.

Gomez-Sanchez, C. E., Zhou, M. Y., Cozza, E. N., Morita, H., Foecking, M. F., and Gomez-Sanchez, E. P. (1997). Aldosterone biosynthesis in the rat brain. *Endocrinology* **138,** 3369–3373.

Griffin, L. D., and Mellon, S. H. (1999). Selective serotonin reuptake inhibitors directly alter activity of neurosteroidogenic enzymes. *Proc. Natl. Acad. Sci. USA* **96,** 13512–13517.

Guennoun, R., Fiddes, R. J., Gouezou, M., Lombes, M., and Baulieu, E. E. (1995). A key enzyme in the biosynthesis of neurosteroids, 3 beta-hydroxysteroid dehydrogenase/delta

5-delta 4-isomerase (3 beta-HSD), is expressed in rat brain. *Brain Res. Mol. Brain Res.* **30**, 287–300

Hammer, F., Compagnone, N. A., Vigne, J.-L., and Mellon, S. H. (1999). Regulation of neurosteroidogenesis: Identification of glial nuclear factors involved in the transcriptional regulation of P450scc. The Endocrine Society, San Diego, CA.

Hara, A., Matsuura, K., Tamada, Y., Sato, K., Miyabe, Y., Deyashiki, Y., and Ishida, N. (1996). Relationship of human liver dihydrodiol dehydrogenases to hepatic bile-acid-binding protein and an oxidoreductase of human colon cells. *Biochem. J.* **313**, 373–376.

Hobkirk, R. (1985). Steroid sulfotransferases and steroid sulfate sulfatases: Characteristics and biological roles. *Can. J. Biochem. Cell Biol.* **63**, 1127–1144.

Honda, S., Harada, N., and Takagi, Y. (1994). Novel exon 1 of the aromatase gene specific for aromatase transcripts in human brain. *Biochem. Biophys. Res. Commun.* **198**, 1153–1160.

Hu, Z. Y., Bourreau, E., Jung-Testas, I., Robel, P., and Baulieu, E. E. (1987). Neurosteroids: Oligodendrocyte mitochondria convert cholesterol to pregnenolone. *Proc. Natl. Acad. Sci. USA* **84**, 8215–8219.

Hutchison, J. B., Wozniak, A., Beyer, C., and Hutchison, R. E. (1996). Regulation of sex-specific formation of oestrogen in brain development: Endogenous inhibitors of aromatase. *J. Steroid Biochem. Mol. Biol.* **56**, 201–207.

Iwamori, M., Moser, H. W., and Kishimoto, Y. (1976). Steroid sulfatase in brain: Comparison of sulfohydrolase activities for various steroid sulfates in normal and pathological brains, including the various forms of metachromatic leukodystrophy. *J. Neurochem.* **27**, 1389–1395.

Jenkins, E. P., Andersson, S., Imperato-McGinley, J., Wilson, J. D., and Russell, D. W. (1992). Genetic and pharmacological evidence for more than one human steroid 5 alpha-reductase. *J. Clin. Invest.* **89**, 293–300.

Jung-Testas, I., Hu, Z. Y., Baulieu, E. E., and Robel, P. (1989). Neurosteroids: Biosynthesis of pregnenolone and progesterone in primary cultures of rat glial cells. *Endocrinology* **125**, 2083–2091.

Kabbadj, K., El-Etr, M., Baulieu, E. E., and Robel, P. (1993). Pregnenolone metabolism in rodent embryonic neurons and astrocytes. *Glia* **7**, 170–175.

Karavolas, H. J., and Hodges, D. R. (1990). Neuroendocrine metabolism of progesterone and related progestins. *Ciba Found. Symp.* **153**, 22–44.

Kato, J., Yamada-Mouri, N., and Hirata, S. (1997). Structure of aromatase mRNA in the rat brain. *J. Steroid. Biochem. Mol. Biol.* **61**, 381–385.

Khanna, M., Qin, K. N., Wang, R. W., and Cheng, K. C. (1995). Substrate specificity, gene structure, and tissue-specific distribution of multiple human 3 alpha-hydroxysteroid dehydrogenases. *J. Biol. Chem.* **270**, 20162–20168.

Kimura, T., and Suzuki, K. (1967). Components of the electron transport system in adrenal steroid hydroxylase. *J. Biol. Chem.* **242**, 485–491.

Knapstein, P., David, A., Wu, C. H., Archer, D. F., FLickinger, G. L., and Touchstone, J. C. (1968). Metabolism of free and sulforconjugated DHEA in brain tissue *in vivo* and *in vitro*. *Steroids* **11**, 885–896.

Kominami, S., Ogawa, N., Morimune, R., De-Ying, H., and Takemori, S. (1992). The role of cytochrome b_5 in adrenal microsomal steroidogenesis. *J. Steroid Biochem. Mol. Biol.* **42**, 57–64.

Krueger, K. E., and Papadopoulos, V. (1990). Peripheral-type benzodiazepine receptors mediate translocation of cholesterol from outer to inner mitochondrial membranes in adrenocortical cells. *J. Biol. Chem.* **265**, 15015–15022.

Labrie, F., Luu-The, V., Lin, S. X., Labrie, C., Simard, J., Breton, R., and Belanger, A. (1997). The key role of 17 beta-hydroxysteroid dehydrogenases in sex steroid biology. *Steroids* **62**, 148–158.

Lakshmi, V., Sakai, R. R., McEwen, B. S., and Monder, C. (1991). Regional distribution of 11 beta-hydroxysteroid dehydrogenase in rat brain. *Endocrinology* **128**, 1741–1748.

Lamacz, M., Tonon, M. C., Smih-Rouet, F., Patte, C., Gasque, P., Fontaine, M., and Vaudry, H. (1996). The endogenous benzodiazepine receptor ligand ODN increases cytosolic calcium in cultured rat astrocytes. *Brain. Res. Mol. Brain Res.* **37**, 290–296.

Lauber, M. E., and Lichtensteiger, W. (1994). Pre- and postnatal ontogeny of aromatase cytochrome P450 messenger ribonucleic acid expression in the male rat brain studied by *in situ* hybridization. *Endocrinology* **135**, 1661–1668.

Lauber, M. E., and Lichtensteiger, W. (1996). Ontogeny of 5 alpha-reductase (type 1) messenger ribonucleic acid expression in rat brain: Early presence in germinal zones. *Endocrinology* **137**, 2718–2730.

Leenders, F., Adamski, J., Husen, B., Thole, H. H., and Jungblut, P. W. (1994). Molecular cloning and amino acid sequence of the porcine 17 beta-estradiol dehydrogenase. *Eur. J. Biochem.* **222**, 221–227.

Lephart, E. D. (1996). A review of brain aromatase cytochrome P450. *Brain Res. Brain Res. Rev.* **22**, 1–26.

Lephart, E. D. (1997). Molecular aspects of brain aromatase cytochrome P450. *J. Steroid Biochem. Mol. Biol.* **61**, 375–380.

Lephart, E. D., Simpson, E. R., McPhaul, M. J., Kilgore, M. W., Wilson, J. D., and Ojeda, S. R. (1992). Brain aromatase cytochrome P-450 messenger RNA levels and enzyme activity during prenatal and perinatal development in the rat. *Brain Res. Mol. Brain Res.* **16**, 187–192.

Lesouhaitier, O., Feuilloley, M., Lihrmann, I., Ugo, I., Fasolo, A., Tonon, M. C., and Vaudry, H. (1996). Localization of diazepam-binding inhibitor-related peptides and peripheral type benzodiazepine receptors in the frog adrenal gland. *Cell Tissue Res.* **283**, 403–412.

Li, X. M., Salido, E. C., Gong, Y., Kitada, K., Serikawa, T., Yen, P. H., and Shapiro, L. J. (1996). Cloning of the rat steroid sulfatase gene (Sts), a non-pseudoautosomal X-linked gene that undergoes X inactivation. *Mamm. Genome* **7**, 420–424.

Lihrmann, I., Plaquevent, J. C., Tostivint, H., Raijmakers, R., Tonon, M. C., Conlon, J. M., and Vaudry, H. (1994). Frog diazepam-binding inhibitor: Peptide sequence, cDNA cloning, and expression in the brain. *Proc. Natl. Acad. Sci. USA* **91**, 6899–6903.

Lin, D., Black, S. M., Nagahama, Y., and Miller, W. L. (1993). Steroid 17 alpha-hydroxylase and 17,20-lyase activities of P450c17: Contributions of serine106 and P450 reductase. *Endocrinology* **132**, 2498–2506.

Lin, D., Sugawara, T., Strauss, J. F. R., Clark, B. J., Stocco, D. M., Saenger, P., Rogol, A., and Miller, W. L. (1995). Role of steroidogenic acute regulatory protein in adrenal and gonadal steroidogenesis. *Science* **267**, 1828–1831.

Lorence, M. C., Murry, B. A., Trant, J. M., and Mason, J. I. (1990). Human 3β-hydroxysteroid dehydrogenase/$\Delta^{5 \to 4}$ isomerase from placenta: Expression in nonsteroidogenic cells of a protein that catalyzes the dehydrogenation/isomerization of C21 and C19 steroids. *Endocrinology* **126**, 2493–2498.

Lund, E. G., Guileyardo, J. M., and Russell, D. W. (1999). cDNA cloning of cholesterol 24-hydroxylase, a mediator of cholesterol homeostasis in the brain. *Proc. Natl. Acad. Sci. USA* **96**, 7238–7243.

Luu-The, V., Labrie, C., Zhao, H. F., Couet, J., Lachance, Y., Simard, J., Leblanc, G., Cote, J., Berube, D., Gagne, R., et al. (1989a). Characterization of cDNAs for human estradiol 17 beta-dehydrogenase and assignment of the gene to chromosome 17: Evidence of two mRNA species with distinct 5′-termini in human placenta. *Mol. Endocrinol.* **3**, 1301–1309.

Luu-The, V., Lachance, Y., Labrie, C., Leblanc, G., Thomas, J. L., Strickler, R. C., and Labrie, F. (1989b). Full length cDNA structure and deduced amino acid sequence of human 3 beta-hydroxy-5-ene steroid dehydrogenase. *Mol. Endocrinol.* **3**, 1310–1312.

MacKenzie, S. M., Clark, C. J., Fraser, R., Gomez-Sanchez, C. E., Connell, J. M., and Davies, E. (2000). Expression of 11beta-hydroxylase and aldosterone synthase genes in the rat brain. *J. Mol. Endocrinol.* **24,** 321–328.

MacLusky, N. J., Philip, A., Hurlburt, C., and Naftolin, F. (1985). Estrogen formation in the developing rat brain: Sex differences in aromatase activity during early post-natal life. *Psychoneuroendocrinology* **10,** 355–361.

Mahendroo, M. S., Cala, K. M., Landrum, D. P., and Russell, D. W. (1997). Fetal death in mice lacking 5alpha-reductase type 1 caused by estrogen excess. *Mol. Endocrinol.* **11,** 917–927.

Mahendroo, M. S., Means, G. D., Mendelson, C. R., and Simpson, E. R. (1991). Tissue-specific expression of human P-450AROM. The promoter responsible for expression in adipose tissue is different from that utilized in placenta. *J. Biol. Chem.* **266,** 11276–11281.

Majewska, M. (1991). Neurosteroids: GABAA-agonistic and GABAA-antagonistic modulators of the GABAA receptor. *In* "Neurosteroids and brain function" (E. Costa and S. M. Paul, Eds.), pp. 109–117. Thieme, New York.

Majewska, M. D. (1992). Neurosteroids: Endogenous bimodal modulators of the GABAA receptor. Mechanism of action and physiological significance. *Prog. Neurobiol.* **38,** 379–395.

Marynick, S. P., Smith, G. B., Ebert, M. H., and Loriaux, D. L. (1977). Studies on the transfer of steroid hormones across the blood-cerebrospinal fluid barrier in the rhesus monkey. *Endocrinology* **101,** 562–567.

Mathieu, M., Mensah-Nyagan, A. G., Vallarino, M., Do-Rego, J. L., Beaujean, D., Luu-The, V., Pelletier, G., and Vaudry, H. (2000). Immunohistochemical localization of 3β-hydroxyseroid dehydrogenase and 5α reductase in the brain of the African lungfish *Protopterus amnectens*. *J. Comp. Neurol,* submitted for publication.

Matteson, K. J., Chung, B. C., Urdea, M. S., and Miller, W. L. (1986a). Study of cholesterol side-chain cleavage (20,22 desmolase) deficiency causing congenital lipoid adrenal hyperplasia using bovine-sequence P450scc oligodeoxyribonucleotide probes. *Endocrinology* **118,** 1296–1305.

Matteson, K. J., Picado-Leonard, J., Chung, B. C., Mohandas, T. K., and Miller, W. L. (1986b). Assignment of the gene for adrenal P450c17 (steroid 17 alpha-hydroxylase/17,20 lyase) to human chromosome 10. *J. Clin. Endocrinol. Metab.* **63,** 789–791.

Means, G. D., Mahendroo, M. S., Corbin, C. J., Mathis, J. M., Powell, F. E., Mendelson, C. R., and Simpson, E. R. (1989). Structural analysis of the gene encoding human aromatase cytochrome P-450, the enzyme responsible for estrogen biosynthesis. *J. Biol. Chem.* **264,** 19385–19391.

Melcangi, R. C., Celotti, F., Castano, P., and Martini, L. (1993). Differential localization of the 5 alpha-reductase and the 3 alpha-hydroxysteroid dehydrogenase in neuronal and glial cultures. *Endocrinology* **132,** 1252–1259.

Melcangi, R. C., Celotti, F., and Martini, L. (1994). Progesterone 5-alpha-reduction in neuronal and in different types of glial cell cultures: Type 1 and 2 astrocytes and oligodendrocytes. *Brain Res.* **639,** 202–206.

Mellon, S. H. (1994). Neurosteroids: Biochemistry, modes of action, and clinical relevance. *J. Clin. Endocrinol. Metab.* **78,** 1003–1008.

Mellon, S. H., and Deschepper, C. F. (1993). Neurosteroid biosynthesis: Genes for adrenal steroidogenic enzymes are expressed in the brain. *Brain Res.* **629,** 283–292.

Mellon, S. H., Kushner, J. A., and Vaisse, C. (1991). Expression and regulation of adrenodoxin and P450scc mRNA in rodent tissues. *DNA Cell Biol.* **10,** 339–347.

Mellon, S. H., and Miller, W. L. (1989). Extraadrenal steroid 21-hydroxylation is not mediated by P450c21. *J. Clin. Invest.* **84,** 1497–1502.

Mensah-Nyagan, A. G., Do-Rego, J. L., Beaujean, D., Luu-The, V., Pelletier, G., and Vaudry, H. (1999). Neurosteroids: Expression of steroidogenic enzymes and regulation of steroid biosynthesis in the central nervous system. *Pharmacol Rev.* **51,** 63–81.

Mensah-Nyagan, A. G., Do-Rego, J. L., Feuilloley, M., Marcual, A., Lange, C., Pelletier, G., and Vaudry, H. (1996a). *In vivo* and *in vitro* evidence for the biosynthesis of testosterone in the telencephalon of the female frog. *J. Neurochem.* **67,** 413–422.

Mensah-Nyagan, A. G., Feuilloley, M., Dupont, E., Do-Rego, J. L., Leboulenger, F., Pelletier, G., and Vaudry, H. (1994). Immunocytochemical localization and biological activity of 3 beta-hydroxysteroid dehydrogenase in the central nervous system of the frog. *J. Neurosci.* **14,** 7306–7318.

Mensah-Nyagan, A. M., Feuilloley, M., Do-Rego, J. L., Marcual, A., Lange, C., Tonon, M. C., Pelletier, G., and Vaudry, H. (1996b). Localization of 17beta-hydroxysteroid dehydrogenase and characterization of testosterone in the brain of the male frog. *Proc. Natl. Acad. Sci. USA* **93,** 1423–1428.

Miller, W. L. (1988). Molecular biology of steroid hormone synthesis. *Endocr. Rev.* **9,** 295–318

Mohandas, T., Geller, R. L., Yen, P. H., Rosendorff, J., Bernstein, R., Yoshida, A., and Shapiro, L. J. (1987). Cytogenetic and molecular studies on a recombinant human X chromosome: Implications for the spreading of X chromosome inactivation. *Proc. Natl. Acad. Sci. USA* **84,** 4954–4958.

Moisan, M. P., Seckl, J. R., and Edwards, C. R. (1990). 11 beta-hydroxysteroid dehydrogenase bioactivity and messenger RNA expression in rat forebrain: Localization in hypothalamus, hippocampus, and cortex. *Endocrinology* **127,** 1450–1455.

Moore, C. C., Mellon, S. H., Murai, J., Siiteri, P. K., and Miller, W. L. (1993). Structure and function of the hepatic form of 11 beta-hydroxysteroid dehydrogenase in the squirrel monkey, an animal model of glucocorticoid resistance. *Endocrinology* **133,** 368–375.

Morfin, R., and Courchay, G. (1994). Pregnenolone and dehydroepiandrosterone as precursors of native 7-hydroxylated metabolites which increase the immune response in mice. *J. Steroid. Biochem. Mol. Biol.* **50,** 91–100.

Morfin, R., Young, J., Corpechot, C., Egestad, B., Sjovall, J., and Baulieu, E. E. (1992). Neurosteroids: Pregnenolone in human sciatic nerves. *Proc. Natl. Acad. Sci. USA* **89,** 6790–6793.

Morohashi, K., Sogawa, K., Omura, T., and Fujii-Kuriyama, Y. (1987). Gene structure of human cytochrome P-450(SCC), cholesterol desmolase. *J. Biochem. (Tokyo)* **101,** 879–887.

Mulder, G. (Ed.). (1981). "The sulfatation of drugs and other compounds" Boca Raton: CRC Press.

Nakamura, Y., Otsuka, H., and Tamaoki, B. (1966). Requirement of a new flavoprotein and a non-heme iron-containing protein in the steroid 11β- and 18-hydroxylase system. *Biochim. Biophys. Acta.* **122,** 34–42.

Normington, K., and Russell, D. W. (1992). Tissue distribution and kinetic characteristics of rat steroid 5 alpha-reductase isozymes. Evidence for distinct physiological functions. *J. Biol. Chem.* **267,** 19548–19554.

Norris, P. J., Hardwick, J. P., and Emson, P. C. (1994). Localization of NADPH cytochrome P450 oxidoreductase in rat brain by immunohistochemistry and *in situ* hybridization and a comparison with the distribution of neuronal NADPH-diaphorase staining. *Neuroscience* **61,** 331–350.

Oftebro, H., Stormer, F. C., and Pedersen, J. L. (1979). The presence of an adrenodoxin-like ferredoxin and cytochrome P-450 in brain mitochondria. *J. Biol. Chem.* **254,** 4331–4334

Ogura, K., Kajita, J., Narihata, H., Watabe, T., Ozawa, S., Nagata, K., Yamazoe, Y., and Kato, R. (1989). Cloning and sequence analysis of a rat liver cDNA encoding hydroxysteroid sulfotransferase. *Biochem. Biophys. Res. Commun.* **165,** 168–174.

Oke, B. O., Suarez-Quian, C. A., Riond, J., Ferrara, P., and Papadopoulos, V. (1992). Cell surface localization of the peripheral-type benzodiazepine receptor (PBR) in adrenal cortex. *Mol. Cell. Endocrinol.* **87,** R1–6.

Omura, T., Sanders, S., Estabrook, R. W., Cooper, D. Y., and Rosenthal, O. (1966). Isolation from adrenal cortex of a non-heme iron protein and a flavoprotein functional as a reduced triphosphopyridine nucleotide-cytochrome P-450 reductase. *Arch. Biochem. Biophys.* **117,** 660.

Onoda, M., and Hall, P. F. (1982). Cytochrome b_5 stimulates purified testicular microsomal cytochrome P-450 (C21 side-chain cleavage). *Biochem. Biophys. Res. Commun.* **108,** 454–460.

Oonk, R. B., Parker, K. L., Gibson, J. L., and Richards, J. S. (1990). Rat cholesterol side-chain cleavage cytochrome P-450 (P-450scc) gene. Structure and regulation by cAMP *in vitro. J. Biol. Chem.* **265,** 22392–22401.

Papadopoulos, V. (1993). Peripheral-type benzodiazepine/diazepam binding inhibitor receptor: Biological role in steroidogenic cell function. *Endocr. Rev.* **14,** 222–240.

Papadopoulos, V., Guarneri, P., Kreuger, K. E., Guidotti, A., and Costa, E. (1992). Pregnenolone biosynthesis in C6-2B glioma cell mitochondria: Regulation by a mitochondrial diazepam binding inhibitor receptor. *Proc. Natl. Acad. Sci. USA* **89,** 5113–5117.

Papadopoulos, V., Mukhin, A. G., Costa, E., and Krueger, K. E. (1990). The peripheral-type benzodiazepine receptor is functionally linked to Leydig cell steroidogenesis. *J. Biol. Chem.* **265,** 3772–3779.

Pardridge, W. M., and Mietus, L. J. (1979a). Regional blood–brain barrier transport of the steroid hormones. *J. Neurochem.* **33,** 579–581.

Pardridge, W. M., and Mietus, L. J. (1979b). Transport of steroid hormones through the rat blood–brain barrier. Primary role of albumin-bound hormone. *J. Clin. Invest.* **64,** 145–154.

Pardridge, W. M., and Mietus, L. J. (1980a). Effects of progesterone-binding globulin versus a progesterone antiserum on steroid hormone transport through the blood–brain barrier. *Endocrinology* **106,** 1137–1141.

Pardridge, W. M., and Mietus, L. J. (1980b). Transport of thyroid and steroid hormones through the blood–brain barrier of the newborn rabbit: Primary role of protein-bound hormone. *Endocrinology* **107,** 1705–1710.

Pardridge, W. M., Moeller, T. L., Mietus, L. J., and Oldendorf, W. H. (1980). Blood–brain barrier transport and brain sequestration of steroid hormones. *Am. J. Physiol.* **239,** E96–102.

Parker, C. R., Jr., Falany, C. N., Stockard, C. R., Stankovic, A. K., and Grizzle, W. E. (1994). Immunohistochemical localization of dehydroepiandrosterone sulfotransferase in human fetal tissues. *J. Clin. Endocrinol. Metab.* **78,** 234–236.

Pasmanik, M., and Callard, G. V. (1985). Aromatase and 5 alpha-reductase in the teleost brain, spinal cord, and pituitary gland. *Gen. Comp. Endocrinol.* **60,** 244–251.

Pasmanik, M., and Callard, G. V. (1988). Changes in brain aromatase and 5 alpha-reductase activities correlate significantly with seasonal reproductive cycles in goldfish (*Carassius auratus*). *Endocrinology* **122,** 1349–1356.

Pawlowski, J., Huizinga, M., and Penning, T. M. (1991). Isolation and partial characterization of a full-length cDNA clone for 3 alpha-hydroxysteroid dehydrogenase: A potential target enzyme for nonsteroidal anti-inflammatory drugs. *Agents Actions* **34,** 289–293.

Pelletier, G., Luu-The, V., and Labrie, F. (1994). Immunocytochemical localization of 5 alpha-reductase in rat brain. *Mol. Cell. Neurosci.* **5,** 394–399.

Pelletier, G., Luu-The, V., and Labrie, F. (1995). Immunocytochemical localization of type I 17 beta-hydroxysteroid dehydrogenase in the rat brain. *Brain Res.* **704,** 233–239.

Peltoketo, H., Isomaa, V., Maentausta, O., and Vihko, R. (1988). Complete amino acid sequence of human placental 17 beta-hydroxysteroid dehydrogenase deduced from cDNA. *FEBS Lett.* **239,** 73–77.

Penning, T. M., Bennett, M. J., Smith-Hoog, S., Schlegel, B. P., Jez, J. M., and Lewis, M. (1997). Structure and function of 3 alpha-hydroxysteroid dehydrogenase. *Steroids* **62,** 101–111.

Poletti, A., Coscarella, A., Negri-Cesi, P., Colciago, A., Celotti, F., and Martini, L. (1998a). 5 alpha-reductase isozymes in the central nervous system. *Steroids* **63,** 246–251.

Poletti, A., Negri-Cesi, P., Rabuffetti, M., Colciago, A., Celotti, F., and Martini, L. (1998b). Transient expression of the 5alpha-reductase type 2 isozyme in the rat brain in late fetal and early postnatal life. *Endocrinology* **139,** 2171–2178.

Price, D. J., and Lotto, R. B. (1996). Influences of the thalamus on the survival of subplate and cortical plate cells in cultured embryonic mouse brain. *J. Neurosci.* **16,** 3247–3255.

Rajkowski, K. M., Robel, P., and Baulieu, E. E. (1997). Hydroxysteroid sulfotransferase activity in the rat brain and liver as a function of age and sex. *Steroids* **62,** 427–436.

Ray, W. J., Bain, G., Yao, M., and Gottlieb, D. I. (1997). CYP26, a novel mammalian cytochrome P450, is induced by retinoic acid and defines a new family. *J. Biol. Chem.* **272,** 18702–18708.

Reddy, V. V. (1979). Estrogen metabolism in neural tissues of rabbits: 17 Beta-hydroxysteroid oxidoreductase activity. *Steroids* **34,** 207–215.

Resko, J. A., Stadelman, H. L., and Norman, R. L. (1979). 17 Beta-hydroxysteroid dehydrogenase activity in the pituitary gland and neural tissue of Rhesus monkeys. *J. Steroid Biochem.* **11,** 1429–1434.

Rheaume, E., Tonon, M. C., Smih, F., Simard, J., Desy, L., Vaudry, H., and Pelletier, G. (1990). Localization of the endogenous benzodiazepine ligand octadecaneuropeptide in the rat testis. *Endocrinology* **127,** 1986–1994.

Robel, P., Bourreau, E., Corpechot, C., Dang, D. C., Halberg, F., Clarke, C., Haug, M., Schlegel, M. L., Synguelakis, M., Vourch, C. *et al.* (1987). Neuro-steroids: 3 beta-hydroxy-delta 5-derivatives in rat and monkey brain. *J. Steroid Biochem.* **27,** 649–655.

Roland, B. L., Li, K. X., and Funder, J. W. (1995). Hybridization histochemical localization of 11 beta-hydroxysteroid dehydrogenase type 2 in rat brain. *Endocrinology* **136,** 4697–4700.

Rose, K. A., Stapleton, G., Dott, K., Kieny, M. P., Best, R., Schwarz, M., Russell, D. W., Bjorkhem, I., Seckl, J., and Lathe, R. (1997). Cyp7b, a novel brain cytochrome P450, catalyzes the synthesis of neurosteroids 7alpha-hydroxy dehydroepiandrosterone and 7alpha-hydroxy pregnenolone. *Proc. Natl. Acad. Sci. USA* **94,** 4925–4930.

Roselli, C. E., Horton, L. E., and Resko, J. A. (1985). Distribution and regulation of aromatase activity in the rat hypothalamus and limbic system. *Endocrinology* **117,** 2471–2417.

Rouet-Smih, F., Tonon, M. C., Pelletier, G., and Vaudry, H. (1992). Characterization of endozepine-related peptides in the central nervous system and in peripheral tissues of the rat. *Peptides* **13,** 1219–1225.

Saenger, P., Klonari, Z., Black, S. M., Compagnone, N., Mellon, S. H., Fleischer, A., Abrams, C. A., Shackelton, C. H., and Miller, W. L. (1995). Prenatal diagnosis of congenital lipoid adrenal hyperplasia. *J. Clin. Endocrinol. Metab.* **80,** 200–205.

Salido, E. C., Li, X. M., Yen, P. H., Martin, N., Mohandas, T. K., and Shapiro, L. J. (1996). Cloning and expression of the mouse pseudoautosomal steroid sulphatase gene (Sts). *Nat. Genet.* **13,** 83–86.

Sanghera, M. K., Simpson, E. R., McPhaul, M. J., Kozlowski, G., Conley, A. J., and Lephart, E. D. (1991). Immunocytochemical distribution of aromatase cytochrome P450 in the rat brain using peptide-generated polyclonal antibodies. *Endocrinology* **129,** 2834–2844.

Sanne, J. L., and Krueger, K. E. (1995). Expression of cytochrome P450 side-chain cleavage enzyme and 3 beta- hydroxysteroid dehydrogenase in the rat central nervous system: A study by polymerase chain reaction and *in situ* hybridization. *J. Neurochem.* **65,** 528–556

Shapiro, L. (1982). Steroid sulfatase deficiency and x-linked ichthyosis. *In* "The Metabolic Basis of Inherited Diseases" (J. B. Stanbury, D. S. Frederickson, J. L. Goldstein, M. S. Brown, Eds.), pp. 1027–1034. McGraw Hill, New York.

Sharp, S., Barker, E. V., Coughtrie, M. W., Lowenstein, P. R., and Hume, R. (1993). Immunochemical characterisation of a dehydroepiandrosterone sulfotransferase in rats and humans. *Eur. J. Biochem.* **211,** 539–548.

Sieghart, W. (1995). Structure and pharmacology of gamma-aminobutyric acid A receptor subtypes. *Pharmacol. Rev.* **47**, 181–234.

Simmons, D. L., and Kasper, C. B. (1989). Quantitation of mRNAs specific for the mixed-function oxidase system in rat liver and extrahepatic tissues during development. *Arch. Biochem. Biophys.* **271**, 10–20.

Slobodyansky, E., Guidotti, A., Wambebe, C., Berkovich, A., and Costa, E. (1989). Isolation and characterization of a rat brain triakontatetraneuropeptide, a posttranslational product of diazepam binding inhibitor: Specific action at the Ro 5-4864 recognition site. *J. Neurochem.* **53**, 1276–1284.

Stapleton, G., Steel, M., Richardson, M., Mason, J. O., Rose, K. A., Morris, R. G., and Lathe, R. (1995). A novel cytochrome P450 expressed primarily in brain. *J. Biol. Chem.* **270**, 29739–29745.

Steckelbroeck, S., Stoffel-Wagner, B., Reichelt, R., Schramm, J., Bidlingmaier, F., Siekmann, L., and Klingmuller, D. (1999). Characterization of 17beta-hydroxysteroid dehydrogenase activity in brain tissue: Testosterone formation in the human temporal lobe. *J. Neuroendocrinol.* **11**, 457–464.

Stewart, P. M., Murry, B. A., and Mason, J. I. (1994). Human kidney 11 beta-hydroxysteroid dehydrogenase is a high affinity nicotinamide adenine dinucleotide-dependent enzyme and differs from the cloned type I isoform. *J. Clin. Endocrinol. Metab.* **79**, 480–484.

Stocco, D. M., and Clark, B. J. (1996). Regulation of the acute production of steroids in steroidogenic cells. *Endocr. Rev.* **17**, 221–244.

Stolz, A., Rahimi-Kiani, M., Ameis, D., Chan, E., Ronk, M., and Shively, J. E. (1991). Molecular structure of rat hepatic 3 alpha-hydroxysteroid dehydrogenase. A member of the oxidoreductase gene family. *J. Biol. Chem.* **266**, 15253–15257.

Swinney, D. C., Ryan, D. E., Thomas, P. E., and Levin, W. (1987). Regioselective progesterone hydroxylation catalyzed by eleven rat hepatic cytochrome P-450 isozymes. *Biochemistry* **26**, 7073–7083.

Tannin, G. M., Agarwal, A. K., Monder, C., New, M. I., and White, P. C. (1991). The human gene for 11 beta-hydroxysteroid dehydrogenase. Structure, tissue distribution, and chromosomal localization. *J. Biol. Chem.* **266**, 16653–16658.

Thigpen, A. E., Davis, D. L., Gautier, T., Imperato-McGinley, J., and Russell, D. W. (1992a). Brief report: The molecular basis of steroid 5 alpha-reductase deficiency in a large Dominican kindred. *N. Engl. J. Med.* **327**, 1216–1219.

Thigpen, A. E., Davis, D. L., Milatovich, A., Mendonca, B. B., Imperato-McGinley, J., Griffin, J. E., Francke, U., Wilson, J. D., and Russell, D. W. (1992b). Molecular genetics of steroid 5 alpha-reductase 2 deficiency. *J. Clin. Invest.* **90**, 799–809.

Thigpen, A. E., Silver, R. I., Guileyardo, J. M., Casey, M. L., McConnell, J. D., and Russell, D. W. (1993). Tissue distribution and ontogeny of steroid 5 alpha-reductase isozyme expression. *J. Clin. Invest.* **92**, 903–910.

Thomas, J. L., Myers, R. P., and Strickler, R. C. (1989). Human placental 3 beta-hydroxy-5-ene-steroid dehydrogenase and steroid 5 → 4-ene-isomerase: Purification from mitochondria and kinetic profiles, biophysical characterization of the purified mitochondrial and microsomal enzymes. *J. Steroid Biochem.* **33**, 209–217.

Thompson, E. A., and Siiteri, P. K. (1973). Studies on the aromatization of C-19 androgens. *Ann. N.Y. Acad. Sci.* **212**, 378–391.

Thompson, E. A., Jr., and Siiteri, P. K. (1974). The involvement of human placental microsomal cytochrome P-450 in aromatization. *J. Biol. Chem.* **249**, 5373–5378.

Toda, K., Simpson, E. R., Mendelson, C. R., Shizuta, Y., and Kilgore, M. W. (1994). Expression of the gene encoding aromatase cytochrome P450 (CYP19) in fetal tissues. *Mol. Endocrinol.* **8**, 210–217.

Tonon, M. C., Smih-Rouet, F., Lamacz, M., Louiset, E., Pelletier, G., and Vaudry, H. (1994). Endozepines: Endogenous ligands for benzodiazepine receptors. *Med./Sci.* **10,** 433–443.

Tremblay, Y., Ringler, G. E., Morel, Y., Mohandas, T. K., Labrie, F., Strauss, J. F. III, and Miller, W. L. (1989). Regulation of the gene for estrogenic 17-ketosteroid reductase lying on chromosome 17cen—q25. *J. Biol. Chem.* **264,** 20458–20462.

Tuinhof, R., Gonzalez, A., Smeets, W. J., Scheenen, W. J., and Roubos, E. W. (1994). Central control of melanotrope cells of *Xenopus laevis. Eur. J. Morphol.* **32,** 307–310.

Tzung, K. W., Ishimura-Oka, K., Kihara, S., Oka, K., and Chan, L. (1994). Structure of the mouse cholesterol 7 alpha-hydroxylase gene. *Genomics* **21,** 244–247.

Ukena, K., Kohchi, C., and Tsutsui, K. (1999). Expression and activity of 3beta-hydroxysteroid dehydrogenase/delta5-delta4-isomerase in the rat *Purkinje* neuron during neonatal life. *Endocrinology* **140,** 805–813.

Ukena, K., Usui, M., Kohchi, C., and Tsutsui, K. (1998). Cytochrome P450 side-chain cleavage enzyme in the cerebellar *Purkinje* neuron and its neonatal change in rats. *Endocrinology* **139,** 137–147.

Usui, E., Okuda, K., Kato, Y., and Noshiro, M. (1994). Rat hepatic 3 alpha-hydroxysteroid dehydrogenase: Expression of cDNA and physiological function in bile acid biosynthetic pathway. *J. Biochem. (Tokyo)* **115,** 230–237.

Voutilainen, R., Ilvesmaki, V., and Miettinen, P. J. (1991). Low expression of 3 beta-hydroxy-5-ene steroid dehydrogenase gene in human fetal adrenals *in vivo*. Adrenocorticotropin and protein kinase C-dependent regulation in adrenocortical cultures. *J. Clin. Endocrinol. Metab.* **72,** 761–767.

Wagner, C. K., and Morrell, J. I. (1996). Distribution and steroid hormone regulation of aromatase mRNA expression in the forebrain of adult male and female rats: A cellular-level analysis using *in situ* hybridization. *J. Comp. Neurol.* **370,** 71–84.

Wang, M. D., Wahlstrom, G., and Backstrom, T. (1997). The regional brain distribution of the neurosteroids pregnenolone and pregnenolone sulfate following intravenous infusion. *J. Steroid Biochem. Mol. Biol.* **62,** 299–306.

Weidenfeld, J., Siegel, R. A., and Chowers, I. (1980). *In vitro* conversion of pregnenolone to progesterone by discrete brain areas of the male rat. *J. Steroid Biochem.* **13,** 961–963.

White, J. A., Beckett-Jones, B., Guo, Y. D., Dilworth, F. J., Bonasoro, J., Jones, G., and Petkovich, M. (1997). cDNA cloning of human retinoic acid-metabolizing enzyme (hP450RAI) identifies a novel family of cytochromes P450. *J. Biol. Chem.* **272,** 18538–18541.

White, J. A., Guo, Y. D., Baetz, K., Beckett-Jones, B., Bonasoro, J., Hsu, K. E., Dilworth, F. J., Jones, G., and Petkovich, M. (1996). Identification of the retinoic acid-inducible all-trans-retinoic acid 4-hydroxylase. *J. Biol. Chem.* **271,** 29922–29927.

Whorwood, C. B., Mason, J. I., Ricketts, M. L., Howie, A. J., and Stewart, P. M. (1995). Detection of human 11 beta-hydroxysteroid dehydrogenase isoforms using reverse-transcriptase-polymerase chain reaction and localization of the type 2 isoform to renal collecting ducts. *Mol. Cell. Endocrinol.* **110,** R7–12.

Wood, A. W., Ryan, D. E., Thomas, P. E., and Levin, W. (1983). Regio- and stereoselective metabolism of two C19 steroids by five highly purified and reconstituted rat hepatic cytochrome P-450 isozymes. *J. Biol. Chem.* **258,** 8839–8847.

Yamano, S., Aoyama, T., McBride, O. W., Hardwick, J. P., Gelboin, H. V., and Gonzalez, F. J. (1989). Human NADPH-P450 oxidoreductase: Complementary DNA cloning, sequence and vaccinia virus-mediated expression and localization of the CYPOR gene to chromosome 7. *Mol. Pharmacol.* **36,** 83–88.

Yanagibashi, K., and Hall, P. F. (1986). Role of electron transport in the regulation of the lyase activity of C21 side-chain cleavage P-450 from porcine adrenal and testicular microsomes. *J. Biol. Chem.* **261,** 8429–8433.

Yanase, T., Sasano, H., Yubisui, T., Sakai, Y., Takayanagi, R., and Nawata, H. (1998). Immunohistochemical study of cytochrome b5 in human adrenal gland and in adrenocortical adenomas from patients with Cushing's syndrome. *Endocr. J.* **45,** 89–95.

Zhang, G., and Miller, W. L. (1996). The human genome contains only two CYP11B (P450c11) genes. *J. Clin. Endocrinol. Metab.* **81,** 3254–3256.

Zhang, P., and Mellon, S. H. (1997). Multiple orphan nuclear receptors converge to regulate rat P450c17 gene transcription: Novel mechanisms for orphan nuclear receptor action. *Mol. Endocrinol.* **11,** 891–904.

Zhang, P., Rodriguez, H., and Mellon, S. H. (1995). Transcriptional regulation of P450scc gene expression in neural and steroidogenic cells: Implications for regulation of neurosteroidogenesis. *Mol. Endocrinol.* **9,** 1571–1582.

Zhou, M. Y., Gomez-Sanchez, E. P., Cox, D. L., Cosby, D., and Gomez-Sanchez, C. E. (1995). Cloning, expression, and tissue distribution of the rat nicotinamide adenine dinucleotide-dependent 11 beta-hydroxysteroid dehydrogenase. *Endocrinology* **136,** 3729–3734.

Zhou, Z., Agarwal, V. R., Dixit, N., White, P., and Speiser, P. W. (1997). Steroid 21-hydroxylase expression and activity in human lymphocytes. *Mol. Cell Endocrinol.* **127,** 11–18.

NEUROSTEROID 7-HYDROXYLATION PRODUCTS IN THE BRAIN

Robert Morfin

Laboratoire de Biotechnologie
Conservatoire National des Arts et Métiers
75003, Paris, France

Luboslav Stárka

Institute of Endocrinology
116 94 Prague 1, Czech Republic

I. Neurosteroid Metabolism in the Brain
II. 7α-Hydroxylation Studies in the Brain
III. 7β-Hydroxylation Studies in the Brain
IV. 7α-Hydroxy-DHEA and 7β-Hydroxy-DHEA as Native Anti-glucocorticoids
V. The Brain and Other Organs
VI. Sex Steroid Metabolism in the Brain
VII. Concluding Remarks
    References

The neurosteroids pregnenolone (PREG) and dehydroepiandrosterone (DHEA) are precursors for both oxidized and hydroxylated metabolites in the brain. Thus, brain production of 7-hydroxylated derivatives is second to that in the liver, and P4507B1-containing hippocampus is the major site for 7α-hydroxylation. Other P450s and/or oxido-reductive mechanisms may be responsible for 7β-hydroxylation. In addition to regulating neurosteroid brain levels, when produced, the 7-hydroxylated derivatives of PREG and DHEA were investigated for antiglucocorticoid-mediated neuroprotective potencies, and both 7α- and 7β-hydroxy-DHEA were efficient in preventing the nuclear uptake of [^{3}H]dexamethasone-activated glucocorticoid receptor in brain cells. Activation of 7α-hydroxylation by increased close contacts of astrocytes and after glucocorticoid treatment suggested that the regulated production of 7α-hydroxysteroids was a key event for the neuroprotection conferred by neurosteroids. © 2001 Academic Press.

I. Neurosteroid Metabolism in the Brain

Because of the formation of pregnenolone (PREG) and dehydroepiandrosterone (DHEA) in the brain independent of the adrenal glands and gonads, they were termed neurosteroids (Corpéchot et al., 1981, 1983). PREG is produced from available cholesterol by oligodendrocytes in the rat brain (Hu et al., 1987), and DHEA production results from a brain-specific peculiar oxidation of PREG (Prasad et al., 1994; Cascio et al., 1998). It is well known that both PREG and DHEA are substrates in the brain for the NAD^+-dependent 3β-hydroxysteroid dehydrogenase responsible for oxidation into progesterone (PROG) (Jung-Testas et al., 1989; Akwa et al., 1993) and androstenedione (Akwa et al., 1993), respectively. Yet, in the brain, PREG and DHEA are both substrates for a reductive pathway where NADPH-dependent cytochromes P450 carry out hydroxylations. Thus, 7α- and 7β-hydroxylations were described with identification of the relevant DHEA and PREG 7-hydroxy-metabolites produced by rat brain microsomes (Akwa et al., 1992; 1993). Mouse brain also produced 7-hydroxylated derivatives of PREG and DHEA (Morfin and Courchay, 1994), and such production

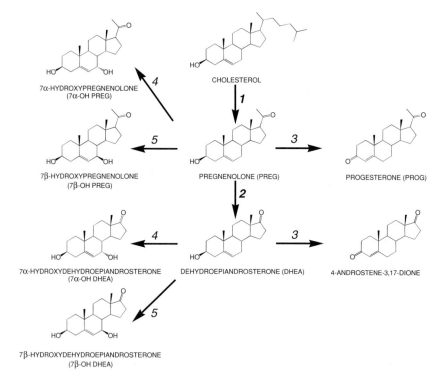

FIG. 1. PREG and DHEA metabolism in brain. Hydroxylations other than 7α and 7β and further metabolism of progesterone and androstenedione are not given. (1) P450scc, (2) $P450_{17\alpha}$,17-20 lyase, (3) 3β-hydroxysteroid dehydrogenase, (4) P4507B1, (5) unknown P450 species.

occurred in other tissues and organs, the brain being second to the liver in production yields (Morfin and Courchay, 1994; Doostzadeh and Morfin, 1996). When outlined (Fig. 1), metabolic pathways of neurosteroids in the brain show that all steroid transformations are irreversible and that the production of 7-hydroxylated metabolites of possible physiological importance also leads to depletion of PREG, DHEA, and their oxidized derivatives.

II. 7α-Hydroxylation Studies in the Brain

PREG and DHEA are both 7α-hydroxylated by NADPH-fortified homogenates of mouse brain in the largest yields when compared with other tissues (Morfin and Courchay, 1994). Incubation of subcellular fractions showed that—in contrast to the thymus, spleen, and heart—brain mitochondria were three to four times more active than microsomes for 7α-hydroxylation of both substrates (Dooostzadeh and Morfin, 1996). Indeed, all of the studies were carried out with brain microsomes, and, to our knowledge, no further investigation of mitochondrial 7α-hydroxylation has been reported. In C57BL/6 mouse brain microsomes, K_M measurements were reported at 1.3 ± 0.1 μM for DHEA and 0.5 ± 0.1 μM for PREG (Doostzadeh and Morfin, 1996). These data are in contrast with those previously reported for rat brain microsomes tested under the same conditions (13.8 ± 3.5 and 4.4 ± 3.2 μM, respectively) (Akwa *et al.*, 1992) and may reflect the difference in species.

From rat, mouse, and human hippocampal transcripts, the cDNA of a novel cytochrome P450 has been sequenced (Stapleton *et al.*, 1995; Wu *et al.*, 1999), and that of mouse was expressed in HeLa cells (Rose *et al.*, 1997). Termed cytochrome P4507B1 (Nelson *et al.*, 1996), that P450 was found to be responsible for 7α-hydroxylation of PREG, DHEA, and other 3β-hydroxysteroids, including estradiol and 25-hydroxycholesterol (Rose *et al.*, 1997; Schwarz *et al.*, 1997). Northern blot analyses provided evidence for P4507B1 localization mainly in the hippocampus (*corpus callosum*) of the adult mouse brain. The human P4507B1 expressed in human kidney 293/T cells was also shown to convert DHEA and 27-hydroxycholesterol into their respective 7α-hydroxylated derivatives (Wu *et al.*, 1999), and its gene was located in chromosome 8 (8q21.3). K_M values for 7α-hydroxylation of PREG and DHEA were measured only with mouse P4507B1 expressed in HeLa cells. In Rose *et al.*'s report (1997), the mouse strain (*Mus musculus*) was not mentioned. Surprisingly, the values reported were similar to those of rat and were one order of magnitude higher than those of C57BL/6 mice (Doostzadeh and Morfin, 1996; Morfin *et al.*, 2000).

TABLE I

COMPARISON OF K_M VALUES FOR 7α-HYDROXYLATION OF PREG AND DHEA AND OF AMINO ACIDS IN THE CYP7B1 PROTEIN OF RAT, *Mus Musculus*, AND C57BL/6 MOUSE STRAIN

K_M (μM) for cell-expressed P4507B1		K_M (μM) in brain microsomes			Amino acid number in P4507B1 sequence			
PREG	DHEA	PREG	DHEA	Species	265	278	432	463
4.0*	13.6*			*Mus musculus*	S	S	R	E
0.3 ± 0.1	1.9 ± 0.3	0.5 ± 0.1	1.3 ± 0.1	C57BL/6	**R**	**P**	**K**	**M**
		4.4 ± 3.2	13.8 ± 3.5	Rat	R	F	K	I

Note. Differences found between *Mus musculus* and C57BL/6 are indicated in **bold**.
*Data reported by Rose *et al.* (1997).

Therefore, and based on the published cDNA sequence of mouse P4507B1, we produced that of C57BL/6 mice (Morfin *et al.*, 2000). Sequence analyses showed only four amino acid changes between *Mus musculus* and the C57BL/6 inbred strain. Expression of the C57BL/6 cDNA in microsomes of transformed yeast provided a tool for comparison with C57BL/6 brain microsomes. Under identical incubation conditions, K_M values of PREG and DHEA 7α-hydroxylation were not significantly different in the two preparations of microsomes. Thus, it is possible that mouse strain differences are reflected at the P4507B1 level and that key amino acid changes modify affinities for the steroid substrates (Table I). Whether this will hold true in mouse and other species, including human, will be addressed by appropriate experiments.

Inhibition studies carried out *in vitro* with mouse brain microsomes showed that increasing inhibitions of PREG and DHEA 7α-hydroxylation were obtained with antipyrine, ketoconazole, metyrapone, and α-naphthoflavone (Dooszadeh *et al.*, 1997a; Dooszadeh and Morfin, 1997). Significant inhibition of DHEA 7α-hydroxylation in brain microsomes was also obtained after *in vivo* treatment (ip) of mice for 1 day with 25 mg/kg metyrapone (Attal-Khémis *et al.*, 1998a). These inhibitors may provide efficient tools for studies of the P450 in the brain responsible for PREG and DHEA 7α-hydroxylation. Other inhibition studies carried out *in vitro* with steroid hormones showed that, aside from PREG and DHEA, epiandrosterone and 5-androstenediol were inhibitors of the 7α-hydroxylation to much larger extents than the Δ^4-3-ketosteroids tested (Dooszadeh *et al.*, 1997a; Dooszadeh and Morfin, 1997). When examined with other reports on 7α-hydroxylation of various steroids in the brain (Akwa *et al.*, 1992; Rose *et al.*, 1997), these findings imply that the P450 responsible for 7α-hydroxylation in the brain binds only 3β-hydroxysteroid substrates.

III. 7β-Hydroxylation Studies in the Brain

In addition to 7α-hydroxylated metabolites, brain microsomes from the rat and the mouse produced 7β-hydroxylated derivatives of PREG and DHEA that were identified by gas chromatography–mass spectrometry and crystallization to constant specific activity (Morfin and Courchay, 1994; Doostzadeh et al., 1996, 1997a; Doostzadeh and Morfin, 1997). In all cases, the production of 7β-hydroxysteroids was 6–12 times less important than that of 7α-hydroxysteroids. In mouse brain microsomes, the K_M values for 7β-hydroxylation of PREG and DHEA were 5.0 ± 0.6 and 4.9 ± 0.2 μM, respectively (Doostzadeh et al., 1996). It is known that, in addition to its 7α-hydroxylating potencies, P4507B1 produces a small amount of 7β-hydroxylated derivatives (Rose et al., 1997), but such production is not extensive enough to explain fully the levels of 7β-hydroxysteroids. Therefore, 7β-hydroxylation of PREG and DHEA may be due to other P450s and/or to other enzymatic conversions. We have proved that the mouse and human P4501A1 were responsible for 7β-hydroxylation of PREG (Doostzadeh et al., 1996, 1997b) and that P4501A1 was present in mouse brain (Doostzadeh et al., 1996). Incubation of yeast-expressed human P4501A1 with PREG provided evidence for 7β-hydroxylation and measurement of K_M at 4.1 ± 0.4 μM and neither P450 1A1 nor other human P450s tested (1A2, 3A4, 3A5, 2C8, 2C9, 2C18, 2C19, 2D6, and 2E1) significantly converted DHEA into 7α- or 7β-hydroxy-DHEA (Doostzadeh et al., 1997b). Therefore, P4501A1 carries out 7β-hydroxylation of PREG exclusively, and 7β-hydroxylation of DHEA may result either from an unidentified P450 or from other enzymatic conversions.

Evidence for the interconversion of 7α-hydroxy-DHEA into 7β-hydroxy-DHEA in rat liver has been reported (Hampl and Stárka, 1969). For such transformations, oxidoreduction of a 3-oxo-DHEA metabolite intermediate was proposed, and works in humans, using transdermal administration of DHEA and 7-oxo-DHEA to male volunteers, showed that blood-borne levels of 7β-hydroxy-DHEA were higher after 7-oxo-DHEA than after DHEA administrration and that the contrary occurred for 7α-hydroxy-DHEA levels (Hampl et al., 2000) (Table II). Therefore, it is possible that oxidation of 7α-hydroxy-DHEA to 7-oxo-DHEA and subsequent reduction into 7β-hydroxy-DHEA occurred in vivo and that part of the 7α-hydroxy-DHEA produced by the 7α-hydroxylase of skin (Khalil et al., 1993) was transformed into 7β-hydroxy-DHEA. Nevertheless, the oxidation step of DHEA could not be obtained in vitro with NADPH-fortified brain microsomes because of the reductive conditions and because no 7-oxo-DHEA could be detected other than the 7α- and 7β-hydroxylated metabolites.

TABLE II
SERUM LEVELS (NMOL/L) OF DHEA, 7α-HYDROXY-DHEA, AND 7β-HYDROXY-DHEA IN SIX NORMAL MALE VOLUNTEERS (AGED 27–68, MEAN 45) BEFORE AND AFTER 5 DAYS OF OINTMENT ON THE ABDOMINAL SKIN WITH 50 MG/DAY DHEA OR 7-OXO-DHEA

Serum levels (nmol/l)	DHEA ointment (50 mg/day)		7oxo-DHEA ointment (50 mg/day)	
	Before	Day 6	Before	Day 6
DHEA	13.0 ± 1.86	19.9 ± 2.08	8.01 ± 1.49	19.6 ± 4.81
7α-Hydroxy-DHEA	1.93 ± 0.35	4.25 ± 0.56	1.1 ± 0.29	1.77 ± 0.30
7α-Hydroxy-DHEA	1.46 ± 0.28	3.99 ± 0.57	1.35 ± 0.28	4.40 ± 0.38

All data (± SEM) were obtained with specific radioimmunoassays and are combined from the report of Hampl *et al.* (2000). All differences measured at Day 6 are significant ($P < 0.05$ for DHEA, $P < 0.01$ for 7α- and 7β-hydroxy-DHEA).

Therefore, interconversion of 7α- and 7β-hydroxy-DHEA in the brain must be assessed in other experiments.

IV. 7α-Hydroxy-DHEA and 7β-Hydroxy-DHEA as Native Anti-glucocorticoids

Much evidence obtained *in vivo* with rodents indicated that DHEA may exert anti-glucocorticoid effects (Kalimi *et al.*, 1994). No such evidence could be obtained *in vitro*, and it was suggested that some of the DHEA metabolites could be responsible for its effects (Kalimi *et al.*, 1994). Because 7α- and 7β-hydroxylation of PREG and DHEA occur with the glucocorticoid receptor in most tissues and organs, it is possible that 7-hydroxylation is the preliminary step necessary for production of native anti-glucocorticoids. Antiglucocorticoid effects of DHEA had been assessed *in vivo* and *in vitro* with rats for counteracting dexamethasone (DEX)-induced thymic and T cell involution (May *et al.*, 1990). Using PREG, 7α-hydroxy-PREG, and 7α-hydroxy-DHEA in addition to DHEA, the same protocol was used in mice for measurement of steroid effects on DEX-induced T-cell apoptosis (Chmielewski *et al.*, 2000). Results indicated that 7α-hydroxy-DHEA was more efficient than DHEA in preventing DEX-induced apoptosis and that PREG and 7α-hydroxy-PREG were not active. Furthermore, additional tests with 7β-hydroxy-DHEA showed no interference with DEX-induced apoptosis. Therefore, to investigate the 7α-hydroxy-DHEA mechanism of anti-glucocorticoid action in the brain, the steroid had to be tested *in vitro* for its interference with [^{3}H]DEX binding in subcellular fractions. The first experiment was carried out with mouse brain cytosol where [^{3}H]DEX binding was assessed ($K_D = 1.2$ nM, $B_{max} = 2.54$ fmol · mg^{-1} protein). Both DEX and corticosterone significantly decreased the binding of 10 nM [^{3}H]DEX, while neither DHEA nor 7α-hydroxy-DHEA was an efficient competitor (Fig. 2A). The second experiment used dissociated intact mouse brain cells cultured in the presence of increasing concentrations of [^{3}H]DEX, according to Mohan and Cleary (1992). Cell nuclei containing the [^{3}H]DEX-activated glucocorticoid receptor were then recovered, extracted, and counted. The same experiment was repeated with cells cultured with a 100-fold excess of DEX, DHEA, 7α-hydroxy-DHEA, 7β-hydroxy-DHEA, or 5α-androstane-3β,7α,17β-triol. DEX competed with [^{3}H]DEX binding, and the resulting nuclear contents in [^{3}H]DEX was decreased. Decrease in nuclear [^{3}H]DEX was also measured when DHEA, 7α-hydroxy-DHEA, or 7β-hydroxy-DHEA was present (Fig. 2B). Computations based on 100% DEX-induced decrease in nuclear [^{3}H]DEX gave $96 \pm 12\%$ for 5α-androstane-3β,7α,17β-triol, $65 \pm 7\%$ for 7β-hydroxy-DHEA, $49 \pm 6\%$ for 7α-hydroxy-DHEA, and $36 \pm 10\%$ for DHEA. Evidence for DHEA potency in preventing the nuclear uptake of the cytosol-containing [^{3}H]DEX-activated glucocorticoid receptor could be explained by the intensive 7α-hydroxylation of DHEA by microsomes, mitochondria, and nuclei of mouse brain cells (Doostzadeh and Morfin, 1996). Other experiments with mouse and rat hepatocytes were more conclusive with 7β- and 7α-hydroxy-DHEA being as potent as DEX for preventing the nuclear uptake of [^{3}H]DEX (Stárka *et al.*, 1998).

Whether the 7α-hydroxy-DHEA-induced decrease in nuclear retention of the [^{3}H]DEX-activated glucocorticoid receptor could be a result of decreased binding to DNA glucocorticoid response elements was tested with use of calf thymus DNA-cellulose and isolated [^{3}H]DEX-activated glucocorticoid receptor from rat liver cytosol (Stárka *et al.*, 1998). Results provided no evidence in support for such a 7α-hydroxy-DHEA-induced decrease (Stárka *et al.*, 1998). A hypothesis that 7α- and 7β-hydroxy-DHEA interfered with the glucocorticoid receptor trafficking events at the nuclear level is being tested.

In mouse brain, the decrease in the nuclear uptake of the [^{3}H]DEX-activated cytosolic glucocorticoid receptor by both 7α-hydroxy-DHEA and 7β-hydroxy-DHEA is in contrast with *in vivo* data showing that 7α-hydroxy-DHEA was efficient in preventing the DEX-induced apoptosis of thymocytes, while 7β-hydroxy-DHEA, when tested, was not (Chmielewski *et al.*, 2000). Even though the thymus and the brain share common microenvironments (Mentlein and Kendall, 2000), different mechanisms in apoptosis take place in lymphocytes and the brain, with an absence in hippocampal neurons of the glucocorticoid-induced DNA cleavage observed in lymphocytes (Masters *et al.*, 1989). Such a difference may contribute to an explanation of the different effects of 7α- and 7β-hydroxy-DHEA. Support for

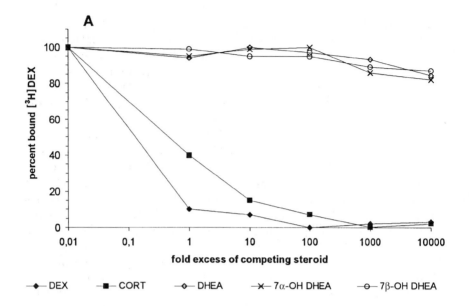
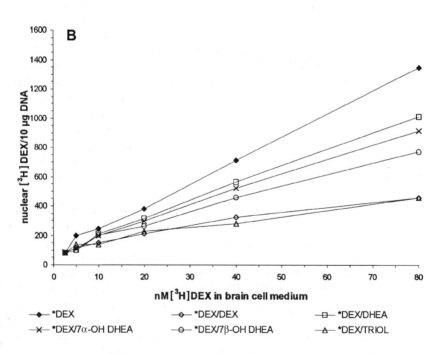

activity of 7β-hydroxy-DHEA as well as its reduced form (androst-5-ene-3β, 7α, 17β-triol) is found in reports showing that then trigger an immune response in mice (Morfin and Courchay, 1994; Loria *et al.*, 1996) but to lower extents than 7α-hydroxy-DHEA (Morfin and Courchay, 1994). Both 7α-hydroxy-DHEA and 7β-hydroxy-DHEA are major DHEA metabolites in the brain, particularly in the hippocampus. Their interference with the nuclear uptake of the activated glucocorticoid receptor may prevent the known glucocorticoid-induced potentiation of toxin-driven damage to primary cultures of fetal rat hippocampal neurons (Sapolsky *et al.*, 1988). The brains of newborn rats contain barely detectable amounts of DHEA 7α-hydroxylase (Akwa *et al.*, 1992), therefore, and even if brain-produced DHEA were available, only minute amounts of 7α-hydroxy-DHEA could be produced for protection from corticosterone. When 7α-hydroxylation was measured in mouse brain microsomes, it increased until adulthood and then decreased with old age (Doostzadeh and Morfin, 1996). To our knowledge, no report on age-related changes in the brain-specific DHEA production is available yet, but aging is known to lower the DHEA blood supply. This and the decreased 7α-hydroxylation may be related to the age-related decrease in neuronal counts and brain performance. After 20-h culture, 5×10^{-6} M DHEA protected HT-22 mouse hippocampal neurons from glutamate toxicity by depleting nuclei of the glutamate-induced accumulation of glucocorticoid receptor (Cardounel *et al.*, 1999). These data do not exclude an effect mediated by 7α-hydroxy-DHEA because of the large concentration of DHEA used and because of its extensive 7α-hydroxylation in hippocampus. Thus, the brain-produced 7α-hydroxy-DHEA and 7β-hydroxy-DHEA may contribute to the neuron protection against glutamate-induced, glucococorticoid-mediated damages and could be termed native neuroprotectors. Other indirect support for neuroprotection was published in 2000 with proof that 7-oxo-DHEA acetate, a putative precursor for 7α- and 7β-hydroxy-DHEA, significantly improved memory retention when administered to treated old mice (Shi *et al.*, 2000).

FIG. 2. Interference of DHEA and 7-hydroxylated derivatives with glucocorticoid mechanism of action. **(A)** Competiton with [^{3}H]dexamethasone (DEX) binding to the glucocorticoid receptor in brain cytosol. Binding of 10 nM [^{3}H]DEX to the receptor in the presence of 0- to 10^4- fold excess of competing steroid was assessed after incubation at 2°C for 24 h by the dextran-charcoal technique. **(B)** Interference with the nuclear uptake of activated glucocorticoid receptor in brain cells. Isolated brain cells were cultured in DMEM/HAMS medium containing increasing concentrations of [^{3}H]DEX at 22°C for 1 h before isolation and counting of nuclei. Nonradioactive steroid tested, according to Mohan and Cleary (1992), was at 100-fold excess of the [^{3}H]DEX. Mean values of three to five experiments are given, and error bars (not exceeding 4%) are not shown for clarity. (CORT, corticosterone; 7α-OH DHEA, 7α-hydroxy-DHEA; 7β-OH DHEA, 7β-hydroxy-DHEA; TRIOL, 5α-androstane-3β,7α,17β-triol.)

V. The Brain and Other Organs

In addition to the brain, 7α- and 7β-hydroxylations take place in numerous tissues and organs of animals and humans, and major productions of the 7-hydroxylated DHEA metabolites occur in the liver. Skin may also be a major producer because of its large area and because of its coverage of specific places, such as around the anal orifice (Morfin *et al.*, 1980; Khalil *et al.*, 1993; Morfin and Courchay, 1994). In mouse liver and in human adipose stromal cells, 7α-hydroxylation was increased after glucocorticoid treatments (Khalil *et al.*, 1994; Attal-Khémis *et al.*, 1998a). These findings provide physiological support for a regulation mechanism of glucocorticoid actions by the 7α-hydroxysteroids produced. From available data, a functional hypothesis may be outlined (Fig. 3). First, effects of 7α-hydroxysteroids must not be systemic but rather paracrine and/or autocrine. Support for this statement is implied from several findings. (1) Blood levels of 7α-hydroxy-DHEA in untreated humans and mice were reported at 1–2 nM (Skinner *et al.*, 1977; Attal-Khémis 1998a,b; Lapcik *et al.*, 1999), which is one order of magnitude lower than that of circulating free DHEA and three orders of magnitude lower than active doses of 7α-hydroxy-DHEA in experimental models (Lafaye *et al.*, 1999; Chmielewski *et al.*, 2000). (2) DHEA or 7α-hydroxy-DHEA does not increase immunity when administered distant from the antigen (Loria *et al.*, 1988; Araneo *et al.*, 1991; Morfin and Courchay, 1994). (3) In human tonsils, 7α-hydroxylation of DHEA was found in stroma cells only, and specific immune responses of activated tonsils B+T cells was triggered by 7α-hydroxy-DHEA and not by DHEA (Lafaye *et al.*, 1999). Second, rat astrocytes cultured at low density with PREG and DHEA produced major yields of PROG and androstenedione and low yields of the 7α-hydroxylated derivatives. With higher densities and close contact of the cells (e.g., in inflammation), PREG and DHEA metabolism was shifted toward 7α-hydroxylated metabolite production exclusively (Akwa *et al.*, 1993). Investigation of the triggering mechanism excluded factors produced by the cells in close contact. Thus, it is possible that cell contacts induced compartment changes in calcium with inhibition of the 3β-hydroxysteroid dehydrogenase and activation of the P450 responsible for 7α-hydroxylation. Support derives from a report on microsomal P450 activation by Ca^{2+} store depletion (Hoebel *et al.*, 1997) and from our finding of a 1.2 times increase of PREG and DHEA 7α-hydroxylation by 0.25–0.5 mM Ca^{2+} in mouse brain microsomes (Doostzadeh and Morfin, 1996). Third, during inflammation or once under close cellular contacts, the increased amounts of 7α-hydroxysteroids produced may reach the cellular concentrations sufficient for autocrine or paracrine protective actions (10^{-6} M) (Lafaye *et al.*, 1999; Chmielewski

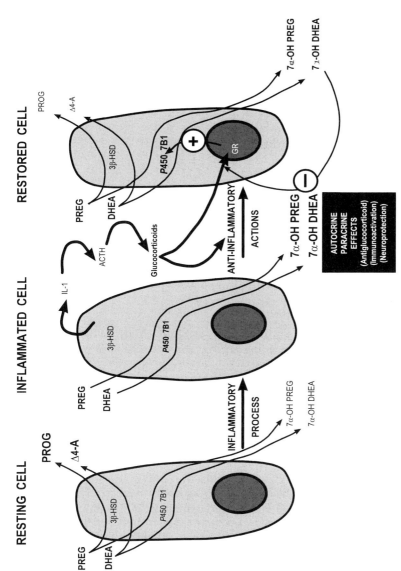

FIG. 3. Functional hypothesis of the timed production, effects, and glucocorticoid-mediated regulation of PREG and DHEA metabolism in cells. (PROG, progesterone; Δ^4-A androstenedione; 7α-OH PREG, 7α-hydroxy-PREG; 7α-OH DHEA, 7α-hydroxy-DHEA; 3βHSD, 3β-hydroxysteroid dehydrogenase; P4507B1, 7α-hydroxylase; GR, glucocorticoic receptor; IL-1, interleukin 1.)

et al., 2000). Production of inflammatory cytokines (interleukin-1, tumor necrosis factor-α) is also known to occur for the triggering of adrenocorticotropic hormone secretion in the pituitary gland and then in turn of glucocorticoids in the adrenal glands. Anti-inflammatory effects are brought to the cells by the increased supply in glucocorticoids together with activation of the 7α-hydroxylating P450 biosynthesis (Khalil *et al.*, 1994; Attal-Khémis *et al.*, 1998a). Yet, the resulting augmented production of 7α-hydroxysteroids may exquisitely balance glucocorticoid actions and protect cells from their unwanted effects.

VI. Sex Steroid Metabolism in the Brain

Testosterone and PROG produced in the gonads are transported in blood and reach the brain, where they are metabolized before they exert proper effects. In contrast, estradiol from the gonads or that produced in the brain after testosterone aromatization binds directly to its receptor, and its metabolism in the brain is second to the estrogenic effects (Polettti *et al.*, 1999). Sex steroids are also directly produced in the brain from the neurosteroid precursors PREG and DHEA (Mellon and Compagnone, 1999), and their metabolism implies production and regulation of active metabolite brain concentrations. Presence in the brain of the irreversible and extensive 5α-reduction of testosterone and PROG leads to their respective 5α-reduced metabolites, which are in turn substrates for the 3α- and 3β-oxidoreductases (Fig. 4). Extensive 7α-hydroxylation of 5α-reduced 3β-hydroxysteroids in the gland pituitary and the brain has been reported (Guiraud *et. al.*, 1979; Warner *et al.*, 1989; Akwa *et al.*,1992; Strömstedt *et al.*,1993), and it was suggested that the 7α-hydroxylation of 3β-hydroxy-5α-pregnan-20-one was a means for regulation of brain concentrations in the anesthetic steroid 3α-hydroxy-5α-pregnan-20-one (Strömstedt *et al.*, 1993). Involvement of 7α-hydroxylation in regulation of 5α-dihydrotestosterone and 5α-androstane-3,17-dione levels could also be suggested in addition to the production of physiologically active 5α-androstane-3β,7α,17β-triol and 7α-hydroxyepiandrosterone metabolites. Such an activity remains putative and is based only on our data showing that 5α-androstane-3β,7α,17β-triol was the best steroid tested for decreasing nuclear uptake of the activated glucocorticoid receptor in mouse brain cells (Fig. 2). Such activity and that of 7α-hydroxyepiandrosterone were also extensively maintained when tested in liver cells. Whether 7α-hydroxysteroid production in the brain contributes to neuroprotection and to the immuno-privileged status of the brain is under investigation.

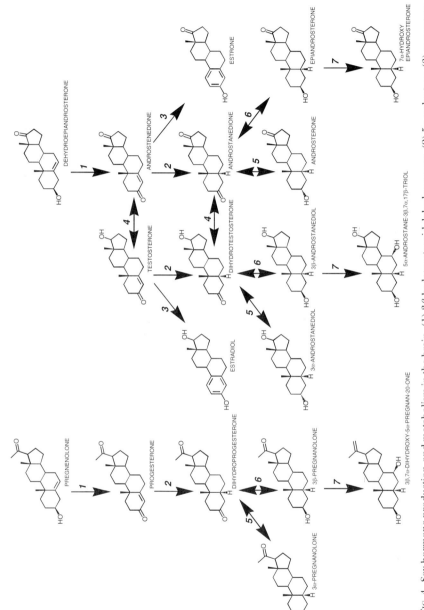

FIG. 4. Sex hormone production and metabolism in the brain. (1) 3β-hydroxysteroid dehydrogenase, (2) 5α-reductase, (3) aromatase (P450aro), (4) 17β-hydroxysteroid oxidoreductase, (5) 3α-hydroxysteroid oxidoreductase, (6) 3β-hydroxysteroid oxidoreductase, and (7) 7α-hydroxylase (P4507B1).

VII. Concluding Remarks

In steroid metabolism, P450s are key enzyme partners leading to the irreversible production of steroid hormones and of physiologically active steroid and sterol derivatives. Thus, cholesterol can be hydroxylated at various positions into bile acid and steroid hormone precursors. Key physiological activity has been established for sterols, steroid hormones, and for most of their metabolites. At the cellular level, irreversible 7α-hydroxylation of 3β-hydroxysteroids produces derivatives for which anti-glucocorticoid, immunity-promoting, and protective activities have been described. In the brain, as "neuroprotectors" and immunity-promoters, 7α-hydroxysteroids could contribute to the panels of cellular protection and defenses. P450s are involved in a metabolic defense mechanism (detoxication), where, in step one, xenobiotics of low molecular weight are irreversibly hydroxylated before conjugation in step two for elimination in urine and bile. For alien proteins, defenses of the organism process through immunity with production of relevant antibodies. Glucocorticoids resulting from three to four P450-mediated transformations of cholesterol, and 7α-hydroxysteroids resulting from two to three P450-mediated hydroxylation steps, inhibit and activate immune processes, respectively. Thus, and in unified view, it is possible that through their wide variety P450s are involved in all defense and protection mechanisms by providing potent hydroxylated derivatives and by contibuting to the elimination of xenobiotics, neurosteroids, and steroid hormones.

Acknowledgments

This collaborative work was made possible by the NATO Grant CRG.CRGP 972957 awarded to R. M. for 1998–2000.

References

Akwa, Y., Morfin, R. F., Robel, P., and Baulieu, E. E. (1992). Neurosteroid metabolism. 7α-Hydroxylation of dehydroepiandrosterone and pregnenolone by rat brain microsomes. *Biochem. J.* **288,** 959–964.

Akwa, Y., Sananès, N., Gouézou, M., Robel, P., Baulieu, E. E., and Le Goascogne, C. (1993). Astrocytes and neurosteroids: Metabolism of pregnenolone and dehydroepiandrosterone. Regulation by cell density. *J. Cell Biol.* **121,** 135–143.

Araneo, B. A., Tad, D., Diegel, M., and Daynes, R. A. (1991). Dihydrotestosterone exerts a depressive influence on the production of interleukin-4 (IL-4), IL-5, and γ-interferon but not IL-2 by activated murine T cells. *Blood* **78,** 688–699.

Attal-Khémis, S., Dalmeyda, V., Michot, J. L., Roudier, M., and Morfin, R. (1998b). Increased total 7α-hydroxy-dehydroepiandrosterone in serum of patients with Alzheimer's disease. *J. Gerontol. Ser. A-Biol. Sci. Med.* **53,** B125–B132.

Attal-Khémis, S., Dalmeyda, V., and Morfin, R. (1998a). Change of 7α-hydroxy-dehydroepiandrosterone levels in serum of mice treated by cytochrome P450-modifying agents. *Life Sci.* **63,** 1543–1553.

Cardounel, A., Regelson, W., and Kalimi, M. (1999). Dehydroepiandrosterone protects hippocampal neurons against neurotoxin-induced cell death: Mechanism of action. *Proc. Soc. Exp. Biol. Med.* **222,** 145–149.

Cascio, C., Prasad, V. V. K., Lin, Y. Y., Lieberman, S., and Papadopoulos, V. (1998). Detection of P450c17-independent pathways for dehydroepiandrosterone (DHEA) biosynthesis in brain glial tumor cells. *Proc. Natl. Acad. Sci. USA* **95,** 2862–2867.

Chmielewski, V., Drupt, F., and Morfin, R. (2000). Dexamethasone-induced apoptosis of mouse thymocytes: Prevention by native 7α-hydroxysteroids. *Immunol. Cell Biol.* **78,** 238–246.

Corpéchot, C., Robel, P., Axelson, M., Sjövall, J., and Baulieu, E. E. (1981). Characterization and measurement of dehydroepiandrosterone sulfate in rat brain. *Proc. Natl. Acad. Sci. USA* **78,** 4704–4707.

Corpéchot, C., Synguelakis, M., Talha, S., Axelson, M., Sjövall, J., Baulieu, E. E., and Robel, P. (1983). Pregnenolone and its sulfate ester in the rat brain. *Brain Res.* **270,** 119–125.

Doostzadeh, J., Urban, P., Pompon, D., and Morfin, R. (1996). Pregnenolone-7β-hydroxylating activties of yeast-expressed mouse cytochrome P450-1A1 and mouse-tissue microsomes. *Eur. J. Biochem.* **242,** 641–647.

Doostzadeh, J., Cotillon, A. C., and Morfin, R. (1997a). Dehydroepiandrosterone 7α- and 7β-hydroxylation in mouse brain microsomes. Effects of cytochrome P450 inhibitors and structure-specific inhibition by steroid hormones. *J. Neuroendocrinol.* **9,** 923–928.

Doostzadeh, J., Flinois, J. P., Beaune, P., and Morfin, R. (1997b). Pregnenolone-7β-hydroxylating activity of human cytochrome P450-1A1. *J. Steroid Biochem. Mol. Biol.* **60,** 147–152.

Doostzadeh, J., and Morfin, R. (1996). Studies of the enzyme complex responsible for pregnenolone and dehydroepiandrosterone 7α-hydroxylation in mouse tissues. *Steroids* **61,** 613–620.

Doostzadeh, J., and Morfin, R. (1997). Effects of cytochrome P450 inhibitors and of steroid hormones on the formation of 7-hydroxylated metabolites of pregnenolone in mouse brain microsomes. *J. Endocrinol.* **155,** 343–350.

Guiraud, J. M., Morfin, R., Ducouret, B., Samperez, S., and Jouan, P. (1979). Pituitary metabolism of 5α-androstane-3β,17β-diol: Intense and rapid conversion into 5α-androstane-3β,6α,17β-triol and 5α-androstane-3β,7α,17β-triol. *Steroids* **34,** 241–248.

Hampl, R., and Stárka, L. (1969). Epimerization of naturally occurring C_{19}-steroid allyl alcohols by rat liver preparations. *J. Steroid Biochem.* **1,** 47–56.

Hampl, R., Lapcík, O., Hill, M., Klak, J., Kasal, A., Novácek, A., Šterzl, I., Šterzl, J., and Stárka, L. (2000). 7-Hydroxydehydroepiandrosterone—A natural antiglucocorticoid and a candidate for steroid replacement therapy? *Physiol. Res.* **49,** S107–S112.

Hoebel, B. G., Kostner, G. M., and Graier, W. F. (1997). Activation of microsomal cytochrome P450 mono-oxygenase by Ca^{2+} store depletion and its contribution to Ca^{2+} entry in porcine aortic endothelial cells. *Br. J. Pharmacol.* **121,** 1579–1588.

Hu, Z. Y., Bourreau, E., Jung-Testas, I., Robel, P., and Baulieu, E. E. (1987). Neurosteroids: Oligodendrocyte mitochondria convert cholesterol to pregnenolone. *Proc. Natl. Acad. Sci. USA* **84,** 8215–8219.

Jung-Testas, I., Hu, Z. Y., Baulieu, E. E., and Robel, P. (1989). Neurosteroids: Biosynthesis of pregnenolone and progesterone in primary cultures of rat glial cells. *Endocrinology* **125,** 2083–2091.

Kalimi, M., Shafagoj, Y., Loria, R. M., Padgett, D., and Regelson, W. (1994). Antiglucocorticoid effects of dehydroepiandrosterone (DHEA). *Mol. Cell. Biochem.* **131,** 99–104.

Khalil, M. W., Strutt, B., Vachon, D., and Killinger, D. W. (1993). Metabolism of dehydroepiandrosterone by cultured human adipose stromal cells: Identification of 7α-hydroxydehydroepiandrosterone as a major metabolite using high performance liquid chromatography and mass spectrometry. *J. Steroid Biochem. Molec. Biol.* **46,** 585–594.

Khalil, M. W., Strutt, B., Vachon, D., and Killinger, D. W. (1994). Effect of dexamethasone and cytochrome P450 inhibitors on the formation of 7α-hydroxydehydroepiandrosterone by human adipose stromal cells. *J. Steroid Biochem. Molec. Biol.* **48,** 545–552.

Lafaye, P., Chmielewski, V., Nato, F., Mazie, J. C., and Morfin, R. (1999). The 7α-hydroxysteroids produced in human tonsils enhance the immune response to tetanus toxoid and *Bordetella pertussis* antigens. *Biochim. Biophys. Acta* **1472,** 222–231.

Lapcik, O., Hampl, R., Hill, M., and Stárka, L. (1999). Immunoassay of 7-hydroxysteroids: 2. Radioimmunoassay of 7alpha-hydroxy-dehydroepiandrosterone. *J. Steroid Biochem. Mol. Biol.* **71,** 231–237.

Loria, R. M., Inge, T. H., Cook, S. S., Szakal, A., and Regelson, W. (1988). Protection against acute lethal infections with the native steroid dehydroepiandrosterone (DHEA). *J. Med. Virol.* **26,** 301–314.

Loria, R. M., Padgett, D. A., and Huynh, P. N. (1996). Regulation of the immune response by dehydroepiandrosterone and its metabolites. *J. Endocrinol.* **150,** S209–S220.

Masters, J. N., Finch, C. E., and Sapolsky, R. M. (1989). Glucocorticoid endangerment of hippocampal neurons does not involve deoxyribonucleic acid cleavage. *Endocrinology* **124,** 3083–3088.

May, M., Holmes, E., Rogers, W., and Poth, M. (1990). Protection from glucocorticoid induced thymic involution by dehydroepiandrosterone. *Life Sci.* **46,** 1627–1631.

Mellon, S. H., and Compagnone, N. A. (1999). Molecular biology and developmental regulation of the enzymes involved in the biosynthesis and metabolism of neurosteroids. *In* "Contemporary Endocrinology: Neurosteroids: A New Regulatory Function" (E. E. Baulieu, P. Robel, and M. Schumacher, Eds.), pp. 27–49. Humana Press, Totowa, NJ.

Mentlein, R., and Kendall, M. D. (2000). The brain and thymus have much in common: A functional analysis of their microenvironments. *Immunology Today* **21,** 133–140.

Mohan, P. F., and Cleary, M. P. (1992). Studies on nuclear binding of dehydroepiandrosterone in hepatocytes. *Steroids* **57,** 244–247.

Morfin, R., and Courchay, G. (1994). Pregnenolone and dehydroepiandrosterone as precursors of active 7-hydroxylated metabolites which increase the immune response in mice. *J. Steroid Biochem. Molec. Biol.* **50,** 91–100.

Morfin, R. F., Leav, I., Orr, J. C., Picart, D., and Ofner, P. (1980). C_{19}-Steroid metabolism by canine prostate, epididymis and perianal glands. Application of the twin-ion technique of gas chromatography–mass spectrometry to establish 7α-hydroxylation. *Eur. J. Biochem.* **109,** 119–127.

Morfin, R., Lafaye, P., Cotillon, A. C., Nato, F., Chmielewski, V., and Pompon, D. (2000). 7α-Hydroxy-dehydroepiandrosterone and immune response. *In* "Neuroimmunomodulation, perspectives at the new millenium" (A. Conti, G. J. M. Maestroni, S. M. McCann, E. M. Sternberg, J. M. Lipton, and C. C. Smith, Eds.), *Annals of the New York Academy of Sciences* **917,** 971–982.

Nelson, D. R., Koymans, L., Kamataki, T., Stegeman, J. J., Feyereisen, R., Waxman, D. J., Waterman, M. R., Gotoh, O., Coon, M. J., Eastbrook, R. W., Gunsalus, I. C., and Nebert, D. W. (1996). The P450 superfamily: Update on new sequences, gene mapping, accession numbers and nomenclature. *Pharmacogenetics* **6,** 1–42.

Poletti, A., Celotti, F., Maggi, R., Melcangi, R. C., Martini, L., and Negri-Cesi, P. (1999). Aspects of hormonal steroid metabolism in the nervous system. *In* "Contemporary Endocrinology: Neurosteroids: A New Regulatory Function" (E. E. Baulieu, P. Robel, and M. Schumacher, eds.), pp. 97–123. Humana Press, Totowa, NJ.

Prasad, V. V. K., Vegesna, S. R., Welsh, M., and Lieberman, S. (1994). Precursors of the neurosteroids. *Proc. Natl. Acad. Sci. USA* **91,** 3220–3223.

Rose, K. A., Stapleton, G., Dott, K., Kieny, M. P., Best, R., Schwarz, M., Russell, D. W., Bjorkhem, I., Seckl, J., and Lathe, R. (1997). Cyp7b, a novel brain cytochrome P450, catalyzes the synthesis of neurosteroids 7α-hydroxy dehydroepiandrosterone and 7α-hydroxy pregnenolone. *Proc. Natl. Acad. Sci. USA* **94,** 4925–4930.

Sapolsky, R. M., Packan, D. R., and Vale, W. W. (1988). Glucocorticoid toxicity in the hippocampus: *In vitro* demonstration. *Brain Res.* **453,** 367–371.

Schwarz, M., Lund, E. G., Lathe, R., Bjorkhem, I., and Russell, D. W. (1997). Identification and characterization of a mouse oxysterol 7α-hydroxylase cDNA. *J. Biol. Chem.* **272,** 23995–24001.

Shi, J., Schulze, S., and Lardy, H. A. (2000). The effect of 7-oxo-DHEA acetate on memory in young and old C57BL/6 mice. *Steroids* **65,** 124–129.

Skinner, S. J. M., Tobler, C. J. P., and Couch, R. A. F. (1977). A radioimmunoassay for 7α-hydroxy-dehydroepiandrosterone in human plasma. *Steroids* **30,** 315–330.

Stapleton, G., Steel, M., Richardson, M., Mason, J. O., Rose, K. A., Morris, R. G. M., and Lathe, R. (1995). A novel cytochrome P450 expressed primarily in brain. *J. Biol. Chem.* **270,** 29739–29745.

Stárka, L., Hill, M., Hampl, R., Malewiak, M. I., Benalycherif, A., Morfin, R., Kolena, J., and Scsukova, S. (1998). Studies on the mechanism of antiglucocorticoid action of 7α-hydroxydehydroepiandrosterone. *Collect. Czech. Chem. Commun.* **63,** 1683–1698.

Strömstedt, M., Warner, M., Banner, C. D., MacDonald, P. C., and Gustafsson, J. Å. (1993). Role of brain cytochrome P450 in regulation of the level of anesthetic steroids in the brain. *Mol. Pharmacol.* **44,** 1077–1083.

Warner, M., Strömstedt, M., Möller, L., and Gustafsson, J. Å. (1989). Distribution and regulation of 5α-androstane-3β,17β-diol hydroxylase in the central nervous system. *Endocrinology* **124,** 2699–2706.

Wu, Z. L., Martin, K. O., Javitt, N. B., and Chiang, J. Y. L. (1999). Structure and functions of human oxysterol 7α-hydroxylase cDNAs and gene CYP7B1. *J. Lipid Res.* **40,** 2195–2203.

NEUROSTEROID ANALYSIS

Ahmed A. Alomary,[*,†,1] Robert L. Fitzgerald,[†,‡] and Robert H. Purdy[*,†,§]

[*]Department of Neuropharmacology, The Scripps Research Institute,
La Jolla, California 92037
[†]Department of Veterans Affairs Medical Center and Veterans Medical Research Foundation,
San Diego, California 92093
[‡]Department of Pathology, University of California–San Diego,
La Jolla, California 92093, and
[§]Department of Psychiatry, University of California–San Diego,
La Jolla, California 92093

I. Introduction
II. Analysis of Neurosteroids by Radioimmunoassay
III. Analysis of Neurosteroids by High-Performance Liquid Chromatography Coupled with Gas Chromatography–Mass Spectrometry
IV. Analysis of Unconjugated Neurosteroids by Gas Chromatography–Mass Spectrometry
V. Derivatization of Neurosteroids
VI. Gas Chromatography Columns for Separation of Neurosteroid Derivatives
VII. Isotopic Dilution for Quantitative Gas Chromatography–Mass Spectrometry
VIII. Data Acquisition Modes in Mass Spectrometry
IX. Mass Spectrometry/Mass Spectrometry
X. Analysis of Neurosteroid Sulfates by Mass Spectrometry
XI. Analysis of Sulfated Steroids by Soft Ionization Mass Spectrometry
XII. Analysis of Steroid by Atmospheric Pressure Chemical Ionization/Mass Spectrometry
XIII. Thermospray Liquid Chromatography/Mass Spectrometry
XIV. Future Analysis of Neurosteroid Sulfates
XV. Conclusions
References

In this chapter, we review techniques used for the analysis of neurosteroids and discuss the advantages and disadvantages of each method. Because radioimmunoassay (RIA) procedures are well known, we focus more on the relatively recent mass spectrometric methods used for analyzing neurosteroids and their sulfates. We also discuss some promising methods that permit the detection of low levels of neurosteroids in small samples with a minimum number of sample pretreatment procedures. Lowering the limits of detection will enable a better understanding of the physiological function of neurosteroids and the mechanism(s) for neurosteroid

[1]Address to whom correspondence should be sent.

regulation of brain function. Moreover, analyzing low levels of neurosteroids more efficiently will increase the throughput, which is important for clinical analysis.

Initially, most neurosteroid analyses were performed by RIA. However, many analyses of neurosteroids are now performed by mass spectrometry. To date, the most sensitive, specific, and accurate method for the simultaneous analysis of several neurosteroids is the method of gas chromatography/electron capture/negative chemical ionization mass spectrometry. This method, with its many variants, is described in detail. © 2001 Academic Press.

I. Introduction

Endogenous steroids are C18, C19, and C21 compounds that have a fused-ring system of three six-membered rings, termed A, B, and C, and one five-membered ring, termed D. In the 1980s, Baulieu and Robel and their co-workers demonstrated, using radioimmunoassay (RIA) procedures, that some steroids, such as dehydroepiandrosterone (DHEA), pregnenolone (PREG), and their sulfates and lipoidal esters are present in higher concentrations in tissue from the central nervous system than in blood. This finding was the initial evidence suggesting that some steroids are synthesized in the brain (see Chapter 1). Such steroids that are biosynthesized from cholesterol in the central and peripheral nervous systems are now universally referred to as neurosteroids. It is implicit in this definition that these neurosteroids are found in larger amounts in brain than in blood. An alternative nomenclature was proposed by Paul and Purdy (1992) to define a category of neuroactive steroids, including both endogenous and synthetic steroids, that alter neuronal activity (Mellon and Compagnone, 1999).

The discovery that neurosteroids alter neuronal activity has driven the need for sensitive, selective, and structurally specific assays to measure their concentration in biological tissues and fluids. To learn whether neurosteroids are synthesized in nervous tissues in quantities sufficient to modulate neuronal activity and to learn if they are distributed uniformly in the brain regions, highly selective and specific analytical methods that can analyze small quantities of neurosteroids, including their sulfates, are required. Many techniques and methods have been used to quantify steroidal compounds. These methods include RIA, gas chromatography–mass spectrometry (GC/MS), high-performance liquid chromatography (HPLC), and liquid chromatography–mass spectrometry (LC/MS). Although these techniques have been successfully used in steroid analysis, quantitative analysis of neurosteroids in small samples has been difficult to achieve because of their low concentrations in nervous tissues. To date, the most sensitive and suitable method capable of simultaneous analysis of several neurosteroids is the one that uses the gas chromatography/electron capture–negative chemical ionization/mass spectrometry (GC/EC–NCI/MS) with selected ion monitoring (SIM), a mode that focuses on a few specific ions bearing structural information.

II. Analysis of Neurosteroids by Radioimmunoassay

RIA remains the method most commonly used in the analysis of steroids. RIA is considered to be a relatively sensitive procedure, depending on the specific activity (millicurie per millimole) of the steroid ligand, and has been used extensively to measure neurosteroids in brain tissue (Purdy et al., 1990; Corpéchot et al., 1993; Murphy, 2000). However, the large number of endogenous steroids that are structurally similar and the insufficient selectivity and specificity of the antibodies that are available for use in these assays, limit the use of RIA for neurosteroid analyses. In the measurement of neurosteroids by RIA, a single steroid is measured per assay. A chromatographic separation by HPLC before RIA greatly improves the specificity of the analysis and is considered to be essential for validation of the RIA method. However, even when combined with HPLC, RIA methods do not offer a reliable criterion for measuring steroids with sufficient specificity for structural identification (Seikman, 1979; Corpéchot et al., 1993). Consequently, reported results for neurosteroids vary significantly (see Chapter 17). Extensive sample purification steps, with inherent product loss, are necessary to avoid quantification errors caused by cross-reactivity or by the presence in tissue extracts of contaminating substances that interfere with the RIA (Purdy et al., 1990, 1991; Morfin et al., 1992; Corpéchot et al., 1993). For example, Korneyev et al. (1993) found it necessary to pass samples extracted from brain tissue through three different HPLC columns to eliminate endogenous substances that interfered with the analysis of PREG by RIA.

III. Analysis of Neurosteroids by High-Performance Liquid Chromatography Coupled with Gas Chromatography–Mass Spectrometry

Cheney et al. (1995) analyzed steroids by performing an HPLC purification step before GC/MS analysis. The steroids were initially characterized by their HPLC retention times compared to those of tritium-labeled recovery standards. The neurosteroids were then characterized by GC/MS analysis of their heptafluorobutyric ester or methoxyamine derivatives. For structural identification, the mass spectra were compared to appropriate reference standards. This approach has high specificity, and the use of SIM increases sensitivity. The detection limit for measuring allopregnanolone achieved in their study was 0.63 pmol (0.2 ng) starting from ~100–300 mg of brain tissue.

IV. Analysis of Unconjugated Neurosteroids by Gas Chromatography–Mass Spectrometry

Several investigators have described the use of GC/MS for the quantitative analysis of neurosteroids. Some of the derivatives are shown in Fig. 1. However, to date, most groups have used extensive sample purification schemes, usually including an HPLC purification step before GC/MS analysis (Cheney et al., 1995; Uzunov et al., 1996; Lière et al., 2000; Vallée et al., 2000) or multiple clean-up steps (Shimada and Yago, 2000). Another drawback of these labor-intensive procedures was the failure to use the appropriate internal standard for each steroid analyzed. This presumably

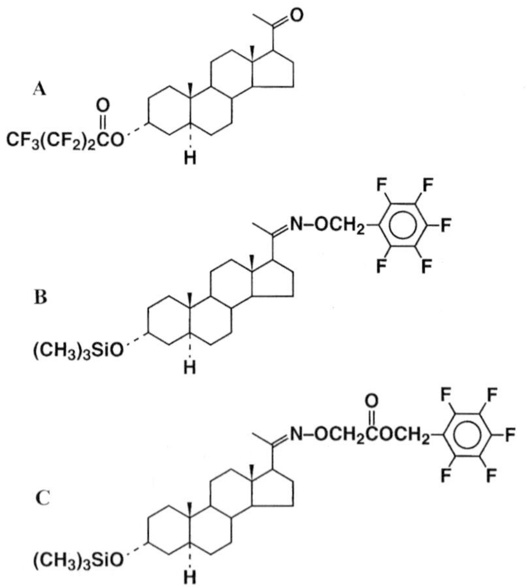

FIG. 1. Structures of three derivatives of allopregnanolone employed in GC/MS: in (**A**) the 3-heptafluorobutyrate, in (**B**) the 20-pentafluorobenzyloxime 3-trimethylsilyl ether, and in (**C**) the 20-pentafluorobenzylcarboxymethoxime 3-trimethylsilyl ether.

was a result of the lack of commercial availability of all of the required deuterated internal standards. It is important to note that deuterated internal standards prepared by exchange-labeling procedures cannot be employed in procedures that require prior purification by HPLC because the deuterated standards are partially separated from the nondeuterated steroids of interest during the HPLC clean-up step.

The amount of steroid that can be detected with a given method depends on many variables. Based on our review of the literature, the manner in which the limit of detection is expressed varies widely. Some authors define their limit of detection as the smallest amount of steroid that can be detected using standards free of a biological matrix. Others express the limit of detection as the smallest amount of steroid that can be reliably detected from a specific amount of a biological matrix (e.g., 250 pg of allopregnanolone from 100 mg of brain tissue). In our opinion, the best way to express sensitivity is the latter method. In addition to good sensitivity, it is essential that the robustness of the assay be demonstrated through the analysis of quality control samples, typically run in duplicate at three concentrations spanning the range of interest for a minimum of five separate analyses.

Neurosteroids have been identified and quantified in rat and human plasma and cerebrospinal fluid (CSF) by Kim *et al.* (2000) using a GC/MS method with a relatively simple sample-preparation scheme. Their procedure was noteworthy because it used only a C18 solid-phase extraction column for prepurification of neurosteroids. For derivatization, they used the method of Hubbard *et al.* (1994), whose product with allopregnanolone is shown in Fig. 1C. This allowed quantification of allopregnanolone, pregnanolone, testosterone, and androsterone in CSF and plasma samples. Kim *et al.* (2000, Fig. 6A) easily separated epiallopregnanolone (which they termed iso-pregnanolone) from its other three isomers. Of these remaining isomers, they were able to partially separate pregnanolone (about 80% resolved) from allopregnanolone and epipregnanolone. Figure 2 shows the separation of the allopregnanolone isomers and PREG using the method of Kim *et al.* (2000), as performed by Z. Wang and E. M. Sellers (personal communication).

Vallée *et al.* (2000) developed a method for quantifying allopregnanolone, DHEA, testosterone, and epiallopregnanolone from rat plasma and brain tissue. The derivative used by these authors is shown in Fig. 1B. This method employed deuterium-labeled internal standards for all of the compounds of interest. It consisted of a simple, solid-phase extraction to isolate neurosteroids from the biological matrix. The limit of quantification from 1 ml of plasma was 100 pg, and the limit from brain tissue was 250 pg. This method was validated in terms of sensitivity, accuracy, and precision, and it can serve as a model for validating quantitative MS methods for neurosteroid analysis. A representative SIM chromatogram is shown in Fig. 3.

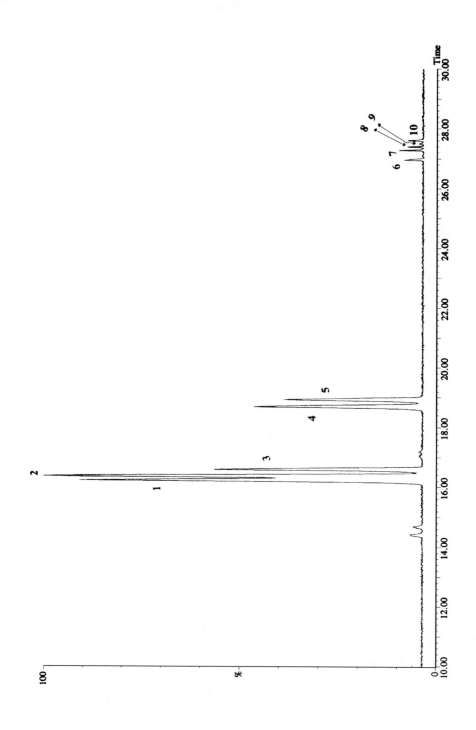

V. Derivatization of Neurosteroids

Steroids are polar compounds that are usually derivatized with a suitable reagent before analysis by GC/MS. Derivatization of polar compounds usually results in decreased polarity and increased volatility. In addition, the mass spectra of the derivatized steroid can be enhanced in terms of sensitivity and specificity. The most common derivatives employed for GC/MS analyses of steroids are silyl ether derivatives. Silyl derivatives can be prepared from alcoholic, phenolic, and carboxyl groups. Enolizable ketones also form silyl derivatives, but yields are not typically as good as for alcoholic or phenolic groups. Although silyl derivatives can be used for polar groups in steroids, silylation is usually employed for hydroxyl groups. Other derivatives are frequently used for carbonyl and carboxyl groups. Oximes and methoximes are the most commonly used derivatives for carbonyl groups. After formation of an oxime derivative with hydroxylamine hydrochloride, it can be further derivatized with a silyl reagent. Dehennin and Scholler (1973) demonstrated that the C3 carbonyl group with a Δ^4-conjugated double bond, as in progesterone or testosterone, forms almost exclusively the 3,5-dienol ether with HFBAA when the reaction is performed in acetone (e.g., Lière *et al.*, 2000). Greatly enhanced sensitivity is obtained with polyfluorinated derivatives and use of NCI GC/MS. Three common polyfluorinated derivatives of allopregnanolone are shown in Fig. 1. Because allopregnanolone has a single hydroxyl group (3α), a monoheptafluorobutyrate is formed (Fig. 1A) using heptafluorobutyric anhydride (Fig. 1B and 1C). Using the procedure of Uzunova *et al.* (1998)—the separation of heptafluorobutyrate of allopregnanolone and its isomers—only three peaks were obtained from the four isomers. However, Romeo and di Michele (personal communication) obtained baseline separation of all allopregnanolone isomers as the HFBAA derivatives as shown in Fig. 4. In addition, the ketone group of allopregnanolone can be reacted to form in one step the pentafluorinated oxime

FIG. 2. The GC/EC–NCI/MS separation and detection of progesterone and its metabolites. **(Peak 1)** epipregnanolone, **(peak 2)** allopregnanolone, **(peak 3)** PREG, **(peak 4)** PREG, **(peak 5)** epiallopregnanolone, **(peak 6)** 5β-dihydroprogesterone, **(peaks 7 and 8)** 5α-dihydroprogestone doublet, **(peaks 9 and 10)** progesterone doublet. One hundred picograms of d_0-derivatized allopregnanolone, PREG, epiallopregnanolone, and epipregnanolone and 1 ng of d_0-derivatized progesterone, 5α-dihydroprogesterone, and 5β-dihydroprogesterone were injected in a single run. The three-step derivatization procedure of Kim *et al.* (2000) was employed. A methylated silicone thin-film capillary column (90510B, Quadrex Co, New Haven, CT; 15 m × 0.25 mm, 0.05 μM film thickness) was used. [Courtesy of W. Zhang and E. M. Sellers, University of Toronto, Toronto, Canada (unpublished)].

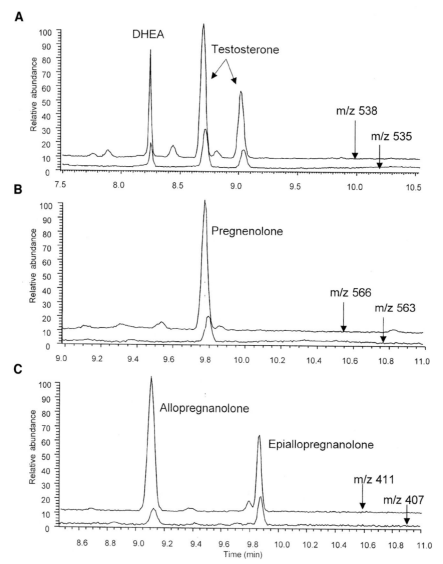

FIG. 3. The GC/MS analysis by selected ion monitoring of neurosteroids extracted from the cortex of the male rat brain. (**A**) DHEA and testosterone were quantified using the m/z 535 ion. The internal standards [^{2}H$_3$]-DHEA and [^{2}H$_3$]testosterone were monitored at m/z 538. (**B**) PREG was quantified using the 563 ion. The internal standard [^{2}H$_4$] PREG was monitored at m/z 563. (**C**) Allopregnanolone and epiallopregnanolone using the m/z 407 ion. The internal standards of [^{2}H$_4$]allopregnanolone and [^{2}H$_4$]epiallopregnanolone were monitored at m/z 411. [from Vallée *et al.* (2000). Quantification of neurosteroids in rat plasma and brain following swim stress and allopregnanolone administration using negative chemical ionization gas chromatography/mass spectrometry. *Anal. Biochem.* **287,** 153–166; reprinted with permission.]

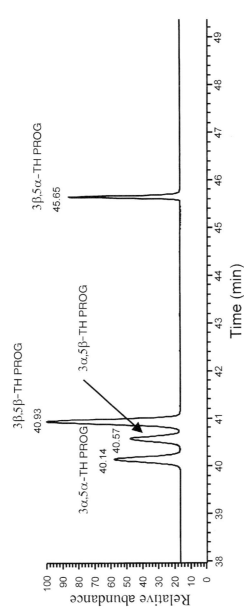

FIG. 4. Selected ion monitoring GC/MS trace (496 m/z) of the four tetrahydroprogesterone 3-heptafluorobutyrate isomers, eluted successively as the derivatives of allopregnanolone ($3\alpha,5\alpha$-TH PROG), pregnanolone ($3\alpha,5\beta$-TH PROG), epipregnanolone ($3\beta,5\beta$-TH PROG), and epiallopregnanolone ($3\beta,5\alpha$-TH PROG). A HP35 MS (35% phenylmethylsilicone) column was employed (30 m × 0.25 mm, 0.15 μm film thickness). [Courtesy of Elena Romeo and Flavia di Michele, Fondazione Santa Lucia, Rome, Italy (unpublished)].

derivative shown in Fig. 1B (Vallée *et al.*, 2000) or in two steps the similar but somewhat larger pentafluorinated derivative shown in Fig. 1C (Kim *et al.*, 2000). It is important to note that with oximes *syn-* and *anti-*isomers are formed as a result of the derivatizing reagent, relative to the lone pair of electrons on the oxime. This can lead to multiple chromatographic peaks, as is observed with testosterone (Fitzgerald and Herold, 1996). When oximes are formed, it is still necessary to derivatize the alcoholic groups. This is done with a silylating reagent, as shown in Fig. 1B and 1C.

Derivatization of unconjugated neurosteroids is essential for increasing the sensitivity by adding functional groups that enhance the ionization efficiency and volatility of the steroid and thus make it more easily detected. For example, derivatization of neurosteroids by reagents with highly electronegative groups improves their detection limit in NCI/MS (Dehennin *et al.*, 1998; Kim *et al.*, 2000; Vallée *et al.*, 2000). Derivatization can also enhance sensitivity of sulfated neurosteroids when analyzed by nano-electrospray/mass spectrometry (nano-ES/MS) procedures. Neurosteroid sulfates can be easily prepared in a single-step reaction in pyridine with the N,N-dimethylformamide complex of sulfur trioxide (Chatman *et al.*, 1999). In another elegant example, the formation of *mono-* or *bis-*oxime derivatives from *mono-* or *di-*ketosteroids with hydroxylamine hydrochloride provided derivatives with detection limits by nano-ESI/MS, which were about 20 times lower than the underivatized neurosteroid (Liu *et al.*, 2000a). In this study, the oximes were found to give abundant positive ions by ES when sprayed from a suitable solvent. Both deuterated neurosteroid sulfates and neurosteroid oximes can easily be prepared in stable crystalline form for use as internal standards.

Derivatization can lead to the shift of fragment ions to higher masses with lower matrix background and to the formation of characteristic fragments. The choice of derivatives to be used involves consideration of several aspects of MS instrumentation, such as the ionization mode, the resolution of the mass spectrometer, and the potential increase in selectivity and/or sensitivity, which can be obtained by using tandem mass spectrometry (MS/MS). In some cases, when structurally related compounds are not separated in the underivatized form, they may be resolved as derivatives.

VI. Gas Chromatography Columns for Separation of Neurosteroid Derivatives

The general rule in column selection is that columns with a polar stationary phase are used to separate polar compounds, while columns with nonpolar stationary phases are used to separate nonpolar compounds. Separation of nonpolar substances on nonpolar columns is a nonselective process, with compounds usually being eluted in the order of increasing boiling point. A variety of capillary columns, including DB-1 (dimethyl polysiloxane), DB-5 (5% phenylmethyl polysiloxane), and DB-17, (50% phenylmethyl polysiloxane) are used in the analysis of steroids by GC/MS.

VII. Isotopic Dilution for Quantitative Gas Chromatography–Mass Spectrometry

For quantitative GC/MS procedures, it is essential that appropriate internal standards are employed to account for variations in extraction efficiency, derivatization, injection volume, and matrix effects. Selection of an appropriate internal standard is critical when performing isotope dilution GC/MS analyses. Ideal internal standards for GC/MS analyses have the same physical and chemical properties as the analyte of interest but are separated by mass. The best internal standards are nonradioactive stable isotopic analogs of the compounds of interest that differ by at least 3, and preferably 4 or 5, atomic mass units. With isotope dilution, the only property that differs between the analyte and the internal standard is a small mass difference, which is readily distinguished by the mass spectrometer. Isotopic dilution procedures are among the most accurate and precise quantitative methods available to analytical chemists. Therefore, in the isotope dilution method, we have an absolute reference (*i.e.*, the response coefficients of the compound and the internal standard are considered to be identical) (Pickup and McPherson, 1976). Isotope dilution is performed by adding a known exact amount of the internal standard, typically a deuterium-labeled analog of the compound of interest, at the initial stage of the analytical process (e.g., to the biological sample). The internal standard is then subjected to the exact conditions as the analyte of interest and thus accounts for any loss of the analyte throughout the entire analytical process. When subject to MS analysis, the amount of the analyte is calculated relative to signal intensity of the internal standard. This is possible because the peak of the labeled standard is shifted to a different position in the spectrum, according to the number and the nature of the atoms that were added in the labeling procedure. The ratio of these two signal intensities is used to calculate their relative proportion. It is essential that the internal standard has the isotopes incorporated in stable positions and that the isotope is contained in the fragments monitored.

VIII. Data Acquisition Modes in Mass Spectrometry

Two common data acquisition modes for GC/MS are scanning and SIM. In the scan mode, a range of mass to charge (m/z) (e.g., from 50 to 700 m/z) is analyzed. In the SIM mode, only ions selected to be characteristics of the analyte of interest are monitored. In general, compounds are initially analyzed in the scan mode to identify all of the ions of interest. Based on the full-scan spectra, ions are then selected for SIM analysis. Good ions for SIM analysis typically are at high m/z and contain a unique fragment of the molecule of interest. For example, it is not good practice to monitor the 73 m/z ion for silylated compounds, as it is characteristic of the derivatizing reagent and therefore not specific for any particular analyte. Because of the increase in dwell time associated with the ions of interest, SIM provides improved sensitivity compared with the scan mode. An example of SIM chromatogram is shown in Fig. 3.

IX. Mass Spectrometry/Mass Spectrometry

In addition to GC/MS and LC/MS, neurosteroids are also analyzed with MS/MS. Typically, MS/MS instrumentation is also connected with LC or GC to form the multihyphenated techniques of LC/MS/MS and GC/MS/MS. These multidimensional analytical systems can be very useful in identifying compounds in complex mixtures and in determining structures of unknown substances.

In the MS/MS technique, the first m/z separation is used to select a single (*precursor*) mass that is characteristic of a given analyte in a mixture. The mass-selected ion(s) is activated to cause it to fall apart to produce fragment (*product*) ions. This is usually done by bombarding the ions with a neutral gas in a process called *collisional activation* or *collision-induced dissociation* (CID). The product ions are then separated according to m/z. The resulting MS/MS spectrum consists only of product ions from the selected precursor. Chemical background and other mixture components are decreased. The chemical noise may decrease more than that of the signal ions, so the signal:noise ratio increases, resulting in greater sensitivity.

X. Analysis of Neurosteroid Sulfates by Mass Spectrometry

Although GC/MS is a suitable method for the analysis of steroids, it is difficult to analyze polar, nonvolatile steroids, such as sulfated steroids, by GC/MS directly without derivatization. GC/MS can be used to analyze sulfated steroids indirectly, where conjugates are hydrolyzed to the parent alcohols, which are then derivatized using a suitable reagent before analysis. Meng *et al.* (1996) described a method of solvolysis of steroid sulfates in tetrahydrofuran with dilute trifluoroacetic acid as a catalyst. A labeled internal standard was used to monitor the efficiency of solvolysis. However, the presence of peroxides in solvents such as ether or tetrahydrofuran is detrimental to efficient recovery of the parent neurosteroids during solvolysis in these solvents. The use of dilute mineral acids, such as hydrochloric or sulfuric acids, as catalysts for solvolysis in ethyl acetate can also result in complete hydrolysis. Direct analysis of sulfated steroids by MS has become common, especially after the advent of soft ionization techniques (Bowers and Sanaullah, 1996; Murry *et al.*, 1996). Soft ionization techniques, such as chemical ionization and electrospray ionization (ESI), can produce pseudo-molecular ions by a gentle process of proton transfer leading to high intensity of intact molecular ions. For molecules that can be vaporized without decomposition, electron impact ionization (EI) is then often used to generate ions for mass analysis. In EI, ionization occurs by electrons accelerated through a potential of 70 volts, which is a highly energetic, or "hard," process and may lead to extensive fragmentation that leaves very little or no trace of a molecular ion.

XI. Analysis of Sulfated Steroids by Soft Ionization Mass Spectrometry

After the advent of fast atom bombardment (FAB), it became possible to analyze intact polar biomolecules, such as sulfated steroids, directly, without the need for solvolysis and derivatization. Several studies have used FAB to analyze steroid sulfates (Liehr *et al.*, 1982; Dumasia *et al.*, 1983; Gaskell *et al.*, 1983; Kingston *et al.*, 1985; Libert *et al.*, 1991). Secondary ion mass spectrometry (SIMS) has also been used to analyze steroid sulfates (Shackleton and Straub, 1982). In FAB and SIMS analyses, the extensive sample purification required can result in limited sensitivity. FAB is not a very "soft ionization" technique; therefore, a softer ionization method, such as ESI, is expected to provide a better opportunity for forming intact molecular ions of sulfated neurosteroids with less fragmentation.

Moreover, the advantage of ESI is a minimal amount of matrix (solvent)-associated chemical noise, making it possible to analyze low levels of steroid sulfates. For example, Bowers and Sanaullah (1996) analyzed steroid glucuronides and sulfates directly by HPLC ESI/MS. Reverse-phase HPLC was capable of resolving all of the isomers studied. The detection limits of 3–25 pg injected on a packed capillary column was achieved in this study. However, no biological analyses were reported. In 1998, Shimada *et al.* used ESI to analyze PREG sulfate (PREGS) in rat brain by SIM of the $[M-H]^-$ ion at m/z 395. Their detection limit using this technique was 1 ng/injection. In 1999, Griffiths *et al.* used nano-ESI/MS for the analysis of steroid sulfates in the rat brain. They were able to analyze deuterated neurosteroid sulfates by nano-ES-MS/MS at a level of 50 pg/mg of brain tissue. They showed that complete structural information can be obtained from 1 ng (3 pmol) of steroid sulfate, while fragment ions of the sulfate esters can be obtained from only 3 pg (10 fmol) of sample. These values correspond to the expected quantities of (PREGS) in about 100 mg and 300 µg of brain, respectively.

Chatman et al. (1999) have also used nano-ESI for neurosteroids analysis. In nano-ESI, the samples are sprayed from pulled capillaries that are usually coated with metal. This allows analysis of very small (e.g., ~1 µl) sample volumes that can be on the order of 10–40 nl/min. This results in decreased sample consumption, more stable spray from a wide pH range and from high buffer concentrations, and less contamination of the instrument. In this study, Chatman et al. (1999) developed a method for extracting unconjugated steroids and steroid sulfates from biofluids. The new extraction method allowed the unconjugated steroids and their sulfate esters to be isolated separately in a two-step procedure using diethyl ether/hexane (90:10, v/v) initially and then chloroform/2-butanol (50:50, v/v) in the second step to extract steroid sulfates. Precursor ion scanning (one of the scan modes used to carry out MS/MS) performed with a triple-quadrupole mass spectrometer was used to quantify the steroids. Deuterated steroids were used as internal standards in the quantification. The limit of detection for steroid sulfates from the biological matrix was 200 amol/µl (~80 fg/µl), with only about 1 µl of sample being injected. Endogenous levels of the unconjugated and sulfated steroids were detected and quantified from physiological samples. Precursor ion scanning reduces the interference from the chemical noise, thus enhancing the signal:noise ratio. The unconjugated steroids were analyzed in the positive mode of nano-ESI, whereas the steroid sulfates were analyzed by negative nano-ESI. Anionic species of steroid sulfates were detected in negative nano-ESI at concentrations lower than the protonated $[M + H]^+$ or $[M - H_2O + H]^+$ species of the unconjugated steroids in positive mode.

More recently, Liu et al. (2001) developed a method based on capillary liquid chromatography (LC)/micro-ESI mass spectrometry. In this study, rat brain was extracted with ethanol and passed through a lipophilic cation and anion exchanger. The sulfated neurosteroids were eluted from the anion exchanger. Oxosteroids in the unconjugated fraction were derivatized to their oximes, which were then sorbed on a bed of the cation exchanger and eluted. The fractions containing the oximes and the sulfates, respectively, were concentrated to 100 µl by micro-solid-phase extraction. The neurosteroid sulfate fraction was analyzed by capillary LC/micro-ESI/MS. Deuterated neurosteroids were used as internal standards. For example, $[^2H_4]$ PREGS was used in the analysis of PREGS, the most abundant neurosteroid sulfate in rat brain. A detection limit of 1 pg on-column was achieved (S/N 3:1) for PREGS. The identity of PREGS from brain was also confirmed by CID of the deprotonated molecule.

XII. Analysis of Steroid by Atmospheric Pressure Chemical Ionization/Mass Spectrometry

Kobayashi *et al.* (1993) used atmospheric pressure chemical ionization (APCI)/MS to study the fragmentation of steroids. They showed that APCI/MS could be applied for determination of the molecular weight of polar nonvolatile thermolabile steroids without derivatization. The steroids used in this study were divided into two groups according to their mass spectral profiles, one having a carbonyl group at C3 together with a conjugated double bond at C4-5 (Group A) and the other with a hydroxyl group at C3 (Group B). In Group A, the predominant peak observed corresponded to the protonated molecular ion $[M+H]^+$. The fragment ion corresponding to the elimination of CH_2OH, $COCH_3$, and/or $COCH_2OH$ from the steroid skeleton appeared as a base peak in some steroids of Group A. In Group B, predominantly $[M+H-H_2O]^+$ and/or $[M+H-2H_2O]^+$ ions were observed, which originated from the loss of water. Other major ions in this group were the protonated molecular ions.

XIII. Thermospray Liquid Chromatography/Mass Spectrometry

Watson *et al.* (1985) analyzed conjugated steroids by coupling LC to MS via a thermospray (Th)-LC/MS interface. In their studies, negative ion Th-mass spectra were recorded for several steroid conjugates. This interface vaporizes the sample before it passes through a small orifice at the tip of a heated stainless steel tube in the ion source. Ionization occurs at the same time, usually promoted by a volatile buffer, such as ammonium acetate.

XIV. Future Analysis of Neurosteroid Sulfates

ESI coupled with Fourier transform ion cyclotron resonance (FT-ICR) MS is one of the most sensitive techniques for the analysis of biological samples. ESI is considered to be the "softest" ionization technique available (Whitehouse *et al.*, 1985); it can produce intact molecular ions, which leads to sensitivity enhancement. ICR is a nondestructive detector that allows multiple detection of the ions, also leading to improvement of sensitivity (Marshall *et al.*, 1985).

Several studies have focused on the improvements of detection limits in FT-ICR MS. For example, ESI coupled with FT-ICR MS has been used to acquire mass spectra from attomole (Valaskovic *et al.*, 1996) and zeptomole ranges (Alomary and Solouki, submitted for publication; Belov *et al.*, 1999). Therefore, ESI coupled with FT-ICR is expected to be the most sensitive and rapid method for the future analysis of neurosteroid sulfates.

XV. Conclusions

Because the concentrations of neurosteroids that are present in brain and plasma are normally very low, several prepurification steps are usually required for analyses of these samples. Therefore, it is necessary to improve detection limits and to introduce new methods that can analyze these samples directly from solution without these labor-intensive prepurification procedures. Most neurosteroid analyses are now performed by RIA or by GC/MS. To date, the most sensitive GC/MS method reported is GC/EC-NCI/ MS, wherein MS is performed in the SIM mode. Neurosteroid sulfates can be directly analyzed by MS without derivatization by using soft ionization methods, such as FAB and ESI, which are currently undergoing further development.

Acknowledgments

This chapter is Publication No. 14076-NP from The Scripps Research Institute. The research was supported by National Institutes of Health Grants AA06420 and AA07456, from the National Institute on Alcohol Abuse and Alcoholism. We are indebted to W. Zhang, Edward M. Sellers, Elena Romeo, and Flavia di Michele for providing chromatograms of their work before publication. We are grateful for the assistance of David A. Herold, Louis Dehennin, and William J. Griffiths in the preparation of this manuscript.

References

Alomary, A., and Solouki, T. (2001). Combined external and internal accumulation to improve detection limit for electrospray ionization Fourier transform ion cyclotron resonance mass spectrometry. *J. Am. Soc. Mass Spectrom.*, submitted for publication.

Belov, M. E., Gorshkov, M. V., Anderson, G. A., Udseth, H. R., and Smith, R. D. (1999). On improving the performance of a FT-ICR mass spectrometer with an external accumulation device, *Proceedings of the 47th ASMS Conference on Mass Spectrometry and Allied Topics,* pp. 767–768.

Bowers, L. D., and Sanaullah (1996). Direct measurement of steroid sulfate and glucuronide conjugates with high-performance liquid chromatography–mass spectrometry. *J. Chromatogr. B.* **687,** 61–68.

Chatman, K., Hollenbeck, T., Hagey, L., Vallée, M., Purdy, R. H., Weiss, F., and Siuzdak, G. (1999). Nanoelectrospray mass spectrometry and precursor ion monitoring for quantitative steroid analysis and attomole sensitivity. *Anal. Chem.* **71,** 2358–2363.

Cheney, D. L., Uzunov, D., Costa, E., and Guidotti, A. (1995). Gas chromatography–mass fragmentometric quantification of 3α-hydroxy-5α-pregnan-20-one (allopregnanolone) and its precursor in blood and brain of adrenalectomized and castrated rats. *J. Neurosci.* **16,** 4641–4650.

Corpéchot, C., Young, J., Calvel, C., Wehrey, J. N., Veltz, G., Touyer, M., Mouren, M., Prasad, V. V. K., Banner, C., Sjövall, J., Baulieu, E. E., and Robel, P. (1993). Neurosteroids: 3α-Hydroxy-5α-pregnan-20-one and its precursor in the brain, plasma, and steroidogenic glands of male and female rats. *Endocrinology* **133,** 1003–1009.

Dehennin, L., Ferry, M., Lafarge, P., Pérès, G., and Lafarge, J.-P. (1998). Oral administration of dehydroepiandrosterone to healthy men: Alteration of the urinary androgen profile and consequences for the detection of abuse in sport by gas chromatography–mass spectrommetry. *Biomed. Mass Spectrom.* **63,** 80–87.

Dehennin, L., and Scholler, R. (1973). Dienol heptafluorobutyrates as derivatives for gas liquid chromatography of steroidal Δ^4-3-ketones determination of the structure of the isomeric dienol esters. *Tetrahedron* **29,** 1591–1594.

Dumasia, M. C., Houghton, E., Bradley, C. V., and Williams, D. H. (1983). Studies related to the metabolism of anabolic steroids in the horse: The metabolism of 1-dehydrosterone and the use of fast atom bombardment mass spectrometry in the identification of steroid conjugates. *Biomed. Mass Spectrom.* **10,** 434–440.

Fitzgerald, R. L., and Herold, D. A. (1996). Serum total testosterone: Immunoassay compared with negative chemical ionization gas chromatography–mass spectrometry. *Clin. Chem.* **42,** 749–755.

Gaskell, S. J., Brownsey, B. G., Brooks, P. W., and Green, B. N. (1983). Fast atom bombardment mass spectrometry of steroid sulphates: Qualitative and quantitative analyses. *Biomed. Mass Spectrom.* **10,** 215–219.

Griffiths, W. J., Liu, S., Yang, Y., Purdy, R. H., and Sjövall, J. (1999). Nano-electrospray tandem mass spectrometry for the analysis of neurosteroid sulphates. *Rapid Commun. Mass Spectrom.* **13,** 1595–1610.

Hubbard, W. C., Bickel, C., and Schleimer, R. P. (1994). Simultaneous quantitation of endogenous levels of cortisone and cortisol in human nasal and bronchoalveolar lavage fluids and plasma via gas chromatography–negative ion chemical ionization mass spectrometry. *Anal. Biochem.* **221,** 109–117.

Kim, Y. S., Zhang, H., and Kim, H. Y. (2000). Profiling neurosteroids in cerebrospinal fluids and plasma by gas chromatography/electron capture negative chemical ionization mass spectrometry. *Anal. Biochem.* **277,** 187–195.

Kingston, E. E., Beynon, J. H., Newton, R. P., and Liehr, J. G. (1985). The differentiation of isomeric biological compounds using collision-induced dissociation of ions generated by fast atom bombardment. *Biomed. Mass Spectrom.* **12,** 525–534.

Kobayashi, Y., Saiki, K., and Watanabe, F. (1993). Characteristics of mass fragmentation of steroids by atmospheric pressure chemical ionization–mass spectrometry. *Biol. Pharm. Bull.* **16,** 1175–1178.

Korneyev, A., Guidotti, A., and Costa, E. (1993). Regional and interspecies differences in brain progesterone metabolism. *J. Neurochem.* **61,** 2041–2047.

Libert, R., Hermans, D., Draye, J. P., Hoof, F. V., Sokal, E., and Hoffmann, E. D. (1991). Bile acids and conjugates identified in metabolic disorders by fast atom bombardment and tandem mass spectrometry. *Clin. Chem.* **37,** 2102.

Liehr, J. G., Beckner, C. F., Baltore, A. M., and Caprioli, R. M. (1982). Fast atom bombardment mass spectrometry of estrogen glucuronides and sulfates. *Steroids* **39,** 599–605.

Lière, P., Akwa, Y., Engerer, S. W., Eychenne, B., Pianos, A., Robel, P., Sjövall, J., Schumacher, M., and Baulieu, E. E. (2000). Validation of an analytical procedure to measure trace amounts of neurosteroids in brain tissue by gas chromatography–mass spectrometry. *J. Chromatogr. B.* **739,** 301–312.

Liu, S., Sjövall, J., and Griffiths, W. J. (2000). Analysis of oxosteroids by nano-electrospray mass spectrometry of their oximes. *Rapid Commun. Mass Spectrom.* **14,** 390–400.

Liu, S., Sjövall, J., and Griffiths, W. J. (2001). Analysis of neurosteroids in brain by nanoscale capillary liquid chromatography/micro-electrospray mass spectrometry. *ASMS* **14,** 390–400.

Marshall, A. G., Wang, T. C. L., and Ricca, T. L. (1985). Tailored excitation for Fourier transform ion cyclotron resonance mass spectrometry. *J. Am. Chem. Soc.* **107,** 7893–7897.

Mellon, S. H., and Compagnone, N. A. (1999). *In* "Neurosteroids: A New Regulatory Function in the Nervous System," ed.), pp. 27–50. Humana Press: Totowa, NJ.

Meng, L. J., Griffiths, W. J., and Sjövall, J. (1996). The identification of novel steroid *N*-acetylglucosaminides in the urine of pregnant women. *J. Steroid Biochem. Molec. Biol.* **58,** 585–598.

Morfin, R., Young, J., Corpechot, C., Egestad, B., Sjövall, J., and Baulieu, E. E. (1992). Neurosteroids: Pregnenolone in human sciatic nerves. *Proc. Natl. Acad. Sci. USA* **89,** 6790–6793.

Murphy, B. E. P. (2000). Determination of progesterone and some of its neuroactive ring A-reduced metabolites in human serum. *J. Steroid. Biochem. Molec. Biol.* **58,** 137–142.

Murry, S., Rendell, N. B., and Taylor, G. W. (1996). Microbore high-performance liquid chromatography–electrospray ionisation mass spectrometry of steroid sulphates. *J. Chromatogr. A.* **738,** 191–199.

Paul, S. M., and Purdy, R. H. (1992). Neuroactive steroids. *FASEB J.* **6,** 2311–2322.

Pickup, J. F., and McPherson, K. (1976). Theoretical considerations in stable isotope dilution mass spectrometry for organic analysis. *Anal. Chem.* **48,** 1885–1890.

Purdy, R. H., Moore, Jr., P. H., Rao, P. N., Hagino, N., Yamaguchi, T., Schmidt, P., Rubinow, D. R., Morrow, A. L., and Paul, S. M. (1990). Radioimmunology of 3α-hydroxy-5α-pregnan-20-one in rat and human plasma. *Steroids* **55,** 290–296.

Purdy, R. H., Morrow, A. L., Moore, P. H., and Paul, S. M. (1991). Stress-induced elevations of γ-aminobutyric acid type A receptor-active steroids in the rat brain. *Proc. Natl. Acad. Sci. USA* **88,** 4553–4557.

Seikman, L. (1979). Determination of steroids by the use of isotope dilution–mass spectrometry: A definitive method in clinical chemistry. *J. Steroid Biochem.* **11,** 117–123.

Shackleton, C. H., and Straub, K. M. (1982). Direct analysis of steroid conjugates: The use of secondary ion mass spectrometry. *Steroids* **40,** 35–51.

Shimada, K., Mukai, Y., and Yago, K. (1998). Studies of neurosteroids. VII. Characterization of pregnenolone, its sulfate and dehydroepiandrosterone in rat brains using liquid chromatography/mass spectrometry. *J. Liq. Chrom. & Rel. Technol.* **21,** 765–775.

Shimada, K., and Yago, K. J. (2000). Studies on neurosteroids X. Determination of pregnenolone and dehydroepiandrosterone in rat brains using gas chromatography–mass spectrometry–mass spectrometry. *Chromatogr. Sci.* **38,** 6–10.

Uzunov, D. P., Cooper, T. B., Costa, E., and Guidotti, A. (1996). Fluoxetine-elicited changes in brain neurosteroid content measured by negative ion mass mass fragmentography. *Proc. Natl. Acad. Sci. USA* **93,** 12599–12604.

Uzunova, V., Sheline, Y., Davis, J. M., Rasmusson, A., Uzunov, D. P., Costa, E., and Guidotti, A. (1998). Increase in the cerebrospinal fluid content of neurosteroids in patients with unipolar major depression who are receiving fluoxetine or fluvoxamine. *Proc. Natl. Acad. Sci. USA* **95,** 3239–3244.

Valaskovic, G. A., Kelleher, N. K., and McLafferty, F. W. (1996). Attomole protein characterization by capillary electrophoresis–mass spectrometry. *Science* **273,** 1199–1202.

Vallée, M., Rivera, J. D., Koob, G. F., Purdy, R. H., and Fitzgerald, R. L. (2000). Quantification of neurosteroids in rat plasma and brain following swim stress and allopregnanolone administration using negative chemical ionization gas chromatography/mass spectrometry. *Anal. Biochem.* **287,** 153–166.

Watson, D., Taylor, G. W., and Murry, S. (1985). Thermospray liquid chromatography negative ion mass spectrometry of steroid sulphate conjugates. *Biomed. Mass Spectrom.* **12,** 610–615.

Whitehouse, C. M., Dreyer, R. N., Yamashita, M., and Fenn, J. B. (1985). Electrospray interface for liquid chromatographs and mass spectrometers. *Anal. Chem.* **57,** 675–679.

ROLE OF THE PERIPHERAL-TYPE BENZODIAZEPINE RECEPTOR IN ADRENAL AND BRAIN STEROIDOGENESIS

Rachel C. Brown[*,†,‡] and Vassilios Papadopoulos[*,†,‡,§,1]

[*]Division of Hormone Research
[†]Department of Cell Biology, and [§]Departments of Pharmacology and Neuroscience
and the [‡]Interdisciplinary Program in Neuroscience
Georgetown University Medical Center
Washington, DC 20007

I. Introduction
II. Peripheral-Type Benzodiazepine Receptor
III. Structure of the PBR Complex
    A. PBR Topography in Mitochondrial Membrane
    B. PBR Modeling
IV. Role of the PBR in Steroidogenesis
    A. Role of the PBR in Hormone-Stimulated Steroidogenesis
    B. Hormonal Regulation of the PBR
    C. PBR-Mediated Cholesterol Transport in Bacteria
    D. Targeted Disruption of the PBR Gene in Steroidogenic Cells
V. The PBR in Adrenal Steroid Biosynthesis
    A. Regulation of Adrenal PBR and Steroidogenesis
VI. Role of the PBR in Brain Neurosteroid Biosynthesis
    A. PBR and DBI Expression in the Brain
    B. The PBR in Rat Glial Cell Steroidogenesis
    C. The PBR in Human Glial Cell Steroidogenesis
VII. Role of the PBR in Pathology
    A. The PBR and Cancer
    B. The PBR in Neuropathological Conditions
VIII. Other Proteins Involved in the Acute Regulation of Steroidogenesis
    A. The PBR and StAR
IX. Conclusions
    References

The peripheral-type benzodiazepine receptor (PBR) has been demonstrated to be critical for steroidogenesis in all steroid-producing tissues. Here, we review the identification and characterization of the PBR, the evidence pointing to its function as a cholesterol pore involved in transporting cholesterol from the cytoplasm of steroid-producing cells into the inner mitochondrial membrane where it is metabolized, and the known mechanisms regulating its function. We present data on the functions of the PBR

---

[1] Author to whom correspondence should be addressed.

in the adrenal gland, a classical steroidogenic tissue, and in the brain, which has only recently been proven to be steroidogenic. Finally, we discuss other potential roles for the PBR in pathological conditions, including cancer, neurodegeneration, and neurotoxicity, and a broader role for the PBR in mediating intracellular cholesterol transport/compartmentalization, which may or may not be linked to steroid biosynthesis.   © 2001 Academic Press.

I. Introduction

The primary point of control in the acute stimulation of steroidogenesis by hormones involves the first step in the steroid biosynthesis pathway, where cholesterol is converted to pregnenolone (PREG) by cytochrome P450 side-chain cleavage (P450scc) and auxiliary electron-transferring proteins, located on the inner mitochondrial membranes of steroidogenic cells (Simpson and Waterman, 1983; Hall, 1985; Kimura, 1986; Jefcoate et al., 1992). Studies have shown that the conversion of cholesterol to PREG by P450scc is not the rate-limiting step in steroid hormone biosynthesis. Rather, it is the supply and transport of the precursor cholesterol to the inner mitochondrial membrane and subsequent loading of cholesterol into the P450scc active site (Simpson and Waterman, 1983; Hall, 1985; Kimura, 1986; Jefcoate et al., 1992). This hormone-dependent mechanism is mediated by cyclic adenosine monophosphate (cAMP). It is thought to be regulated by a cytoplasmic protein and localized in the mitochondrion, where it regulates intramitochondrial cholesterol transport. This cholesterol transport consists of three steps: (1) transfer of cholesterol from the outer leaflet of the outer mitochondrial membrane to the inner leaflet of the outer membrane, (2) translocation from the inner leaflet of the outer membrane to the outer leaflet of the inner membrane through the aqueous intramembranal space, and (3) transfer and loading of cholesterol to P450scc present in the inner leaflet (matrix side) of the inner mitochondrial membrane (Fig. 1). Since 1990, a component of this cholesterol transport mechanism has been identified and characterized as the peripheral-type benzodiazepine receptor (PBR) (Papadopoulos, 1993).

II. Peripheral-Type Benzodiazepine Receptor

Benzodiazepines have long been prescribed for their anxiolytic, anticonvulsant, and hypnotic actions. It has been well established that the major

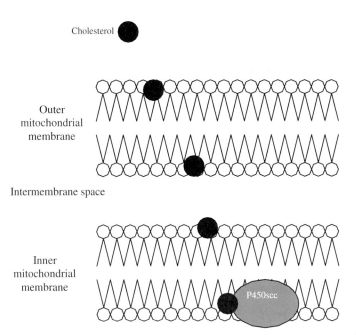

FIG. 1. Cholesterol movement from the cytoplasm to the inner mitochondrial membrane. Cholesterol is stored in intracellular lipid droplets and the plasma membrane. For cholesterol to be metabolized into PREG and other steroid hormones, it must move to the mitochondria, incorporate into the outer leaflet of the outer mitochondrial membrane, and translocate to the inner leaflet of the outer mitochondrial membrane. Cholesterol, a hydrophobic molecule, must then cross the aqueous intermembrane space, incorporate into the outer leaflet of the inner mitochondrial membrane, and finally translocate to the inner leaflet of the inner mitochondrial membrane where it can be converted to PREG by P450scc.

pharmacological effects of benzodiazepines are mediated by γ-aminobutyric acid $(GABA)_A$ receptors in the central nervous system (CNS) (Haefely et al., 1975; Costa and Guidotti, 1979). However, a search for specific benzodiazepine binding sites outside of the CNS found another class of binding sites, first observed in the kidney (Braestrup and Squires, 1977) and later determined to be present in all tissues, including the CNS (Verma and Snyder, 1989; Gavish et al., 1992; Papadopoulos, 1993; Parola et al., 1993). This class of binding sites is referred to as the peripheral-type benzodiazepine recognition site, or receptor (PBR), due to its initial discovery in peripheral tissues using the benzodiazepine diazepam (Valium^{TM}) as a ligand. However, the name does not really apply to the protein because PBR is also present in the CNS. Furthermore, there are many classes of compounds that bind to the receptor with affinities higher than benzodiazepines. A new name

TABLE I
PHARMACOLOGICAL PROFILES OF THE PERIPHERAL-TYPE BENZODIAZEPINE RECEPTOR AND THE $GABA_A$ RECEPTOR

	PBR	$GABA_A$	References
Diazepam	+++	+++	Papadopoulos *et al.*, 1996
4'-Chlorodiazepam	+++	–	Papadopoulos, 1993; Parola, 1993
Clonazepam	–	++++	Papadopoulos *et al.*, 1996
Flumazenil	–	++++	Papadopoulos *et al.*, 1996
Alpidem	++++		Krueger, 1993
Zolpidem	–	++++	Krueger, 1993
Isoquinolines (PK 11195)	+++	No binding	Le Fur, 1983; Papadopoulos, 1993; Parola, 1993
Indole acetamides	++++	No binding	Papadopoulos, 1993; Parola, 1993

for the protein, based on its function, may be needed but, for historical purposes, we will continue to use PBR.

PBR and $GABA_A$ receptors exhibit distinct pharmacological profiles. Although both receptors bind diazepam with relatively high affinity, they exhibit very different binding specificities (Papadopoulos, 1993; Parola *et al.*, 1993), as shown in Table I. In addition to differences in drug specificity, it has been well established that the $GABA_A$ receptor is composed of 50–55 kDa protein subunits and expresses chloride channel activity in synaptosomes. The PBR is an 18-kDa protein functioning as a channel for cholesterol, located on mitochondria (Anholt *et al.*, 1985; Basile and Skolnick, 1986), and more specifically on the outer mitochondrial membrane (Anholt *et al.*, 1986). Later work showed that the PBR is not exclusive to the mitochondria (Oke *et al.*, 1992; Hardwick *et al.*, 1999; Brown *et al.*, 2000a). Diazepam-binding inhibitor (DBI), the endogenous ligand for the PBR, has been identified and characterized. DBI is a 10-kDa protein originally purified from the brain owing to its ability to displace diazepam from $GABA_A$ receptors (Guidotti *et al.*, 1983). DBI is present in a variety of tissues but is highly expressed in steroidogenic cells, where it stimulates steroid formation (Papadopoulos *et al.*, 1991; Papadopoulos, 1993), mediated by interactions with the PBR (Papadopoulos *et al.*, 1997a).

III. Structure of the PBR Complex

An 18-kDa isoquinoline-binding protein has been identified as the PBR (Doble *et al.*, 1987). This protein has been purified (Antikiewicz-Michaluk *et al.*, 1988; Riond *et al.*, 1989), and the corresponding cDNA cloned from rat

(Sprengel, 1989), bovine (Parola *et al.*, 1991), mouse (Garnier *et al.*, 1994b), and human (Riond *et al.*, 1991; Chang *et al.*, 1992). Expression studies with cDNA probes demonstrate that the 18-kDa protein contains the binding domains for PBR ligands, although the presence of other proteins important for PBR binding cannot be excluded. Primary amino acid sequence analysis indicates that the PBR contains a predominance of hydrophobic amino acids predicting five transmembrane domains (Sprengel, 1989; Riond *et al.*, 1991; Chang *et al.*, 1992; Parola *et al.*, 1993; Garnier *et al.*, 1994b).

As noted, the PBR was identified and characterized initially by its high affinity for two distinct classes of compounds, the benzodiazepines and the isoquinolines. High-affinity isoquinoline binding is diagnostic of the PBR, but the affinity of benzodiazepines for the PBR is species specific, ranging from high affinity (rodents) to low affinity (bovine) (Papadopoulos, 1993; Parola *et al.*, 1993). These species differences may result from structural differences in the 18-kDa protein or differences in the various components comprising the PBR complex in the mitochondrial membrane. We isolated and sequenced a 626 base pair cDNA, specifying an open reading frame of 169 amino acid residues, from MA-10 Leydig cells (Garnier *et al.*, 1994b). Expression of PBR cDNA in mammalian cells resulted in an increase in the density of both benzodiazepine- and isoquinoline-binding sites. To determine whether the increased drug binding is the result of the 18-kDa PBR protein alone, or to other constitutively expressed components of the receptor, an *in vitro* system was developed using recombinant PBR protein (Garnier *et al.*, 1994b). Isolated maltose-binding protein (MBP)-PBR recombinant fusion protein was incorporated into liposomes, formed using lipids found in steroidogenic outer mitochondrial membranes. The MBP-PBR fusion protein, but not MBP alone, maintained its ability to bind isoquinolines, but not benzodiazepines. Addition of mitochondrial extracts to the liposomes restored benzodiazepine binding. The protein responsible for this effect was purified and identified as the 34-kDa, voltage-dependent anion channel (VDAC), which by itself, does not express any drug binding. VDAC is functionally associated with the PBR and is part of the benzodiazepine-binding site in the PBR (Papadopoulos, 1998) (Fig. 2).

These findings provide functional evidence for previous observations on 30- to 35-kDa proteins nonspecifically labeled using isoquinolines and benzodiazepines (Gavish *et al.*, 1992; Papadopoulos, 1993; Parola *et al.*, 1993). The observation that the 18-kDa PBR was isolated as a complex with the 34-kDa VDAC and the inner mitochondrial membrane adenine nucleotide carrier (ADC) (McEnery *et al.*, 1992) suggested that the PBR is not a single protein receptor but a multimeric complex. Studies in our laboratory using a yeast two-hybrid system have pulled out several PBR-associated

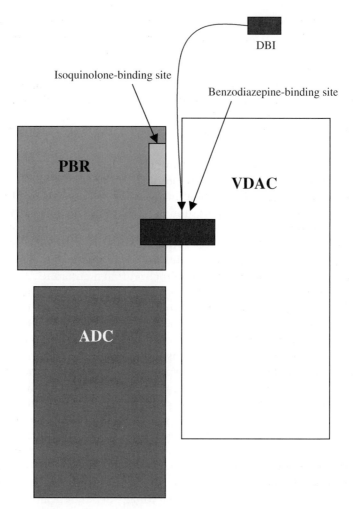

FIG. 2. The PBR complex. PBR–ligand interactions require the formation of a complex of proteins. The 18-kDa-PBR protein contains the isoquinolone binding site. However, benzodiazepine/DBI binding requires the presence of the VDAC in addition to the PBR. Furthermore, the complex also contains the adenosine nucleotide carrier (ADC), which is found on the inner side of the outer mitochondrial membrane.

proteins (PAPs) (Papadopoulos, 1998; Li *et al.*, 2000). PAP7, specifically, has been demonstrated to be localized to steroidogenic tissues and to play a role in modulating steroid synthesis, further supporting the hypothesis that PBR-mediated regulation of steroid biosynthesis occurs via a complex of proteins.

A. PBR Topography in Mitochondrial Membrane

Native Leydig cell mitochondrial preparations were examined by transmission electron microscopic and atomic force microscopic (AFM) procedures to investigate the topography and organization of the PBR. Mitochondria were immunolabeled with an anti-PBR antiserum coupled to gold-labeled secondary antibodies. The results obtained indicate that the PBR protein is organized in clusters of four to six molecules in the hormone-unstimulated state (Papadopoulos and Guarneri, 1994). On many occasions, the molecular clusters formed a pore. Because the PBR protein is functionally associated with the pore-forming VDAC protein, which is preferentially located at the contact sites of the two mitochondrial membranes, these results suggest that the mitochondrial PBR complex may function as a pore. Treating cells with human chorionogonadotropin (hCG), which stimulates steroid biosynthesis and increases PBR binding, causes a reorganization of the mitochondrial membrane at these sites, inducing the appearance of large clusters, varying from 15 to 25 gold particles. AFM analyses of these sites further demonstrated membrane reorganization (Vidic *et al.*, 1995; Boujrad *et al.*, 1996). This effect is specific to hCG because inhibitors of hCG-stimulated steroid biosynthesis also block the changes in PBR topography (Boujrad *et al.*, 1996). Therefore, hormones that stimulate steroid biosynthesis also induce the rapid reorganization of mitochondrial membranes, favoring the formation of contact sites that may facilitate the transfer of cholesterol from the outer to the inner mitochondrial membrane. An increase in the formation of contact sites between the mitochondrial membranes with hCG treatment was previously reported (Stevens *et al.*, 1985). Free cholesterol from the outer mitochondrial membrane seems to transfer freely via the contact sites to the inner membrane where P450scc is located, as demonstrated for intramitochondrial transport of phospholipids (Simbeni *et al.*, 1991).

B. PBR Modeling

Based on the amino acid sequence of the human and mouse PBR protein, a three-dimensional model of this receptor was developed using molecular dynamics simulations (Bernassau *et al.*, 1993; Culty *et al.*, 1999). According to this model, the five transmembrane domains of the PBR are modeled as five α-helices that span one phospholipid layer of the outer mitochondrial membrane. Further studies in yeast have demonstrated that the PBR can span the entire membrane bilayer (Joseph-Liauzun *et al.*, 1998). This receptor model was tested as a cholesterol carrier, and it was shown that the PBR

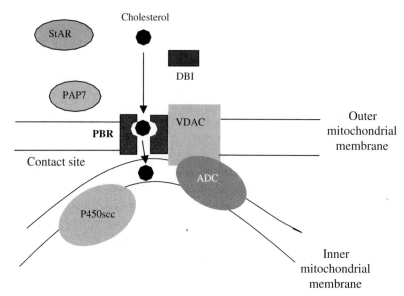

FIG. 3. A working model of PBR function. The PBR, VDAC, and ADC form a complex at contact sites, potentially with a number of other proteins, including PAP7. Cholesterol from cellular stores moves to the outer mitochondrial membrane in a process that may be mediated by StAR. Cholesterol is incorporated into the outer mitochondrial membrane by uptake and transport through PBR. In this manner, cholesterol can cross the outer mitochondrial bilayer, bypass the aqueous intermembrane space (which is eliminated at contact sites), and incorporate into the inner mitochondrial membrane. Cholesterol can then be converted to PREG by P450scc, located on the inner side of the inner mitochondrial membrane.

could accommodate a cholesterol molecule and function as a channel. It was suggested that the receptor's function is to carry cholesterol molecules from the outer lipid monolayer to the inner lipid monolayer of the outer membrane, acting as a "shield," to hide the cholesterol from the hydrophobic membrane inner medium. Considering that the PBR complex forms at contact sites, this cholesterol movement could end in the inner mitochondrial membrane, where P450scc is located. This theoretical model further supports experimental data on the role of the PBR in intramitochondrial cholesterol transport (Fig. 3).

IV. Role of the PBR in Steroidogenesis

Two important observations indicate that the PBR plays a role in steroidogenesis: First, PBRs are found primarily on outer mitochondrial membranes;

and second, PBRs are extremely abundant in steroidogenic cells (Gavish *et al.,* 1992; Papadopoulos, 1993; Parola *et al.,* 1993). We have reported that a spectrum of ligands, which bind to the PBR with a wide range of affinity, will stimulate steroid biosynthesis in various cell systems (Mukhin *et al.,* 1989; Parola *et al.,* 1993). The relationship between the affinities of these compounds for the PBR and the concentrations of each compound required to stimulate steroidogenesis was examined and showed an excellent correlation, suggesting that these drugs, via binding to the PBR, stimulate steroid biosynthesis. However, the stimulatory effect of the PBR ligands was not additive to the stimulation by hormones and cAMP (Krueger and Papadopoulos, 1990). Considering the mitochondrial localization of the PBR, we then examined the direct effect of PBR ligands on mitochondrial steroid formation. PBR ligands were found to stimulate production of PREG by isolated mitochondria (Papadopoulos *et al.,* 1990). The stimulatory effect of PBR ligands on intact mitochondria was not observed with mitoplast preparations (mitochondria devoid of the outer membrane) in agreement with the outer mitochondrial localization of the receptor (Papadopoulos *et al.,* 1990). In these studies, we concluded that PBRs are implicated in the acute stimulation of adrenocortical and Leydig cell steroidogenesis by mediating the entry into and/or distribution of cholesterol within mitochondria.

To further elucidate the exact point at which PBR ligands activate PREG formation in mitochondria, we measured the amount of cholesterol distributed in the inner and outer mitochondrial membranes before and after treatment with PBR ligands. PBR-ligand-induced stimulation of PREG formation was found to be the result of PBR-mediated translocation of cholesterol from the outer to the inner mitochondrial membrane (Krueger and Papadopoulos, 1990). The abundance of PBRs in steroidogenic tissues, together with tissue-specific cholesterol transport, points to the PBR being the regulator of the rate-determining cholesterol supply for steroid biosynthesis. Different laboratories have corroborated these observations (Ritta and Calandra, 1989; Papadopoulos *et al.,* 1992) and extended them to placental (Barnea *et al.,* 1989) and ovarian granulosa cells (Amsterdam and Suh, 1991). Moreover, a similar mechanism has been shown to regulate brain glial cell neurosteroid biosynthesis (Brown *et al.,* 2000a).

A. ROLE OF THE PBR IN HORMONE-STIMULATED STEROIDOGENESIS

We examined the possibility that the PBR participates in hormone-stimulated steroidogenesis. In searching for a PBR ligand that might affect hormone-stimulated steroid production, we found that flunitrazepam, a benzodiazepine that binds to the PBR with high nanomolar affinity, inhibited hormone- and cAMP-stimulated steroidogenesis (Papadopoulos *et al.*, 1991). This drug also inhibited mitochondrial PREG formation, which was found to result from a reduction in cholesterol transport to the inner mitochondrial membrane. Binding studies revealed a single class of PBR binding sites for flunitrazepam. These studies suggested that hormone-induced steroidogenesis involves, in part, the participation of PBRs.

B. Hormonal Regulation of the PBR

Long-term treatment of MA-10 Leydig cells with hCG has no effect on PBR binding or message levels (Boujrad *et al.*, 1994). However, addition of hCG causes a very rapid increase in PBR binding capacity, which gradually returns to basal levels. Scatchard analyses reveal that, in addition to the known high-affinity benzodiazepine binding site, hCG treatment induced the appearance of a second, higher-affinity binding site for benzodiazepines. Steroid synthesis occurred within a similar time frame. Addition of flunitrazepam abolished the hCG-induced rapid stimulation of steroid synthesis, which suggests that hormones can alter PBR to activate cholesterol delivery to the inner mitochondrial membrane and thus alter subsequent steroid formation. In a constitutive steroid-producing cell line, steroid synthesis takes place independent of hormonal control (Ward *et al.*, 1960). R2C cells are derived from rat Leydig cell tumors and maintain their *in vitro* capacity to synthesize steroids in a hormone-independent manner (Freeman, 1987). One can predict that constitutive steroidogenesis is driven by the unregulated expression of the hormonal mechanism that controls steroid production. Binding assays on intact R2C cells reveal the presence of a single class of PBR binding sites with an affinity 10 times higher ($K_d = 0.5$ nM) than the MA-10 PBR ($K_d = 5$ nM) (Garnier *et al.*, 1994a). Photolabeling of R2C and MA-10 cell mitochondria with a photoactive PBR ligand labels an 18-kDa protein. Moreover, a PBR synthetic ligand was able to increase steroid production in isolated mitochondria from R2C cells. These data demonstrate that ligand binding to the higher-affinity PBR binding site in mitochondria is involved in maintaining R2C cell constitutive steroidogenesis.

C. PBR-Mediated Cholesterol Transport in Bacteria

When the aforementioned studies are considered, the hypothesis emerges that the PBR functions as a channel-like protein for cholesterol. This channel is thought to span the outer mitochondrial membrane at contact sites and allow cholesterol to move to the inner mitochondrial membrane, where P450scc metabolizes it. To test this hypothesis, we looked at bacteria, which constitute a model system without endogenous cholesterol, endogenous PBR protein, or ligand binding (Li and Papadopoulos, 1998). We transfected *Escherichia coli* with mouse PBR cDNA and induced PBR expression. This resulted in protein expression and a time- and temperature-dependent uptake of radiolabeled cholesterol, but not of other radiolabeled steroids. Treatment of cholesterol-loaded, protein-induced bacterial membranes with PK 11195 resulted in the release of cholesterol, suggesting that cholesterol is captured by the PBR expressed in the bacterial membranes that, upon ligand binding, release cholesterol. Thus, the PBR serves as a channel by which cholesterol can freely enter and remain stored within the membrane without being incorporated into the lipid bilayer. PBR ligands may control the opening of the channel, mediating the movement of cholesterol across membranes.

Amino acid deletion analyses identified sequences 5–20 and 41–51 of the PBR-predicted amino acid sequence as the drug-ligand-binding domains and identified the carboxy-terminal sequence 153–'69 as responsible for cholesterol uptake. Site-directed mutagenesis identified two amino acids interacting with cholesterol, Y153 and R156 (Li and Papadopoulos, 1998). We postulated the existence of a common cholesterol recognition–interaction amino acid consensus pattern-L/V-$(X)_{1-5}$-Y-$(X)_{1-5}$-R/K-. Indeed, we found this consensus pattern in a number of proteins that are shown to interact with cholesterol (Li and Papadopoulos, 1998). Furthermore, treatment of MA-10 Leydig cells with this cholesterol-binding domain will decrease steroid biosynthesis by binding up free cholesterol within the cells and preventing its transport into the mitochondria.

These data suggest that PBR expression confers the ability to take up and release cholesterol upon ligand activation. They also suggest that the PBR has two distinct cholesterol- and ligand-binding sites. The finding that the PBR mediates cholesterol transport across membranes suggests that the PBR has a more general role in intracellular cholesterol transport, trafficking and compartmentalization, related to the presence of the PBR in specific cell membranes. In this regard, the PBR has been localized to the plasma membrane (Oke *et al.*, 1992) and to the nucleus in aggressive human tumor cells, which is further discussed below.

D. TARGETED DISRUPTION OF THE PBR GENE IN STEROIDOGENIC CELLS

To manipulate levels of PBR expression, we developed a molecular approach based on the disruption of the PBR gene in R2C cells by homologous recombination (Papadopoulos *et al.*, 1997b). By using gene targeting, PBR-mutant (negative) cell lines were generated and were found to produce minimal amounts of progesterone (PROG) as compared with wild-type cells (Papadopoulos *et al.*, 1997b; Amri *et al.*, 1999). The addition of a hydrosoluble derivative of cholesterol, or transfection of the cells with PBR-cDNA-rescued steroid production by the PBR-mutant R2C cells, indicating that the cholesterol transport mechanism was impaired in the absence of the 18-kDa protein. Furthermore, we noted that the PBR-mutant cells grew much more slowly than control R2C cells (Amri *et al.*, 1999), which supports other observations indicating a possible role of the PBR in cell proliferation (Papadopoulos, 1993; Hardwick, 1999; Brown *et al.*, 2000a).

In agreement with these findings, Kelly-Herskovitz *et al.* (1998) reported that transfection of MA-10 cells with an antisense knockout vector decreased PK 11195 ligand-binding capacity by 40%. However, this group did not note any effect of the reduction of PBR levels on the rate of cell proliferation. It is possible that the remaining PBR in these cells is sufficient to maintain a normal rate of growth. Efforts to generate a PBR gene knockout mouse model have failed because the animals died at an early embryonic stage, suggesting that the PBR is involved in functions vital to embryonic development. Thus, other means have been developed to modulate PBR expression and function *in vivo*.

V. The PBR in Adrenal Steroid Biosynthesis

A. REGULATION OF ADRENAL PBR AND STEROIDOGENESIS

The ability of the adrenal gland to synthesize steroids in response to adrenocorticotropic hormone (ACTH) is developmentally regulated. PBR and DBI mRNAs both are expressed during embryonic development, beginning around embryonic day 16 (ED16), peaking by ED18, and gradually declining during the later part of gestation until postnatal day 6 (PD6). At PD6, the levels of DBI are almost undetectable, and levels of PBR are very low (Burgi *et al.*, 1999). Although the adrenal gland of the adult rat is ACTH sensitive, adrenals from neonatal rats show a period of greatly reduced sensitivity to ACTH that gradually increases during the first 2 weeks of life (Arai and Widmaier, 1993). Thus, neonatal rats have lower circulating corticosterone levels after exposure to stress (Arai and Widmaier, 1993). The expression of the PBR in the adrenal glands of neonatal and adult rats as well as adrenal sensitivity to ACTH *in vivo* and *in vitro* have been examined (Zilz *et al.*, 1999). The ontogeny of both 18-kDa PBR protein expression and PBR-ligand-binding capacity were found to parallel directly that of ACTH-inducible steroidogenesis in isolated rat adrenal cells and in rats injected with ACTH (Zilz *et al.*, 1999), suggesting that PBR is critical for the stress-hyporesponsive characteristics of the neonatal adrenal gland (Dallman, 2000). In addition, adrenal PBR from neonates had a three- to fivefold higher affinity for PK 11195 than adult adrenal PBR, which may partly explain the relatively high constitutive steroidogenesis that is characteristic of the neonatal rat adrenal cell. Therefore, the stress-hyporesponsive period observed in neonatal rats may result from decreased PBR expression. This is consistent with the hypothesis that the PBR is an absolute prerequisite for adrenocortical steroid biosynthesis.

Glucocorticoid excess has broad pathologic potential, including neurotoxicity and immunosuppression. Glucocorticoid synthesis is regulated by ACTH, which acts by accelerating the transport of cholesterol to the mitochondria for steroidogenesis. Searching for a pharmacologic way to regulate PBR expression and glucocorticoid biosynthesis, we examined the effect of *Ginkgo biloba* extract EGb 761 and isolated ginkgolides. Leaf extracts from *Ginkgo biloba* have been used in traditional medicine. EGb 761, a standardized extract, is described as having beneficial effects on vigilance, memory, and cognitive functions associated with aging and senility, dementia, mood changes, and the ability to cope with daily stressors (DeFeudis, 1998). This substance also exerts an "anti-stress" effect in rodents that differs from that of antidepressants or anxiolytics (Rapin *et al.*, 1994). Given the pathogenic potential of glucocorticoid excess, we examined whether EGb 761 and its bioactive terpenoid ginkgolide B (GKB) exert their beneficial effects by controlling glucocorticoid levels. In adult rats, treatment with EGb 761 or GKB decreased serum corticosterone levels (Amri *et al.*, 1996). Because glucocorticoid secretion is associated with a negative feedback mechanism in the hypothalamus, the GKB-induced decrease in glucocorticoid levels resulted in an increased ACTH release, which in turn induced expression of the hormone-dependent steroidogenic acute regulatory protein (StAR) (Stocco and Clark, 1996).

Although EGb 761 and GKB did not affect the P450scc enzyme, they dramatically decreased expression of the 18-kDa PBR protein, expression of PBR mRNA, and the number of adrenal PBR-binding sites. The extracts did not affect the binding characteristics of kidney PBR, which suggests that they specifically reduced PBR mRNA stability or DNA transcription in the adrenal cortex (Amri *et al.*, 1996).

In an attempt to determine the specific effects of EGb 761 and GKB, adrenocortical cells were isolated from rats treated with EGb 761, GKB, or saline (Amri *et al.*, 1997). Cells isolated from animals treated with GKB had an 80% reduction of corticosterone production in response to ACTH stimulation as compared to controls. Analyses of metabolically radiolabeled proteins showed that, in cells from control and drug-treated animals, ACTH induced the same level of StAR synthesis. In addition, EGb 761 and GKB specifically altered the synthesis of an 18-kDa protein identified as PBR. In summary, these results suggest that EGb 761 and GKB exert specific effects on adrenocortical cells by inhibiting PBR expression, limiting the transport of cholesterol to P450scc, and subsequently reducing the ACTH-stimulated corticosterone production. Marcilhac *et al.* (1998) have confirmed these results.

Weizman, *et al.* (1997a) used pregnancy as a model to evaluate the impact of hormonal changes on adrenal and gonadal PBR expression. Pregnancy is accompanied by a marked increase in serum levels of PROG and estrogen. This group found an increase in PBR density in the ovary during pregnancy. Simultaneously, adrenal PBR density was downregulated, probably to prevent overreactivity of the hypothalamic-pituitary-adrenal axis (Weizman *et al.*, 1997a). Indeed, high levels of circulating cortisol may lead to hypertension, hyperglycemia, and impairment of immune function, which should be avoided during pregnancy (Weizman *et al.*, 1997a). In another study, the same group examined the effects of long-term administration of PBR ligands on adrenal PBR (Weizman *et al.*, 1997b). The treatment resulted in increased PBR density but did not change circulating corticosterone levels. This is not surprising because the adrenal cortex is already extremely rich in PBRs, and increases in PBR levels may not have additional effects. From the findings reported to date, it seems that a minimal level of PBRs is required to maintain the flow of cholesterol into the mitochondria. If the level of PBRs falls below this critical level, then the flow of cholesterol and the rate of steroid biosynthesis decrease. This may lead to changes in the outer mitochondrial membrane pore formation, release of cytochrome c, and apoptosis, as shown by Papadopoulos *et al.* (1999). This conclusion was reinforced by separate studies performed by Siripurpong *et al.* (1997) and in our laboratory (Amri *et al.*, 1999). Various stimuli have been shown to modulate PBR levels in various tissues, including the adrenal gland. Using various stress models, Siripurpong *et al.* (1997) showed that stress did not change adrenal PBR mRNA levels. However, treatment of rats with dexamethasone caused a 30% decrease in adrenal PBR gene expression. In a similar *in vitro* study, we found that dexamethasone decreased PBR mRNA levels by 30% in both Y-1 adrenal and MA-10 Leydig tumor cells. In addition, these changes translated into decreased PBR densities (as determined by radioligand binding studies) but were not accompanied by a significant decrease in steroid biosynthesis. Despite the 30% decrease in PBR density, the amount of PBRs remaining was more than was necessary to maintain the flow of cholesterol and the rate of steroid biosynthesis in response to hormones.

VI. Role of the PBR in Brain Neurosteroid Biosynthesis

The specific interactions of steroid with binding sites at neuronal membranes and the ability of various steroids to modulate brain function (Lambert *et al.*, 1995) have prompted the investigation of the steroidogenic potential of CNS structures. The pioneering work of Baulieu *et al.* (Baulieu and Robel, 1990) demonstrated that glial cells could convert cholesterol to PREG and give rise to neuroactive steroid metabolites. It has been shown that oligodendrocytes and Schwann cells, the myelinating cells of the CNS and peripheral nervous system (PNS), respectively, and several glioma cell lines have the ability to convert cholesterol to PREG, the first step in steroid biosynthesis.

A. PBR AND DBI EXPRESSION IN THE BRAIN

A study has examined the expression of mRNA for both PBR and its endogenous ligand, DBI, over the course of pre- and postnatal development (Burgi *et al.*, 1999). Levels of PBR mRNA are detectable by ED12 in the rat nervous system and peak by ED18–20. Highest levels of expression are seen in the cerebral cortex, caudate putamen, and choroid plexus at ED18. By ED22, PBR mRNA is undetectable except in the diencephalon. The PBR was not detected in the postnatal period. Levels of DBI expression are much higher, with high levels seen as early as ED11. High levels are maintained in the neuroepithelium and the trigeminal ganglia throughout the period studied (ED11–PD6). In other parts of the nervous system, DBI mRNA levels begin to decline by ED16 and are low by PD6. These data suggest that PBR–DBI interactions and, potentially, neurosteroidogenesis may be important during development. Several studies have suggested that neurosteroids may be important molecules for axonal path finding during development of the hippocampus and neocortex (Brinton, 1994; Compagnone and Mellon, 1998). It remains to be determined what the *in vivo* role of PBR is during neurodevelopment.

B. THE PBR IN RAT GLIAL CELL STEROIDOGENESIS

We examined neurosteroid synthesis using the C6-2B subclone of the rat C6 glioma cell line. These cells express PBR immunoreactivity (Cascio *et al.*, 1999) and are able to synthesize PREG from various substrates, including cholesterol (Guarneri *et al.*, 1992). To examine the activity of

P450scc present in these cells, hydroxylated analogs of cholesterol, which will freely cross the mitochondrial membrane, were used, thus bypassing PBR-mediated cholesterol transport and directly interacting P450scc. Three different hydroxylated cholesterols stimulated mitochondrial production of PREG three- to fivefold (Cascio *et al.*, 1999). PREG formation could be inhibited by aminoglutethimide, a specific inhibitor for P450scc (Papadopoulos *et al.*, 1992; Papadopoulos and Guarneri, 1994). These findings suggest a functional analogy between adrenal and glial P450scc.

It is important to note that P450scc activity is related to the oligodendrocyte differentiation process (Jung-Testas *et al.*, 1989). Cholesterol accumulation in the brain is also related to differentiation (Shah, 1993) and coincides with the rate of myelinization by oligodendrocytes (Norton and Poduslo, 1973). Interestingly, all three activities reach their maximum in the rat at 20 days of age, and cholesterol accumulation in the brain declines after maturation of CNS structures. These findings demonstrate a temporal relationship among cholesterol accumulation, steroid synthesis, myelinization, and nerve synapse formation, which is still being investigated. Our studies have shown a correlation between PBR expression, steroid biosynthesis, and oligodendrocyte differentiation, which suggest that the PBR may be important in regulating development and differentiation processes (Cascio *et al.*, 2000).

The regulation of neurosteroid biosynthesis is in stark contrast to peripheral regulatory mechanisms. There are no clearly demonstrated hormone-stimulated mechanisms known to act in the brain to regulate neurosteroidogenesis. We have examined the PBR in brain tissue, primary glial cultures, and C6-2B glioma cells. Subcellular fractionation indicated that the majority of the PBR is localized in the mitochondrial fraction (Krueger and Papadopoulos, 1991a,b). Ligand-binding studies indicated that rat brain mitochondria contains approximately 1 pmol/mg protein PK 11195 binding sites. In contrast, mitochondria from primary glial cultures and C6-2B glioma cells exhibited a very high density for PBR binding sites (25–50 pmol/mg of protein). This single class of binding sites had an apparent K_d of 4 nM, similar to that found in Leydig cells (Papadopoulos *et al.*, 1990). In addition, the glioma receptor expressed identical pharmacological profile to the adrenocortical and testicular Leydig cell PBR (Cascio *et al.*, 1999). These findings demonstrate that, within the CNS, the PBR is primarily localized in glial cells. In addition, the mitochondrial localization and density of the PBR in glia suggest that it may serve a function similar to that seen in peripheral steroidogenic tissues.

We investigated whether PBR ligands affect PREG formation in C6-2B glioma cell mitochondria. At nanomolar concentrations, PBR ligands induced a twofold stimulation of PREG formation (Papadopoulos *et al.*, 1992). A similar increase was obtained with anxiolytic benzodiazepines that bind to both classes of benzodiazepine receptors, whereas clonazepam, a ligand selective for the $GABA_A$ receptor, had no effect on steroid synthesis at any concentration tested (Papadopoulos *et al.*, 1992). In these studies, exogenous cholesterol was not supplied to the mitochondria, suggesting that PBR facilitates the transport of cholesterol from the outer mitochondrial membrane, which is then metabolized by P450scc to PREG. In our studies, glial cell mitochondria have a rate of PREG formation 10 times slower than adrenocortical or Leydig cell mitochondria (Cascio *et al.*, 1999). However, in all three models, PBR drug ligands stimulated the rate of PREG formation by two- to threefold.

Biosynthesis of [^{3}H]cholesterol and [^{3}H]PREG was demonstrated in C6-2B glioma cell cultures to occur within seconds on addition of the precursor [^{3}H]mevalonactone (Guarneri *et al.*, 1992). Addition of 10 nM Ro5 4864, a specific PBR ligand, resulted in 141% and 205% increase in cholesterol and PREG formation, respectively (Guarneri *et al.*, 1992). This effect of Ro5 4864 was dose and time dependent and demonstrates that PBR ligands stimulate steroid formation in cultured glial cells. In addition, PBR drug ligands were found to increase rat forebrain PREG synthesis *in vivo* (Costa *et al.*, 1994) and to elicit antineophobic and anticonflict actions, presumably via their PBR-mediated steroidogenic effect and the subsequent action of the synthesized neuroactive steroids on the $GABA_A$ receptor.

C. THE PBR IN HUMAN GLIAL CELL STEROIDOGENESIS

We have demonstrated that human glioma cell lines make neurosteroids *de novo* from a radiolabeled cholesterol precursor (Brown *et al.*, 2000b). We blocked endogenous cholesterol synthesis with the 3-hydroxy-3-methylglutaryl-CoA reductase inhibitor Mevastatin and incubated the cells with [^{3}H]mevalonactone. By blocking the metabolism of PREG with specific enzyme inhibitors, we attempted to maximize PREG levels. High-performance liquid chromatography separation of the samples indicated that the MGM-3 cell line, isolated from a glioblastoma multiforme tumor, synthesizes PREG *de novo*. However, another glioblastoma multiforme cell line, MGM-1, does not make PREG *de novo*. Immunocytochemical localization using anti-PBR antiserum demonstrated that the MGM-3 cell line has the PBR localized to the cytoplasmic compartment, and presumably to mitochondria, while MGM-1 cells have the PBR primarily localized around the nucleus (Brown *et al.*, 2000a). Radioligand-binding analyses demonstrated that, while the two cell lines have similar binding affinities for PK 11195 (MGM-3 $K_d = 2.15 \pm 0.6$ nM vs MGM-1 $K_d = 1.19 \pm 0.2$ nM), the distribution of these binding sites is differentially localized in the two cell lines, with MGM-1 cells having a larger proportion of their total binding sites in the nucleus. The largely nuclear localization of the PBR in these cells renders them unable to make PREG from cholesterol, although they do have P450scc mRNA and protein.

In summary, the aforementioned studies have demonstrated that, in addition to being crucial for adrenal and gonadal steroidogenesis, the PBR is also central to neurosteroidogenesis. Furthermore, PBR function, and the regulation of the supply of cholesterol for metabolism by P450scc, remains the only known mechanism by which to regulate PREG formation in the brain.

VII. Role of the PBR in Pathology

The bulk of this chapter has focused on the critical role of the PBR in the acute regulation of steroidogenesis. However, we now address a more recent development in our understanding of PBR function within the cell.

A. THE PBR AND CANCER

Work in our laboratory has demonstrated a correlation between PBR expression and aggressive phenotype of human breast cancer cell lines (Hardwick et al., 1999). Human breast cancer cells with high aggressivity rating have a much higher expression of PBR protein and mRNA. Further examination showed that MDA-231 cells, an aggressive cell line, have PBR binding sites localized primarily to the perinuclear area (Hardwick et al., 1999). Human breast tumor biopsies show a similar nuclear localization for PBR immunoreactivity. Furthermore, treatment of these breast cancer cells with PK 11195, a specific PBR ligand, stimulates cell proliferation. This increase in proliferation was associated with nuclear cholesterol uptake. We have found a similar PBR-induced proliferation in a human glioma cell line with nuclear PBR immunoreactivity (Brown et al., 2000a). Nuclear PBR location was confirmed in human brain tumor biopsies and was found to be specific to a glioblastoma multiforme tumor, although an astrocytoma expressed cytoplasmic PBR immunoreactivity (Brown et al., 2000a). Other studies have shown an upregulation in cell proliferation in gliomas after treatment with PBR ligands (Ikezaki and Black, 1990) as well as a general upregulation of PBR expression in gliomas (Cornu et al., 1992; Miettinen et al., 1995). These results suggest that, in cells with nuclear PBR, the receptor appears to be involved in regulating cell proliferation, potentially by mediating nuclear cholesterol transport, as in aggressive human breast cancer cells (Hardwick *et al.*, 1999). The presence of DBI in the glioma cell lines further suggests that the PBR is functional (Brown *et al.*, 2000a). These results indicate that the PBR may serve a function in regulating cellular proliferation in aggressive tumors.

B. The PBR in Neuropathological Conditions

The PBR is upregulated in a number of neuropathologies, including gliomas (Cornu *et al.*, 1992; Miettinen *et al.*, 1995) and Alzheimer's disease (McGeer *et al.*, 1998). Furthermore, injection of a neurotoxicant, trimethyltin, causes an increase in [^{3}H]PK 11195 binding in areas of neurodegeneration, such as the hippocampus, primary olfactory cortex, posteriomedial cortical amygdaloid nucleus, subiculum, and entorhinal cortex, that persists for more than a month after initial injury (Guilarte *et al.*, 1995). This phenomenon is also seen with neurotoxic insults from domoic acid or MPTP (Kuhlmann and Guilarte, 1997, 1999). Further studies have demonstrated that activated microglia and reactive astrocytes express the PBR in the nucleus, as opposed to the classical, mitochondrial localization for the receptor, after neurotoxicological insult, suggesting that PBR expression may be a marker of neurotoxic insult and injury (Kuhlmann and Guilarte, 2000). It remains to be seen what mechanism targets the PBR to novel compartments, such as the nucleus or the plasma membrane (Oke *et al.*, 1992), and what is the ultimate function of the protein in these areas.

The PBR has also been implicated by Lacor *et al.* (1999) in nerve degeneration and regeneration. This group showed that PBR density and the levels of active DBI metabolites were increased after peripheral nerve injury. Once regeneration of the peripheral nerve was complete, PBR and DBI levels dropped back to normal. When regeneration did not occur, PBR and DBI remained elevated. This study concluded that there is a role for the PBR and DBI in peripheral nerve regeneration and that neurosteroidogenesis may exert trophic effects in cases of nerve injury.

VIII. Other Proteins Involved in the Acute Regulation of Steroidogenesis

The PBR is present in the mitochondria in a complex of a number of different proteins. These other proteins may be very important in helping to regulate cholesterol transport across the mitochondrial membranes. We have already discussed PAPs; now we focus on StAR, another protein thought to be involved in acute regulation of steroid production.

A. The PBR and StAR

The PBR is not the only protein involved in the acute regulation of steroidogenesis. One of the other major candidates is steroidogenic acute regulatory protein, or StAR. StAR was first identified as a 30-kDa phosphoprotein isolated from stimulated adrenocortical cells from mouse and rat as well as from rat corpus luteum cells and mouse Leydig cells (Stocco, 1999). StAR has been linked to steroidogenesis by its association with mitochondria and its ability to confer steroidogenic ability to unstimulated MA-10 cells or COS-1 monkey kidney tumor cells when overexpressed (Stocco, 1999). Furthermore, mutations in the StAR gene have been found in a potentially lethal clinical disorder known as congenital lipoid adrenal hyperplasia (lipoid CAH) (Lin et al., 1995). This disease is characterized by a loss of steroid hormone synthesis due to the lack of delivery of cholesterol to P450scc (Stocco, 2000). A StAR knockout mouse has been generated (Caron et al., 1997), and the animals exhibit many of the same symptoms as seen in human patients with lipoid CAH, including external female genitalia in both genetic males and females and severe defects in adrenal steroids. All of these animals die within a few weeks if they are not treated with corticosteroid replacement. However, in these animals, although the gonads and the adrenal glands do not synthesize steroids in the full amount of steroids, they still make testosterone, estrogen, and PROG. The steroidogenic ability of the placenta of these knockout animals, which is derived from embryonic cells and is required to make PROG for the pregnancy to go to full term, does not seem to be affected. Therefore, StAR may be more important in regulating steroidogenesis in the adrenals and gonads. Similar attempts to create a PBR knockout mouse have been unsuccessful because interfering with the expression of the PBR during development results in an embryonic lethal phenotype. Experiments are in process to develop a tissue-specific inducible knockout of PBR to determine its specific role in steroidogenesis.

The mechanism by which StAR regulates cholesterol transport into mitochondria is not well understood. Initially, it was thought that intramitochondrial cholesterol transport required the importation of StAR into the mitochondria (Stocco and Clark, 1996). This has been shown not to be the case because deleting the mitochondrial targeting sequence on the precursor protein has no effect on the ability of StAR to stimulate steroidogenesis (Arkane et al., 1996; Wang et al., 1998). StAR's role may be involved in transporting cholesterol from intracellular stores and presenting it to the complex in the outer mitochondrial membrane (Stocco, 2000).

The role of StAR in the brain is only beginning to be examined. Initial reports were unable to find StAR in either human or mouse brain (Clark et al., 1995; Sugawara et al., 1995), but more recent studies using reverse transcription–polymerase chain reaction techniques have found StAR transcripts in the cerebral cortex, hippocampus, dentate gyrus, olfactory bulb, and *Purkinje* cells in the cerebellum (Furukawa *et al.*, 1998). It remains to be seen whether the StAR protein is made in the brain and whether it plays an important role in regulating brain steroidogenesis. The final picture will most likely involve interplay among PBR, StAR, and any number of other known and as yet unidentified proteins (Fig. 3).

IX. Conclusions

Acute regulation of steroidogenesis is a critical process in the regulation of homeostasis and neuronal excitability. It is now clear that this process is largely mediated through the control of the supply of cholesterol to the inner mitochondrial membrane and P450scc. The PBR plays a critical role in regulating cholesterol transport into the mitochondrial matrix and therefore all further steroid hormone biosynthesis. This function is conserved between tissues and across species, including insects, where the PBR and DBI were shown to be involved in the synthesis of ecdysteroids (Snyder and Van Antwerpen, 1998). It is also becoming increasingly apparent that the actions of the PBR will not be limited to steroidogenesis but may also be involved in development, differentiation, and a number of pathological conditions via a more general role in cholesterol compartmentalization within the cell. Understanding the function of the PBR will be crucial for developing further treatments for these clinical conditions as well as for understanding basic cellular function.

Acknowledgments

This chapter was supported by grants from the National Institutes of Health (ES-07747, HD-37031), the National Science Foundation (IBN-9728261), and the Institut Henri Beaufour, France. R. C. B. was supported by a National Science Foundation predoctoral fellowship.

References

Amri, H., Ogwegbu, S. O., Boujrad, N., Drieu, K., and Papadopoulos, V. (1996). *In vivo* regulation of the peripheral-type benzodiazepine receptor and glucocorticoid synthesis by the *Ginkgo biloba* extract EGb 761 and isolated ginkgolides. *Endocrinology* **137,** 5707–5718.

Amri, H., Drieu, K., and Papadopoulos, V. (1997). *Ex vivo* regulation of adrenal cortical cell steroid and protein synthesis, in response to adrenocorticotropic hormone stimulation, but the *Ginkgo biloba* extract EGb 761 and isolated ginkgolides. *Endocrinology* **138**, 5415–5426.

Amri, H., Li, H., Culty, M., Gaillard, J. L., Teper, G., and Papadopoulos, V. (1999). The peripheral-type benzodiazepine receptor and adrenal steroidogenesis. *Curr. Opin. Endocrinol. Diabetes* **6**, 179–184.

Amsterdam, A., and Suh, B. S. (1991). An inducible functional peripheral benzodiazepine receptor in mitochondria of steroidogenic granulosa cells. *Endocrinology* **128**, 503–510.

Anholt, R. R. H., DeSouza, E. B., Oster-Granite, M. L., and Snyder, S. H. (1985). Peripheral-type benzodiazepine receptors: Autoradiographic localization in whole-body sections of neonatal rats. *J. Pharm. Exp. Ther.* **233**, 517–526.

Anholt, R. R. H., Pedersen, P. L., DeSouza, E. B., and Snyder, S. H. (1986). The peripheral-type benzodiazepine receptor: Localization of the mitochondrial outer membrane. *J. Biol. Chem.* **261**, 576–583.

Antikiewicz-Michaluk, L., Mukhin, A. G., Guidotti, A., and Krueger, K. E. (1988). Purification and characterization of a protein associated with peripheral-type benzodiazepine binding sites. *J. Biol. Chem.* **263**, 17317–17321.

Arai, M., and Widmaier, E. P. (1993). Steroidogenesis in isolated adrenocortical cells during development in rats. *Mol. Cell. Endocr.* **92**, 91–97.

Arkane, F., Sugawara, T., Nichino, H., Liu, Z., Holt, H. A., Pain, D., Stocco, D. M., Miller, W. L., and Strauss, III, J. F. (1996). Steroidogenic acute regulatory protein (StAR) retains activity in the absence of its mitochondrial import sequence: Implications for the mechanism of StAR action. *Proc. Natl. Acad. Sci. USA* **93**, 13731–13736.

Barnea, E. R., Fares, F., and Gavish, M. (1989). Modulatory action of benzodiazepines on human term placental steroidogenesis *in vitro*. *Mol. Cell. Endocrinol.* **64**, 155–159.

Basile, A. S., and Skolnick, P. (1986). Subcellular localization of "peripheral-type" binding sites for benzodiazepines in rat brain. *J. Neurochem.* **46**, 305–308.

Baulieu, E. E., and Robel, P. (1990). Neurosteroids: A new brain function? *J. Steroid Biochem. Mol. Biol.* **37**, 395–403.

Bernassau, J. M., Reverat, J. L., Ferrara, P., Caput, D., and Lefu, G. (1993). A 3D model of the peripheral benzodiazepine receptor and its implication in intra mitochondrial cholesterol transport. *J. Mol. Graph.* **11**, 236–245.

Boujrad, N., Gaillard, J.-L., Garnier, M., and Papadopoulos, V. (1994). Acute action of choriogonadotropin in Leydig tumor cells. Induction of a higher affinity benzodiazepine receptor related to steroid biosynthesis. *Endocrinology* **135**, 1576–1583.

Boujrad, N., Vidic, B., and Papadopoulos, V. (1996). Acute action of choriogonadotropin on Leydig tumor cells: Changes in the topography of the mitochondrial peripheral-type benzodiazepine receptor. *Endocrinology* **137**, 5727–5730.

Braestrup, C., and Squires, R. F. (1977). Specific benzodiazepine receptors in rat brain characterized by high-affinity [^{3}H]diazepam binding. *Proc. Natl. Acad. Sci. USA* **74**, 3805–3809.

Brinton, R. (1994). The neurosteroid 3α-hydroxy-5α-pregnan-20-one induces cytoarchitectural regression in cultured fetal hippocampal neurons. *J. Neurosci. Res.* **14**, 2763–2774.

Brown, R. C., Cascio, C., and Papadopoulos, V. (2000). Neurosteroid biosynthesis in cell lines from human brain: Regulation of dehydroepiandrosterone formation by oxidative stress and β-amyloid peptide. *J. Neurochem.* **74**, 847–859.

Brown, R. C., Degenhardt, B., Kotoula, M., and Papadopoulos, V. (2000). Location dependent role of the peripheral-type benzodiazepine receptor in proliferation and steroid biosynthesis in human glioma cells. *Cancer Lett.* **156**, 125–132.

Caron, K. M., Soo, S. C., Wetsel, W. C., Stocco, D. M., Clark, B. J., and Parker, K. L. (1997). Targeted disruption of the mouse gene encoding steroidogenic acute regulatory protein provides insights into congenital lipoid adrenal hyperplasia. *Proc. Natl. Acad. Sci. USA* **94,** 11540–11545.

Cascio, C., Guarneri, P., Li, H., Brown, R. C., Amri, H., Boujrad, N., Kotoula, M., Vidic, B., Drieu, K., and Papadopoulos, V. (1999). Peripheral-type benzodiazepine receptor. Role in the regulation of steroid and neurosteroid biosynthesis. *In* "Contemporary Endocrinology: Neurosteroids: A New Regulatory Function in the Nervous System" (E. E. Baulieu and M. Schumacher, Eds.), pp. 75–96. Humana Press, Totowa, N J.

Cascio, C., Brown, R. C., Liu, Y., Han, Z., Hales, D. B., and Papadopoulos, V. (2000). Pathways of dehydroepiandrosterone formation in developing rat brain glia. *J. Steroid Biochem. Mol. Biol.* **889,** 181–190.

Chang, Y. J., McCabe, R. T., Rennert, H., Budarf, M. L., Sayegh, R., Emmanuel, B. S., Skolnick, P., and Strauss, J. F. (1992). The human "peripheral-type" benzodiazepine receptor: Regional mapping of the gene and characterization of the receptor expressed from cDNA. *DNA Cell Biol.* **11,** 471–480.

Clark, B. J., Soo, S. C., Caron, K. M., Ikeda, Y., Parker, K. L., and Stocco, D. M. (1995). Hormonal and developmental regulation of the steroidogenic acute regulatory (StAR) protein. *Mol. Endocrinol.* **9,** 1346–1355.

Compagnone, N. A., and Mellon, S. H. (1998). Dehydroepiandrosterone: A potential signaling molecule for neocortical organization during development. *Proc. Natl. Acad. Sci. USA* **95,** 4678–4683.

Cornu, P., Benavides, J., Scatton, B., Hauw, J. J., and Philippon, J. (1992). Increase in omega 3 (peripheral-type benzodiazepine) binding site densities in different types of human brain tumors. A quantitative autoradiography study. *Acta Neurochir.* **119,** 146–152.

Costa, E., Cheney, D. L., Grayson, D. R., Korneyev, A., Longone, P., Pani, L., Romeo, E., Zivkovich, E., and Guidotti, A. (1994). Pharmacology of neurosteroid biosynthesis. *Ann. N.Y. Acad. Sci.* **746,** 223–242.

Costa, E., and Guidotti, A. (1979). Molecular mechanism in the receptor actions of benzodiazepines. *Ann. Rev. Pharmacol. Toxicol.* **19,** 531–545.

Culty, M., Li, H., Boujrad, N., Amri, H., Vidic, B., Bernassau, J. M., Reversat, J. L., and Papadopoulos, V. (1999). *In vitro* studies on the role of the peripheral-type benzodiazepine receptor in steroidogenesis. *J. Steroid Biochem. Mol. Biol.* **69,** 123–130.

DeFeudis, F. V. (1998). "*Ginkgo biloba* extract (EGb 761): From chemistry to the clinic." Wiebaden: Ullestein Medical.

Doble, A., Ferris, O., Burgevin, M. C., Menager, J., Uzan, A., Dubroeucq, M. C., Renault, C., Gueremy, C., and Le Fur, G. (1987). Photoaffinity labeling of peripheral-type benzodiazepine binding sites. *Mol. Pharmacol.* **31,** 42–49.

Farges, R., Joseph-Liauzun, E., Shire, D., Caput, D., Le Fur, G., Loison, G., and Ferrara, P. (1993). Molecular basis for the different binding properties of benzodiazepines to human and bovine peripheral-type benzodiazepine receptors. *FEBS Lett.* **335,** 305–308.

Freeman, D. A. (1987). Constitutive steroidogenesis in the R2C Leydig tumor cell line is maintained by the adenosine $3':5'$-cyclic monophosphate-independent production of a cyclohexamide-sensitive factor that enhances mitochondrial pregnenolone biosynthesis. *Endocrinology* **120,** 124–132.

Furukawa, A., Miyatake, A., Ohnishi, T., and Ichikawa, Y. (1998). Steroidogenic acute regulatory protein (StAR) transcripts constitutively expressed in the adult rat central nervous system: Colocalization of StAR, cytochrome P-450scc (CYP X1A1), and 3β-hydroxysteroid dehydrogenase in the rat brain. *J. Neurochem.* **71,** 2231–2238.

Garnier, M., Boujrad, N., Ogwuegbu, S. O., Hudson, J. R., and Papadopoulos, V. (1994). The polypeptide diazepam binding inhibitor and a higher affinity peripheral-type benzodiazepine receptor sustain constitutive steroidogenesis in the R2C Leydig tumor cell line. *J. Biol. Chem.* **269,** 22105–22112.

Garnier, M., Dimchev, A. B., Boujrad, N., Price M. J., Musto, N. A., and Papadopoulos, V. (1994). In vitro reconstitution of a functional peripheral-type benzodiazepine receptor from mouse Leydig tumor cells. *Mol. Pharm.* **45,** 201–211.

Gavish, M., Katz, Y., Bar-Ami, S., and Weizman, R. (1992). Biochemical, physiological, and pathological aspects of the peripheral benzodiazepine receptor. *J. Neurochem.* **58,** 1589–1601.

Guilarte, T. R., Kuhlmann, A. C., O'Callaghan, J. P., and Miceli, R. C. (1995). Enhanced expression of peripheral benzodiazepine receptors in trimethyltin-exposed rat brain: A biomarker of neurotoxicity. *Neurotoxicology* **16,** 441–450.

Haefely, W., Kulcsar, A., Mohler, H., Pieri, L., Polc, P., and Schaffner, R. (1975). Possible involvement of GABA in the central actions of benzodiazepine derivatives. In "Advances in Biochemical Psychopharmacology" (E. Costa and P. Greengard, Eds.), pp. 131–151. Raven Press, NY.

Hall, P. F. (1985). Trophic stimulation of steroidogenesis: In search of the elusive trigger. *Recent Prog. Horm. Res.* **41,** 1–39.

Hardwick, M., Fertikh, D., Culty, M., Li, H., Vidic, B., and Papadopoulos, V. (1999). Peripheral-type benzodiazepine receptor (PBR) in human breast cancer: Correlation of breast cancer cell aggressive phenotype with PBR expression, nuclear localization, and PBR-mediated cell proliferation and nuclear transport of cholesterol. *Cancer Res.* **59,** 831–842.

Ikezaki, K., and Black, K. L. (1990). Stimulation of cell growth and DNA synthesis by peripheral benzodiazepine. *Cancer Lett.* **49,** 115–120.

Jefcoate, C. R., McNamara, B. C., Artemenko, I., and Yamazaki, T. (1992). Regulation of cholesterol movement to mitochondrial cytochrome P450scc in steroid hormone synthesis. *J. Steroid Biochem. Mol. Biol.* **43,** 751–767.

Joseph-Liauzun, E., Delmas, P., Shire, D., and Ferrara, P. (1998). Topological analysis of the peripheral benzodiazepine receptor in yeast mitochondrial membranes supports a five-transmembrane structure. *J. Biol. Chem.* **273,** 2146–2152.

Jung-Testas, I., Hu, Z. Y., Baulieu, E. E., and Robel, P. (1989). Neurosteroids: Biosynthesis of pregnenolone and progesterone in primary cultures of rat glial cells. *Endocrinology* **125,** 2083–2091.

Kelly-Herskovitz, E., Weizman, R., Spanier, I., Leschiner, S., Lahav, M., Weisinger, G., and Gavish, M. (1998). Effects of peripheral-type benzodiazepine receptor antisense knockout on MA-10 Leydig cell proliferation and steroidogenesis. *J. Biol. Chem.* **273,** 5478–5483.

Kimura, T. (1986). Transduction of ACTH signal from plasma membrane to mitochondria in adrenocortical steroidogenesis. Effects of peptide, phospholipid and calcium. *J. Steroid Biochem.* **25,** 711–716.

Krueger, K. E., and Papadopoulos, V. (1990). Peripheral-type benzodiazepine receptors mediate translocation of cholesterol from outer to inner mitochondrial membranes in adrenocortical cells. *J. Biol. Chem.* **265,** 15015–15022.

Krueger, K. E., and Papadopoulos, V. (1991a). Molecular and functional characterization of peripheral-type benzodiazepine receptors in glial cells. In "Biological Psychiatry" (G. Racagni, Ed.), Vol. 1, pp. 744–746. Elsevier Science NY.

Krueger, K. E., and Papadopoulos, V. (1991b). The peripheral-type benzodiazepine receptor: Cell biological role and pharmacological significance. In "Transmitter Amino Acid Receptors: Structures, Transduction and Models for Drug Development," FIDIA Research Foundation Symposium Series, Vol. 6, pp. 153–166. Thieme, NY.

Kuhlmann, A. C., and Guilarte, T. R. (1997). The peripheral benzodiazepine receptor is a sensitive indicator of domoic acid neurotoxicity. *Brain Res.* **751,** 281–288.

Kuhlmann, A. C., and Guilarte, T. R. (1999). Regional and temporal expression of the peripheral benzodiazepine receptor in MPTP neurotoxicity. *Toxicol. Sci.* **48,** 107–116.

Kuhlmann, A. C., and Guilarte, T. R. (2000). Cellular and subcellular localization of peripheral benzodiazepine receptors after trimethyltin neurotoxicity. *J. Neurochem.* **74,** 1694–1704.

Lacor, P., Gandolfo, P., Tonon, M.-C., Brault, E., Dalibert, I., Schumacher, M., Benavides, J., and Ferzaz, B. (1999). Regulation of the expression of peripheral benzodiazepine receptors and their endogenous ligands during rat sciatic nerve degeneration and regeneration: A role for PBR in neurosteroidogenesis. *Brain Res.* **815,** 70–80.

Lambert, J. J., Belelli, D., Hill-Venning, C., and Peters, J. A. (1995). Neurosteroids and GABA$_A$ receptor function. *Trends Pharm. Sci.* **16,** 295–303.

Li, H., and Papadopoulos, V. (1998). Peripheral-type benzodiazepine receptor function in cholesterol transport. Identification of a putative cholesterol recognition/interaction amino acid sequence and consensus pattern. *Endocrinology* **139,** 4991–4997.

Li, H., Degenhardt, B., Tobin, D., Tasken, K., and Papadopoulos, V. (2000). Novel element in the hormonal regulation of steroidogenesis. PAP7: A peripheral-type benzodiazepine receptor-associated and a kinase anchoring protein. *Endocrine Society Abstracts.*

Lin, D., Sugawara, T., Strauss III, J. F., Clark, B. J., Stocco, D. M., Saenger, P., Rogol, A., and Miller, W. L. (1995). Role of steroidogenic acute regulatory protein in adrenal and gonadal steroidogenesis. *Science* **267,** 1828–1831.

Marcilhac, A., Dakine, N., Bourhim, N., Guillaume, V., Grino, M., Drieu, K., and Oliver, C. (1998). Effect of chronic administration of *Ginkgo biloba* extract or ginkgolide on the hypothalamic-pituitary-adrenal axis in the rat. *Life Sci.* **62,** 2329–2340.

McEnery, M. W., Sowman, A. M., Trifiletti, R. R., and Snyder, S. H. (1992). Isolation of the mitochondrial benzodiazepine receptor: Association with the voltage-dependent anion channel and the adenine nucleotide carrier. *Proc. Natl. Acad. Sci. USA* **89,** 3170–3174.

McGeer, E. G., Singh, E. A., and McGeer, P. L. (1998). Peripheral-type benzodiazepine binding in Alzheimer disease. *Alzheim. Dis. Assoc. Disord.* **2,** 331–336.

Miettinen, H., Kononen, J., Haapasalo, H., Helen, P., Sallinen, P., Harjuntausta, T., Helin, H., and Alho, H. (1995). Expression of peripheral-type benzodiazepine receptor and diazepam binding inhibitor in human astrocytomas: Relationship to cell proliferation. *Cancer Res.* **15,** 2691–2695.

Mukhin, A. G., Papadopoulos, V., Costa, E., and Kreuger, K. E. (1989). Mitochondrial benzodiazepine receptors regulate steroid biosynthesis. *Proc. Natl. Acad. Sci. USA* **86,** 9813–9816.

Norton, W. T., and Poduslo, S. E. (1973). Myelinization in rat brain: Changes in myelin composition during brain maturation. *J. Neurochem.* **21,** 759–773.

Oke, B. O., Suarez-Quian, C. A., Riond, J., Ferrara, P., and Papadopulos, V. (1992). Cell surface localization of the peripheral-type benzodiazepine receptor (PBR) in adrenal cortex. *Mol. Cell. Endocrinol.* **87,** R1–R6.

Papadopoulos, V. (1993). Peripheral-type benzodiazepine/diazepam binding inhibitor receptor: Biological role in steroidogenic cell function. *Endocr. Rev.* **14,** 222–240.

Papadopoulos, V. (1998). Structure and function of the peripheral-type benzodiazepine receptor in steroidogenic cells. *Proc. Soc. Exp. Biol. Med.* **217,** 130–142.

Papadopoulos, V., and Guarneri, P. (1994). Regulation of C6 glioma cell steroidogenesis by adenosine 3′,5′-cyclic monophosphate. *Glia* **10,** 75–78.

Papadopoulos, V., Mukhin, A. G., Costa, E., and Krueger, K. E. (1990). The peripheral-type benzodiazepine receptor is functionally linked to Leydig cell steroidogenesis. *J. Biol. Chem.* **265,** 3772–3779.

Papadopoulos, V., Berkovich, A., Krueger, K. E., Costa, E., and Guidotti, A. (1991). diazepam binding inhibitor (DBI) and its processing products stimulate mitochondrial steroid biosynthesis via an interaction with mitochondrial benzodiazepine receptors. *Endocrinology* **129,** 1481–1488.

Papadopoulos, V., Guarneri, P., Krueger, K. E., Guidotti, A., and Costa, E. (1992). Pregnenolone biosynthesis in C6 glioma cell mitochondria: Regulation by a diazepam binding inhibitor mitochondrial receptor. *Proc. Natl. Acad. Sci. USA* **89,** 5113–5117.

Papadopoulos, V., Amri, H., Boujrad, N., Cascio, C., Culty, M., Garnier, M., Hardwick, M., Li, H., Vidic, B., Brown, A. S., Reversat, J. L., Bernassau, J. M., and Drieu, K. (1997a). Peripheral benzodiazepine receptor in cholesterol transport and steroidogenesis. *Steroids* **62,** 21–28

Papadopoulos, V., Amri, H., Li, H., Boujrad, N., Vidic, B., and Garnier, M. (1997b). Targeted disruption of the peripheral-type benzodiazepine receptor gene inhibits steroidogenesis in the R2C Leydig tumor cell line. *J. Biol. Chem.* **272,** 32129–32135.

Papadopoulos, V., Dharmarajan, A. M., Li, H., Culty, M., Lemay, M., and Sridaran, R. (1999). Mitochondrial peripheral-type benzodiazepine receptor expression: Correlation with gonadotropin-releasing hormone (GnRH) agonist-induced apoptosis in the corpus luteum. *Biochem. Pharm.* **58,** 1389–1393.

Parola, A. L., Stump, D. G., Pepper, D. J., Krueger, K. E., Regan, J. W., and Laird, H. E. (1991). Cloning and expression of a pharmacologically unique bovine peripheral-type benzodiazepine receptor isoquinoline binding protein. *J. Biol. Chem.* **266,** 14082–14087.

Parola, A. L., Yamamura, H. I., and Laird, H. E. (1993). Peripheral-type benzodiazepine receptors. *Life Sci.* **52,** 1329–1342.

Rapin, J. R., Lamproglou, I., Drieu, K., and DeFeudis, F. V. (1994). Demonstration of the "antistress" activity of an extract of *Ginkgo biloba* (EGb 761) using a discrimination learning task. *Gen. Pharmacol.* **25,** 1009–1016.

Riond, J., Vita, N., Le Fur, G., and Ferrara, P. (1989). Characterization of a peripheral-type benzodiazepine binding site in the mitochondrial of Chinese hamster ovary cells. *FEBS Lett.* **245,** 238–244.

Riond, J., Mattei, M. G., Kaghad, M., Dumont, X., Guillemot, J. C., Le Fur, G., Caput, D., and Ferrara, P. (1991). Molecular cloning and chromosomal localization of a human peripheral-type benzodiazepine receptor. *Eur. J. Biochem.* **195,** 305–311.

Ritta, M. N., and Calandra, R. S. (1989). Testicular interstitial cells as targets for peripheral benzodiazepines. *Neuroendocrinology* **49,** 262–266.

Shah, S. N. (1993). Cholesterol metabolism in brain. *Adv. Structural Biology* **2,** 171–189.

Simbeni, R., Pon, L., Zinser, E., Paltauf, F., and Daum, G. (1991). Mitochondrial membrane contact sites of yeast. Characterization of lipid components and possible involvement in intramitochondrial translocation of phospholipids. *J. Biol. Chem.* **266,** 10047–10049.

Simpson, E. R., and Waterman, M. R. (1983). Regulation by ACTH of steroid hormone biosynthesis in the adrenal cortex. *Can. J. Biochem. Cell. Biol.* **61,** 692–707.

Siripurpong, P., Harnyuttanakorn, P., Chindaduangratana, C., Kotchabhakdi, N., Wichyanuwat, P., and Casalotti, S. O. (1997). Dexamethasone, but not stress, induces measurable changes of mitochondrial benzodiazepine receptor mRNA levels in rats. *Eur. J. Pharmacol.* **331,** 227–235.

Snyder, M. J., and Van Antwerpen, R. (1998). Evidence for a diazepam-binding inhibitor benzodiazepine receptor-like mechanism in ecdysteroidogenesis by the insect prothoracic gland. *Cell Tissue Res.* **294,** 161–168.

Sprengel, R., Werner, P., Seeburg, P. H., Mukhin, A. G., Santi, M. R., Grayson, D. R., Guidotti, A., and Krueger, K. E. (1989). Molecular cloning and expression of cDNA encoding a peripheral-type benzodiazepine receptor. *J. Biol. Chem.* **264,** 20415–20421.

Stevens, V. L., Tribble, D. L., and Lambeth, J. D. (1985). Regulation of mitochondrial compartment volumes in rat adrenal cortex by ether stress. *Arch. Biochem. Biophys.* **242,** 324–327.

Stocco, D. M. (1999). Steroidogenic acute regulatory (StAR) protein: What's new? *BioEssays* **21,** 768–775.

Stocco, D. M. (2000). The role of the StAR protein in steroidogenesis: Challenges for the future. *J. Endocrinol.* **164,** 247–253.

Stocco, D. M., and Clark, B. J. (1996). Regulation of the acute production of steroids in steroidogenic cells. *Endocr. Rev.* **17,** 221–244.

Sugawara, T., Holt, J. A., Driscoll, D., Strauss, III, J. F., Lin, D., Miller, W. L., Patterson, D., Clancy, K. P., Hart, I. M., Clark, B. J., and Stocco, D. M. (1995). Human steroidogenic acute regulatory protein: Functional activity in COS-1 cells, tissue-specific expression, and mapping of the gene to 8p11·2 and a pseudogene to chromosome 13. *Proc. Natl. Acad. Sci. USA* **92,** 4778–4782.

Verma, A., and Snyder, S. H. (1989). Peripheral type benzodiazepine receptors. *Ann. Rev. Pharmacol. Toxicol.* **29,** 307–322.

Vidic, B., Boujrad, N., and Papadopoulos, V. (1995). Hormone-induced changes in the topography of the mitochondrial peripheral-type benzodiazepine receptor. *Scanning* **17,** V34–35.

Wang, X. J., Liu, Z., Eimerl, S., Weiss, A. M., Orly, J., and Stocco, D. M. (1998). Effect of truncated forms of the steroidogenic acute regulatory (StAR) protein on intramitochondrial cholesterol transfer. *Endocrinology* **139,** 3903–3912.

Ward, J. A., Krantz, S., Medeloff, J., and Halriwanger, E. (1960). Interstitial cell tumor of the testes: Report of two cases. *J. Clin. Endocrinol. Metab.* **22,** 1622–1629.

Weizman, R., Dagan, E., Snyder, S. H., and Gavish, M. (1997a). Impact of pregnancy and lactation on GABA$_A$ receptor and central-type and peripheral-type benzodiazepine receptors. *Brain Res.* **752,** 302–314.

Weizman, R., Leschiner, S., Schlegel, W., and Gavish, M. (1997b). Peripheral-type benzodiazepine receptor ligands and serum steroid hormones. *Brain Res.* **772,** 203–208.

Zilz, A., Li, H., Castello, R., Papadopoulos, V., and Widmaier, E. (1999). Developmental expression of the peripheral-type benzodiazepine receptor and the advent of steroidogenesis in rat adrenal glands. *Endocrinology* **140,** 859–864.

FORMATION AND EFFECTS OF NEUROACTIVE STEROIDS IN THE CENTRAL AND PERIPHERAL NERVOUS SYSTEM

Roberto Cosimo Melcangi, Valerio Magnaghi, Mariarita Galbiati, and Luciano Martini

Department of Endocrinology, University of Milan, 20133, Milan, Italy

I. Introduction
II. 5α-Reductase and 3α-Hydroxysteroid Dehydrogenase System
   A. General Consideration
   B. Cellular Localization in the Central Nervous System
   C. Regulation in the Central Nervous System
III. Effect of Glial–Neuronal Interactions on the Formation of Neuroactive Steroids
IV. Effects of Neuroactive Steroids on Glial Cells of the Central Nervous System
V. Formation of Neuroactive Steroids in the Peripheral Nervous System
VI. Effects of Neuroactive Steroids on the Peripheral Nervous System
   A. Effects on Peripheral Myelin
   B. Effects on Schwann Cells
VII. Conclusion
   References

This chapter summarizes several observations that emphasize the importance of neuroactive steroids in the physiology of the central and peripheral nervous systems. A new, and probably important, concept is emerging: Neuroactive steroids not only modify neuronal physiology but also intervene in the control of glial cell functions. The data presented here underscore that (1) the mechanism of action of the various steroidal molecules may involve both classical (progesterone and androgens) and nonclassical steroid receptors [γ-aminobutyric acid type A ($GABA_A$) receptor], (2) in many instances, the actions of hormonal steroids are not due to their native molecular forms but to their 5α- and 3α,5α-reduced metabolites, (3) several neuroactive steroids exert dramatic actions on the proteins proper of the peripheral myelin (e.g., glycoprotein Po and peripheral myelin protein 22), and (4) the effects of steroids and of their metabolites might have clinical significance in cases in which the rebuilding of the peripheral myelin is needed (e.g., aging, peripheral injury). © 2001 Academic Press.

I. Introduction

In the 1990s, many observations suggested the importance of neuroactive steroids in the control of several functions of the nervous system. In particular, it has been shown that the nervous system is able to synthesize steroid hormones and possesses the enzymatic capability to convert several physiological steroids into neuroactive metabolites (see for review Celotti *et al.*, 1992; Melcangi *et al.*, 1999a). These metabolites may occasionally be more effective than their corresponding parent compounds, or they may have totally different biological actions; neuroactive steroids may exert their actions either through classical or nonclassical receptors, which are localized both in the neuronal and in the glial compartments. In particular, the importance of glial cells as a target of neuroactive steroids is emphasized by the presence, in the oligodendrocytes, the astrocytes, and the Schwann cells, of classical intracellular receptors for many families of hormonal steroids [e.g., receptors for glucocorticoids (GR), mineralocorticoids (MR), androgens (AR), estrogens (ER), and progesterone (PR)] (Vielkind *et al.*, 1990; Jung-Testas *et al.*, 1992, 1996; Langub and Watson, 1992; Wolff *et al.*, 1992; Santagati *et al.*, 1994; Magnaghi *et al.*, 1999). Moreover, it must be mentioned that astrocytes (Bovolin *et al.*, 1992; Hosli *et al.*, 1997) and Schwann cells (Melcangi *et al.*, 1999b) possess $GABA_A$ receptors and, consequently, may respond to neuroactive steroids that are able to interact with this neurotransmitter receptor (see for review Celotti *et al.*, 1992; Melcangi *et al.*, 1999a).

This chapter analyzes only one of the major pathways converting hormonal steroids, namely the 5α-reductase- 3α-hydroxysteroid dehydrogenase (5αR-3αHSD) enzymatic system. For brevity, aromatase (i.e., the enzyme that converts androgens into estrogens) is not considered. Moreover, only the effects of neuroactive steroids on the central and peripheral glial elements and on their specific products (e.g., myelin membranes) are considered.

II. 5α-Reductase and 3α-Hydroxysteroid Dehydrogenase System

A. General Consideration

Several observations have shown that the enzymatic complex formed by the 5αR and the 3αHSD is found not only in the classical peripheral steroid target structures (e.g., the prostate, the epididymis) that respond to androgens, but also is present in the nervous system (see for review, Celotti *et al.*,

1992; Melcangi *et al.*, 1999a). This enzymatic system is very versatile, in the sense that every steroid possessing the Δ^4-3-ketosteroid configuration may be first 5α-reduced and, subsequently, 3α-hydroxylated. In particular, testosterone (T) can be converted into dihydrotestosterone (DHT) and, subsequently, into 5α-androstane-3α, 17β-diol (3α-diol); progesterone (PROG) into dihydroprogesterone (5α-DH PROG) and, subsequently, into tetrahydroprogesterone (3α,5α-TH PROG); deoxycorticosterone (DOC) into dihydrodeoxycorticosterone (5α-DH DOC) and finally into tetrahydrodeoxycorticosterone (3α,5α-TH DOC); corticosterone into dihydrocorticosterone (DHC) (see for review Melcangi *et al.*, 1999a). The existence, in peripheral structures, of more than one 5αR isozyme was postulated years ago on the basis of studies using various inhibitors (Motta *et al.*, 1986; Zoppi *et al.*, 1992) and different substrates (Zoppi *et al.*, 1992). More recently, two isoforms of the 5αR (called type 1 and type 2) have been cloned, in human, in rat, and in monkey (Andersson *et al.*, 1989; Andersson *et al.*, 1991; Labrie *et al.*, 1992; Normington and Russell, 1992; Russell and Wilson, 1994; Levy *et al.*, 1995). In humans, the type 1 5αR gene, on chromosome 5, is composed of five exons and four introns and produces a protein of 259 amino acids. The type 2 5αR gene, located on chromosome 2, has a similar structure, but the resulting protein is composed of 254 amino acids. The structures of the enzymatic proteins, determined from their respective cDNAs, show a limited degree of homology (about 47%) and a predicted molecular weight of 28–29 kDa. Despite the fact that the two major isoforms of the 5αR (type 1 and type 2) catalyze the same reaction (e.g., T to DHT, PROG to 5α-DH PROG), they possess different biochemical and possibly functional properties. In rats, the affinity of T for the type 1 isoform is about 15- to 20-fold lower than that determined for the type 2 isoform. The difference in affinity is evident also in the case of the human enzymes, even if it is less marked. Both in rat and in human, the capability of reducing the substrate is much higher in the case of the type 1 isoform. The two isoforms have a different pH optimum: The type 1 isoform is active in a wide range of pH (from 5 to 8), while the type 2 5αR possesses a narrow pH optimum around 5, with a very low activity at pH 7.5. The two isozymes also show a differential sensitivity to synthetic inhibitors, such as finasteride, which blocks preferentially the human type 2 isozyme (Thigpen and Russell, 1992).

The enzyme 3αHSD, also known as 3α-hydroxy steroid oxidoreductase, may be considered the second element of the 5αR/3αHSD system. At variance with the two isoforms of the 5αR, this enzyme appears to be able to catalyze the controlled reaction both in the oxidative and in the reductive directions. The amounts of 5α-reduced/3α-hydroxylated compounds present at any given moment in a cell are therefore dependent on the equilibrium of this reaction. The first 3αHSD enzyme to be purified and cloned was the one present in the rat liver cytosol (Cheng et al., 1991; Pawlovski et al., 1991; Cheng et al., 1994). The cDNA of this enzyme codes for a protein of 322 amino acids with an estimated molecular weight of 37 kDa and catalyzes the oxidoreduction of steroids using either NADP/NADPH or NAD/NADH as a cofactor. The three-dimensional structure and other relevant features of the rat liver enzyme have been reported (Bennett et al., 1996; Penning, 1996). This enzyme shares 84% sequence similarity with the human liver dihydrodiol dehydrogenase, which is considered the human counterpart of the rat 3αHSD (Cheng, 1992; Penning et al., 1996). The human gene coding for this enzyme possesses more than 47 kb and contains, in the 5'-flanking region, consensus sequences for AP-1 and Oct -1 as well as multiple copies of perfect and imperfect steroid responsive elements for the ER, and for the GR/PR (Penning, 1996; Cheng, 1992; Penning et al., 1996).

B. CELLULAR LOCALIZATION IN THE CENTRAL NERVOUS SYSTEM

The majority of the earlier studies on the 5αR activity present in the brain have been performed on the hypothalamus, due to the pre-eminent endocrine role of this structure (see for review Celotti et al., 1992; Melcangi et al., 1999a). However, later analyses have indicated that the formation of DHT is not limited to the hypothalamus but also occurs in several other brain areas, including the cerebral cortex (see for review Celotti et al., 1992; Melcangi et al., 1999a). It has also been reported that the conversion of T into DHT is several times higher in brain structures mainly composed of white matter (e.g., corpus callosum, midbrain tegumentum) (Krieger et al., 1983; Melcangi et al., 1988a,b; Sholl et al., 1989). Further studies have linked this activity to the presence of the myelin, and purified myelin membranes have been shown to possess a 5αR activity about eight times higher than that of brain homogenates (Melcangi et al., 1988a,b; 1989). The physiological meaning of the presence of the 5αR in the myelin is still obscure, but the hypothesis that 5α-reduced steroids locally formed in the myelin might play a role in the process of myelination is certainly attractive. Such a hypothesis is indirectly supported by data showing that there are peaks of 5αR activity in the rat central nervous system (CNS) (Massa et al., 1975; Degtiar et al., 1981) and in the purified CNS myelin in the first weeks of life (Melcangi et al., 1988b), at the time of the initiation of the process of myelinization (Norton and Poduslo, 1973). Moreover, the cells that manufacture the myelin, the oligodendrocytes, possess some 5αR activity (Melcangi et al., 1988b, 1993, 1994a).

The distribution of the 5αR activity in the different cell types of the rat brain has been analyzed using either T or PROG as the substrates (Melcangi *et al.*, 1993, 1994a). In particular, the ability to metabolize T was first studied in freshly isolated cell preparations (Melcangi *et al.*, 1990a); subsequent studies were performed in cultures of neurons, of type 1 astrocytes (A1), of type 2 astrocytes, and of oligodendrocytes (Melcangi *et al.*, 1993, 1994a). The two different groups of experiments have provided similar data. In particular, it has been observed that, using T as substrate, fetal rat neurons possess significantly higher amounts of 5αR activity than neonatal oligodendrocytes and astrocytes (Melcangi *et al.*, 1993). Among glial cells, type 2 astrocytes possess a considerable 5αR activity, while A1 are almost devoid of such activity. On the contrary, the enzyme 3αHSD appears to be mainly localized in A1 (Melcangi *et al.*, 1993). The compartmentalization of these two strictly correlated enzymes, 5αR and 3αHSD, in separate CNS cell populations suggests the simultaneous participation of neurons and glial cells in the 5α-reductive metabolism of T and other delta 4-3 ketosteroids. It must be pointed out that neurons seem to be the only brain cell population that, in addition to the 5αR, also contain the enzyme aromatase and, consequently, are able to transform androgens into estrogens (Negri-Cesi *et al.*, 1992). However, in 1999, Garcia-Segura *et al.* observed that aromatase may also be expressed in astrocytes after brain injury.

The formation of the metabolites of PROG has also been evaluated in cultures of neurons, oligodendrocytes, A1, and type 2 astrocytes (Melcangi *et al.*, 1994a). Similar to what was observed using T, and when PROG was used as substrate, neurons were shown to possess significantly higher activity than any glial cell analyzed; these convert PROG with very similar and relatively low yields. The 3αHSD activity (which was analyzed using 5α-DH PROG as substrate and which gives origin to 3α,5α-TH PROG) appeared predominantly concentrated in A1 (Melcangi *et al.*, 1994a). However, a consistent formation of 3α,5α-TH PROG was also present in oligodendrocytes; the amounts of 3α,5α-TH PROG measured in the cultures of these cells were significantly lower than those found in A1 but significantly higher than those formed in cultures of type 2 astrocytes and neurons. These data agree, in general, with the previous findings in which androgens (T and DHT) were used as the labeled substrates (Melcangi *et al.*, 1993). However, it is important to point out that, as a result of the different affinity of the 5αR for PROG and for T (see for review Celotti *et al.*, 1992; Melcangi *et al.*, 1999a), it appears that the formation of 5α-DH PROG was about two times higher than that of DHT.

It is important to recall that the conversion of PROG into 5α-DH PROG and 3α,5α-TH PROG occurring in neurons and/or in glial cells may represent an important step in the mechanism of action of the parent steroid in the brain (see for review Melcangi *et al.*, 1999a) PROG, either derived from the circulation or directly synthesized in the CNS, influences several physiological processes in the glia as well as in neurons (see for review Melcangi *et al.*, 1999a). In glial cells, which have been shown to possess estrogen-inducible PROG nuclear receptors (Jung-Testas *et al.*, 1991; 1992), PROG inhibits cell growth; the hormone also affects astrocyte and oligodendrocyte morphology (Jung-Testas *et al.*, 1994) and alters the expression of specific cell markers [myelin basic protein (MBP) in oligodendrocytes, glial fibrillary acidic protein (GFAP) in astrocytes, etc.]. Further details on the importance of 5α- and $3\alpha,5\alpha$-derivatives of PROG and T on the physiology of the glial cells are presented in Section IV.

Not only differentiated CNS cells possess the 5αR-3αHSD system. Considerable enzymatic activities for converting steroid hormones are also present in undifferentiated cells, as shown in our studies performed on undifferentiated stem cells originating from the mouse striatum (Melcangi *et al.*, 1996a).

Thus far, few studies have been specifically dedicated to analyzing the distribution of the two 5αR isozymes cloned in the brain. In adult human (postmortem samples of the cerebellum, hypothalamus, pons, and medulla oblongata), the combined results of immunoblotting with a 5αR polyclonal antisera and of RNA blot hybridization have indicated that apparently only the type 1 enzyme is present (Thigpen *et al.*, 1993; Eicheler *et al.*, 1994). However, in another study, the immunostaining of autopsy specimens has also shown the presence of the type 2 enzyme in the pyramidal cells of the cerebral cortex (Eicheler *et al.*, 1994). The expression/production of the two isoforms has been studied on the whole brain of the rat. Some authors have shown that specific mRNAs coding for both isoforms (Normington and Russell, 1992) are detectable using Northern analyses on the total RNA obtained from the whole brain of adult male rats; however, we and other authors have demonstrated that, in the whole adult rat brain, the type 1 isoform is largely preponderant [Lephart, 1993 (see for review Melcangi *et al.*, 1998a)]. In contrast, a study performed on the whole rat brain at different stages of development (including adulthood) has shown the presence of high levels of type 1 mRNA at all ages examined, while the expression of the type 2 isoform appeared to be linked to specific times, being maximal in the perinatal period and almost undetectable in adulthood (see for review Melcangi *et al.*, 1998a).

Several metabolic studies suggest that the rat CNS possesses at least two 3αHSD. It has, indeed, been shown that there are two enzymes—one cytosolic and one microsomal. Both types of enzymes appear to be endowed with oxidoreductase capabilities, but they might act in opposite directions (Cheng *et al.*, 1994; Krieger and Scott, 1984). The cytosolic form has been purified from the rat brain to apparent homogeneity (Penning *et al.*, 1985), and preliminary attempts to clone it produced clones very similar to those of the liver enzyme (Penning *et al.*, 1996). The purified brain enzyme is a monomer, with an apparent molecular weight of about 31 kDa, which shows a preference for NADPH and high affinity for DHT; the enzyme shows a specific activity about 100-fold higher than that of the 5αR, indicating that *in vivo* it could rapidly transform a large portion of the 5α-reduced compounds formed by the 5αR (Penning *et al.*, 1985). Karavolas and Hodges (1990) have isolated a cytosolic enzyme from the rat hypothalamus that shows an affinity for 3α,5α-TH PROG much lower than that for 5αDH PROG, indicating that the reductive activity might be its prevailing function. The brain cytosolic enzyme(s), like that of the liver, is(are) inhibited by all major classes of nonsteroidal anti-inflammatory drugs (Penning *et al.*, 1985, 1996). Much less information is available on the microsomal 3αHSD. This enzyme has not been purified yet; it appears to be NADH linked and to possess a 300-fold lower affinity for 5α-DH PROG than for 3α,5α-TH PROG, suggesting that the reaction it controls may proceed in the oxidative direction (Karavolas and Hodges, 1990).

C. REGULATION IN THE CENTRAL NERVOUS SYSTEM

The physiological control of the 5αR in the brain is still a debated issue. The data available were obtained mainly when the existence of the two isoforms of the enzyme was unknown. It is generally accepted that, in the brain, the enzymatic system formed by the 5αR-3αHSD is not sexually dimorphic. Moreover, extensive data indicate that this system is not regulated by sex steroids because castration and substitution therapies are unable to influence its activity (see for review Celotti *et al.*, 1992). This has been shown both in the whole rat brain and in specific CNS areas. Only a few studies appear to disagree with this conclusion. An increase in 5αR activity has been observed after orchidectomy in the basolateral amygdala of the rhesus monkey (Roselli *et al.*, 1987) but not in several other brain structures (suprachiasmatic nucleus, supraoptic nucleus, lateral hypothalamus, basolateral nuclei of the septum, and caudate nucleus) (see for review Celotti *et al.*, 1992).

Also, neural inputs seem to be ineffective in regulating the activity of the 5αR at least at the hypothalamic level. This was shown in the rat by abolishing, with appropriate pharmacological manipulations, inputs reaching the hypothalamus from other brain centers. The use of reserpine, atropine, *p*-chlorophenylalanine, morphine, and naloxone has excluded, respectively, the participation of adrenergic, serotoninergic, cholinergic, and opioid mediators. The final demonstration of the lack of participation of inputs transported from extrahypothalamic neurons in the control of the hypothalamic levels of the 5αR was obtained performing total hypothalamic deafferentations; the 5αR activity remained unchanged in the isolated hypothalamus (Celotti et al., 1983).

Some experiments directed toward analyzing the possible control of the enzymatic complex 5αR-3αHSD have been performed *in vitro* using cultures of mixed glial cells (Melcangi et al., 1992a). Glial cells exhibit on their surface a large variety of receptors, whose activation brings about the increase (e.g., β_1 adrenergic, adenosine, PGE_1 receptors) or the decrease (e.g., α_2 adrenergic and somatostatin receptors) of the formation of cyclic adenosine monophosphate (cAMP); other types of glial receptors (e.g., muscarinic, α_1 adrenergic, glutamate, histamine, serotonin, bradykinin receptors), use the phosphatidylinositol and/or the arachidonic acid intracellular signaling pathways (McCarthy et al., 1988; Hansson, 1989). On this basis, the approach selected was that of using molecules able to activate the PKC [(e.g., the phorbol ester, 12-O-tetradecanoyl-phorbol-13-acetate (TPA)] to mimic the effects of the activation of the phosphatidylinositol pathway, and a cAMP analog (8-Br-cAMP) to duplicate the effects of the activation of the adenylcyclase. The 5αR and the 3αHSD activities present in the cultures of the glial cells have been evaluated at different time intervals after treatment, using labeled T and DHT as specific substrates for the two enzymes. The results obtained indicate that, in glial cell cultures, the formation of DHT is not modified by the addition of phorbol esters, indicating that PKC is probably not involved in the intracellular signaling system controlling the enzyme 5αR in these cells. On the contrary, a statistically significant increase of the 5αR activity over control levels has been observed after incubation with 8-Br-cAMP (Melcangi et al., 1992a). The effect of the cAMP analog appears to be specific for the 5αR because the activity of the 3αHSD did not show any variation. As mentioned, there is abundant evidence indicating that the glia receives a lot of nervous inputs that use cAMP as the second messenger (McCarthy et al., 1988; Hansson, 1989). Among these, β_1-adrenergic and α_2-adrenergic are particularly relevant. On this basis, one may hypothesize that all inputs using the cAMP pathway might play a role in modulating the 5αR activity present in glial cells.

Another important aspect that has emerged is that growth factors originating in the astroglial cells may influence the activity of the enzymes 5αR and 3αHSD in neuronal cell populations. In this context, it is interesting to recall that transforming growth factor β_1 (TGFβ1) has been shown to be able to stimulate the formation of DHT in genital skin fibroblasts (Wahe et al., 1993), and that, in various cell systems, TGFβ1 expression is controlled by sex steroids (see, for review, Hering et al., 1995). We have observed that TGFβ1 is able to decrease the 5αR activity converting T into DHT in GT1-1 cells (a cell line derived from a hypothalamic LHRH-producing tumor, induced by genetically targeted tumorigenesis in a female transgenic mouse).

A similar phenomenon was observed when co-incubating GT1 cells with A1 and measuring the 5αR in the GT1 cells (Cavarretta *et al.*, 1999). This parallelism may suggest that TGFβ1 is one of the factors transferred from A1 to GT1-1 cells and blocking the activity of the 5αR. In the same experimental protocol, the formation of 5α-DH PROG was highly stimulated (rather than inhibited) by the presence of A1 in GT1-1; however, 5α-DH PROG formation remained unchanged after exposure of GT1-1 cells to TGFβ1 (Cavarretta *et al.*, 1999). These data suggest that the 5αR activity that transforms PROG and T in GT1 cells is two different enzymes; they also show that TGFβ1 is not involved in transferring the stimulatory effect exerted by the glial elements (see also Section III).

The possible androgenic control on the gene expression of the two isozymes has been analyzed *in vitro* on cultured hypothalamic neurons as well as *in vivo*, exposing the animals *in utero* to the androgen antagonist flutamide (see for review Melcangi *et al.*, 1998a). T treatment greatly induced the expression of the 5αR type 2 gene in cultured hypothalamic neurons, which normally do not express this isozyme. *In vivo* treatment with flutamide counteracted the expression of the type 2 gene occurring, at time of birth, in the whole brains of male neonates, whose genotype was determined by evaluating the expression of the male specific gene SRY (see for review Melcangi *et al.*, 1998a). When the same phenomenon was analyzed in the neonatal female brain, the effect of flutamide was not present, suggesting, for the first time, a sexual dimorphism of the 5αR system in the brain; these data also lead to the hypothesis that factors other than androgens might control the expression of the 5αR type 2 in the female brain. There was no effect of T or flutamide on the expression of the type 1 gene in the brain of either sex.

III. Effect of Glial–Neuronal Interactions on the Formation of Neuroactive Steroids

For many years, glial cells have been believed to provide only mechanical support to neurons. This concept is obviously no longer acceptable. The existence of important functional glia–neuron relationships has become apparent. The interaction of glial cells with neurons plays an important role in neuronal migration, neurite outgrowth, and axonal guidance during neural development (Rakic, 1971; Banker, 1980; Gasser and Hatten, 1990). Moreover, it has been demonstrated that glial cells synthesize, and possibly release, an array of bioactive agents, such as neurotransmitters, growth factors, prostaglandins, and neurosteroids (see for review LoPachin and Aschner, 1993), which are likely to exert specific influences on neuronal activity. In contrast, glial–neuronal interactions are certainly not one- way because neurons can interfere with the proliferation and maturation of glial elements. For example, neuronal activity has been shown to upregulate the expression of a specific glial marker, the GFAP (Steward *et al.*, 1991). Thus, the concept has emerged that neurons and glia form in the CNS a functional unit, in which each single element exerts effects on the others. As mentioned (see Section IIB), the enzyme $5\alpha R$ is present both in neurons (where it shows its higher concentrations) and in the glia (Melcangi *et al.*, 1993, 1994a), while the enzyme $3\alpha HSD$ is almost exclusively present in A1 (Melcangi *et al.*, 1993, 1994a). The peculiar cellular localization of these two enzymes in the brain has prompted us to investigate whether neurons and glial cells might interact, via humoral messages, to control these two enzymes.

Two approaches have been used to examine the interactions between A1 and neurons: (1) a co-culture system in which the two types of cells remain physically separated but that allows the free transfer of secretory products from one to the other type of co-cultured cells; and (2) the addition of conditioned medium (CM) of neurons to cultures of A1 and of CM of astrocytes to cultures of neurons. The data have consistently shown that co-culture with neurons, or exposure to the neuronal CM, stimulates the $5\alpha R$ and the $3\alpha HSD$ activities in A1. In contrast, there was no effect of astrocyte secretory products on the enzymatic activity of neurons (Melcangi *et al.*, 1994b). It appears, then, that one or more soluble substances secreted by neurons are able to stimulate the activities of the two enzymes in A1. Because the physiological role of the $5\alpha R$-$3\alpha HSD$ system in glial cells is largely unknown, it is difficult to speculate on the biological significance of a neuronal mechanism that stimulates the metabolism of steroid hormones in A1 cells. However, one should recall that the formation of DHT is a mechanism for potentiating androgenic actions, while the subsequent conversion to 3α-diol is generally considered to be a mechanism of steroid catabolism because 3α-diol is much less effective than T as an androgen, and does not bind to the AR (see for review Celotti *et al.*, 1992). Consequently, neuronal influences on A1 cells may increase (by acting on the $5\alpha R$), or decrease (by acting on the $3\alpha HSD$) the androgenic potential. However, the physiological meaning of this phenomenon remains to be established. Caution regarding this hypothesis may come from observations indicating that 3α-diol might interact with the $GABA_A$ receptor (Frye *et al.*, 1996a,b), and, consequently, might exert some thus far unknown anabolic effect.

The same protocol (co-culture system and transfer of CM of A1) has been used to analyze the possible effects of astrocytic secretions on GT1-1 cells, the neuronal cell line that synthesizes and secretes LHRH. We have found that the co-culture with A1, or the exposure to the CM of A1, strongly stimulates the release of LHRH from GT1-1 cells (Melcangi *et al.*, 1995).

TGFβ appears to be the factor secreted from astrocytes to release LHRH because the activity of CM in which A1 had been grown was fully antagonized by a TGFβ-neutralizing antibody. Moreover, direct exposure to TGFβ1, like the addition of CM of A1, stimulated LHRH release from GT1-1 cells (Melcangi *et al.*, 1995). In parallel experiments, it was shown that TGFβ1, as well as CM of A1, also modifies LHRH gene expression in GT1-1 cells (Galbiati *et al.*, 1996). These effects are not surprising, as we have also observed that the G+1−1 cells express the messengers coding for the two most important TGFβ receptors, TGFβRI and TGFβRII, and consequently may be considered a target for the action of members of the TGFβ family (Messi *et al.*, 1999). GT1-1 cells also appear to be steroid sensitive because they express ER, AR, PR, and GR (see for review Herbison, 1998) and possess 5αR and 3αHSD activities (Melcangi *et al.*, 1997a; Cavarretta *et al.*, 1999). As mentioned in Section IIC, we have observed that, in a co-culture system, A1 alter in opposite directions these enzymatic activities in GT1-1 cells, depending of the substrate used; a significant decrease of the formation of DHT vs a significant increase in the formation of 5α-DH PROG was observed (Melcangi *et al.*, 1997a; Cavarretta *et al.*, 1999). These observations are intriguing because they suggest that the substrate itself may influence 5αR activity. The differential effects on the formation of DHT and 5α-DH PROG cannot be explained on the basis of an effect on either one of the two isoforms of the 5αR so far cloned. In our opinion, these data may be better explained by postulating the existence of a third 5αR isoform, which has not yet been cloned. Several other observations made in our laboratory, both *in vitro* and *in vivo*, support such a hypothesis. For example, pluripotential CNS stem cells derived from mice striatum, when induced in culture to differentiate into glial cells, start to form DHT and 5α-DH PROG at different times (Melcangi *et al.*, 1996a). The formation of 5α-DH PROG peaks on Day 10, while that of DHT increases only after 14 days of differentiation. The CM of C6 glioma and of 1321N1 human astrocytoma cell lines are unable to modify the formation of DHT from T in A1, while inducing a statistically significant decrease in the formation of 5α-DH PROG from PROG in the same type of cultures (Melcangi *et al.*, 1998b). The exposure of C6 cells to CM of rat fetal neurons stimulates the formation of DHT but not that of 5α-DH PROG (Melcangi *et al.*, 1998b).

IV. Effects of Neuroactive Steroids on Glial Cells of the Central Nervous System

As mentioned, glial cells may be considered a target for the action of neuroactive steroids because they possess classical and nonclassical steroid receptors. Researchers in several laboratories have tried to analyze the effects of distinct hormonal steroids on different parameters of glial cells. In particular, a large body of evidence has been obtained on the effect that steroids and other principles exert on an important and specific marker of the astrocytes, the cytoskeletal protein named GFAP. Since the late-1990s, some data have shown that GFAP may be under the control of a wide variety of humoral principles that include, in addition to cytokines and growth factors, hormonal steroids (see for review Laping *et al.*, 1994; Garcia-Segura *et al.*, 1996). The major observations in the literature on the effect exerted by neuroactive steroids on GFAP are summarized in Table I. It appears that important influences of gonadal steroids on the expression and synthesis of GFAP are clearly evident; however, not one of the experiments thus far performed has addressed the question of whether steroidal molecules like PROG and T act directly on the astrocytes or whether the preliminary conversion into their active metabolites is requested for these effects to appear. A systematic study of the effects of sex steroids and of their metabolites (P, 5α-DH PROG, $3\alpha,5\alpha$-TH PROG, T, DHT and 3α-diol) on the gene expression of GFAP in astrocytic cultures was consequently performed in our laboratory (Melcangi *et al.*, 1996b). A1 cultures were exposed to the various steroids for 2, 6, and 24 h, and GFAP mRNA was measured by Northern blot analyses. A significant elevation in GFAP mRNA was observed after exposure to either PROG or 5α-DH PROG; the effect of 5α-DH PROG appeared more promptly (at 2 h) than that of PROG (at 6 h). This result suggests that the effect of PROG might be linked to its conversion into 5α-DH PROG; this hypothesis has been confirmed by showing that the addition of finasteride (a blocker of the 5αR) is able to abolish the effect of PROG. At late intervals, exposure to 5α-DH PROG caused a decrease rather than an increase in GFAP gene expression (Melcangi *et al.*, 1996b). This inhibitory effect appears similar to the one evoked by the direct exposure to $3\alpha,5\alpha$-TH PROG; consequently, $3\alpha,5\alpha$-TH PROG might be the metabolite derived from 5α-DH PROG, and it may be exerting an inhibitory effect on GFAP mRNA. This concept is supported by the observation that A1 are extremely rich in the enzyme 3αHSD, which converts 5α-DH PROG into $3\alpha,5\alpha$-TH PROG (Melcangi *et al.*, 1994a). Together, these observations seem to suggest that two metabolites of PROG, 5α-DH PROG and $3\alpha,5\alpha$-TH PROG, exert different effects on GFAP mRNA levels present in A1 and that differential intracellular signaling pathways may be involved. PROG and 5α-DH PROG may bind to their PR (Rupprecht *et al.*, 1993, 1996) which are present in the astrocytes (Jung-Testas *et al.*, 1992). $3\alpha,5\alpha$-TH PROG which does not bind to this receptor, might act via a GABAergic pathway. $3\alpha,5\alpha$-TH PROG is indeed a potent ligand for the $GABA_A$ receptor (Majewska *et al.*, 1986; Paul and Purdy, 1992; Puia *et al.*, 1990). On the basis of these observations, it is possible that the inhibitory effect exerted by $3\alpha,5\alpha$-TH PROG (and probably by 5α-DH PROG after its conversion into $3\alpha,5\alpha$-TH PROG) on GFAP expression might be mediated by the activation of the $GABA_A$ receptors. In contrast to what we had observed with PROG and its derivatives, neither T nor 3α-diol have appeared to be able to change GFAP expression at any time of exposure. DHT produced a significant decrease of GFAP mRNA only after 24 h (Mclcangi et al., 1996b). The efficacy of DHT might be explained by the affinity of DHT for the AR which is higher than that of T (see for review Celotti et al., 1992; Melcangi et al., 1999a).

As shown in Table I, in addition to sex steroids corticosteroids are also able to influence GFAP gene expression. In this case, we have evaluated whether these steroids act directly on the astrocytes or whether their preliminary conversion into their 5α-reduced metabolites is necessary (Melcangi et al., 1997b). In particular, we have considered corticosterone, DHC, DOC, 5α-DH DOC, and $3\alpha,5\alpha$-TH DOC. The data obtained have indicated that GFAP mRNA levels are increased by exposure to corticosterone for 6 and 24 h, while DHC was ineffective. The minerocorticoid DOC was also ineffective, while 5α-DH DOC strongly inhibited GFAP gene expression after 6 h of exposure (Melcangi et al., 1997b). $3\alpha,5\alpha$-TH DOC, was ineffective. It is interesting to note that the stimulatory effects of corticosterone, a steroid with mixed glucominerocorticoid activities, on GFAP expression are not duplicated by the pure minerocorticoid derivatives used. This fact suggests that the observed effect may be due to the gluco- rather than to the minerocorticoid properties of corticosterone. In this connection it is relevant point out that the synthesis of GFAP may represent an early response of the astrocytes and of the microglia to brain injury and/or lesioning (see for review Laping et al., 1994). Consequently, one might extrapolate from the present data that stress might influence GFAP synthesis also via the activation of the adrenal cortex.

Glucocorticoids are also able to influence other important functional parameters of the astrocytes. For instance, it has been demonstrated that glucocorticoids like methylprednisolone and dexamethasone enhance astrocytic calcium signaling, increasing both resting cytosolic calcium levels and the extent and the amplitude of Ca^{2+} wave propagation (Simard et al., 1999). This effect seems to be mediated via activation of the cytosolic GR, as the antagonist of this receptor RU 486 is able to inhibit the effects of methylprednisolone. Glucocorticoids are also able to influence the metabolism of glutamate in the astrocytes because glucocorticoids increase the activity of both the glutamine synthetase and glutamate dehydrogenase present in these kinds of cells [Hardin-Pouzet et al., 1996 (see for review Vardimon et al., 1999)]. On both enzymes, the control of the glucocorticoids occurs at transcriptional levels. In the case of glutamine synthetase, the regulatory

TABLE I
Effects of Steroids on GFAP

Steroid	Change	Tissue	Reference
Estradiol	↓ mRNA and protein levels	Hippocampus of castrated male rat	Day et al. (1993). Neuroscience **55**, 435–443.
Estradiol	↓ Number of immunoreactive cells	Hippocampus of brain-injured female rat	Garcia-Estrada et al. (1993). Brain Res. **628**, 271–278.
Estradiol	↑ Surface density of immunoreactive cells	Hippocampus of castrated female rat	Luquin et al. (1993). J. Neurobiol. **24**, 913–924.
Estradiol	↓ Immunoreactivity	Hippocampus of hypogonadal rat	McQueen (1994). J. Endocrinol. **143**, 411–415.
Estradiol	↑ Extension of immunoreactive cell processes	Hippocampal slice of castrated male rat	Del Cerro et al. (1995) Glia **14**, 65–71.
Estradiol (physiological $[K^+]_0$)	↑ Extension of immunoreactive cell processes	Hippocampal slice of castrated male rat	Del Cerro et al. (1996). Glia **18**, 293–305.
Estradiol	↑ mRNA levels	Rat cortical astrocytic cultures	Stone et al. (1998). Endocrinology **139**, 3202–3209.
Estradiol	↓ mRNA levels	Co-culture rat cortical astrocytes–neurons	Stone et al. (1998). Endocrinology **139**, 3202–3209.
T	↑ Immunoreactivity	Hippocampus of castrated male rat	Day et al. (1993). Neuroscience **55**, 435–443.
T	↓ Number of immunoreactive cells	Hippocampus of brain-injured male rat	Garcia-Estrada et al. (1993). Brain Res. **628**, 271–278.
T	↓ Immunoreactivity	Hippocampus of hypogonadal rat	McQueen, (1994). J. Endocrinol. **143**, 411–415.
T	↑ Extension of immunoreactive cell processes	Hippocampal slice of castrated male rat	Del Cerro et al. (1995). Glia **14**, 65–71.

T (high [K^+]0)	↓ Extension of immunoreactive cell processes	Hippocampal slice of castrated male rat	Del Cerro *et al.* (1996). *Glia* **18**, 293–305
T	↑ mRNA levels	Hypothalamus of female rat	Chowen *et al.* (1995). *Neuroscience* **69**, 519–532.
T	↑ mRNA levels	Hypothalamus of castrated male rat	Chowen *et al.* (1995). *Neuroscience* **69**, 519–532.
T	↑ Immunoreactivity	Interpeduncular nucleus of castrated male rat	Hajos *et al.* (1999). *Neuroreport* **10**, 2229–2233.
DHT	↑ Immunoreactivity	Hippocampus of castrated male rat	Day *et al.* (1993). *Neuroscience* **55**, 435–443.
PROG	↓ Number of immunoreactive cells	Hippocampus of brain-injured female rat	Garcia-Estrada *et al.* (1993). *Brain Res.* **628**, 271–278.
PROG	↑ Surface density of immunoreactive cells	Hippocampus of castrated female rat	Luquin *et al.* (1993). *J. Neurobiol.* **24**, 913–924.
Pregnenolone	↑ Extension of immunoreactive cell processes	Hippocampal slice of castrated male rat	Del Cerro *et al.* (1995). *Glia* **14**, 65–71.
Pregnenolone (high [K^+]0)	↓ Extension of immunoreactive cell processes	Hippocampal slice of castrated male rat	Del Cerro *et al.* (1996). *Glia* **18**, 293–305.
Pregnenolone	↓ Surface density of immunoreactive cell processes	Cortex, amygdala, and thalamus of aged male rat	Legrand and Alonso (1998). *Brain Res.* **802**, 125–133.
Corticosterone	↓ mRNA levels	Hippocampus of male rat	Nichols *et al.* (1990). *Mol. Brain Res.* **7**, 1–7.
Corticosterone	↑ mRNA and protein levels	Rat cortical astrocytic cultures	Rozovsky *et al.* (1995). *Endocrinology* **136**, 2066–2073.
Corticosterone	↓ mRNA and protein levels	Coculture rat cortical astrocytes–neurons	Rozovsky *et al.* (1995). *Endocrinology* **136**, 2066–2073.

regions of the gene contain a glucocorticoid response element capable of binding the GR and thus conferring responsiveness to the hormone, which can be evaluated by attaching a reporter gene (see for review Vardimon et al., 1999). In the case of glutamate dehydrogenase, it has been demonstrated that the responsive elements activated by hydrocortisone are localized in the $-557/+1$ region of the promoter (Hardin-Pouzet et al., 1996).

The effects of neuroactive steroids have been also evaluated in ammonia-induced astrocyte swelling in culture. It has been demonstrated by Bender and Norenberg (1998) that neuroactive steroids like, for instance, $3\alpha,5\alpha$-TH PROG and pregnenolone sulfate, at nanomolar concentrations diminished induced swelling. These observations might be very useful for new therapeutic approaches to acute hyperammonemic syndromes and other associated pathological conditions.

Little information is available on the direct effects of corticosteroids on the oligodendrocytes (i.e., on the cells producing the myelin). A dual influence of corticosteroids on the process of myelinization emerges from the observation that adrenalectomy, performed in the rat on the 11th day of life, increases the dry weights of the myelin measured on Day 63 (Meyer and Fairman, 1985); however, adrenalectomy, performed on the 14th day of life, appears to be able to reduce, rather than to increase, the amounts of myelin isolated from the whole brain on Days 21 and 22 (Preston and McMorris, 1984). Also, an important component of central myelin membranes, the MBP, is affected by corticoids. Treatment with dexamethasone, performed *in vivo* for 7 consecutive days in neonatal rats, is able to significantly decrease in the cerebrum (at the 20th and 30th days of life), and in the cerebellum (at the 10th day of life) the relative abundance of MBP mRNA (Tsuneishi et al., 1991). However, using a cell-free translation system programmed with synthetic messages, it has been observed that dexamethasone, cortisone, and hydrocortisone stimulate the translation of MBP mRNA (Verdi et al., 1989, Verdi and Campagnoni, 1990). Experiments performed in our laboratory have taken in consideration the effects of corticosterone, DHC, DOC, 5α-DH DOC, and $3\alpha,5\alpha$-TH DOC on the gene expression of MBP present in cultures of rat oligodendrocytes (Melcangi et al., 1997b). The data obtained have indicated that only 5α-DH DOC is effective and diminishes the mRNA levels of MBP. This is reminiscent of that observed in the case of GFAP (Melcangi et al., 1997b).

Some observations have also been performed analyzing the effect of gonadal steroids on some other parameters of oligodendrocyte. It has been demonstrated that both PROG and estradiol are able to increase the expression of MBP in oligodendrocytes present in primary cultures of glial cells (Jung-Testas et al., 1992, 1994). Moreover, using a monoclonal antibody, which is able to recognize an oligodendrocyte-specific cell surface antigen, it has been demonstrated that T is able to accelerate oligodendrocyte maturation in several brain areas (Kafitz *et al.*, 1992). Furthermore, *in vivo* treatment with T in juvenile male zebra finches is able to increase the degree of myelination in the forebrain and in the cerebellum (Kafitz *et al.*, 1992).

V. Formation of Neuroactive Steroids in the Peripheral Nervous System

The enzymatic complex $5\alpha R$-$3\alpha HSD$ is present not only in the CNS but also in the peripheral nervous system (PNS) (see for review Melcangi, Martini *et al.*, 1999a); there are, however, quantitative differences. For instance, it has been shown that, in the rat sciatic nerve, the formation of DHT is at least equal to that found in the central white matter, and higher than that found in the cerebral cortex. However, in contrast with what happens in the CNS, where 3α-diol is made in minute amounts (Melcangi *et al.*, 1993), in the sciatic nerve, this steroid is formed in amounts similar to those of DHT (Melcangi *et al.*, 1990b, 1992b). At variance with what it was observed in the CNS, in which the $5\alpha R$ activity is strongly associated with myelin membranes (Melcangi *et al.*, 1988a,b, 1989), analyses of the formation of the 5α-reduced metabolites of T in the myelin obtained from the sciatic nerve of adult male rats have indicated that the purification of the myelin is associated with a marked decrease in the $5\alpha R$ activity (Melcangi *et al.*, 1992b). These data do not appear to be consistent with a possible association of the enzyme with the myelin membranes of the PNS, even if caution appears to be necessary because of differences in the process of isolation of the myelin from the central and the peripheral structures.

The effects of aging on the activities of the enzymes $5\alpha R$ and of $3\alpha HSD$ present in the peripheral nerves have been evaluated. In particular, using, respectively, T and DHT as the substrates, we have quantitated these enzymatic activities in the sciatic nerve of aged (20-month-old) male rats and compared the values obtained with those found in adult (3-month-old) male rats (Melcangi *et al.*, 1990b, 1992b). The data obtained indicate that aging deeply impairs the 5α-reduction of T in the sciatic nerve. This observation, initially performed using tissue fragments (Melcangi *et al.*, 1990b), was confirmed in further experiments in which nerve homogenates and purified PNS myelin were used (Melcangi *et al.*, 1992b). On the contrary, aging seems not to affect the $3\alpha HSD$ activity of the PNS (Melcangi *et al.*, 1990b, 1992b).

The presence of $5\alpha R$ and $3\alpha HSD$ activities has been evaluated in cultures of Schwann cells obtained from the sciatic nerve of the neonatal rat

(Melcangi et al., 1998c). It has been observed that, using PROG as the substrate, Schwann cells (Melcangi et al., 1998c) possess a 5αR activity that is similar to that found in fetal central neurons (Melcangi et al., 1993, 1994a) and higher than that present in the oligodendrocytes (Melcangi et al., 1993, 1994a). On the contrary, Schwann cells possess a 3αHSD activity lower than that of the oligodendrocytes (Melcangi et al., 1993, 1994a). Guennoun et al. (1997) have shown that not only Schwann cells, but also peripheral sensory neurons, can metabolize PROG into 5α-DH PROG; however, unlike Schwann cells, dorsal root ganglia neurons seem not to be able to produce 3α,5α-TH PROG. Interestingly, the formation of 5α-DH PROG is higher in neurons grown in the presence of Schwann cells than in neurons grown alone (Guennoun et al., 1997).

VI. Effects of Neuroactive Steroids on the Peripheral Nervous System

A. Effects on Peripheral Myelin

Since the late 1990s, several observations obtained in our and others laboratories have indicated the importance of neuroactive steroids in controlling PNS physiology.

The effects of neuroactive steroids have been evaluated both in adult and in aged male rats. In particular, we have directed our attention to the proteins proper of the peripheral myelin (see for review, Melcangi et al., 2000a). Using *in situ* hybridization and/or Northern blot analyses, we have compared the levels of the messengers for the glycoprotein Po (Po) and the peripheral myelin protein 22 (PMP22) in the sciatic nerves of 3-month-old and of 22- to 24-month-old male rats. The results have indicated that the mRNA levels of these two myelin proteins are significantly decreased in the sciatic nerves of aged male rats (Melcangi et al., 1998c,d, 1999b). Subsequent experiments have indicated that, in the sciatic nerve of male rats, not only Po mRNA levels but also Po protein levels are decreased during the aging process (Melcangi et al., 1998d). These observations are in line with, and provide a possible explanation for, several observations that indicate that the morphological and functional aspects of peripheral myelin are profoundly modified during aging (see for review Melcangi et al., 2000a).

Another relevant aspect of aging is represented by progressive changes in the hormonal milieu, which include many alterations of the synthesis, metabolism, and transport of sex steroids (Finch et al., 1984; Melcangi et al., 1990b, 1992b; Wise et al., 1991; Wang et al., 1993). On this basis, and keeping in mind that the sciatic nerve possesses the messengers for AR and PR

(Magnaghi *et al.*, 1999), we have analyzed whether PROG and its physiological metabolites could counteract the drop of Po and PMP22 levels found in the sciatic nerve of aged animals (Melcangi *et al.*, 1998c,d, 1999b). To this purpose, we have treated old male rats with PROG, 5α-DH PROG, or 3α,5α-TH PROG. The data obtained have indicated that only 5α-DH PROG was able to significantly increase the levels of the Po messenger (Melcangi *et al.*, 1998c) and protein (Melcangi *et al.*, 1998d) in the sciatic nerve; PROG and 3α,5α-TH PROG were apparently inactive (Melcangi *et al.*, 1998c, 1999b). The fact that, in the *in vivo* experiments, PROG, at variance with 5α-DH PROG, was poorly active in inducing an increase of Po gene expression might be ascribed to the fact that, as previously mentioned, the process of 5αR in the whole sciatic nerve decreases in a significant way during aging (Melcangi *et al.*, 1990b, 1992b); consequently, in the aged animals used in this study, very little 5α-DH PROG could actually be formed from PROG. It is noteworthy that, in parallel experiments, PROG, 5α-DH PROG, or 3α,5α-TH PROG were totally ineffective in modifying the gene expression of PMP22 present in the sciatic nerve of aged male rats (Melcangi *et al.*, 1999b). The effect on Po in old animals seems to be a prerogative of progestagens, as androgens like T and its 5α-reduced metabolite DHT were unable to modify Po gene expression in the sciatic nerve of old animals (Melcangi *et al.*, 1998d).

As just mentioned, 5α-DH PROG is able to stimulate Po mRNA levels in the sciatic nerve of aged male rats. To evaluate whether such an effect was a peculiarity occurring in senescent animals or whether it was a more generalized phenomenon, we have analyzed the effect of PROG, 5α-DH PROG, and 3α,5α-TH PROG in normal adult male rats (Melcangi *et al.*, 1999b). The data obtained have indicated that *in vivo* treatment with PROG, 5α-DH PROG, or 3α,5α-TH PROG is effective in increasing Po gene expression in the sciatic nerve of adult animals; 5α-DH PROG was significantly more effective than the other two steroids. In a parallel experiment, in which we have evaluated the effects of the three steroids on the gene expression of PMP22 in the sciatic nerve of adult male rats (Melcangi *et al.*, 1999b), we found that the mRNA levels of this myelin protein were only slightly stimulated by PROG or 5α-DH PROG but were significantly increased by 3α,5α-TH PROG. More recent experiments performed in our laboratory have evaluated the effects of PROG and its metabolites (5α-DH PROG and 3α,5α-TH PROG) on the proteins of the peripheral myelin after nerve transection (Melcangi *et al.*, 2000b). In this context, it is important to remember that peripheral nerves are able to regenerate following damage; however, their functional recovery depends on several factors, including the type of injury and the distance over which axons must regrow (Sunderland, 1991). Several experimental procedures have been tried to improve the regeneration and the functional recovery of lesioned or transected nerves. For instance, modifying the environment close to the lesion has been attempted (see for review Melcangi et al., 2000b). Moreover, several morphological and physiological studies have demonstrated that transected peripheral nerves are better able to regenerate when a tubular support of silicone or of other synthetic materials is provided (the so-called entubulation repair) and the interstump gap does not exceed 10 mm (Archibald et al., 1995, Butì et al., 1996).

We have used the transection and entubulation technique to study whether the repair of the sciatic nerve of adult male rat might be facilitated by PROG, 5α-DH PROG, and 3α,5α-TH PROG. Using this technique, we have employed the effects on the gene expressions of Po and PMP22 as markers of the repair process. The data obtained have shown that both PROG and 5α-DH PROG are able to increase, in a statistically significant fashion, the low levels of messenger for Po present in the distal portion from the cut of the sciatic nerve. On the contrary, 3α,5α-TH PROG was ineffective (Melcangi et al., 2000b). The fact that PROG and 5α-DH PROG, in this experimental protocol, are able to stimulate Po mRNA levels is in agreement with the effects exerted by these steroids on the basal levels of Po present in the intact sciatic nerve of adult animals (Melcangi et al., 1998c,d, 1999b). The fact that, in this experimental paradigm, 3α,5α-TH PROG was not able to stimulate Po gene expression is rather surprising because, as mentioned before, this steroid was effective on the basal levels of Po present in the intact sciatic nerve of adult male rats (Melcangi et al., 1999b). We have mentioned that 3α,5α-TH PROG is able to directly interact with the $GABA_A$ receptor, or after its retro-conversion into 5α-DH PROG, with the PR; on the basis of these considerations, we are tempted to hypothesize that, in the experimental procedure applied, a possible decrease of the $GABA_A$ receptor and/or of the 3αHSD activity might be present in the transected sciatic nerve. Partial support for the hypothesis of the occurrence of a decrease of $GABA_A$ receptor after nerve transection comes from parallel experiments in which we have analyzed, in the distal portion from the cut of the sciatic nerve, the mRNA levels of PMP22. In agreement with what we had observed in intact adult male rats (Melcangi et al., 1999b), PROG and 5α-DH PROG have not been found able to stimulate the gene expression of this myelin protein. However, at variance with what we had previously observed in normal adult rats (Melcangi et al., 1999b), the treatment with 3α,5α-TH PROG was not able to increase the mRNA levels of PMP22 in the transected sciatic nerve. It is important to recall that the inefficacy of 3α,5α-TH PROG in stimulating Po and PMP22 mRNA levels in the transected sciatic nerve is reminiscent of what we have observed in aged male rats, in which a decrease of the $GABA_A$ receptor may occur (Melcangi et al., 1998c, 1999b). The positive effects we have obtained with PROG and 5α-DH PROG are reminiscent of the data of

Koenig et al. (1995), who have demonstrated that PROG and pregnenolone are able to counteract the decrease of the amounts of myelin membranes induced by a cryolesion in the sciatic nerve of the mouse. Obviously, our results appear to be more impressive because we have used a more severe challenge.

After having shown the effects of PROG, 5α-DH PROG, and 3α,5α-TH PROG in the control of Po mRNA levels, we have taken into consideration the possibility that other classes of sex steroids might exert similar effects in adult animals (Magnaghi et al., 1999). To this purpose, we have analyzed whether castration might influence the gene expression of Po in the PNS of adult male rats; we have measured the mRNA levels for this myelin protein, by Northern blot analyses, in the sciatic nerve of 6-month-old animals that had been castrated 3 months before; these levels were compared to those present in normal 6-month-old controls. The results obtained have indicated that orchidectomy clearly decreases (about 40%) Po mRNA levels in the sciatic nerve (Magnaghi et al., 1999). Subsequent experiments have been performed to investigate whether the systemic administration of T or DHT to castrated animals (during the last 32 days before sacrifice) might counteract the drop of the levels of Po induced, in the sciatic nerve, by orchidectomy. The results obtained indicate that the *in vivo* treatment with DHT, but not with T, is able to increase the levels of the messenger of this myelin protein (Magnaghi et al., 1999). The lack of effect of T in these experiments is not surprising because the activity of the 5αR, which converts T into DHT, is significantly decreased in the sciatic nerve following castration (Magnaghi et al., 1999). Consequently, it is possible that very low levels of DHT could be locally formed in castrated male rats treated with T. It is known that DHT is considered to be the active form of T.

B. Effects on Schwann Cells

The experiments performed *in vivo* in male rats could not identify where the effects of the various steroids were taking place because both the neuronal and the glial components are simultaneously present. We have turned to pure cultures of rat Schwann cells to analyze whether the effects of steroids on peripheral myelin proteins might be exerted directly on the glial compartment. For this purpose, we have first analyzed the effects of PROG, 5α-DH PROG, and 3α,5α-TH PROG in the control of the gene expressions of Po and PMP22 in cultures of rat Schwann cells. We have observed that within 2 h PROG and 5α-DH PROG induce a significant increase of Po gene expression; this effect decreases at later intervals (6 h) (Melcangi et al., 1998c). There was no effect on PMP22 gene expression after treatment with either steroid. However, $3\alpha,5\alpha$-TH PROG proved to be very effective in stimulating both Po (after 6 h exposure) and PMP22 (after 24 h of exposure) gene expressions (Melcangi *et al.*, 1998c, 1999b). The responses to steroids of cultured Schwann cells appeared to be identical to those observed in the sciatic nerve of adult animals subjected to *in vivo* treatments. The fact that PROG stimulates Po gene expression in Schwann cell cultures (Melcangi *et al.*, 1998c) has been confirmed by Désarnaud *et al.* (1998) using a different experimental set up (Schwann cells transiently transfected with a reporter construct in which the luciferase expression is controlled by the promoter region of the Po gene). Unfortunately, in the experiments of Désarnaud *et al.* (1998), the effects of 5α-DH PROG and $3\alpha,5\alpha$-TH PROG have not been evaluated. However, at variance with our observations, Désarnaud *et al.* (1998) have found that PROG is also able to stimulate the gene expression of PMP22, acting on promoter 1, but not on promoter 2, of the corresponding gene. Because, as mentioned previously, these experiments have been performed using Schwann cells transiently transfected with a reporter construct, this discrepancy is probably due to the different experimental models applied. In this connection, it is also relevant that experiments by Chan *et al.* (1998) have demonstrated that PROG is able to accelerate the time of initiation and to enhance the rate of myelin synthesis in a co-culture of dorsal root ganglia neurons and of Schwann cells; also, in this case, the effects of the 5α-reduced metabolites of PROG have not been evaluated.

It is clear from all of these results that different steroid molecules may modify the synthesis of these two important proteins of the peripheral myelin. We have proposed a hypothesis on the mechanisms of action of the various steroids on the gene expression of these two peripheral myelin proteins (Melcangi *et al.*, 2000b); this is schematically represented in Fig. 1. In the case of Po, it is possible that the effects of PROG and 5α-DH PROG might directly involve the PR, which, as we have demonstrated, is present both in the sciatic nerve and in Schwann cells (Magnaghi *et al.*, 1999). One might also postulate that PROG might act mainly following transformation into 5α-DH PROG, in view of the higher efficacy of 5α-DH PROG over PROG; in this connection, it is important to recall that the 5αR is present in the sciatic nerve and in Schwann cells in culture (Melcangi *et al.*, 1990b, 1992b, 1998c; Magnaghi *et al.*, 1999). Furthermore, because the activity of the 3αHSD is bidirectional, it is possible to assume that the efficacy of $3\alpha,5\alpha$-TH PROG, which is not able to directly bind to the PR, may result from a retro-conversion of this steroid into 5α-DH PROG, with subsequent binding to the PR (Rupprecht *et al.*, 1993, 1996). However, one cannot exclude a direct effect of this steroid via the interaction with the GABA$_A$ receptor, because, as we have mentioned, $3\alpha,5\alpha$-TH PROG has been reported

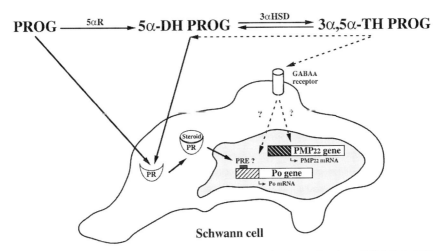

FIG. 1. Schematic representation of the possible mechanisms of action of PROG, 5α-DH PROG, and 3α,5α-TH PROG on the gene expression of Po and PMP22.

to interact actively with this receptor (see for review Celotti *et al.*, 1992; Melcangi *et al.*, 1999a) and, as described in Section I, we have shown that the messengers for several subunits ($\alpha 2$, $\beta 3$, $\beta 1$, $\beta 2$, and $\beta 3$) of the $GABA_A$ receptor are expressed both in the sciatic nerve and in Schwann cell cultures (Melcangi *et al.*, 1999b). An effect of PROG and 5α-DH PROG (and possibly of 3α,5α-TH PROG, after retro-conversion to 5α-DH PROG), through the PR is supported by the results of a computer analysis we have performed, which has permitted identification of some putative PROG-responsive elements (PRE) on the Po promoter (Magnaghi *et al.*, 1999).

As mentioned, the control of PMP22 seems to be different from that of Po. In fact, only 3α,5α-TH PROG proved to be effective in modulating PMP22 mRNA levels; a retro-conversion of 3α,5α-TH PROG to 5α-DH PROG cannot be postulated to explain the action of this steroid because 5α-DH PROG was ineffective on the expression of this protein (Melcangi *et al.*, 1999b). The possibility then remains that 3α,5α-TH PROG acts on PMP22 expression via an interaction with the $GABA_A$ receptor, even if the final proof of the interaction of 3α,5α-TH PROG with the subunits of the $GABA_A$ receptor that we have identified both in the sciatic nerve and in the Schwann cells is still lacking.

To explain the effects exerted by androgens, and in particular by DHT, on the gene expression of Po, we have been forced to hypothesize that the gene expression of this protein is stimulated by androgen-dependent mechanisms acting on Schwann cells in an indirect fashion. This is because

Schwann cells do not express the AR (Magnaghi et al., 1999). For example, androgens might act through the neuronal component, which does contain AR (Magnaghi et al., 1999). In this context, it is important to remember that the gene expression of Po is controlled by axonal signals; these trigger a wave of cAMP formation that modulates the synthesis of various transcription factors that eventually lead to the transcription and translation of the Po gene (LeBlanc et al., 1992).

This hypothesis, however, could be slightly modified on the basis of the results obtained *in vitro* using rat Schwann cell cultures. Also, in this case, the treatment with T or with 3α-diol was ineffective, while that with DHT was able to stimulate Po gene expression (Magnaghi et al., 1999; Melcangi et al., 2000b); this is obviously in agreement with our previous observations *in vivo* (Magnaghi et al., 1999). However, it is very difficult to explain this effect because, as mentioned, AR mRNA is not present in Schwann cells (Magnaghi et al., 1999). To explain the effects that DHT exerts on the gene expression of Po acting directly in Schwann cells, we have tested the hypothesis that DHT might be able to activate the gene expression of this myelin protein by acting through a steroid receptor other than the AR. Because 5α-DH PROG, a steroid that interacts with the PR, may activate Po gene expression (Melcangi et al., 1998c), we have postulated that DHT might interact with the PR, and activate a PRE. The data we have obtained indicate that, in a human neuroblastoma cell line co-transfected with the hPR_B and with a reporter plasmid containing a PRE in the context of the mouse mammary tumor virus promoter upstream of the luciferase gene, DHT is able to exert a transcriptional activity via the human PR (Magnaghi et al., 1999). Even if such a hypothesis deserves further experimental consideration, these observations suggest the possibility of an interesting cooperation and, possibly of a receptorial competition, between PROG and androgens on Po gene expression. We have further evaluated the actions of androgens on the synthesis of proteins of the peripheral myelin by analyzing the effect of T, DHT, or 3α-diol on the mRNA levels of PMP22 in Schwann cells. The data obtained indicate that, while T and DHT are ineffective at any time considered (e.g., 2, 6, and 24 h), 3α-diol treatment was able, after 6 h exposure, to significantly increase the gene expression of PMP22 (Melcangi et al., 2000b). These data differ from the results obtained evaluating the effects of AR on Po gene expression in cultures of Schwann cells. The fact that 3α-diol is the only derivative of T able to stimulate PMP22 gene expression appears to be rather interesting because of its similarity with the efficacy of $3\alpha,5\alpha$-TH PROG (Melcangi et al., 1999b), the other 3α-hydroxy-5α-reduced derivatives we have tested. It is important to recall that it has been proposed that 3α-diol, which does not bind to the AR, might interact with the $GABA_A$ receptor (Frye et al., 1996a,b); consequently, these observations further support the

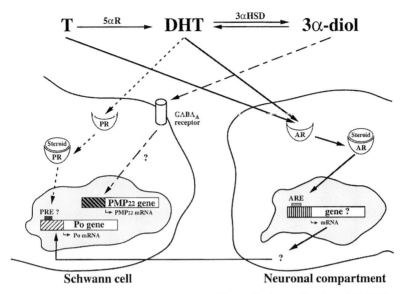

FIG. 2. Schematic representation of the possible mechanisms of action of T, DHT, and 5α-androstane-3α,17β-diol on the gene expression of Po and PMP22.

concept that the PMP22 mRNA levels are stimulated in Schwann cells via the $GABA_A$ receptor. The hypothesized mechanisms of control of Po and PMP22 by androgenic molecules are represented in Fig. 2.

Since the late-1990s, several studies have been performed in different laboratories to evaluate the possible effects of neuroactive steroids on different physiological parameters of Schwann cells. For instance, it has been shown that the synthetic glucocorticoid dexamethasone is able to strongly enhance in Schwann cell cultures the mitogenic activity exerted by axolemma-enriched fraction (Neuberger et al., 1994). In the same experimental model, sex steroids (e.g., PROG, T, 17β-estradiol), as well as aldosterone, do not exert any co-mitogenic action (Neuberger et al., 1994). At partial variance with these observations, it has been shown that estrogens are able to promote Schwann cell proliferation in culture in the presence of agents elevating intracellular cAMP (Jung-Testas et al., 1993). The effects of estrogen and PROG on the proliferation of Schwann cells have been also studied in cultures of segments of the rat sciatic nerve obtained from adult or newborn male and female rats (Svenningsen and Kanje, 1999). In these experiments, it has been observed that these two sex steroids are able to enhance [^{3}H]thymidine incorporation into the Schwann cells but that this effect is dependent on the sex and the age of the animals. In fact, estrogens are effective on Schwann cell proliferation in segments from adult male and newborn rats but do not exert any effect on segments from adult female rats; PROG, on the contrary, increases Schwann cell proliferation in segments obtained from adult female and newborn rats. The effects of estrogens and PROG were blocked by their respective receptor antagonists (Svenningsen and Kanje, 1999).

VII. Conclusion

Altogether, the data presented here clearly show the importance of the formation and of the effects of several neuroactive steroids in the glial component of the nervous system. In particular, some important points should be stressed: (1) the mechanism of action of the various steroidal molecules may involve both classical (PR and AR) and nonclassical steroid receptors ($GABA_A$ receptor); (2) in some instances, the action of hormonal steroids is not linked to their native molecular forms but to their metabolism into 5α- and $3\alpha,5\alpha$-reduced derivatives; (3) several neuroactive steroids exert important actions on the proteins proper of the peripheral myelin (e.g., Po and PMP22); and (4) the effects of steroids and of their metabolites on Po and PMP22 might have clinical significance in cases in which the rebuilding of the peripheral myelin is needed (e.g., aging, peripheral injury).

References

Andersson, S., Bishop, R. W., and Russell, D. W. (1989). Expression and regulation of steroid 5α-reductase, an enzyme essential for male sexual differentiation. *J. Biol. Chem.* **264,** 16249–16255.

Andersson, S., Berman, D. M., Jenkins, E. P., and Russell, D. W. (1991). Deletion of steroid 5α-reductase 2 gene in male pseudohermaphroditism. *Nature* **354,** 159–161.

Archibald, S. J., Shefner, J., Krarup, C., and Madison, R. D. (1995). Monkey median nerve repaired by nerve graft or collagen nerve guide tube. *J. Neurosci.* **15,** 4109–4123.

Banker, G. A. (1980). Trophic interactions between astroglial cells and hippocampal neurons in culture. *Science* **209,** 809–810.

Bender, A. S., and Norenberg, M. D. (1998). Effect of benzodiazepines and neurosteroids on ammonia-induced swelling in cultured astrocytes. *J. Neurosci. Res.* **54,** 673–680.

Bennett, M. J., Schlegel, B. P., Lez, J. M., Penning, T. M., and Lewis, M. (1996). Structure of 3α-hydroxysteroid/dihydrodiol dehydrogenase complexed with $NADP^+$. *Biochemistry* **33,** 10702–10711.

Bovolin, P., Santi, M. R., Puia, G., Costa, E., and Grayson, D. (1992). Expression patterns of γ-aminobutyric acid type a receptor subunit mRNAs in primary cultures of granule neurons and astrocytes from neonatal rat cerebella. *Proc. Natl. Acad. Sci. USA* **89,** 9344–9348.

Butì, M., Verdú, E., Labrador, R. O., Vilches, J. J., Fores, J., and Navarro, X. (1996). Influence of physical parameters of nerve chambers on peripheral nerve regeneration and reinnervation. *Exp. Neurol.* **137,** 26–33.

Cavarretta, I., Magnaghi, V., Ferraboschi, P., Martini, L., and Melcangi, R. C. (1999). Interactions between type 1 astrocytes and LHRH-secreting neurons (GT1-1 cells): Modification of steroid metabolism and possible role of TGFβ1. *J. Steroid Biochem. Mol. Biol.* **71,** 41–47.

Celotti, F., Negri-Cesi, P., Limonta, P., and Melcangi, C. (1983). Is the 5α-reductase of the hypothalamus and of the anterior pituitary neurally regulated? Effects of hypothalamic deafferentations and of centrally acting drugs. *J. Steroid Biochem.* **19,** 229–234.

Celotti, F., Melcangi, R. C., and Martini, L. (1992). The 5α-reductase in the brain: Molecular aspects and relation to brain function. *Front. Neuroendocrinol.* **13,** 163–215.

Chan, J. R., Phillips Ii, L. J., and Glaser, M. (1998). Glucocorticoids and progestins signal the initiation and enhance the rate of myelin formation. *Proc. Natl. Acad. Sci. USA* **95,** 10459–10464.

Cheng, K. C. (1992). Molecular cloning of rat liver 3α-hydroxysteroid dehydrogenase and identification of structure related proteins from rat lung and kidney. *J. Steroid Biochem. Molec. Biol.* **43,** 1083–1088.

Cheng, K. C., White, P. C., and Quin, K. N. (1991). Molecular cloning and expression of rat liver 3α-hydroxysteroid dehydrogenase. *Mol. Endocrinol.* **5,** 823–828.

Cheng, K. C, Lee, J., Khanna, M., and Quin, K. N. (1994). Distribution and ontogeny of 3α-hydroxysteroid dehydrogenase in the rat brain. *J. Steroid Biochem. Mol. Biol.* **50,** 85–89.

Degtiar, V. G., Loseva, B., and Isatchenkov, P. (1981). *In vitro* metabolism of androgens in hypothalamus and pituitary from infantile and adolescent rats of both sexes. *Endocrinol. Exp.* **15,** 181–190.

Désarnaud, F., Do Thi, A. N., Brown, A. M., Lemke, G., Suter, U., Baulieu, E.-E., and Schumacher, M. (1998). Progesterone stimulates the activity of the promoters of peripheral myelin protein-22 and protein zero genes in Schwann cells. *J. Neurochem.* **71,** 1765–1768.

Eicheler, W., Tuohimaa, P., Vilja, P., Adermann, K., Forssmann, W. G., and Aumüller, G. (1994). Immunocytochemical localization of human 5α-reductase 2 with polyclonal antibodies in androgen target and non-target human tissues. *J. Histochem. Cytochem.* **42,** 664–675.

Finch, C. E., Felicio, L. S., Mobbs, C. V., and Nelson, J. F. (1984). Ovarian and steroidal influences on neuroendocrine aging processes in female rodents. *Endocrine Rev.* **5,** 467–497.

Frye, C. A., Van Keuren, K. R., and Erkine, M. S. (1996a). Behavioral effects of 3alpha-androstanediol. I: Modulation of sexual receptivity and promotion of GABA-stimulated chloride flux. *Behav. Brain Res.* **79,** 109–118.

Frye, C. A., Duncan, J. E., Basham, M., and Erkine, M. S. (1996b). Behavioral effects of 3alpha-androstanediol. II: Hypothalamic and preoptic area actions via a GABAergic mechanism. *Behav. Brain Res.* **79,** 119–130.

Galbiati, M., Zanisi, M., Messi, E., Cavarretta, I., Martini, L., and Melcangi, R. C. (1996). Transforming growth factor-β and astrocytic conditioned medium influence luteinizing hormone-releasing hormone gene expression in the hypothalamic cell line GT1. *Endocrinology* **137,** 5605–5609.

Garcia-Segura, L. M., Chowen, J. A., and Naftolin, F. (1996). Endocrine glia: Roles of glial cells in the brain actions of steroid and thyroid hormone and in the regulation of hormone secretion. *Front. Neuroendocrinol.* **17,** 180–211.

Garcia-Segura, L. M., Wozniak, A., Azcoitia, I., Rodriguez, J. R., Hutchison, R. E., and Hutchison, J. B. (1999). Aromatase expression by astrocytes after brain injury: Implications for local estrogen formation in brain repair. *Neuroscience* **89,** 567–578.

Gasser, U. E., and Hatten, M. E. (1990). Neuron–glia interactions of rat hippocampal cells *in vitro*: Glial-guided neuronal migration and neuronal regulation of glial differentiation. *J. Neurosci.* **10,** 1276–1285.

Guennoun, R., Schumacher, M., Robert, F., Delespierre, B., Gouézou, M., Eychenne, B., Akwa, Y., Robel, P., and Baulieu, E. E. (1997). Neurosteroids: Expression of functional 3β-hydroxysteroid dehydrogenase by rat sensory neurons and Schwann cells. *Eur. J. Neurosci.* **9,** 2236–2247.

Hansson, E. (1989). Co-existence between receptors, carriers, and second messengers on astrocytes grown in primary cultures. *Neurochem. Res.* **14,** 811–819.

Hardin-Pouzet, H., Giraudon, P., Belin, M. F., and Didier-Bazes, M. (1996). Glucocorticoid upregulation of glutamate dehydrogenase gene expression *in vitro* in astrocytes. *Mol. Brain Res.* **37,** 324–328.

Herbison, A. E. (1998). Multimodal influence of estrogen upon gonadotropin-releasing hormone neurons. *Endocrine Rev.* **19,** 302–330.

Hering, S., Surig, D., Freystadt, D., Schatz, H., and Pfeiffer, A. (1995). Regulation of transforming growth factor β by sex steroids. *Horm. Metab. Res.* **27,** 345–351.

Hosli, E., Otten, U., and Hosli, L. (1997). Expression of GABA$_A$ receptors by reactive astrocytes in explant and primary cultures of rat CNS. *Int. J. Dev. Neuroscience* **15,** 949–960.

Jung-Testas, I., Renoir, J. M., Gasc, J. M., and Baulieu, E.-E. (1991). Estrogen-inducible progesterone receptor in primary cultures of rat glial cells. *Exp. Cell Res.* **193,** 12–19.

Jung-Testas, I., Renoir, J. M., Bugnard, H., Greene, G. L., and Baulieu, E.-E. (1992). Demonstration of steroid receptors and steroid action in primary cultures of rat glial cells. *J. Steroid Biochem. Mol. Biol.* **41,** 815–821.

Jung-Testas, I., Schumacher, M., Bugnard, H., and Baulieu, E.-E. (1993). Stimulation of rat Schwann cell proliferation by estradiol: Synergism between estrogen and cAMP. *Dev. Brain Res.* **72,** 282–290.

Jung-Testas, I., Schumacher, M., Robel, P., and Baulieu, E.-E. (1994). Actions of steroid hormones and growth factors on glial cells of the central and peripheral nervous system. *J. Steroid Biochem. Mol. Biol.* **48,** 145–154.

Jung-Testas, I., Schumacher, M., Robel, P., and Baulieu, E.-E. (1996). Demonstration of progesterone receptors in rat Schwann cells. *J. Steroid Biochem. Mol. Biol.* **58,** 77–82.

Kafitz, K. W., Herth, G., Bartsch, U., Guttinger, H. R., and Schachner, M. (1992). Application of testosterone accelerates oligodendrocyte maturation in brain of zebra finches. *Neuroreport* **3,** 315–318.

Karavolas, H. J., and Hodges, D. (1990). Neuroendocrine metabolism of progesterone and related progestins. *In* "Steroids and Neuronal Activity" (D. Chadwick and K. Widdows Eds.), pp. 22–55. Ciba Foundation Symposium 153, John Wiley & Sons, Chichester.

Koenig, H. L., Schumacher, M., Ferzas, B., Do Thi, A. N., Ressouches, A., Guennoun, R., Jung-Testas, I., Robel, P., Akwa, Y., and Baulieu, E.-E. (1995). Progesterone synthesis and myelin formation by Schwann cells. *Science* **268,** 1500–1503.

Krieger, N. R., Scott, R. G., and Jurman, M. E. (1983). Testosterone 5α-reductase in rat brain. *J. Neurochem.* **40,** 1460–1464.

Krieger, N. R., and Scott, R. G. (1984). 3α-hydroxysteroid dehydrogenase in rat brain. *J. Neurochem.* **42,** 887–890.

Labrie, F., Sugimoto, Y., Luu-The, V., Simard, J., Lachance, Y., Bachvarov, D., Leblanc, G., Durocher, F., and Paquet, N. (1992). Structure of human type 2 5α-reductase gene. *Endocrinology* **131,** 1571–1573.

Langub, M. C., and Watson, R. E., Jr. (1992). Estrogen receptor-immunoreactive glia, endothelia, and ependyma in guinea pig preoptic area and median eminence: Electron microscopy. *Endocrinology* **130,** 364–372.

Laping, N. J., Teter, B., Nichols, N. R., Rozovsky, I., and Finch, C. E. (1994). Glial fibrillary acidic protein: Regulation by hormones, cytokines, and growth factors. *Brain Pathol.* **1,** 259–275.

Leblanc, A. C., Windebank, A. J., and Poduslo, J. F. (1992). Po gene expression in Schwann cells is modulated by an increase of cAMP which is dependent on the presence of axons. *Mol. Brain Res.* **12,** 31–38.

Lephart, E. D. (1993). Brain 5α-reductase: cellular, enzymatic, and molecular perspectives and implications for biological function. *Mol. Cell Neurosci.* **4,** 473–484.

Levy, M. A., Brandt, M., Sheedy, K. M., Holt, D. A., Heaslip, J. I., Trill, J. J., Ryan, P. J., Morris, R. A., Garrison, L. M., and Bergsma, D. J. (1995). Cloning, expression and functional characterization of type 1 and type 2 steroid 5α-reductases from cynomolgus monkey: Comparison with human and rat isoenzymes. *J. Steroid Biochem. Mol. Biol.* **52,** 307–319.

LoPachin, R. M., Jr., and Aschner, M. (1993). Glial–neuronal interactions: Relevance to neurotoxic mechanisms. *Toxicol. Appl. Pharmacol.* **118,** 141–158.

Magnaghi, V., Cavarretta, I., Zucchi, I., Susani, L., Rupprecht, R., Hermann, B., Martini, L., and Melcangi, R. C. (1999). Po gene expression is modulated by androgens in the sciatic nerve of adult male rats. *Mol. Brain Res.* **70,** 36–44.

Majewska, M. D., Harrison, N. L., Schwartz, R. D., Barker, J. L., and Paul, S. M. (1986). Steroid hormone metabolites are barbiturate-like modulators of the GABA receptor. *Science* **232,** 1004–1007.

Massa, R., Justo, S., and Martini, L. (1975). Conversion of testosterone into 5α-reduced metabolites in the anterior pituitary and in the brain of maturing rats. *J. Steroid Biochem.* **6,** 567–571.

McCarthy, K. D., Salm, A., and Lerea, L. S. (1988). Astroglial receptors and their regulation of intermediate filament protein phosphorilation. *In* "Glial Cell Receptors" (H. K. Kimelberg Ed.), pp. 1–22. Raven Press, New York.

Meyer, J. S., and Fairman, K. R. (1985). Early adrenalectomy increases myelin content of the rat brain. *Dev. Brain Res.* **17,** 1–9.

Melcangi, R. C., Celotti, F., Ballabio, M., Poletti, A., Castano, P., and Martini, L. (1988a). Testosterone 5α-reductase activity in the rat brain is highly concentrated in white matter structures and in purified myelin sheaths of axons. *J. Steroid Biochem.* **31,** 173–179.

Melcangi, R. C., Celotti, F., Ballabio, M., Castano, P., Poletti, A., Milani, S., and Martini, L. (1988b). Ontogenetic development of the 5α-reductase in the rat brain: Cerebral cortex, hypothalamus, purified myelin and isolated oligodendrocytes. *Dev. Brain Res.* **44,** 181–188.

Melcangi, R. C., Celotti, F., Ballabio, M., Carnaghi, R., Poletti, A., and Martini, L. (1989). Effect of postnatal starvation on the 5α-reductase activity of the brain and of the isolated myelin membranes. *Exp. Clin. Endocrinol.* **94,** 253–261.

Melcangi, R. C., Celotti, F., Ballabio, M., Castano, P., Massarelli, R., Poletti, A., and Martini, L. (1990a). 5α-reductase activity in isolated and cultured neuronal and glial cells of the rat. *Brain Res.* **516,** 229–236.

Melcangi, R. C., Celotti, F., Ballabio, M., Poletti, A., and Martini, L. (1990b). Testosterone metabolism in peripheral nerves: Presence of the 5α-reductase-3α-hydroxysteroid-dehydrogenase enzymatic system in the sciatic nerve of adult and aged rats. *J. Steroid Biochem.* **35,** 145–148.

Melcangi, R. C., Celotti, F., Castano, P., and Martini, L. (1992a). Intracellular signalling systems controlling the 5α-reductase in glial cell cultures. *Brain Res.* **585,** 411–415.

Melcangi, R. C., Celotti, F., Castano, P., and Martini, L. (1992b). Is the 5α-reductase-3α-hydroxysteroid dehydrogenase complex associated with the myelin in the peripheral nervous system of young and old male rats? *Endocrine Reg.* **26,** 119–125.

Melcangi, R. C., Celotti, F., Castano, P., and Martini, L. (1993). Differential localization of the 5α-reductase and the 3α-hydroxysteroid dehydrogenase in neuronal and glial cultures. *Endocrinology* **132,** 1252–1259.

Melcangi, R. C., Celotti, F., and Martini, L. (1994a). Progesterone 5α-reduction in neurons, astrocytes and oligodendrocytes. *Brain Res.* **639,** 202–206.

Melcangi, R. C., Celotti, F., and Martini, L. (1994b). Neurons influence the metabolism of testosterone in cultured astrocytes via humoral signals. *Endocrine* **2,** 709–713.

Melcangi, R. C., Galbiati, M., Messi, E., Piva, F., Martini, L., and Motta, M. (1995). Type 1 astrocytes influence luteinizing hormone-releasing hormone release from the hypothalamic cell line GT1-1: Is transforming growth factor-β the principle involved? *Endocrinology* **136,** 679–686.

Melcangi, R. C., Froelichsthal, P., Martini, L., and Vescovi, A. L. (1996a). Steroid metabolizing enzymes in pluripotential progenitor CNS cells: Effect of differentiation and maturation. *Neuroscience* **72,** 467–475.

Melcangi, R. C., Riva, M. A., Fumagalli, F., Magnaghi, V., Racagni, G., and Martini, L. (1996b). Effect of progesterone, testosterone and their 5α-reduced metabolites on GFAP gene expression in type 1 astrocytes. *Brain Res.* **711,** 10–15.

Melcangi, R. C., Galbiati, M., Messi, E., Magnaghi, V., Cavarretta, I., Riva, M. A., and Zanisi, M. (1997a). Astrocyte–neuron interactions *in vitro*: Role of growth factors and steroids on LHRH dynamics. *Brain Res. Bull.* **44,** 465–469.

Melcangi, R. C., Magnaghi, V., Cavarretta, I., Riva, M. A., and Martini, L. (1997b). Corticosteroid effects on gene expression of myelin basic protein in oligodendrocytes and of glial fibrillary acidic protein in type 1 astrocytes. *J. Neuroendocrinol.* **9,** 729–733.

Melcangi, R. C., Poletti, A., Cavarretta, I., Celotti, F., Colciago, A., Magnaghi, V., Motta, M., Negri-Cesi, P., and Martini, L. (1998a). The 5α-reductase in the central nervous system: Expression and modes of control. *J. Steroid Biochem. Mol. Brain Res.* **65,** 295–299.

Melcangi, R. C., Cavarretta, I., Magnaghi, V., Ballabio, M., Martini, L., and Motta, M. (1998b). Crosstalk between normal and tumoral brain cells. Effect on sex steroid metabolism. *Endocrine* **8,** 65–71.

Melcangi, R. C., Magnaghi, V., Cavarretta, I., Martini, L., and Piva, F. (1998c). Age-induced decrease of glycoprotein Po and myelin basic protein gene expression in the rat sciatic nerve. Repair by steroid derivatives. *Neuroscience* **85,** 569–578.

Melcangi, R. C., Magnaghi, V., Cavarretta, I., Riva, M. A., Piva, F., and Martini, L. (1998d). Effects of steroid hormones on gene expression of glial markers in the central and peripheral nervous system: Variations induced by aging. *Exp. Gerontol.* **33,** 827–836.

Melcangi, R. C., Magnaghi, V., and Martini, L. (1999a). Steroid metabolism and effects in central and peripheral glial cells. *J. Neurobiol.* **40,** 471–483.

Melcangi, R. C., Magnaghi, V., Cavarretta, I., Zucchi, I., Bovolin, P., D'Urso, D., and Martini, L. (1999b). Progesterone derivatives are able to influence peripheral myelin protein 22 and Po gene expression: Possible mechanisms of action. *J. Neurosci. Res.* **56,** 349–357.

Melcangi, R. C., Magnaghi, V., and Martini, L. (2000a). Aging in peripheral nerves: Regulation of myelin protein genes by steroid hormones. *Prog. Neurobiol.* **60,** 291–308.

Melcangi, R. C., Magnaghi, V., Galbiati, M., Ghelarducci, B., Sebastiani, L., and Martini, L. (2000b). The action of steroid hormones on peripheral myelin proteins: A possible new tool for the rebuilding of myelin? *J. Neurocytol.* **23,** 327–339.

Messi, E., Galbiati, M., Magnaghi, V., Zucchi, I., Martini, L., and Melcangi, R. C. (1999). Transforming growth factor β2 is able to modify mRNA levels and release of luteinizing hormone-releasing hormone in a immortalized hypothalamic cell line (GT1-1). *Neuroscience Lett.* **270,** 165–168.

Motta, M., Zoppi, S., Brodie, A. M., and Martini, L. (1986). Effect of 1,4,6-androstatriene-3,17-dione (ATD), 4-hydroxy-4-androstene-3,17-dione (4-OH-A) and 4-acetoxy-4-androstene-3,17-dione (4-Ac-A) on the 5α-reduction of androgens in the rat prostate. *J. Steroid Biochem.* **25,** 593–600.

Negri-Cesi, P., Melcangi, R. C., Celotti, F., and Martini, L. (1992). Aromatase activity in cultured brain cells: Difference between neurons and glia. *Brain Res.* **589,** 327–332.

Neuberger, T. J., Kalimi, O., Regelson, W., Kalimi, M., and De Vries, G. H. (1994). Glucocorticoids enhance the potency of Schwann cell mitogens. *J. Neurosci. Res.* **38,** 300–313.

Normington, K., and Russell, D. W. (1992). Tissue distribution and kinetic characteristics of rat steroid 5α-reductase isozymes. Evidence for distinct physiological functions. *J. Biol. Chem.* **267,** 19548–19554.

Norton, W. T., and Poduslo, S. E. (1973). Myelination in the rat brain: Changes in myelin composition during brain maturation. *J. Neurochem.* **21,** 759–773.

Paul, S. M., and Purdy, R. H. (1992). Neuroactive steroids. *FASEB J.* **6,** 2311–2322.

Pawlovski, J. E, Huizinga, M., and Penning, T. M. (1991). Cloning and sequencing of the cDNA for rat liver 3α-hydroxysteroid/dihydrodiol dehydrogenase. *J. Biol. Chem.* **266,** 8820–8825.

Penning, T. M. (1996). 3α-hydroxysteroid dehydrogenase: Three dimensional structure and gene regulation. *J. Endocrinol.* **150,** 175–187.

Penning, T., Sharp, R. B., and Krieger, N. R. (1985). Purification properties of 3α-hydroxysteroid dehydrogenase from rat brain cytosol: Inhibition by nonsteroidal anti-inflammatory drugs and progestins. *J. Biol. Chem.* **260,** 15266–15272.

Penning, T., Pawlowski, J. E., Schlegel, B. P, Jez, J.M., Lin, H.-K., Hoog, S. S., Bennett, M. J., and Lewis, M. (1996). Mammalian 3α-hydroxysteroid dehydrogenases. *Steroids* **61,** 508–523.

Preston, S. L., and McMorris, F. A. (1984). Adrenalectomy of rats results in hypomyelination of the central nervous system. *J. Neurochem.* **42,** 262–267.

Puia, G., Santi, M., Vicini, S., Pritchett, D. B., Purdy, R. H., Paul, S. M., Seeburg, P. H., and Costa, E. (1990). Neurosteroids act on recombinant human $GABA_A$ receptors. *Neuron* **4,** 759–765.

Rakic, P. (1971). Neuron–glia relationship during granule cell migration in developing cerebellar cortex: A golgi and electron microscopic study in macacus rhesus. *J. Comp. Neurol.* **141,** 282–312.

Roselli, C. E., Stadelman, H., Horton, L. E., and Resko, J. A. (1987). Regulation of androgen metabolism and luteinizing hormone-releasing hormone content in discrete hypothalamic and limbic areas of male rhesus macaques. *Endocrinology* **120,** 97–106.

Rupprecht, R., Hauser, C. A. E., Trapp, T., and Holsboer, F. (1996). Neurosteroids: Molecular mechanisms of action and psychopharmacological significance. *J. Steroid Biochem. Mol. Biol.* **56,** 163–168.

Rupprecht, R., Reul, J. M. H. M., Trapp, T., Van Steensel, B., Wetzel, C., Damm, K., Zieglgansberger, W., and Holsboer, F. (1993). Progesterone receptor-mediated effects of neuroactive steroids. *Neuron* **11,** 523–530.

Russell, D. W., and Wilson, J. D. (1994). Steroid 5α-reductase: Two genes/two enzymes. *Ann. Rev. Biochem.* **63,** 25–61.

Santagati, S., Melcangi, R. C., Celotti, F., Martini, L., and Maggi, A. (1994). Estrogen receptor is expressed in different types of glial cells in culture. *J. Neurochem.* **63,** 2058–2064.

Sholl, S. A., Goy, R. W., and Kim, K. . (1989). 5α-reductase, aromatase, and androgen receptor levels in the monkey brain during fetal development. *Endocrinology* **124,** 627–634.

Simard, M., Couldwell, W. T., Zhang, W., Song, H., Liu, S., Cotrina, M. L., Goldman, S., and Nedergaard, M. (1999). Glucocorticoids—Potent modulators of astrocytic calcium signaling. *Glia* **28,** 1–12.

Steward, O., Torre, E. R., Tomasulo, R., and Lothman, E. (1991). Neuronal activity up-regulates astroglial gene expression. *Proc. Natl. Acad. Sci. USA* **88,** 6819–6823.

Sunderland, S. (1991). A classification of nerve injury. *In* "Nerve Injuries and Their Repair. A Critical Appraisal" (S. Sunderland Ed.), pp. 221–232. Churchill, Edinburgh.

Svenningsen, A. F., and Kanje, M. (1999). Estrogen and progesterone stimulate Schwann cell proliferation in a sex- and age-dependent manner. *J. Neurosci. Res.* **57,** 124–130.

Thigpen, A. E., and Russell, D. W. (1992). Four-amino acid segment in steroid 5α-reductase 1 confers sensitivity to finasteride, a competitive inhibitor. *J. Biol. Chem.* **267,** 8577–8583.

Thigpen, A. E., Silver, R. I., Guileyardo, J. M., Casey, M. L., McConnel, J. D., and Russell, D. W. (1993). Tissue distribution and ontogeny of steroid 5α-reductase isozyme expression. *J. Clin. Invest.* **92,** 903–910.

Tsuneishi, S., Takada, S., Motoike, T., Ohashi, T., Sano, K., and Nakamura, H. (1991). Effects of dexamethasone on the expression of myelin basic protein, proteolipid protein, and glial fibrillary acidic protein genes in developing rat brain. *Dev. Brain Res.* **61,** 117–123.

Vardimon, L., Ben-Dror, I., Avisar, N., Oren, A., and Shiftan, L. (1999). Glucocorticoid control of glial gene expression. *J. Neurobiol.* **40,** 513–527.

Verdi, J. M., and Campagnoni, A. T. (1990). Translational regulation by steroids. *J. Biol. Chem.* **265,** 20314–20320.

Verdi, J. M., Kampf, K., and Campagnoni, A. T. (1989). Translational regulation of myelin protein synthesis by steroids. *J. Neurochem.* **52,** 321–324.

Vielkind, U., Walencewicz, A., Levine, J. M., and Bohn, M. C. (1990). Type II glucocorticoid receptors are expressed in oligodendrocytes and astrocytes. *J. Neurosci. Res.* **27,** 360–373.

Wahe, M., Antonipillai, I., and Horton, R. (1993). Effects of transforming growth factor β and epidermal growth factor on steroid 5α-reductase activity in genital skin fibroblasts. *Mol. Cell Endocrinol.* **98,** 55–59.

Wang, C., Leung, A., and Sinha-Hikim, A. P. (1993). Reproductive aging in the male brown-norway rat/ a model for human. *Endocrinology* **133,** 2773–2781.

Wise, P. M., Scarbrough, K., Larson, G. H., Lloyd, J. M., Weiland, N. G., and Chiu, S. (1991). Neuroendocrine influences on aging of the female reproductive system. *Front. Neuroendocrinol.* **12,** 323–356.

Wolff, J. E., Laterra, J., and Goldstein, G. W. (1992). Steroid inhibition of neural microvessel morphogenesis *in vitro*: Receptor mediation and astroglial dependence. *J. Neurochem.* **58,** 1023–1032.

Zoppi, S., Lechuga, M., and Motta, M. (1992). Selective inhibition of the 5α-reductase of the rat epididymis. *J. Steroid Biochem. Mol. Biol.* **42,** 509–514.

NEUROSTEROID MODULATION OF RECOMBINANT AND SYNAPTIC GABA$_A$ RECEPTORS

Jeremy J. Lambert, Sarah C. Harney, Delia Belelli, and John A. Peters

Department of Pharmacology and Neuroscience, Ninewells Hospital and Medical School
Dundee University, Dundee, DD1 9SY Scotland

I. Introduction
II. Transmitter-Gated Ion Channels and Neurosteroid Selectivity
   A. Glycine Receptors
   B. Neuronal Nicotinic Receptors
   C. 5-Hydroxytryptamine$_3$ (5-HT$_3$) Receptors
   D. Ionotropic Glutamate Receptors
III. Influence of GABA$_A$-Receptor Subunit Composition on Neurosteroid Action
   A. α Subunits
   B. β Subunits
   C. γ Subunits
   D. δ and ε Subunits
IV. Mechanism of Neurosteroid Modulation of GABA$_A$ Receptors
V. Neurosteroid Modulation of Inhibitory Synaptic Transmission
   A. Phosphorylation Influences Neurosteroid Effects on Synaptic Transmission
VI. Structure–Activity Relationships for Steroids at the GABA$_A$ Receptor
   A. Steroids with Increased Oral Bioavailability
   B. Water-Soluble Steroids
VII. Multiple Steroid Binding Sites on the GABA$_A$ Receptor
VIII. Concluding Remarks
   References

Certain pregnane steroids are now established as potent, positive allosteric modulators of the γ-aminobutyric acid type A (GABA$_A$) receptor. These compounds are known to be synthesized in the periphery by endocrine glands, such as the ovaries and the adrenal glands, and can rapidly cross the blood–brain barrier. Therefore, such steroids could act as endogenous modulators of the major inhibitory receptor in the mammalian central nervous system. However, the demonstration that certain neurons and glia can synthesize the pregnane steroids (i.e., neurosteroids) additionally suggests that they may serve a paracrine role by influencing GABA$_A$-receptor function through their local release in the brain itself. Here, we demonstrate that these neurosteroids are highly selective and extremely potent modulators of the GABA$_A$ receptor. The subunit composition of the GABA$_A$ receptor may influence the actions of the neurosteroids, particularly when considering concentrations of these agents thought to occur physiologically, which may underlie their reported differential effects at certain inhibitory synapses. However, recent work suggests that the phosphorylation status of either the synaptic $GABA_A$ receptor or its associated proteins may also influence neurosteroid sensitivity; these findings are discussed. Upon administration, the neurosteroids exhibit clear behavioral effects, including sedation, anticonvulsant actions, and behaviors predictive of anxiolysis; when given at high doses, they induce general anesthesia. Numerous synthetic steroids have been synthesized in an attempt to therapeutically exploit these properties, and these data are reviewed in this chapter. However, targeting the brain enzymes that synthesize and metabolize the neurosteroids may offer a new approach to exploit this novel endocrine–paracrine neurotransmitter interaction. © 2001 Academic Press.

I. Introduction

In 1941, Hans Selye first described the rapid sedative and anesthetic effects of certain pregnane steroids. However, it was not until approximately 40 years had passed that a possible molecular mechanism emerged. Harrison and Simmonds (1984) demonstrated that the synthetic steroidal anesthetic alphaxalone (3α-hydroxy-5α-pregnane-11,20-dione) selectively enhanced the interaction of γ-aminobutyric acid (GABA) with the $GABA_A$ receptor. Given that $GABA_A$ receptors mediate much of the "fast" inhibitory synaptic transmission in the central nervous system (CNS) this observation provided a logical mechanism to explain the rapid central depressant effects of these steroids.

The $GABA_A$ receptor is an important therapeutic target for a number of structurally diverse compounds, including a variety of general anesthetic agents (e.g., isoflurane, thiopentone, propofol, etomidate) and the benzodiazepines (Sieghart, 1995; Belelli et al., 1999) (Fig. 1). In common with these agents, the pregnane steroids exhibit anxiolytic, anticonvulsant, analgesic, sedative, and, at relatively high doses, anesthetic actions (Lambert et al., 1995; Gasior et al., 1999). The $GABA_A$ receptor is a member of the transmitter-gated ion channel family, formed from five subunits drawn from a repertoire that includes: α_{1-6}, β_{1-3}, γ_{1-3}, δ, ε, π, and θ (Barnard et al., 1998; Barnard, 2001). Importantly, these subunits have a distinct distribution throughout the CNS (Pirker et al., 2000), and the receptor subunit composition influences both the physiological and the pharmacological properties of the receptor (Barnard et al., 1998). Furthermore, experiments using transgenic mice engineered to express benzodiazepine-insensitive $GABA_A$ receptors have demonstrated that some of the behavioral characteristics

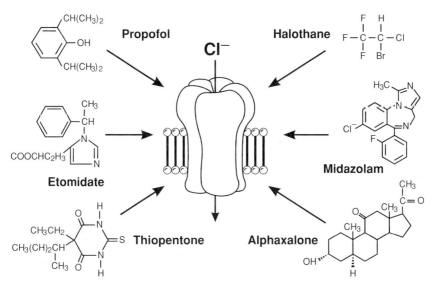

FIG. 1. A diagrammatic representation of the GABA$_A$ receptor, illustrating some of the varied structures that are known to act as positive allosteric modulators of this receptor.

of this class of compounds reside with distinct receptor isoforms (e.g., the sedative and anxiolytic actions being mediated by α_1- and α_2-containing receptors, respectively) (Rudolph *et al.*, 1999; Crestani *et al.*, 2000; McKernan *et al.*, 2000; Sieghart, 2000).

The initial electrophysiological experiments with alphaxalone (Harrison and Simmonds 1984; Cottrell *et al.*, 1987) were soon extended to a number of endogeneous pregnane steroids, and some of these, including 3α-hydroxy-5α-pregnan-20-one ($3\alpha,5\alpha$-TH PROG), 3α-hydroxy-5β-pregnan-20-one ($3\alpha,5\beta$-TH PROG), and $3\alpha,21$-dihydroxy-5α-pregnan-20-one ($3\alpha,5\alpha$-TH DOC), were found to be active at the GABA$_A$ receptor at concentrations as low as 1–3 nM (Lambert *et al.*, 1995) (Fig. 2). Such concentrations are well within the physiological range suggesting that these compounds may act as endogeneous modulators of the GABA$_A$ receptor. These GABA-active steroids are produced by peripheral endocrine glands, such as the adrenals and ovaries (Robel *et al.*, 1999; Poletti *et al.*, 1999). However, the demonstration that certain glial and neuronal cells within the CNS itself can both synthesize and metabolize such compounds raises the possibility that the activity of the major inhibitory circuitry in the brain could be "fine-tuned" by these locally produced steroids.

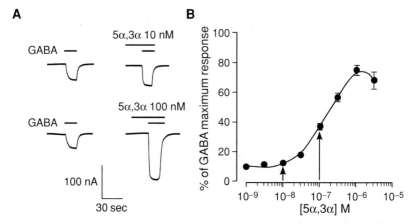

FIG. 2. Allopregnanolone potently enhances GABA-activated chloride currents. (**A**) Traces representing GABA-evoked currents recorded from an oocyte expressing $\alpha_1\beta_2\gamma_2$ GABA$_A$ receptors and their enhancement by 10 nM and 100 nM $3\alpha,5\alpha$-TH PROG ($5\alpha,3\alpha$). (**B**) Relationship between the concentration of $5\alpha,3\alpha$ (logarithmic scale) and the GABA-evoked current expressed as a percentage of the maximum response to GABA. The arrows indicate the possible "physiological range of steroid concentrations."

II. Transmitter-Gated Ion Channels and Neurosteroid Selectivity

The behavioral profile of steroids such as $3\alpha,5\alpha$-TH PROG is similar to those of other compounds identified as positive allosteric modulators of the GABA$_A$ receptor. Therefore, the question arises as to whether the behavioral actions of such steroids are mediated exclusively through this receptor. Here, we consider neurosteroid selectivity across additional members of the transmitter-gated ion channel family. The discussion is mainly restricted to pregnane steroids known to potently enhance GABA$_A$-receptor function.

A. GLYCINE RECEPTORS

In the brain stem and spinal cord, glycine is an important neurotransmitter; it produces neuronal depression by activating anion-selective, strychnine-sensitive glycine receptors. The glycine receptor is composed of five transmembrane-crossing subunits drawn from a palette of a β subunit and α_{1-4} subunits (Betz *et al.*, 2001). These subunits exhibit considerable sequence homology and a common (predicted) membrane topology to those of the $GABA_A$ receptor. Furthermore, certain general anesthetics and central depressants enhance function at both $GABA_A$ and glycine receptors (Pistis et al., 1997; Belelli et al., 1999; Thompson and Wafford, 2001). However, such similarities do not extend to neurosteroid modulation, as even micromolar concentrations of pregnane steroids such as alphaxalone and $3\alpha,5\alpha$-TH PROG are inactive at both native and recombinant glycine receptors (Harrison and Simmonds, 1984; Barker et al., 1987; Wu et al., 1990; Pistis et al., 1997) (see also Table I). Not all pregnane steroids are inactive at the glycine receptor, as we find that the water-soluble steroid minaxolone (2β-ethoxy-3α-hydroxy-11α-dimethylamino-5α-pregnan-20-one) produces a large potentiation of glycine-evoked currents recorded from oocytes preinjected with rat spinal cord mRNA. However, this effect requires concentrations of the anesthetic some 20-fold greater than those required to produce an equivalent effect at the $GABA_A$ receptor (see Table I).

TABLE I
SELECTIVITY OF ACTION OF ALPHAXALONE AND MINAXALONE

Receptor	Alphaxalone	Minaxolone
$GABA_A$	2.2 ± 0.3 μM	0.5 ± 0.1 μM
($\alpha_1\beta_2\gamma_2$: EC_{50})	($78 \pm 3\%$)	($93 \pm 5\%$)
Glycine	60 μM	11 ± 1 μM
($\alpha_1\beta$: EC_{50})	No effect	($89 \pm 4\%$)
AMPA	60 μM	100 μM
(IC_{50})	No effect	No effect
NMDA	30 μM	30 μM
(IC_{50})	No effect	No effect
Nicotinic	5 ± 1 μM	19 ± 3 μM
($\alpha_4\beta_2$: IC_{50})		
Nicotinic	13 ± 2 μM	11 ± 1 μM
(α_7 : IC_{50})		
5-HT_3	~50 μM	8 ± 1 μM
(h5-HT_{3A} : IC_{50})		

All experiments were performed on oocytes voltage-clamped at –60 mV. The sources of receptor were $GABA_A$, human $\alpha_1\beta_2\gamma_2$; glycine, human α_1 rat β for alphaxalone and rat spinal cord mRNA for minaxalone; kainate and N-methyl-D-aspartate, rat cerebellar mRNA; nicotinic, rat $\alpha_4\beta_2$ and chick α_7 5-HT_3 human 5-HT_{3A}. All experiments on $GABA_A$ and glycine receptors used the EC_{10} concentration of the natural agonist. For the other receptors, the appropriate EC_{50} was used. For $GABA_A$ and glycine receptors, the steroid EC_{50} and the maximal potentiation produced (expressed as a percentage of the maximum response to GABA or glycine) are given in parentheses. For kainate, NMDA, nicotinic and 5-HT_3 receptors, the IC_{50} values are given, where appropriate.

B. NEURONAL NICOTINIC RECEPTORS

Neuronal nicotinic receptors are composed of α and β subunits (α_{2-9}, β_{2-4}), which can combine to form hetero-oligomeric and homo-oligomeric (α_{7-9}) nicotinic receptors that exhibit distinct physiological and pharmacological properties (Albuquerque et al., 1997). Neuronal nicotinic receptors have been implicated in analgesia, anxiety, memory acquisition, synaptic plasticity, and neuronal excitotoxicity (Albuquerque et al., 1997), and certain receptor isoforms are extremely sensitive to some general anesthetics (Evers and Steinbach, 1997). However, a comparison of the actions of pregnane steroids at $GABA_A$ and neuronal nicotinic receptors reveals the latter to be relatively insensitive (Buisson and Bertrand, 1999; Paradiso et al., 2000). Hence, early studies demonstrated that high micromolar concentrations of alphaxalone were required to inhibit chromaffin cell nicotinic receptors, whereas nanomolar concentrations of this anesthetic are active at $GABA_A$ receptors (Cottrell et al., 1987). Similarly, relatively high concentrations of alphaxalone and minaxolone are required to inhibit nicotine-induced currents mediated by neuronal (α_7 homomeric or $\alpha_4\beta_2$ heteromeric receptors; see Table I). In addition, the orientation of the hydroxyl group at the 3-position is known to be critically important for both the behavioral and the $GABA_A$ receptor effects of the anesthetic pregnane steroids (3α-hydroxy active; 3β-hydroxy inactive; see Section VI). However, the behaviorally inert 3β-ol diastereomer of alphaxalone, betaxalone (3β-hydroxy-5α-pregnane-11,20-dione), is equieffective at neuronal nicotinic receptors (Cottrell et al., 1987) (see Table I). The relative insensitivity of nicotinic receptors to steroids coupled with poor correlation between the structure-activity relationships for behavior and nicotinic receptor inhibition suggest that these proteins are unlikely to constitute the major locus for mediating the behavioral actions of the pregnane steroids.

C. 5-HYDROXYTRYPTAMINE$_3$ (5-HT$_3$) RECEPTORS

5-HT$_3$ receptors are composed of five subunits, with each subunit thought to adopt the characteristic four-transmembrane topology that is common to $GABA_A$-glycine-, and nicotinic-receptor subunits (Davies et al., 1999). To date, only two distinct receptor subunits (5-HT$_{3A}$ and 5-HT$_{3B}$) have been isolated and, upon expression, they can form either homomeric (5-HT$_{3A}$) or heteromeric (5-HT$_{3A}$ + 5-HT$_{3B}$) receptors (Davies et al., 1999). The 5-HT$_{3B}$ subunit was isolated only recently and hence the majority of pharmacological studies have been performed using either 5-HT$_3$ receptors that are native to certain cell lines and neurons, or 5-HT$_{3A}$ homo-oligomeric receptors expressed in either cell lines or *Xenopus laevis* oocytes. The 5-HT$_{3A}$ subunit has been isolated from different species (e.g., human, mouse, rat, ferret) and, importantly, these species homologues exhibit distinct pharmacological properties that result from small differences in the primary amino acid sequence of these subunits (e.g., Hope *et al.*, 1999). Behavioral studies with selective 5-HT$_3$-receptor antagonists suggest that the receptor may be implicated in anxiety, cognition, and addictive behaviors, but to date the clinical use of such antagonists is restricted to the prevention of emesis and nausea induced by cytotoxic drugs, radiation, or by general anesthetics.

Human recombinant 5-HT$_{3A}$ homomeric receptors are relatively insensitive to alphaxalone, with inhibition of 5-HT-evoked currents occurring only with high micromolar concentrations of the anesthetic (Table II). Furthermore, the behaviorally inactive diastereomer betaxalone was equieffective in this respect (see Table II). Similarly, $3\alpha,5\alpha$-TH PROG is reported to be a relatively weak antagonist of the 5-HT$_3$ receptor (Rupprecht and Holsboer, 1999). Minaxalone is more effective than alphaxalone as a 5-HT$_3$-receptor

TABLE II
INFLUENCE OF SUBUNIT COMPOSITION OF GABA$_A$ RCEPTOR ON
GABA-MODULATORY EFFECTS OF $3\alpha,5\alpha$-TH PROG

Human recombinant receptor combination	EC$_{50}$	E_{max}
$\alpha_1\beta_1$	380 ± 10 nM	143 ± 2%
$\alpha_1\beta_1\gamma_1$	559 ± 22 nM	62 ± 8%
$\alpha_1\beta_1\gamma_{2L}$	89 ± 6 nM	69 ± 4%
$\alpha_1\beta_1\gamma_3$	294 ± 36 nM	74 ± 5%
$\alpha_1\beta_2\gamma_{2L}$	177 ± 2 nM	75 ± 4%
$\alpha_1\beta_3\gamma_{2L}$	195 ± 36 nM	72 ± 4%
$\alpha_2\beta_1\gamma_{2L}$	146 ± 11 nM	66 ± 6%
$\alpha_3\beta_1\gamma_{2L}$	74 ± 1 nM	67 ± 7%
$\alpha_4\beta_1\gamma_{2L}$	317 ± 25 nM	72 ± 6%
$\alpha_5\beta_1\gamma_{2L}$	302 ± 38 nM	81 ± 2%
$\alpha_6\beta_1\gamma_{2L}$	220 ± 12 nM	131 ± 6%
$\alpha_6\beta_2\gamma_{2L}$	350 ± 29 nM	108 ± 5%
$\alpha_6\beta_3\gamma_{2L}$	264 ± 33 nM	90 ± 9%
$\alpha_1\beta_1\varepsilon$	N.D.	15 ± 2%

All parameters are calculated from steroid concentration–effect relationships obtained from oocytes expressing human recombinant GABA$_A$ receptors. The EC$_{50}$ is defined as the concentration of steroid that produces an enhancement of the GABA(EC$_{10}$)-evoked current to 50% of the maximum potentiation produced by that steroid. The E_{max} is the maximum potentiation of the GABA (EC$_{10}$)-evoked current produced by the steroid expressed as a percentage of the GABA maximum.

antagonist, although the active concentration range for this effect is still an order of magnitude greater than that required for modulation of the GABA$_A$ receptor (see Table II). The incorporation of the 5-HT$_{3B}$ subunit to form a hetero-oligomeric receptor (5-HT$_{3A}$ + 5-HT$_{3B}$) has a considerable effect on certain biophysical properties of the receptor (e.g., increasing the single-channel conductance) but, to date, has not been demonstrated to greatly influence the antagonist pharmacology of this receptor (Davies et al., 1999). Nevertheless, it would be of interest to reinvestigate the actions of the neurosteroids on heteromeric 5-HT$_3$ receptors. Finally, it should be noted that some steroids (e.g., 17β-estradiol, estrone) are extremely potent (nanomolar), agonist-dependent inhibitors of murine 5-HT$_3$ receptors (Steele and Martin, 1999).

D. Ionotropic Glutamate Receptors

Glutamate is the major "fast" excitatory neurotransmitter in the mammalian CNS. These effects are mediated by glutamate-activating cation selective ion channels that have been broadly classified into three main subtypes based on their selectivity for the agonists N-methyl-D-aspartate (NMDA), DL-α-amino-3-hydroxy-5-methyl-4-isopropionic acid (AMPA), and kainate (Dingledine et al., 1999). Glutamate receptors are multisubunit proteins, although there is still debate as to whether neuronal receptors are composed of four or five subunits. For each of the receptor classes (NMDA, AMPA, and kainate), subtypes exist as a consequence of the combination of distinct subunit isoforms, which exhibit distinct physiological and pharmacological properties and a heterogeneous distribution throughout the mammalian CNS (Dingledine et al., 1999). Initially, glutamate-receptor subunits were thought to have a membrane topology similar to that of nicotinic, glycine, and GABA$_A$ receptors. However, it is now apparent that ionotropic glutamate receptor subunits have a distinctive topology from that of the cysteine-loop receptors and constitute a separate family (Dingledine et al., 1999). We have utilized the Xenopus laevis oocyte expression system (injected with rat cerebellar mRNA) to investigate the actions of alphaxalone and minaxalone on expressed glutamate receptors (see Table I). Using this paradigm, we found that even micromolar concentrations of these anesthetic steroids had no effect on kainate-evoked currents. These findings were in agreement with our earlier observations demonstrating kainate-evoked currents recorded from rat hippocampal neurons to be insensitive to alphaxalone (Lambert et al., 1990). Similarly, micromolar concentrations of alphaxalone or minaxolone had no effect on NMDA-evoked currents recorded from oocytes (see Table I) and hippocampal neurons (Lambert et al., 1990).

In general, pregnane steroids that act to enhance $GABA_A$-receptor-mediated responses at nanomolar concentrations have been found to have little or no effect on ionotropic glutamate receptors, even at micromolar concentrations (Gibbs *et al.*, 1999). Finally, at concentrations generally greater than those required for $GABA_A$- and glycine-receptor antagonism, pregnenolone sulfate acts to enhance NMDA-mediated responses with no effect on kainate and AMPA receptors (Gibbs *et al.*, 1999).

III. Influence of $GABA_A$-Receptor Subunit Composition on Neurosteroid Action

Radioligand binding and chloride flux studies performed in various brain regions have indicated that neuroactive steroids could discriminate between different $GABA_A$-receptor isoforms (Gee *et al.*, 1988; Prince and Simmonds, 1993; Olsen and Sapp, 1995). More recently, whole-cell clamp electrophysiological studies have demonstrated conclusively that neurosteroids such as $3\alpha,5\alpha$-TH PROG act differentially at synaptic $GABA_A$ receptors in different brain regions. However, whether this heterogeneity is the result of the expression of distinct $GABA_A$-receptor isoforms or it is caused by other factors, such as phosphorylation or local steroid metabolism (Pinna *et al.*, 2000), is not clear. Furthermore, studies investigating the dependence of neurosteroid action on the subunit composition of the $GABA_A$ receptor have not provided an unequivocal picture (e.g., Lambert *et al.*, 1995). For clarity, this review focuses on electrophysiological experiments.

A. α Subunits

The isoform of the α subunit within the heteromeric $GABA_A$ receptor has a major effect on both the binding and the function of the benzodiazepine class of compounds (Luddens *et al.*, 1995; Smith and Olsen 1995; Sieghart, 2000). By contrast, the influence of the α isoform on the neurosteroid pharmacology is more modest (see for review Lambert *et al.*, 1995). Furthermore, the presence of an α subunit is not a prerequisite for neurosteroid sensitivity, as $3\alpha,5\alpha$-TH PROG and alphaxalone are active at recombinant receptors composed of only β_1- and γ_2-receptor subunits (Maitra and Reynolds, 1999). We have used the *Xenopus laevis* oocyte expression system to determine the influence of the α isoform on the potency (EC_{50}) and maximal (E_{max}) GABA-modulatory effects of $3\alpha,5\alpha$-TH PROG (see Table II). Essentially in agreement with previous studies, these experiments illustrate that the maximal GABA-modulatory effects of the neurosteroid acting at $\alpha_{1-5}\beta_1\gamma_2$ receptors is not influenced by the isoform of the α subunit (an ~six- to sevenfold increase of the current induced by an EC_{10} concentration of GABA (see Table II), although, for the $\alpha_6\beta_1\gamma_2$ receptor, the steroid somehow increases the GABA response by ~12-fold (i.e., above the apparent GABA maximum). Evaluation of the EC_{50} reveals only a three- to fourfold difference across the α isoforms (74–317 nM, see Table II).

Although the effect of the α isoform on the steroid EC_{50} is modest, the α_4 and α_5 containing receptors are significantly less sensitive to low concentrations (10–100 nM) of $3\alpha,5\alpha$-TH PROG compared with receptors incorporating α_1, α_2, α_3, or α_6 subunits. This reduced sensitivity may be of physiological significance, as levels of this neurosteroid in plasma normally fluctuate between 3 and 10 nM but rise to 30–60 nM following mild stress and may reach 100 nM just before parturition (Paul and Purdy, 1992). The relative insensitivity of the α_4-containing receptor is of particular interest given the increased expression of this subunit that is reported to occur in the hippocampus on progesterone withdrawal (Smith et al., 1998a,b). This effect on $GABA_A$-receptor expression appears to be caused by withdrawal of the progesterone metabolite $3\alpha,5\alpha$-TH PROG, rather than by falling levels of progesterone *per se* (Smith et al., 1998a,b). Hippocampal neurons isolated from such animals exhibit physiological and pharmacological properties consistent with those reported for recombinant $GABA_A$ receptors containing the α_4 subunit. Hence, GABA-evoked currents recorded from such neurons are relatively brief in duration, are insensitive to lorazepam, and are characteristically enhanced by benzodiazepine antagonists and inverse agonists (Wafford et al., 1996; Smith et al., 1998a,b). In addition, the hippocampal $GABA_A$ receptors of these progesterone-withdrawn animals are insensitive to physiological (10 nM) levels of $3\alpha,5\alpha$-TH PROG, a feature that would be consistent with the reduced effect of low concentrations of this steroid acting at recombinant receptors expressing α_4 subunits (see Table II).

B. β Subunits

The anesthetic etomidate and the anticonvulsant loreclezole preferentially modulate β_2- and β_3- over β_1- containing receptors (Belelli et al., 1999). However, neuroactive steroids such as alphaxalone, $3\alpha,5\alpha$-TH PROG, and $3\alpha,5\alpha$-TH DOC do not differentiate among the β-subunit isoforms when expressed in hetero-oligomeric receptors ($\alpha_1\beta_x\gamma_2$; where $x = 1, 2,$ or 3) (see Hadingham et al., 1993; Sanna et al., 1997; see also Table II)].

C. γ SUBUNITS

The presence of a γ subunit is essential for benzodiazepines to act as high-affinity allosteric modulators of the $GABA_A$ receptor, with the isoform (1–3) of the γ subunit greatly influencing the benzodiazepine pharmacology of the receptor (Luddens et al., 1995). By contrast, a γ subunit is not required for steroid modulation of the $GABA_A$ receptor (Puia et al., 1990; Shingai et al., 1991). Indeed, comparison of the action of 3α,5α-TH PROG at recombinant $α_1β_1$ and $α_1β_1γ_2$ $GABA_A$ receptors reveals that omission of the γ subunit increases the maximal potentiation produced by the steroid, although the steroid EC_{50} was ∼ fourfold greater for the $α_1β_1$ receptor (see Table II). The isoform of the γ subunit had no effect on the maximal potentiation produced by the steroid, although, in comparison to $γ_2$ containing receptors (i.e., $α_1β_1γ_{2L}$), the EC_{50} for 3α,5α-TH PROG was ∼ 3.3- and 6.3- fold greater for $γ_3$- and $γ_1$- containing receptors respectively (see Table II) (c.f. Maitra and Reynolds, 1999). Hence, neurons expressing $γ_1$ subunits might be expected to be less sensitive to "physiological" concentrations of the neurosteroids. Interestingly, in the hypothalamus, the allosteric actions of certain anabolic steroids (which appear to act at a distinct site from the pregnane steroids) differentiate between $γ_1$- (medial preoptic area) and $γ_2$- (ventromedial nucleus) expressing neurons. The relatively high concentration of 1 μM 3α,5α-TH PROG is active at synaptic $GABA_A$ receptors of both neuronal types (Jorge-Rivera et al., 2000). Clearly, it would be of interest to investigate the actions of lower concentrations of this steroid in this brain region (see Section V).

D. δ AND ε SUBUNITS

$GABA_A$ receptors containing the δ subunit are benzodiazepine insensitive and also have been reported to be insensitive to the GABA-enhancing actions of the pregnane steroids (Zhu et al., 1996). Indeed, for cerebellar granule cells, the loss of steroid sensitivity that occurs with development has been attributed to the increased expression of this subunit (Zhu et al., 1996). However, a study that compared the pharmacological properties of cell lines engineered to express $α_4β_3δ$, or $α_4β_3γ_2$ $GABA_A$ receptors demonstrated both alphaxalone and 3α,5α-TH PROG to produce a much greater enhancement of GABA-evoked currents mediated by the former receptor (Brown et al., 2001). In apparent agreement with this study, in δ knockout mice, the anesthetic actions of 3α,5α-TH PROG and alphaxalone are reduced, whereas those of other $GABA_A$-receptor modulators, such as propofol, etomidate, and pentobarbitone, are not (Mihalek et al., 1999).

Similarly, in such mice, the anxiolytic effect of the synthetic pregnane steroid ganaxolone (3α-hydroxy, 3β-methyl-5α-pregnan-20-one) are abolished (Mihalek *et al.*, 1999). A number of laboratories have found expression of the δ subunit to be problematic (Brown *et al.*, 2001). However, it will be important to clarify whether the δ subunit suppresses pregnane steroid modulation (Zhu *et al.*, 1996) or does not (Brown *et al.*, 2001).

Similar to the δ subunit, the incorporation of the ε subunit into α- and β-subunit-containing receptors produces a $GABA_A$ receptor that is benzodiazepine insensitive (Davies *et al.*, 1997; Whiting *et al.*, 1997). The influence of the ε subunit on neurosteroid modulation is controversial, as the incorporation of this subunit, together with α and β subunits, has been reported to produce both a steroid-insensitive and -sensitive receptor (Davies *et al.*, 1997; Whiting *et al.*, 1997). We find that 3α,5α-TH PROG has no GABA-modulatory effect on the $\alpha_1\beta_1\varepsilon$ receptor isoform when expressed in oocytes (see Table II). However, quantifying the effects of steroids is complicated, as $GABA_A$ receptors that contain an ε subunit exhibit spontaneous chloride channel openings in the absence of GABA (Neelands *et al.*, 1999). Indeed, we find that 3α,5α-TH PROG greatly enhances the background chloride current mediated by constitutively active $\alpha_1\beta_1\varepsilon$ receptors, demonstrating that, although devoid of GABA-modulatory actions, 3α,5α-TH PROG can bind to this receptor isoform.

IV. Mechanism of Neurosteroid Modulation of $GABA_A$ Receptors

Experiments that investigated the influence of alphaxalone on the GABA-induced increase of membrane current noise in mouse spinal neurons suggested that this anesthetic acted to enhance $GABA_A$-receptor function by principally prolonging the mean open time of the GABA-activated chloride ion channel (Barker *et al.*, 1987). Single-channel experiments arrived at the same conclusion for the related neurosteroids 3α,5α-TH PROG or 3α,5β-TH PROG and additionally confirmed that these depressant steroids had no effect on the GABA-gated single-channel conductance (Callachan *et al.*, 1987; Lambert *et al.*, 1987). These studies also revealed that, at concentrations in excess of those required for GABA modulation, these depressant steroids had a second action, to directly activate the $GABA_A$ receptor (Callachan *et al.*, 1987; Lambert *et al.*, 1987).

Subsequently, a detailed quantitative kinetic analysis of the GABA-modulatory actions of neuroactive steroids was performed on mouse spinal neurons grown in cell culture. Using subsaturating concentrations of GABA, three kinetically distinct open states of the GABA-gated ion channel were revealed of brief, intermediate, and long duration (MacDonald *et al.*, 1989b; MacDonald and Olsen, 1994). Under these conditions, the neuroactive steroids enhanced the actions of GABA by promoting the occurrence of open states of intermediate and long duration, with a concomitant reduction in the appearance of openings of brief duration (Twyman and MacDonald, 1992; MacDonald and Olsen, 1994). This pertubation of channel kinetics is similar to that produced by anesthetic barbiturates (MacDonald *et al.*, 1989a), although the steroids also increase the frequency of single-channel openings (Twyman and MacDonald, 1992). Whether this latter effect is the result of an increased probability of GABA gating the chloride channel or whether it is a consequence of the GABA-mimetic effect of the steroids is not known.

The aforementioned studies were performed using relatively low concentrations of GABA. Evidence is now emerging that the excitability of some neurons may be influenced in part by a tonic background of GABA; hence, these studies may be of physiological relevance (Brickley *et al.*, 1996). However, when considering the effects of steroids on synaptic transmission, it appears that, at least for some synapses, the postsynaptic $GABA_A$ receptors are briefly exposed to relatively high concentrations of GABA (Mody *et al.*, 1994; Edwards, 1995). Therefore, an investigation of the influence of neuroactive steroids on the currents induced by rapidly applied high concentrations of GABA may be more pertinent in revealing how the steroid-induced pertubation of channel kinetics influences inhibitory synaptic transmission. The rapid and brief application of saturating concentrations of GABA to nucleated membrane patches excised from cerebellar granule cells induces currents that decay biphasically (Zhu and Vicini, 1997). Similarly, the decay of some miniature inhibitory postsynaptic currents (mIPSCs), which result from the activation of synaptically located $GABA_A$ receptors by a single vesicle of GABA, also exhibit a bi-exponential decay. For both exogenous and synaptic GABA, the fast-time constant is thought to result from channels oscillating between bound open and closed conformations, with the slow component reflecting receptors entering and exiting various desensitized states (Jones and Westbrook, 1995). The neuroactive steroid $3\alpha,5\alpha$-TH DOC prolongs the slow time constant of decay of GABA-evoked currents recorded from nucleated patches (Zhu and Vicini, 1997). It is postulated that this steroid slows the recovery of $GABA_A$ receptors from desensitization and as receptors exiting desensitization, may reconduct, this action would effectively prolong the GABA-evoked current. In support of this mechanism, $3\alpha,5\alpha$-TH DOC, in the presence of a saturating concentration of GABA, increased the probability of the channel being in the open state by augmenting the number of late channel openings (Zhu and Vicini, 1997). The prolongation of GABAergic synaptic currents by such steroids may result from this mechanism.

V. Neurosteroid Modulation of Inhibitory Synaptic Transmission

In rat hippocampal neurons, steroidal positive allosteric modulators of the $GABA_A$ receptor, such as $3\alpha,5\alpha$-TH PROG, act to prolong the decay of evoked IPSCs but have little effect on IPSC amplitude or rise time (Harrison et al., 1987a). Both evoked and spontaneous IPSCs may result from the asynchronous release of GABA from multiple release sites, complicating the interpretation of the actions of the neurosteroids on synaptic transmission (Mody et al., 1994; Williams et al., 1998). Tetrodotoxin blocks voltage-activated sodium channels and as a consequence will inhibit presynaptic action potential discharge. Inhibitory synaptic events recorded in the presence of this toxin are generally thought to result from the vesicular release of a single quanta of GABA [although some cerebellar *Purkinje* neuron mIPSCs may be multiquantal (Llano et al., 2000)], which activates a relatively small number of $GABA_A$ receptors at a single synapse, giving rise to the mIPSC (see Mody et al., 1994). Hence, in studies using *in vitro* brain slices, analysis has concentrated on the effects of depressant steroids on mIPSCs or on spontaneously occurring IPSCs that appear to be insensitive to tetrodotoxin (i.e., are action potential independent and therefore are probably monoquantal), rather than interpreting their effects on multiquantal IPSCs. These studies demonstrate that steroids such as $3\alpha,5\beta$-TH PROG or $3\alpha,5\alpha$-TH PROG and $3\alpha,5\alpha$-TH DOC prolong the mIPSC decay time recorded from neurons in the hippocampus (CA1 pyramidal and dentate granule cell neurons), cerebellum (*Purkinje* neurons), and hypothalamus (oxytocin-containing neurons of the supraoptic nucleus), with little or no effect on mIPSC amplitude or rise time (Brussaard et al., 1997, 1999; Cooper et al., 1999; Harney et al., 1999, 2000; Brussaard and Herbison, 2000; Fancsik et al., 2000). By contrast, $3\alpha,5\alpha$-TH PROG increases the amplitude of sIPSCs recorded from neurons located in the ventromedial nucleus and the medial preoptic area of the hypothalamus (Jorge-Rivera et al., 2000).

Interestingly, GABAergic mIPSCs in different brain regions do not appear to be uniformally sensitive to the neurosteroids with low nanomolar concentrations of steroids, such as $3\alpha,5\beta$-TH PROG or $3\alpha,5\alpha$-TH DOC, prolonging the mIPSC decay of rat hippocampal CA1 and cerebellar *Purkinje* neurons (Cooper et al., 1999; Harney et al., 1999; Harney, 2000), whereas micromolar concentrations of such steroids are required to similarly influence hypothalamic mIPSCs (Brussaard et al., 1997, 1999). The explanation for these apparent differences could, of course, be trivial (e.g., a differential access of the neurosteroid to the synapse). However, in the rat hippocampus (20 days old), CA1 pyramidal neuron mIPSCs are prolonged by nanomolar concentrations of $3\alpha,5\beta$-TH PROG, whereas those of the dentate granule neurons (recorded from the same *in vitro* brain slice) are less sensitive

(Harney *et al.*, 1999; Harney, 2000). Furthermore, the steroid sensitivity of dentate granule neuron $GABA_A$ receptors appears to be developmentally regulated, as mIPSCs recorded from brain slices made from 10-day-old rats are more sensitive than those made from 20-day-old animals (Cooper *et al.*, 1999). The subunit composition of dentate granule cell $GABA_A$ receptors are known to undergo developmental changes during this time and therefore it is conceivable that such changes may influence the steroid sensitivity of synaptic $GABA_A$ receptors (Fritschy *et al.*, 1994; Hollrigel and Soltesz, 1997; Kapur *et al.*, 1999). However, alternative explanations, including local steroid metabolism (Pinna *et al.*, 2000), and the influence of phosphorylation on steroid modulation of $GABA_A$ receptors should also be considered.

As mentioned, certain anabolic steroids have been shown to influence $GABA_A$-receptor function (Jorge-Rivera *et al.*, 2000). Comparison of their structures with those of the pregnane steroids makes it unlikely that these steroids act through a common site. The effects of the anabolic steroids on GABAergic transmission are highly dependent on the neuronal type, with 17α-methyltestosterone, stanozolol, and nandrolone enhancing the amplitude and prolonging the decay of spontaneously occurring IPSCs (sIPSCs) (i.e., in the presence of tetrodotoxin) recorded from neurons located in one part of the hypothalamus (the ventromedial nucleus), but inhibiting their amplitude in another (medial preoptic area) (see Jorge-Rivera *et al.*, 2000). By contrast, 3α,5α-TH PROG at the relatively high concentration of 1μM enhances the amplitude of sIPSCs in both neuronal types (Jorge-Rivera *et al.*, 2000). The differential effect of the anabolic steroids on these neurons seems to be dependent on the subunit composition of the $GABA_A$-receptors. The major $GABA_A$ receptor isoform in the ventromedial nucleus is thought to contain α_2, β_3, and γ_2 subunits, whereas, in the medial preoptic area, α_2, β_3, and γ_1 subunits predominate (Wisden *et al.*, 1992; Fenelon *et al.*, 1995). In agreement, 17α-methyltestosterone enhanced GABA-evoked currents recorded from HEK293 cells expressing recombinant α_2, β_3, and γ_2 subunits, whereas it inhibited currents recorded from cells expressing α_2, β_3 and γ_1 subunits (Jorge-Rivera *et al.*, 2000). Although not tested for this particular subunit combination, in *Xenopus* oocyte expression studies, comparison of $\alpha_1\beta_1\gamma_{2L}$ and $\alpha_1\beta\gamma_1$ receptors, revealed the latter to be less sensitive to physiological concentrations of 3α,5α-TH PROG (see Table II).

The observation that the steroid sensitivity of dentate granule cell $GABA_A$ receptors changes with development demonstrates that neurosteroid effects on inhibitory synaptic transmission can be plastic (Cooper *et al.*, 1999). A dramatic example of neurosteroid plasticity occurs in the hypothalamus. This study investigated the influence of 3α,5α-TH PROG on the decay phase of sIPSCs obtained from hypothalamic magnocellular oxytocin neurons (located in the supraoptic nucleus) at different stages of the reproductive cycle of the rat (Brussaard *et al.*, 1997, 1999; Brussaard and Herbison, 2000). These neurons are known to secrete the hormone oxytocin during parturition and lactation. $3\alpha,5\alpha$-TH PROG produces a prolongation of sIPSCs recorded from both virgin animals and animals 1 day before parturition. However, upon parturition, the circulating levels of $3\alpha,5\alpha$-TH PROG dramatically decrease and in tandem the sIPSCs become steroid insensitive, although these synaptic currents do exhibit a prolonged decay. The changes to both neurosteroid sensitivity and synaptic decay are long lasting, with these properties only reverting to those of preparturition several weeks after the cessation of lactation (Brussaard *et al.*, 1999). The changes in synaptic decay and steroid sensitivity are accompanied by an increase in the ratio of α_2 to α_1 $GABA_A$-subunit mRNA in these neurons. Should these changes be reflected at the level of the expressed protein, then it is conceivable that an altered subunit composition of the synaptic $GABA_A$ receptors may be responsible for the observed alteration of sIPSC decay kinetics (Brussaard *et al.*, 1997). Whether such a subunit switch would also produce neurosteroid-insensitive $GABA_A$ receptors is not clear. Indeed, the GABA-enhancing effect of $3\alpha,5\alpha$-TH PROG is similar for α_1- and α_2- containing recombinant receptors, although a reduced metabolite of this steroid (5α-pregnane-$3\alpha,20\alpha$-diol) is less potent at the latter (Belelli *et al.*, 1996). However, the studies on recombinant receptors used relatively slow and prolonged means of applying the agonist in comparison to the efficient and brief presentation of GABA known to occur during vesicular release and hence it may be inappropriate to compare the two approaches. Furthermore, the properties of synaptically located α_2-containing receptors could be functionally distinct from those of recombinant receptors expressed in a non-neuronal host cell. Alternatively, these neurons may express additional subunits (e.g., the ε subunit) that may explain the neurosteroid insensitivity.

Independent of the underlying mechanism, this clear example of neurosteroid plasticity is of physiological importance because the GABAergic input to the magnocellular neurons is known to play an essential role in regulating hormonal release and hence these changes in the properties of the $GABA_A$ receptors may underpin the timed release of oxytocin that is required for parturition and lactation (Brussaard *et al.*, 1999; Brussaard and Herbison, 2000).

A. Phosphorylation Influences Neurosteroid Effects on Synaptic Transmission

It is now evident that phosphorylation/dephosphorylation of $GABA_A$-receptor subunits, or their associated proteins, represents an important regulatory mechanism whereby the internal biochemistry of the neuron can influence inhibitory receptor function, turnover, and assembly (Smart et al., 2001). Preliminary evidence suggested that phosphorylation processes may also influence the interaction of neurosteroids with the $GABA_A$ receptor (Gyenes et al., 1994; Leidenheimer and Chapell, 1997). More recently, a role for phosphorylation in regulating the interaction of neurosteroids with synaptic $GABA_A$ receptors has emerged (Brussaard et al., 2000; Fancsik et al., 2000; Harney, 2000). Inhibitors of protein kinase C have been reported to abolish or reduce the effects of $3\alpha,5\alpha$-TH PROG or $3\alpha,5\beta$-TH PROG on the decay of sIPSCS or mIPSCs recorded from hypothalamic magnocellular neurons in the supraoptic nucleus and hippocampal dentate granule neurons, respectively (Fancsik et al., 2000; Harney, 2000). However, phorbol ester activators of protein kinase C did not influence the steroid effect, suggesting that normally the protein target for this enzyme is optimally phosphorylated (Fancsik et al., 2000; Harney, 2000). In the hypothalamus, inhibitors of protein kinase G had no effect—and inhibitors of protein kinase A, little effect—on neurosteroid modulation of synaptic $GABA_A$ receptors (Fancsik et al., 2000). By contrast, in dentate granule neurons, the intracellular application of a specific protein kinase A peptide greatly attenuated the effect of low concentrations of $3\alpha,5\beta$-TH PROG to prolong the decay of mIPSCs (Harney, 2000). In these examples, it is not yet known whether the site of phosphorylation is the $GABA_A$ receptor itself or an associated protein. However, irrespective of the locus of phosphorylation, these data suggest an additional mechanism whereby the sensitivity of the inhibitory synapse to an endogenous modulator can be regulated. Furthermore, such a mechanism may contribute to the differential sensitivity of neurons to the neurosteroids previously described.

Finally, a novel effect of $3\alpha,5\alpha$-TH PROG—its influence on the actions of a cell-surface-located, G-protein-coupled receptor that signals through protein kinase C—has been reported (Brussaard et al., 2000). Oxytocin, acting on magnocellular neurons located in the supraoptic nucleus of the hypothalmus decreases the amplitude of GABA-mediated sIPSCs, an effect that can be mimicked by activating protein kinase C directly with phorbol esters. However, in magnocellular neurons that express synaptic $GABA_A$ receptors that are sensitive to $3\alpha,5\alpha$-TH PROG (e.g., from rats that are either in a juvenile stage, or in a late stage of pregnancy), this effect of oxytocin, or indeed the effects of stimulating protein kinase C directly, are prevented by pretreatment with the neurosteroid (Brussaard et al., 2000). As described previously, the synaptic $GABA_A$ receptors of these neurons become steroid insensitive upon parturition and, in concordance at this stage, the steroid can no longer inhibit the effects of oxytocin/protein kinase C on these receptors. Hence, it would appear that allosteric regulation of the $GABA_A$ receptor is a prerequisite for the neurosteroid to block the protein-kinase-C-mediated modulation of the synaptic receptors. Collectively, these data suggest that the neurosteroid produces a conformational change to the receptor that makes the phosphorylation site(s) on the receptor protein inaccessible, or alternatively disrupts an important $GABA_A$ receptor–protein interaction that is dependent on protein kinase C (Brussaard et al., 2000). It will be interesting to investigate whether neurosteroids can influence the actions of additional G-protein-coupled receptors and, if so, the effects in different brain regions. Such studies are required to determine whether this newly elucidated interaction is restricted to the hypothalamus or whether it is representative of a more general and additional mechanism whereby the neurosteroids can influence synaptic signaling and neuronal excitability.

VI. Structure–Activity Relationships for Steroids at the $GABA_A$ Receptor

Early studies analyzing the structural requirements for steroid modulation of the $GABA_A$ receptor deduced that optimal activity is associated with structures containing a 5α- or 5β-reduced pregnane (or androstane) skeleton, an α-hydroxyl substituent at C3 of the steroid A-ring, and a keto group at either C20 of the pregnane steroid side chain or C17 of the androstane ring system (Harrison et al., 1987b; Gee et al., 1988; Peters et al., 1988). Those are features of the naturally occurring steroids $3\alpha,5\alpha$-TH PROG, $3\alpha,5\beta$-TH PROG, $3\alpha,5\alpha$-TH DOC, and androsterone (3α-hydroxy-5α-androstan-17-one). More extensive explorations of the structure–activity relationship using a wide range of synthetic steroids have revealed that a saturated, or closed, steroid ring system is not imperative for activity (Rodgers-Neame et al., 1992; Hawkinson et al., 1994). In addition, 20-keto reduced analog of $3\alpha,5\alpha$-TH PROG and $3\alpha,5\beta$-TH PROG (i.e., pregnanediols) retain activity as partial agonists at the steroid site, with potencies and efficacies that vary depending on *cis* or *trans* fusion of the A- and B- rings and the orientation (α or β), of the 20-hydroxyl moiety (McCauley et al., 1995; Belelli et al., 1996). Consistent with the replacement of the hydrogen-bond-accepting keto group by a hydrogen-bond-donating hydroxyl as the cause of such changes, the modulatory potency of a 3α-hydroxy-5α-androstane bearing a 17β-methoxy group greatly exceeds that of the corresponding 17β-diol compound (Anderson et al., 2000). Such refinements, along with the effects of chemical substitutions, and in some instances-epimerisation, at the C2, C3, C5, C10, C11, and C17 positions of the steroid ring system and C21 of acetyl side chain have been reviewed in detail

(Lambert *et al.*, 2001). This discussion is limited to developments with particular practical or theoretical significance.

A. STEROIDS WITH INCREASED ORAL BIOAVAILABILITY

One limitation to the potential use of pregnane steroids as, for example, anticonvulsant, anxiolytic, or hypnotic drugs results from their low oral bioavailabilty and short half-life (Gasior *et al.*, 1999). Rapid hepatic metabolism, via conjugation or oxidation of the 3α-hydroxyl group, which is crucial for steroid action at the GABA$_A$ receptor, underlies these pharmacokinetic features (Gasior *et al.*, 1999). Steroids protected from rapid metabolism may be prepared by the introduction of simple alkane and alkyl halide moieties at the 3β position. A well-studied example is provided by ganaxolone, a 3β-methyl substituted analog of 3α,5α-TH PROG, which largely retains the potency and efficacy of the parent compound (Carter *et al.*, 1997). Orally administered ganaxolone, unlike 3α,5α-TH PROG, demonstrates anticonvulsant activity against acute, chemically induced, seizures in rats (Carter *et al.*, 1997). Moreover, the compound is effective in rodent kindling models (Carter *et al.*, 1997). Ganaxolone demonstrates undiminished efficacy over time against chemically induced seizures in rats in a chronic dosing regimen, indicating that significant tolerance to the drug itself does not develop (Reddy and Rogawski, 2000). However, ganaxolone does cause cross-tolerance to diazepam (Reddy and Rogawski, 2000), and the development of self-tolerance has been observed for the sedative effect of other synthetic steroids in mice (Marshall *et al.*, 1997). In small-scale clinical trials, ganaxolone has been shown, as add-on therapy, to reduce the incidence of refractory seizures in children (Monaghan *et al.*, 1999). The introduction of the 3β-methyl group does not influence activity at the GABA$_A$ receptor, as 3α,5α-TH PROG and ganaxolone are approximately equipotent in this respect (Carter *et al.*, 1997).

The introduction of a 3β-trifluoromethyl group also provides a degree of metabolic protection (Gasior *et al.*, 1999). In 5α-pregnanes (e.g., 3α-hydroxy-3β-trifluoromethyl-5α-pregnan-20-one), this modification appears to be associated with partial agonism at the steroid site because efficacy is reduced in comparison to 3α,5α-TH PROG, and positive allosteric regulation by the latter is antagonized by the 3β-trifluoromethyl derivative (Hawkinson *et al.*, 1996). Steroids with limited efficacy, including the preganediols, could, in principle, offer advantages over full agonists in certain clinical settings. In 5β-pregnanes, 3β-substitution apparently does not result in a reduction in efficacy (Hogenkamp *et al.*, 1997). Examples of 3β-substituted 5β-pregnanes that have been promoted for clinical trial include a 19-nor derivative of

3α-hydroxy-3β-trifluoromethyl-5β-pregnan-20-one advocated for the treatment of insomnia and the 20-hydroxy derivative of the latter proposed as a novel anxiolytic (Gasior *et al.*, 1999).

B. WATER-SOLUBLE STEROIDS

Solubility in water is clearly a desirable property of drugs that are to be administered *via* the intravenous route. Indeed, the demise of the one-time clinically useful, but water-insoluble, steroid intravenous general anesthetic alphaxalone can be traced to the propensity of the vehicle Chremphor EL to provoke rare anaphalactoid reactions in humans (and dogs) rather than a deficiency in steroid anesthesia per se (Sear, 1997). Substantial effort has therefore been devoted to the development of water-soluble steroids that might retain the desirable anesthetic profile of alphaxalone.

Several water-soluble steroids that retain both activity at the $GABA_A$ receptor and the ability to induce general anesthesia have been described. An example is provided by minaxolone, wherein the incorporation of the 11α- dimethyl amino group results in water solubility (Phillips *et al.*, 1979). In addition to potentiating the action of GABA at $GABA_A$ receptors minaxolone, unlike alphaxalone, acts as a positive allosteric modulator of strychnine-sensitive glycine receptors, albeit with a less pronounced effect (Shepherd *et al.*, 1996) (see Table I). Modulation of $GABA_A$-receptor activity by pregnane steroids rendered water soluble by the introduction of a 2β-morpholinyl group [e.g., 21-chloro-3α-hydroxy-2β-(4-morpholinyl)-5α-pregnan-20-one methanesulphonate (ORG 20599), 3α-hydroxy-2β-(2,2-dimethyl-morpholin-4-yl)-5α-pregnane-11,20-dione (ORG 21465)] has also been reported (Hill-Venning *et al.*, 1996; Anderson *et al.*, 1997). It is clear that the steroid binding site of the $GABA_A$ receptor can tolerate rather bulky substituents at the 2β-position, as even a 2,6-dibutyl morpholinyl derivative of ORG 21465 is accommodated without loss of potency (Anderson *et al.*, 1997). Unfortunately, although water-soluble steroidal anesthetics demonstrated promise in preclinical studies, clinical testing has revealed either pharmacokinetic or side-effect profiles that are unacceptable in humans, thereby halting, at least to date, clinical development of this class of agents as general anesthetic compounds.

VII. Multiple Steroid Binding Sites on the $GABA_A$ Receptor

As detailed previously, $GABA_A$-receptor heterogeneity may provide for binding sites at which steroids display differing affinities and/or efficacies and the GABA-modulatory and GABA-mimetic activities of the steroids are known to be differentially affected by the subunit composition of the receptor. A study examining the effect of synthetic enantiomers of $3\alpha,5\alpha$-TH PROG and $3\alpha,5\beta$-TH PROG provides additional evidence for such heterogeneity in steroid action (Covey et al., 2000). In assays that include displacement of [^{35}S]tert-butylbicyclophosphorothionate binding from rat brain membranes, potentiation of GABA-evoked currents recorded from rat hippocampal neurons, and loss of the righting reflex in tadpoles and mice, a pronounced and consistent enantioselectivty of action is observed for the naturally occurring $3\alpha,5\alpha$-TH PROG vs the synthetic ent-$3\alpha,5\alpha$-TH PROG (Wittmer et al., 1996; Covey et al., 2000). A similar correlation between GABA-modulatory and anesthetic potency has been established for 5α-androstane enantiomers bearing a 17β-carbonitrile substituent (Wittmer et al., 1996; Covey et al., 2000) and for the enantiomers of the benz[e]indene BI-1 (Zorumski et al., 1996). In the absence of structural data, these observations provide the most convincing evidence for a direct interaction between certain pregnane steroids and the GABA$_A$ receptor because enantiomers act dissimilarly only in a chiral (e.g., protein) environment. It is thus significant that the degree of enantioselectivity observed for $3\alpha,5\beta$-TH PROG vs ent-$3\alpha,5\beta$-TH PROG and 17β-carbonitrile-substituted 5β-androstane enantiomers is vastly reduced across the same battery of assays. As expected from many previous studies (Lambert et al., 2001), little diastereoselectivity was observed between the $3\alpha,5\alpha$-TH PROG and $3\alpha,5\beta$-TH PROG pair of compounds (Covey et al., 2000). These data can be interpreted in at least two ways: (1) the binding pocket on the GABA$_A$ receptor cannot accommodate ent-$3\alpha,5\alpha$-TH PROG or (2) the pair $3\alpha,5\beta$-TH PROG/ent-$3\alpha,5\beta$-TH PROG bind to a site distinct from that recognizing $3\alpha,5\alpha$-TH PROG and ent-$3\alpha,5\alpha$-TH PROG.

It has been reported that the anabolic steroids 17α-methyltestosterone, stanozolol, and nandrolone also exert a rapid, nongenomic modulation of GABA$_A$-receptor activity (Jorge-Rivera et al., 2000). In view of the selectivity of these anabolic steroids for particular isoforms of the GABA$_A$ receptor, coupled with their structures not possessing the crucial 3α-hydroxyl group, it would appear most unlikely that they share a common site of action with the pregnane compounds discussed previously.

VIII. Concluding Remarks

Neurosteroids such as $3\alpha,5\alpha$-TH PROG are now well established as potent, selective modulators of the GABA$_A$ receptor. These compounds may be synthesized peripherally in endocrine glands and transported into the brain or, indeed, they may be made locally in the brain itself. Evidence from a variety of approaches now suggests that under certain physiological and pathophysiological conditions these steroids will attain levels sufficient to enhance the function of the brain's major inhibitory neurotransmitter receptor, the $GABA_A$ receptor, and, as a consequence, enhance synaptic inhibition. Given the heterogeneous distribution of $GABA_A$ receptor isoforms throughout the mammalian CNS, it is of importance to establish whether all inhibitory pathways are similarly influenced. We have examined in some detail the influence of receptor subunit composition on the GABA-modulatory actions of $3\alpha,5\alpha$-TH PROG. Although these studies demonstrate most isoforms of the $GABA_A$ receptor to be sensitive to this neurosteroid, they also reveal that "physiological concentrations" (3–100 nM) of $3\alpha,5\alpha$-TH PROG are discriminatory. Hence, these findings suggest that endogenous neurosteroids selectively influence particular inhibitory pathways. Consistent with this proposal, the few reports to date on the effects of neurosteroids on GABAergic synaptic transmission reveal such heterogeneity. However, although this selectivity may be caused by the subunit composition of the synaptic $GABA_A$ receptors, experiments demonstrate that the phosphorylation status of either the receptors per se, or their associated proteins, must also be considered.

Irrespective of the molecular locus of this synaptic specificity, further studies on neurosteroid modulation of inhibitory transmission may ultimately permit a better understanding of their behavioral effects. Behaviorally, $3\alpha,5\alpha$-TH PROG exhibits a similar profile to the benzodiazepines, including anxiolytic, anticonvulsant, and sedative actions. For the benzodiazepines, progress is being made in assigning these behaviors to particular $GABA_A$-receptor isoforms and therefore, by inference, to certain anatomical locations and inhibitory pathways (e.g., the sedative actions of diazepam appear to be mediated mainly by receptors incorporating the α_1 subunit, whereas α_2-subunit-containing receptors are required for the anxiolytic actions of this compound) (Crestani et al., 2000; McKernan et al., 2000; Rudolph et al., 1999). Whether these receptor isoforms also mediate the sedative and anxiolytic actions of the neurosteroids remains to be determined.

A number of studies have investigated the relationship between steroid structure and $GABA_A$ receptor activity. By inference, these data may reveal something of the chemical environment the steroid experiences when bound to the $GABA_A$ receptor protein. In addition, these studies have demonstrated that many of the physicochemical properties of endogenous steroids, such as $3\alpha,5\alpha$-TH PROG, that make them unsuitable for clinical development (e.g., poor solubility, low oral availability, short half-life) can be overcome by relatively straightforward chemical modification. Furthermore, neurosteroids have been identified that have a reduced maximal efficacy at the GABA$_A$ receptor in comparison to 3α,5α-TH PROG and have been classified as "partial agonists" of the neurosteroid binding site. Theoretically, such compounds would be expected to have a limited behavioral repertoire in comparison with steroid "full agonists" and, at least in animals, this seems to be the case (e.g., anxiolytic properties without sedation) (Hawkinson *et al.*, 1994). Collectively, findings with synthetic steroids have encouraged their clinical development as hypnotics, anxiolytics, anticonvulsants, and general anesthetics (Gasior *et al.*, 1999). It will be of interest to establish whether such compounds, based on the structure of an endogeneous modulator, offer any clinical advantage over those synthetic compounds available to date (e.g., the benzodiazepines). However, the discovery of novel compounds that selectively interact with the brain enzymes that synthesize, or metabolize, the neurosteroids may offer a new approach to therapeutically exploit this novel endocrine–paracrine neurotransmitter interaction (Uzunova *et al.*, 1998; Griffin and Mellon, 1999).

Acknowledgments

Some of the work reported here was supported by an EC Bioscience and Health Grant BMH4-CT97-2359 and by financial support of the Commision of the European Communities, RTD program "Quality of Life and Management of Living Resources," QLK1-CT-2000-00179. D.B. is an MRC senior fellow, and S.C.H. was supported by an MRC studentship.

References

Albuquerque, E. X., Alkondon, M., Pereira, E. F., Castro, N. G., Schrattenholz, A., Barbosa, C. T., Bonfante-Cabarcas, R., Aracava, Y., Eisenberg, H. M., and Maelicke, A. (1997). Properties of neuronal nicotinic acetylcholine receptors: Pharmacological characterization and modulation of synaptic function. *J. Pharmacol. Exp. Ther.* **280,** 1117–1136.
Anderson, A., Boyd, A. C., Byford, A., Campbell, A. C., Gemmell, D. K., Hamilton, N. M., Hill, D. R., Hill-Venning, C., Lambert, J. J., Maidment, M. S., May, V., Marshall, R. J., Peters, J. A., Rees, D. C., Stevenson, D., and Sundaram, H. (1997). Anesthetic activity of novel water-soluble 2 beta-morpholinyl steroids and their modulatory effects at GABA$_A$ receptors. *J. Med. Chem.* **40,** 1668–1681.
Anderson, A., Boyd, A. C., Clark, J. K., Fielding, L., Gemmell, D. K., Hamilton, N. M., Maidment, M. S., May, V., McGuire, R., McPhail, P., Sansbury, F. H., Sundaram, H., and Taylor, R. (2000). Conformationally constrained anesthetic steroids that modulate GABA(A) receptors. *J. Med. Chem.* **43,** 4118–4125.
Barker, J. L., Harrison, N. L., Lange, G. D., and Owen, D. G. (1987). Potentiation of gamma-aminobutyric-acid-activated chloride conductance by a steroid anaesthetic in cultured rat spinal neurones. *J. Physiol. (London)* **386,** 485–501.

Barnard, E. A. (2001). The molecular architecture of GABA$_A$ receptors. *In* "Handbook of Experimental Pharmacology" (H. Mohler, Ed.), Vol. 150, pp. 79–99. Springer Verlag, Berlin.

Barnard, E. A., Skolnick, P., Olsen, R. W., Mohler, H., Sieghart, W., Biggio, G., Braestrup, C., Bateson, A. N., and Langer, S. Z. (1998). International Union of Pharmacology: XV. Subtypes of gamma-aminobutyric acid$_A$ receptors: Classification on the basis of subunit structure and receptor function. *Pharmacol. Rev.* **50,** 291–313.

Belelli, D., Lambert, J. J., Peters, J. A., Gee, K. W., and Lan, N. C. (1996). Modulation of human recombinant GABA$_A$ receptors by pregnanediols. *Neuropharmacology* **35,** 1223–1231.

Belelli, D., Pistis, I., Peters, J. A., and Lambert, J. J. (1999). General anesthetic action at transmitter-gated inhibitory amino acid receptors. *Trends Pharmacol. Sci.* **20,** 496–502.

Betz, H., Harvey, R., and Schloss, P. (2001). Structures, diversity and pharmacology of glycine receptors and transporters. *In* "Handbook of Experimental Pharmacology" (H. Mohler, Ed.), Vol. 150, pp. 375–401. Springer, Berlin

Brickley, S. G., Cull-Candy, S. G., and Farrant, M. (1996). Development of a tonic form of synaptic inhibition in rat cerebellar granule cells resulting from persistent activation of GABA$_A$ receptors. *J. Physiol* **497,** (3) 753–759.

Brown, N. K. J., Bonnert, T. P., and Wafford, K. A. (2001). Pharmacology of a novel $\alpha_4\beta_3\delta$ GABA$_A$ receptor cell line. *Br. J.Pharmacol.* In press.

Brussaard, A. B., Devay, P., Leyting-Vermeulen, J. L., and Kits, K. S. (1999). Changes in properties and neurosteroid regulation of GABAergic synapses in the supraoptic nucleus during the mammalian female reproductive cycle. *J. Physiol. (London)* **516,** (2) 513–524.

Brussaard, A. B., Kits, K. S., Baker, R. E., Willems, W. P., Leyting-Vermeulen, J. W., Voorn, P., Smit, A. B., Bicknell, R. J., and Herbison, A. E. (1997). Plasticity in fast synaptic inhibition of adult oxytocin neurons caused by switch in GABA(A) receptor subunit expression. *Neuron* **19,** 1103–1114.

Brussaard, A. B., and Herbison, A. E. (2000). Long-term plasticity of postsynaptic GABA$_A$-receptor function in the adult brain: Insights from the oxytocin neurone. *Trends Neurosci.* **23,** 190–195.

Brussaard, A. B., Wossink, J., Lodder, J. C., and Kits, K. S. (2000). Progesterone-metabolite prevents protein kinase C-dependent modulation of gamma-aminobutyric acid type A receptors in oxytocin neurons. *Proc. Natl. Acad. Sci. USA* **97,** 3625–3630.

Buisson, B., and Bertrand, D. (1999). Steroid modulation of nicotinic achetylcholine receptor. *In* "Neurosteroids: A New Regulatory Function in the Nervous System" (E. Baulieu, P. Robel, and M. Schumacher, Eds.), pp. 207–223. Humana Press, Totowa, NJ.

Callachan, H., Cottrell, G. A., Hather, N. Y., Lambert, J. J., Nooney, J. M., and Peters, J. A. (1987). Modulation of the GABA$_A$ receptor by progesterone metabolites. *Proc. R. Soc. Lond B. Biol. Sci.* **231,** 359–369.

Carter, R. B., Wood, P. L., Wieland, S., Hawkinson, J. E., Belelli, D., Lambert, J. J., White, H. S., Wolf, H. H., Mirsadeghi, S., Tahir, S. H., Bolger, M. B., Lan, N. C., and Gee, K. W. (1997). Characterization of the anticonvulsant properties of ganaxolone (CCD 1042; 3alpha-hydroxy-3beta-methyl-5alpha-pregnan-20-one), a selective, high-affinity, steroid modulator of the gamma-aminobutyric acid(A) receptor. *J. Pharmacol. Exp. Ther.* **280,** 1284–1295.

Cooper, E. J., Johnston, G. A., and Edwards, F. A. (1999). Effects of a naturally occurring neurosteroid on GABA$_A$ IPSCs during development in rat hippocampal or cerebellar slices. *J. Physiol. (London)* **521,** (2) 437–449.

Cottrell, G. A., Lambert, J. J., and Peters, J. A. (1987). Modulation of GABA$_A$ receptor activity by alphaxalone. *Br. J. Pharmacol.* **90,** 491–500.

Covey, D. F., Nathan, D., Kalkbrenner, M., Nilsson, K. R., Hu, Y., Zorumski, C. F., and Evers, A. S. (2000). Enantioselectivity of pregnanolone-induced γ-aminobutyric acid$_A$ receptor modulation and anesthesia. *J. Pharmacol. Exp. Ther.* **293,** 1009–1016.

Crestani, F., Martin, J. R., Mohler, H., and Rudolph, U. (2000). Resolving differences in GABA$_A$ receptor mutant mouse studies. *Nat. Neurosci.* **3,** 1059.

Davies, P. A., Hanna, M. C., Hales, T. G., and Kirkness, E. F. (1997). Insensitivity to anaesthetic agents conferred by a class of GABA(A) receptor subunit. *Nature* **385,** 820–823.

Davies, P. A., Pistis, M., Hanna, M. C., Peters, J. A., Lambert, J. J., Hales, T. G., and Kirkness, E. F. (1999). The 5-HT$_{3B}$ subunit is a major determinant of serotonin-receptor function. *Nature* **397,** 359–363.

Dingledine, R., Borges, K., Bowie, D., and Traynelis, S. F. (1999). The glutamate receptor ion channels. *Pharmacol. Rev.* **51,** 7–61.

Edwards, F. A. (1995). Anatomy and electrophysiology of fast central synapses lead to a structural model for long-term potentiation. *Physiol. Rev.* **75,** 759–787.

Evers, A. S., and Steinbach, J. H. (1997). Supersensitive sites in the central nervous system. Anesthetics block brain nicotinic receptors. *Anesthesiology* **86,** 760–762.

Fancsik, A., Linn, D. M., and Tasker, J. G. (2000). Neurosteroid modulation of GABA IPSCs is phosphorylation dependent. *J. Neurosci.* **20,** 3067–3075.

Fenelon, V. S., Sieghart, W., and Herbison, A. E. (1995). Cellular localization and differential distribution of GABA$_A$ receptor subunit proteins and messenger RNAs within hypothalamic magnocellular neurons. *Neuroscience* **64,** 1129–1143.

Fritschy, J. M., Paysan, J., Enna, A., and Mohler, H. (1994). Switch in the expression of rat GABA$_A$-receptor subtypes during postnatal development: An immunohistochemical study. *J. Neurosci.* **14,** 5302–5324.

Gasior, M., Carter, R. B., and Witkin, J. M. (1999). Neuroactive steroids: Potential therapeutic use in neurological and psychiatric disorders. *Trends Pharmacol. Sci.* **20,** 107–112.

Gee, K. W., Bolger, M. B., Brinton, R. E., Coirini, H., and McEwen, B. S. (1988). Steroid modulation of the chloride ionophore in rat brain: Structure–activity requirements, regional dependence and mechanism of action. *J. Pharmacol. Exp. Ther.* **246,** 803–812.

Gibbs, T. T., Yaghoubi, N., Weaver, C., Park-Chung, J., Russek, S., and Farb, D. H. (1999). Modulation of ionotropic glutamate receptors by neuroactive steroids. *In* "Neurosteroids: A New Regulatory Function in the Nervous System" (E. Baulieu, P. Robel, and M. Schumacher, Eds.), pp. 167–190. Human Press, Totowa, NJ.

Griffin, L. D., and Mellon, S. H. (1999). Selective serotonin reuptake inhibitors directly alter activity of neurosteroidogenic enzymes. *Proc. Natl. Acad. Sci. USA* **96,** 13512–13517.

Gyenes, M., Wang, Q., Gibbs, T. T., and Farb, D. H. (1994). Phosphorylation factors control neurotransmitter and neuromodulator actions at the gamma-aminobutyric acid type A receptor. *Mol. Pharmacol.* **46,** 542–549.

Hadingham, K. L., Wingrove, P. B., Wafford, K. A., Bain, C., Kemp, J. A., Palmer, K. J., Wilson, A. W., Wilcox, A. S., Sikela, J. M., and Ragan, C. I. (1993). Role of the beta subunit in determining the pharmacology of human gamma- aminobutyric acid type A receptors. *Mol. Pharmacol.* **44,** 1211–1218.

Harney, S. (2000). Neurosteroid modulation of GABA$_A$ receptor-mediated synaptic currents. (PhD Thesis, University of Dundee, Scotland.)

Harney, S., Frenguelli, B., and Lambert, J. J. (1999). Neurosteroid modulation of GABA$_A$ receptor-mediated miniature inhibitory post-synaptic currents in the rat hippocampus. *Br. J. Pharmacol.* **126,** 7P.

Harrison, N. L., and Simmonds, M. A. (1984). Modulation of the GABA receptor complex by a steroid anesthetic. *Brain Res.* **323,** 287–292.

Harrison, N. L., Vicini, S., and Barker, J. L. (1987a). A steroid anesthetic prolongs inhibitory postsynaptic currents in cultured rat hippocampal neurons. *J. Neurosci.* **7,** 604–609.

Harrison, N. L., Majewska, M. D., Harrington, J. W., and Barker, J. L. (1987b). Structure–activity relationships for steroid interaction with the gamma-aminobutyric acid$_A$ receptor complex. *J. Pharmacol. Exp. Ther.* **241,** 346–353.

Hawkinson, J. E., Kimbrough, C. L., Belelli, D., Lambert, J. J., Purdy, R. H., and Lan, N. C. (1994). Correlation of neuroactive steroid modulation of [^{35}S]*t*-butyl-bicyclophosphorothionate and [^{3}H]flunitrazepam binding and gamma-aminobutyric acid$_A$ receptor function. *Mol. Pharmacol.* **46,** 977–985.

Hawkinson, J. E., Drewe, J. A., Kimbrough, C. L., Chen, J. S., Hogenkamp, D. J., Lan, N. C., Gee, K. W., Shen, K. Z., Whittemore, E. R., and Woodward, R. M. (1996). 3α-Hydroxy-3β-trifluoromethyl-5α-pregnan-20-one (Co 2- 1970): A partial agonist at the neuroactive steroid site of the γ-aminobutyric acid$_A$ receptor. *Mol. Pharmacol.* **49,** 897–906.

Hill-Venning, C., Peters, J. A., Callachan, H., Lambert, J. J., Gemmell, D. K., Anderson, A., Byford, A., Hamilton, N., Hill, D. R., Marshall, R. J., and Campbell, A. C. (1996). The anesthetic action and modulation of GABA$_A$ receptor activity by the novel water-soluble aminosteroid Org 20599. *Neuropharmacology* **35,** 1209–1222.

Hogenkamp, D. J., Tahir, S. H., Hawkinson, J. E., Upasani, R. B., Alauddin, M., Kimbrough, C. L., Acosta-Burruel, M., Whittemore, E. R., Woodward, R. M., Lan, N. C., Gee, K. W., and Bolger, M. B. (1997). Synthesis and *in vitro* activity of 3 beta-substituted-3 alpha-hydroxypregnan-20-ones: Allosteric modulators of the GABA$_A$ receptor. *J. Med. Chem.* **40,** 61–72.

Hollrigel, G. S., and Soltesz, I. (1997). Slow kinetics of miniature IPSCs during early postnatal development in granule cells of the dentate gyrus. *J. Neurosci.* **17,** 5119–5128.

Hope, A. G., Belelli, D., Mair, I. D., Lambert, J. J., and Peters, J. A. (1999). Molecular determinants of (+)-tubocurarine binding at recombinant 5-hydroxytryptamine$_{3A}$ receptor subunits. *Mol. Pharmacol.* **55,** 1037–1043.

Jones, M. V., and Westbrook, G. L. (1995). Desensitized states prolong GABA$_A$ channel responses to brief agonist pulses. *Neuron* **15,** 181–191.

Jorge-Rivera, J. C., McIntyre, K. L., and Henderson, L. P. (2000). Anabolic steroids induce region- and subunit-specific rapid modulation of GABA$_A$ receptor-mediated currents in the rat forebrain. *J. Neurophysiol.* **83,** 3299–3309.

Kapur, J., and Macdonald, R. L. (1999). Postnatal development of hippocampal dentate granule cell gamma-aminobutyric acid$_A$ receptor pharmacological properties. *Mol. Pharmacol.* **55,** 444–452.

Lambert, J. J., Peters, J. A., and Cottrell, G. A. (1987). Actions of synthetic and endogenous steroids on the GABA$_A$ receptor. *Trends Pharmacol. Sci.* **8,** 224–227.

Lambert, J. J., Peters, J. A., Harney, S., and Belelli, D. (2001). Steroid modulation of GABA$_A$ receptors. *In* "Handbook of Experimental Pharmacology" (H. Mohler, Eds.), Vol. 150, pp. 117–140. Springer Verlag, Berlin.

Lambert, J. J., Peters, J. A., Sturgess, N. C., and Hales, T. G. (1990). Steroid modulation of the GABA$_A$ receptor complex: Electrophysiological studies. *Ciba Found. Symp.* **153,** 56–71.

Lambert, J. J., Belelli, D., Hill-Venning, C., and Peters, J. A. (1995). Neurosteroids and GABA$_A$ receptor function. *Trends Pharmacol. Sci.* **16,** 295–303.

Leidenheimer, N. J., and Chapell, R. (1997). Effects of PKC activation and receptor desensitization on neurosteroid modulation of GABA(A) receptors. *Brain Res. Mol. Brain Res.* **52,** 173–181.

Llano, I., Gonzalez, J., Caputo, C., Lai, F. A., Blayney, L. M., Tan, Y. P., and Marty, A. (2000). Presynaptic calcium stores underlie large-amplitude miniature IPSCs and spontaneous calcium transients. *Nat. Neurosci.* **3,** 1256–1265.

Luddens, H., Korpi, E. R., and Seeburg, P. H. (1995). GABA$_A$/benzodiazepine receptor heterogeneity: Neurophysiological implications. *Neuropharmacology* **34,** 245–254.

Macdonald, R. L., and Olsen, R. W. (1994). GABA$_A$ receptor channels. *Ann. Rev. Neurosci.* **17,** 569–602.

Macdonald, R. L., Rogers, C. J., and Twyman, R. E. (1989a). Barbiturate regulation of kinetic properties of the GABA$_A$ receptor channel of mouse spinal neurones in culture. *J. Physiol. (London)* **417,** 483–500.

Macdonald, R. L., Rogers, C. J., and Twyman, R. E. (1989b). Kinetic properties of the GABA$_A$ receptor main conductance state of mouse spinal cord neurones in culture. *J. Physiol. (London)* **410,** 479–499.

Maitra, R., and Reynolds, J. N. (1999). Subunit dependent modulation of GABA$_A$ receptor function by neuroactive steroids. *Brain Res.* **819,** 75–82.

Majewska, M. D., Harrison, N. L., Schwartz, R. D., Barker, J. L., and Paul, S. M. (1986). Steroid hormone metabolites are barbiturate-like modulators of the GABA receptor. *Science* **232,** 1004–1007.

Marshall, F. H., Stratton, S. C., Mullings, J., Ford, E., Worton, S. P., Oakley, N. R., and Hagan, R. M. (1997). Development of tolerance in mice to the sedative effects of the neuroactive steroid minaxolone following chronic exposure. *Pharmacol. Biochem. Behav.* **58,** 1–8.

McCauley, L. D., Liu, V., Chen, J. S., Hawkinson, J. E., Lan, N. C., and Gee, K. W. (1995). Selective actions of certain neuroactive pregnanediols at the gamma-aminobutyric acid type A receptor complex in rat brain. *Mol. Pharmacol.* **47,** 354–362.

McKernan, R. M., Rosahl, T. W., Reynolds, D. S., Sur, C., Wafford, K. A., Atack, J. R., Farrar, S., Myers, J., Cook, G., Ferris, P., Garrett, L., Bristow, L., Marshall, G., Macaulay, A., Brown, N., Howell, O., Moore, K. W., Carling, R. W., Street, L. J., Castro, J. L., Ragan, C. I., Dawson, G. R., and Whiting, P. J. (2000). Sedative but not anxiolytic properties of benzodiazepines are mediated by the GABA(A) receptor alpha1 subtype. *Nat. Neurosci.* **3,** 587–592.

Mihalek, R. M., Banerjee, P. K., Korpi, E. R., Quinlan, J. J., Firestone, L. L., Mi, Z. P., Lagenaur, C., Tretter, V., Sieghart, W., Anagnostaras, S. G., Sage, J. R., Fanselow, M. S., Guidotti, A., Spigelman, I., Li, Z., DeLorey, T. M., Olsen, R. W., and Homanics, G. E. (1999). Attenuated sensitivity to neuroactive steroids in gamma-aminobutyrate type A receptor delta subunit knockout mice. *Proc. Natl. Acad. Sci. USA* **96,** 12905–12910.

Mody, I., De Koninck, Y., Otis, T. S., and Soltesz, I. (1994). Bridging the cleft at GABA synapses in the brain. *Trends Neurosci.* **17,** 517–525.

Monaghan, E. P., McAuley, J. W., and Data, J. L. (1999). Ganaxolone: A novel positive allosteric modulator of the GABA(A) receptor complex for the treatment of epilepsy. *Expert. Opin. Investig. Drugs* **8,** 1663–1671.

Neelands, T. R., Fisher, J. L., Bianchi, M., and Macdonald, R. L. (1999). Spontaneous and gamma-aminobutyric acid (GABA)-activated GABA(A) receptor channels formed by epsilon subunit-containing isoforms. *Mol. Pharmacol.* **55,** 168–178.

Olsen, R. W., and Sapp, D. W. (1995). Neuroactive steroid modulation of GABA$_A$ receptors. *Adv. Biochem. Psychopharmacol.* **48,** 57–74.

Paradiso, K., Sabey, K., Evers, A. S., Zorumski, C. F., Covey, D. F., and Steinbach, J. H. (2000). Steroid inhibition of rat neuronal nicotinic alpha4beta2 receptors expressed in HEK 293 cells. *Mol. Pharmacol.* **58,** 341–351.

Paul, S. M., and Purdy, R. H. (1992). Neuroactive steroids. *FASEB J.* **6,** 2311–2322.

Peters, J. A., Kirkness, E. F., Callachan, H., Lambert, J. J., and Turner, A. J. (1988). Modulation of the GABA$_A$ receptor by depressant barbiturates and pregnane steroids. *Br. J. Pharmacol.* **94,** 1257–1269.

Phillips, G. H., Ayres, B. E., Bailey, E. J., Ewan, G. B., Looker, B. E., and May, P. J. (1979). Water-soluble steroidal anaesthetics. *J. Steroid Biochem.* **11,** 79–86.

Pinna, G., Uzunova, V., Matsumoto, K., Puia, G., Mienville, J. M., Costa, E., and Guidotti, A. (2000). Brain allopregnanolone regulates the potency of the GABA(A) receptor agonist muscimol. *Neuropharmacology* **39,** 440–448.

Pirker, S., Schwarzer, C., Wieselthaler, A., Sieghart, W., and Sperk, G. (2000). GABA(A) receptors: Immunocytochemical distribution of 13 subunits in the adult rat brain. *Neuroscience* **101,** 815–850.

Pistis, M., Belelli, D., Peters, J. A., and Lambert, J. J. (1997). The interaction of general anaesthetics with recombinant $GABA_A$ and glycine receptors expressed in *Xenopus laevis* oocytes: A comparative study. *Br. J. Pharmacol.* **122,** 1707–1719.

Poletti, A., Celotti, F., Maggi, R., Melcangi, R., Martini, L., and Negri-Cesi, P. (1999). Aspects of hormonal steroid metabolism in the nervous system. *In* "Neurosteroids: A New Regulatory Function in the Nervous System" (E. Baulieu, P. Robel, and M. Schumacher, Eds.), pp. 97–124. Humana Press, Totowa, NJ.

Prince, R. J., and Simmonds, M. A. (1993). Differential antagonism by epipregnanolone of alphaxalone and pregnanolone potentiation of [^{3}H]flunitrazepam binding suggests more than one class of binding site for steroids at $GABA_A$ receptors. *Neuropharmacology* **32,** 59–63.

Puia, G., Santi, M. R., Vicini, S., Pritchett, D. B., Purdy, R. H., Paul, S. M., Seeburg, P. H., and Costa, E. (1990). Neurosteroids act on recombinant human $GABA_A$ receptors. *Neuron* **4,** 759–765.

Reddy, D. S., and Rogawski, M. A. (2000). Chronic treatment with the neuroactive steroid ganaxolone in the rat induces anticonvulsant tolerance to diazepam but not to itself. *J. Pharmacol. Exp. Ther.* **295,** 1241–1248.

Robel, P., Scumacher, M., and Baulieu, E. (1999). Neurosteroids: From definition and biochemistry to physiopathological function. *In* "Neurosteroids: A New Regulatory Function in the Nervous System" (E. Baulieu, P. Robel, M. Schumacher, Eds.), pp. 1–26. Humana Press, Totowa, NJ.

Rodgers-Neame, N. T., Covey, D. F., Hu, Y., Isenberg, K. E., and Zorumski, C. F. (1992). Effects of a benz[e]indene on gamma-aminobutyric acid-gated chloride currents in cultured postnatal rat hippocampal neurons. *Mol. Pharmacol.* **42,** 952–957.

Rudolph, U., Crestani, F., Benke, D., Brunig, I., Benson, J. A., Fritschy, J. M., Martin, J. R., Bluethmann, H., and Mohler, H. (1999). Benzodiazepine actions mediated by specific gamma-aminobutyric acid(A) receptor subtypes [published erratum appears in *Nature* 2000 Apr 6;404(6778):629]. *Nature* **401,** 796–800.

Rupprecht, R., and Holsboer, F. (1999). Neuroactive steroids: Mechanisms of action and neuropsychopharmacological perspectives. *Trends Neurosci.* **22,** 410–416.

Sanna, E., Murgia, A., Casula, A., and Biggio, G. (1997). Differential subunit dependence of the actions of the general anesthetics alphaxalone and etomidate at gamma-aminobutyric acid type A receptors expressed in *Xenopus laevis* oocytes. *Mol. Pharmacol.* **51,** 484–490.

Sear, J. (1997). Steroids. *In* "Textbook of Intravenous Anesthesia" (P. White, Eds.), pp. 153–169. Williams and Wilkins, Baltimore, MD.

Selye, H. (1941). Anaesthetic effects of steroid hormones. *Proc. Soc. Exp. Biol. Med.* **46,** 116–121.

Shepherd, S., Peters, J. A., and Lambert, J. J. (1996). The interaction of intravenous anaesthetics with rat inhibitory and excitatory amino acid receptors expressed in *Xenopus laevis* oocytes. *Br. J. Pharmacol.* **119,** 364P.

Shingai, R., Sutherland, M. L., and Barnard, E. A. (1991). Effects of subunit types of the cloned $GABA_A$ receptor on the response to a neurosteroid. *Eur. J. Pharmacol.* **206,** 77–80.

Sieghart, W. (1995). Structure and pharmacology of gamma-aminobutyric acid$_A$ receptor subtypes. *Pharmacol. Rev.* **47,** 181–234.

Sieghart, W. (2000). Unraveling the function of GABA(A) receptor subtypes. *Trends Pharmacol. Sci.* **21,** 411–413.

Smart, T., Thomas, P., Brandon, N., and Moss, S. (2001). Heterologous regulation of $GABA_A$ receptors: Protein phosphorylation. *In* "Handbook of Experimental Pharmacology" (H. Mohler, Eds.), Vol. 150, pp. 195–226. Springer, Berlin, Germany.

Smith, G. B., and Olsen, R. W. (1995). Functional domains of GABA$_A$ receptors. *Trends Pharmacol. Sci.* **16,** 162–168.
Smith, S. S., Gong, Q. H., Hsu, F. C., Markowitz, R. S., French-Mullen, J. M., and Li, X. (1998a). GABA(A) receptor alpha4 subunit suppression prevents withdrawal properties of an endogenous steroid. *Nature* **392,** 926–930.
Smith, S. S., Gong, Q. H., Li, X., Moran, M. H., Bitran, D., Frye, C. A., and Hsu, F. C. (1998b). Withdrawal from 3alpha-OH-5alpha-pregnan-20-one using a pseudopregnancy model alters the kinetics of hippocampal GABA$_A$-gated current and increases the GABA$_A$ receptor alpha4 subunit in association with increased anxiety. *J. Neurosci.* **18,** 5275–5284.
Steele, J., and Martin, I. (1999). Interaction of oestrone and 17β-oestradiol with the 5HT$_3$ receptor. *Br. J. Pharmacol.* **126,** 15P.
Thompson, S. A., and Wafford, K. A. (2001). Mechanism of action of general anaesthetics—New information from molecular pharmacology. *Cur. Op. Pharmacol.* **1,** 78–83.
Twyman, R. E., and Macdonald, R. L. (1992). Neurosteroid regulation of GABA$_A$ receptor single-channel kinetic properties of mouse spinal cord neurons in culture. *J. Physiol. (London)* **456,** 215–245.
Uzunova, V., Sheline, Y., Davis, J. M., Rasmusson, A., Uzunov, D. P., Costa, E., and Guidotti, A. (1998). Increase in the cerebrospinal fluid content of neurosteroids in patients with unipolar major depression who are receiving fluoxetine or fluvoxamine. *Proc. Natl. Acad. Sci. USA* **95,** 3239–3244.
Wafford, K. A., Thompson, S. A., Thomas, D., Sikela, J., Wilcox, A. S., and Whiting, P. J. (1996). Functional characterization of human gamma-aminobutyric acid$_A$ receptors containing the alpha4 subunit. *Mol. Pharmacol.* **50,** 670–678.
Whiting, P. J., McAllister, G., Vassilatis, D., Bonnert, T. P., Heavens, R. P., Smith, D. W., Hewson, L., O'Donnell, R., Rigby, M. R., Sirinathsinghji, D. J., Marshall, G., Thompson, S. A., Wafford, K. A., and Vasilatis, D. (1997). Neuronally restricted RNA splicing regulates the expression of a novel GABA$_A$ receptor subunit conferring atypical functional properties. *J. Neurosci.* **17,** 5027–5037.
Williams, S. R., Buhl, E. H., and Mody, I. (1998). The dynamics of synchronized neurotransmitter release determined from compound spontaneous IPSCs in rat dentate granule neurones *in vitro*. *J. Physiol. (London)* **510,** (2) 477–497.
Wisden, W., Laurie, D. J., Monyer, H., and Seeburg, P. H. (1992). The distribution of 13 GABA$_A$ receptor subunit mRNAs in the rat brain. I. Telencephalon, diencephalon, mesencephalon. *J. Neurosci.* **12,** 1040–1062.
Wittmer, L. L., Hu, Y., Kalkbrenner, M., Evers, A. S., Zorumski, C. F., and Covey, D. F. (1996). Enantioselectivity of steroid-induced gamma-aminobutyric acid$_A$ receptor modulation and anesthesia. *Mol. Pharmacol.* **50,** 1581–1586.
Wu, F. S., Gibbs, T. T., and Farb, D. H. (1990). Inverse modulation of gamma-aminobutyric acid-glycine-induced currents by progesterone. *Mol. Pharmacol.* **37,** 597–602.
Zhu, W. J., Wang, J. F., Krueger, K. E., and Vicini, S. (1996). Delta subunit inhibits neurosteroid modulation of GABA$_A$ receptors. *J. Neurosci.* **16,** 6648–6656.
Zhu, W. J., and Vicini, S. (1997). Neurosteroid prolongs GABA$_A$ channel deactivation by altering kinetics of desensitized states. *J. Neurosci.* **17,** 4022–4031.
Zorumski, C. F., Wittmer, L. L., Isenberg, K. E., Hu, Y., and Covey, D. F. (1996). Effects of neurosteroid and benz[e]indene enantiomers on GABA$_A$ receptors in cultured hippocampal neurons and transfected HEK-293 cells. *Neuropharmacology* **35,** 1161–1168.

GABA$_A$-RECEPTOR PLASTICITY DURING LONG-TERM EXPOSURE TO AND WITHDRAWAL FROM PROGESTERONE

Giovanni Biggio,* Paolo Follesa,* Enrico Sanna,* Robert H. Purdy,†
and Alessandra Concas*

*Department of Experimental Biology "Bernardo Loddo," University of Cagliari
09123 Cagliari, Italy
†Department of Psychiatry, Veterans Affairs Medical Center Research Foundation, University
of California, San Diego, California 92161, and Department of Neuropharmacology
The Scripps Research Institute, La Jolla, California

I. Introduction
II. Effects of Long-Term Exposure to PROG on GABA$_A$-Receptor Gene Expression and Function *in Vitro*
    A. GABA$_A$-Receptor Gene Expression
    B. GABA$_A$ Receptor Function
III. Effects of PROG Withdrawal on GABA$_A$-Receptor Gene Expression and Function *in Vitro*
IV. Effects of Long-Term Exposure to and Subsequent Withdrawal of PROG in Pseudo-pregnancy
V. Effects of Long-Term Exposure to and Subsequent Withdrawal of PROG in Pregnancy
    A. Neuroactive Steroid Concentrations in the Brain
    B. GABA$_A$-Receptor Density and Function in the Cortex and Hippocampus
    C. GABA$_A$-Receptor Gene Expression in the Cortex and Hippocampus
    D. GABA$_A$-Receptor Gene Expression in Hypothalamic Magnocellular Neurons
VI. Oral Contraceptives and GABA$_A$-Receptor Plasticity
VII. Mechanism of the Effect of Long-Term PROG Exposure on GABA$_A$-Receptor Plasticity
VIII. Conclusions
    References

The subunit composition of native γ-aminobutyric acid type A (GABA$_A$) receptors is an important determinant of the role of these receptors in the physiological and pharmacological modulation of neuronal excitability and associated behavior. GABA$_A$ receptors containing the α_1 subunit mediate the sedative-hypnotic effects of benzodiazepines (Rudolph *et al.*, 1999; McKernan *et al.*, 2000), whereas the anxiolytic effects of these drugs are mediated by receptors that contain the α_2 subunit (Löw *et al.*, 2000). In contrast, GABA$_A$ receptors containing the α_4 or α_6 subunits are insensitive to benzodiazepines (Barnard *et al.*, 1998). Characterization of the functions of GABA$_A$-receptors thus requires an understanding of the mechanisms by which the receptor subunit composition is regulated. The expression of specific $GABA_A$-receptor subunit genes in neurons is affected by endogenous and pharmacological modulators of receptor function. The expression of $GABA_A$-receptor subunit genes is thus regulated by neuroactive steroids both *in vitro* and *in vivo*. Such regulation occurs both during physiological conditions, such as pregnancy, and during pharmacologically induced conditions, such as pseudo-pregnancy and long-term treatment with steroid derivatives or anxiolytic-hypnotic drugs. Here, we summarize results obtained by our laboratory and by other groups pertaining to the effects of long-term exposure to, and subsequent withdrawal from, progesterone and its metabolite $3\alpha,5\alpha$-tetrahydroprogesterone on both the expression of $GABA_A$-receptor subunits and $GABA_A$-receptor function. © 2001 Academic Press.

I. Introduction

The observation that both the expression of γ-aminobutyric acid type A ($GABA_A$)-receptor subunit genes and $GABA_A$-receptor function are affected by long-term administration of sedative-hypnotic, anxiolytic, and anticonvulsant drugs has suggested that the mechanisms responsible for these changes may also contribute to the physiological modulation of $GABA_A$ receptors by endogenous compounds. Neurosteroids are steroid derivatives that are synthesized *de novo* from cholesterol in the central nervous system (CNS) (Hu *et al.*, 1987; Le Goascogne *et al.*, 1987; Mathur *et al.*, 1993; Mellon and Deshepper, 1993; Prasad *et al.*, 1994), some of which modulate $GABA_A$-receptor function with potencies and efficacies similar to or greater than those of benzodiazepines and barbiturates (Harrison and Simmonds, 1984; Majewska *et al.*, 1986; Majewska, 1992). These molecules have thus been suggested to be endogenous modulators of $GABA_A$-receptor-mediated neurotransmission. The progesterone (PROG) metabolite 3α-hydroxy-5α-pregnan-20-one [also known as $3\alpha,5\alpha$-tetrahydroprogesterone ($3\alpha,5\alpha$-TH PROG) or allopregnanolone] induces opening of the $GABA_A$-receptor-associated Cl^- channel at nanomolar concentrations *in vitro*, and, when administered systemically, induces pharmacological and behavioral effects similar to those elicited by anxiolytic, anticonvulsant, and hypnotic drugs that modulate $GABA_A$-receptor-mediated transmission (Majewska, 1992; Lambert *et al.*, 1995). The anxiolytic and anticonvulsant properties of PROG are mostly attributable to its conversion to $3\alpha,5\alpha$-TH PROG (Bitran *et al.*, 1993, 1995; Freeman *et al.*, 1993; Picazo and Fernández-Guasti, 1995; Kokate *et al.*, 1999; Reddy and Rogawski, 2000).

Changes in the peripheral or central production of PROG and consequent changes in the synaptic concentrations of $3\alpha,5\alpha$-TH PROG might therefore contribute to regulation of $GABA_A$-receptor-mediated synaptic activity and to emotional states associated with physiological conditions, such as stress, pregnancy, the menstrual cycle, and menopause. PROG and $3\alpha,5\alpha$-TH PROG also might be important in a variety of neurological and psychiatric disorders characterized by changes in affective behavior, sleep pattern, and neuronal excitability. Indeed, a selective decrease in the concentrations of $3\alpha,5\alpha$-TH PROG in plasma and cerebrospinal fluid has been detected in individuals with major depression, and these concentrations returned to normal values after treatment (Romeo *et al.*, 1998; Uzunova *et al.*, 1998; Ströhle *et al.*, 1999). Moreover, physiological and pharmacologically induced fluctuations in the plasma or brain concentration of $3\alpha,5\alpha$-TH PROG are associated with changes in $GABA_A$-receptor plasticity and function (Weiland and Orchinik, 1995; Concas *et al.*, 1998; Follesa *et al.*, 1998; Smith *et al.*, 1998a,b). Thus, fluctuations in the secretion of PROG and $3\alpha,5\alpha$-TH PROG that accompany various physiological and pathological conditions, together with the ability of neurons to synthesize $3\alpha,5\alpha$-TH PROG from PROG (Follesa *et al.*, 2000), are likely to be important determinants of the regulation of $GABA_A$-receptor gene expression and function. Such regulation may contribute to changes in neuronal excitability and to the development of emotional and affective disorders often associated with these conditions. For example, the rapid decrease in the peripheral and central concentrations of PROG that occurs during the menstrual cycle may contribute to mental symptoms associated with premenstrual syndrome, and the marked decrease in the secretion of steroid hormones that occurs during menopause may be important in the development of such symptoms in postmenopausal women (Schmidt *et al.*, 1994; Wang *et al.*, 1996; Rapkin *et al.*, 1997; Bicikova *et al.*, 1998; Genazzani *et al.*, 1998).

II. Effects of Long-Term Exposure to PROG on $GABA_A$-Receptor Gene Expression and Function *in Vitro*

A. $GABA_A$-RECEPTOR GENE EXPRESSION

Long-term treatment of chick neurons in culture with pregnanolone [$3\alpha,5\beta$-tetrahydroprogesterone ($3\alpha,5\beta$-TH PROG)] or $3\alpha,5\alpha$-TH PROG induces both homologous and heterologous uncoupling of the various recognition sites associated with $GABA_A$ receptors (Friedman *et al.*, 1993, 1996), effects that are similar, but not identical, to those apparent after chronic treatment of cultured neurons with benzodiazepines, barbiturates, or GABA (Roca *et al.*, 1989, 1990). Long-term exposure of neurons to $3\alpha,5\beta$-TH PROG, to GABA, to the benzodiazepine flunitrazepam, or to pentobarbital thus induced complete uncoupling of barbiturate and benzodiazepine sites, partial uncoupling of GABA and benzodiazepine sites, and different extents of uncoupling of steroid and benzodiazepine sites. Moreover, the heterologous uncoupling induced by 3α,5β-TH PROG was inhibited by concomitant treatment with a GABA antagonist, whereas homologous uncoupling was resistant to the antagonist. The type and extent of uncoupling induced by positive modulators of $GABA_A$ receptors therefore appear to depend on the specific modulator. Accordingly, the state of enhanced seizure susceptibility induced by PROG withdrawal in rats is accompanied by a marked increase in the anticonvulsant potency of steroid derivatives and a decrease in that of diazepam or valproic acid (Reddy and Rogawski, 2000).

Yu and Ticku (1995a,b) also showed that chronic treatment of mouse cerebral cortical neurons with 3α,5α-TH PROG did not affect the basal binding of the benzodiazepines [^{3}H]flunitrazepam, [^{3}H]Ro 15-1788, or [^{3}H]Ro 15-4513 to neuronal membranes, the dissociation constant (K_d), or the maximal number of binding sites (B_{max}) for [^{3}H]flunitrazepam binding to intact neurons. However, long-term treatment with 3α,5α-TH PROG induced uncoupling between GABA, barbiturate, and neurosteroid sites and the benzodiazepine site. Although the median effective concentration (EC_{50}) values of these ligands were not substantially altered, their maximal response (E_{max}) values were reduced after chronic 3α,5α-TH PROG treatment. The binding of [^{3}H]GABA and that of t-[^{35}S]butylbicyclophosphorothionate (TBPS) were also reduced after long-term exposure of neurons to 3α,5α-TH PROG, effects that were attributable to decreases in the B_{max} value of the low-affinity GABA binding site (the high-affinity site being unaffected) and in that of the TBPS binding site. Furthermore, chronic treatment with 3α,5α-TH PROG reduced the efficacy of GABA in functional assays, an effect that was attributable to a decrease in the E_{max} value (with no effect on the EC_{50}). The same treatment also reduced both the E_{max} value of diazepam with regard to potentiation of GABA-induced $^{36}Cl^-$ influx as well as the maximal inhibitory response (I_{max}) of the benzodiazepine receptor inverse agonist DMCM with regard to inhibition of GABA-induced $^{36}Cl^-$ influx; again, the respective EC_{50} and IC_{50} values were not affected. Long-term exposure to 3α,5α-TH PROG also reduced the E_{max} value of 3α,5α-TH PROG with regard to potentiation of GABA-induced $^{36}Cl^-$ influx without affecting the EC_{50} value. Finally, electrophysiological measurements with cortical neurons in the whole-cell mode revealed that chronic 3α,5α-TH PROG treatment reduced the GABA-induced current by 78%, an effect that was associated with a reduced potentiation of this current by pentobarbital or 3α,5α-TH PROG. These observations thus support the notion that the potentiation of GABA action by pentobarbital or neurosteroids is attenuated by long-term exposure of neurons to 3α,5α-TH PROG.

The decrease in the numbers of GABA and TBPS binding sites and the reduction in $GABA_A$-receptor function induced by long-term exposure of cultured neurons to $3\alpha,5\alpha$-TH PROG are associated with changes in the abundance of mRNAs encoding specific $GABA_A$-receptor subunits (Yu et al., 1996a). Such changes had been proposed as a possible mechanism for the downregulation of $GABA_A$ receptors observed in cultured chick neurons during chronic treatment with $3\alpha,5\beta$-TH PROG (Friedman et al., 1993) and demonstrated in mammalian neurons after chronic treatment with $3\alpha,5\alpha$-TH PROG (Yu et al., 1996a). Thus, changes in the expression of genes for the various $GABA_A$-receptor subunits and the consequent synthesis of new receptors might be a mechanism by which the sensitivity of neurons to positive and negative modulators of $GABA_A$ receptors is altered by long-term exposure to neuroactive steroids. Such changes that result in the production of $GABA_A$ receptors with uncoupled steroid-, benzodiazepine-, and GABA-recognition sites and a reduced sensitivity to $3\alpha,5\alpha$-TH PROG and to other positive modulators might be of clinical relevance.

To characterize further the contributions to $GABA_A$-receptor modulation of neurosteroids produced in the CNS from precursors, such as PROG, synthesized in the periphery, we have examined the effects of long-term exposure to PROG with cultures of rat cerebellar granule cells. These cultures are composed mostly (95%) of neurons, with the remaining cell types including glial cells. We first examined whether the granule cell cultures synthesize the mRNAs for 5α-reductase and 3α-hydroxysteroid oxidoreductase (3αHSD), the enzymes required for the conversion of PROG to $3\alpha,5\alpha$-TH PROG. An RNase protection assay detected both 5α-reductase and 3αHSD transcripts in the cultured cells (Fig. 1). *In situ* hybridization also revealed the presence of 5α-reductase mRNA within the cell bodies of cerebellar granule cells (Follesa et al., 2000). The abundance of 5α-reductase mRNA was not affected by exposure of the cultures to 1 μM PROG for 5 days (Follesa et al., 2000). Consistent with the presence of these enzyme mRNAs in the cerebellar granule cells, exposure of cultures to 1 μM PROG for 5 days resulted in a large increase in the amount of $3\alpha,5\alpha$-TH PROG in both the cells and culture medium (Table I), suggesting that the granule cells are the major source of $3\alpha,5\alpha$-TH PROG in this primary culture system.

An RNase protection assay revealed that exposure of the cultured granule cells to 1 μM PROG for 5 days resulted in a marked decrease in the abundance of transcripts encoding both γ_2 subunits (γ_2S and γ_2L) of the $GABA_A$ receptor (Fig. 2). Similar treatment of these cells with 1 μM $3\alpha,5\alpha$-TH PROG reduced the abundance of γ_2S and γ_2L subunit transcripts by $27 \pm 3\%$ and $40 \pm 8\%$, respectively ($P < 0.01$) (Follesa et al., 2000). Long-term exposure of cultures to PROG also significantly reduced the abundance of mRNAs encoding the α_1, α_3, and α_5 subunits, but it had no effect on that of the α_4-, β_1-, or β_2-subunit mRNAs (Fig. 2). The conversion of PROG to $3\alpha,5\alpha$-TH PROG by the cultured cells was blocked (Table I) by the 5α-reductase

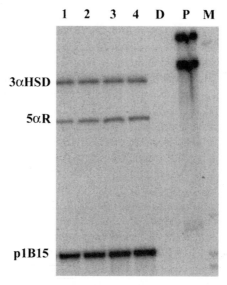

FIG. 1. Detection of 5α-reductase and 3α-hydroxysteroid oxidoreductase mRNAs in rat cerebellar granule cell cultures. (**Lanes 1–4**) Cells were cultured for 8 days, after which total RNA was extracted and subjected to an RNase protection assay with probes specific for 5α-reductase (5αR), 3α-hydroxysteroid oxidoreductase (3αHSD), and cyclophilin (p1B15; internal standard) mRNAs. (**Lane D**) digested probes. (**Lane P**) Probes alone. (**Lane M**) Molecular size markers.

inhibitor finasteride (Rittmaster, 1994; Azzolina *et al.*, 1997). This inhibitor also blocked the PROG-induced decreases in the amounts of γ_2L and γ_2S transcripts in the granule cell cultures (Fig. 3).

Our observation that finasteride prevented both the conversion of PROG to 3α,5α-TH PROG and the effect of PROG on the abundance of GABA$_A$-receptor subunit mRNAs in granule cell cultures provided direct evidence that a compound produced by neuronal metabolism of PROG, rather than PROG itself, modulates GABA$_A$-receptor plasticity. Consistent with this notion, 3α,5α-TH PROG, but not PROG, exhibits a positive allosteric modulatory action at GABA$_A$ receptors (Majewska *et al.*, 1986; Wu *et al.*, 1990; Friedman *et al.*, 1993).

Our demonstration that PROG modulates GABA$_A$-receptor gene expression in cultured neurons is consistent with the results of both previous *in vivo* studies of pregnant and pseudo-pregnant rats (Fenelon and Herbison, 1996; Brussaard *et al.*, 1997; Concas *et al.*, 1998; Follesa *et al.*, 1998; Smith *et al.*, 1998a,b) and *in vitro* studies (Yu *et al.*, 1996a). Moreover, our data are also consistent with the general observation that chronic treatment with positive allosteric modulators that act at different sites of the GABA$_A$ receptor

TABLE I

EFFECTS OF PROG AND FINASTERIDE ON THE AMOUNT OF $3\alpha,5\alpha$-TH PROG IN CELLS AND CONDITIONED MEDIUM OF CEREBELLAR GRANULE CELL CULTURES

	$3\alpha,5\alpha$-TH PROG			
	Cells		Medium	
Treatment	(% of control)	(ng/g protein)	(% of control)	(ng/ml medium)
Control	100 ± 9	0.60 ± 0.05	100 ± 5	0.56 ± 0.03
Progesterone	$563 \pm 123^*$	3.40 ± 0.70	$269 \pm 34^*$	1.50 ± 0.19
Finasteride + progesterone	$106 \pm 12^\dagger$	0.63 ± 0.07	$102 \pm 14^\dagger$	0.57 ± 0.08
Finasteride	70 ± 5	0.42 ± 0.03	68 ± 7	0.38 ± 0.04

From Follesa *et al.* (2000). Allopregnanolone synthesis in cerebellar granule cells: Roles in regulation of GABA$_A$ receptor expression and function during progesterone treatment and withdrawal. *Mol. Pharmacol.* 57, 1262–1270; adapted with permission.

Note. Cultures were incubated for 5 days in the absence or presence of 1 μM progesterone or 1 μM finasteride as indicated, after which the amount of $3\alpha,5\alpha$-TH PROG in cells and medium was measured by radioimmunoassay. Data are expressed as absolute amounts and as a percentage of control (solvent) and are means $\pm$ SEM of values from four independent experiments ($n = 11$–16).

$^*P < 0.01$ vs control.
$^\dagger P < 0.01$ vs PROG.

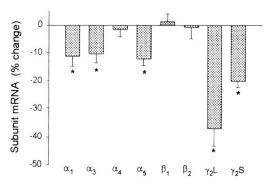

FIG. 2. Effects of long-term exposure of cerebellar granule cell cultures to PROG on the abundance of GABA$_A$-receptor-subunit mRNAs. Cultured cerebellar granule cells were incubated for 5 days in the presence of 1 μM progesterone, after which total RNA was extracted and the amounts of receptor subunit mRNAs were measured by RNase protection assay. Data are expressed as the percentage change in transcript abundance compared with values for control (vehicle-treated) cells and are means $\pm$ SEM ($n = 13$–23) of values from three to four independent experiments. ($^*P < 0.01$ vs control cells.) [From Follesa *et al.* (2000). Allopregnanolone synthesis in cerebellar granule cells: Roles in regulation of GABA$_A$ receptor expression and function during progesterone treatment and withdrawal. *Mol. Pharmacol.* 57, 1262–1270; reprinted with permission.]

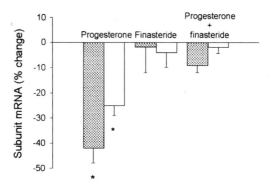

FIG. 3. Effect of finasteride on the prog-induced decrease in the abundance of $GABA_A$-receptor γ_2-subunit transcripts. Cultured cerebellar granule cells were incubated for 5 days in the absence or presence of 1 μM progesterone or 1 μM finasteride as indicated, after which total RNA was extracted and subjected to RNase protection assay for determination of the amounts of γ_2L (shaded bars) and γ_2S (open bars) transcripts. Data are expressed as the percentage change relative to the values for control (vehicle-treated) cells, and are means ± SEM ($n = 9$) of values from three independent experiments. (*$P < 0.01$ control cells.) [From Follesa et al. (2000). Allopregnanolone synthesis in cerebellar granule cells: Roles in regulation of $GABA_A$ receptor expression and function during progesterone treatment and withdrawal. Mol. Pharmacol. 57, 1262–1270; reprinted with permission.]

results in downregulation of the receptor by reducing the abundance of specific receptor subunit mRNAs (Roca et al., 1989, 1990; Morrow et al., 1990; Montpied et al., 1991; Mhatre and Ticku, 1992; Mhatre et al., 1993; Holt et al., 1996, 1997; Impagnatiello et al., 1996; Yu et al., 1996a; Follesa et al., unpublished observation).

B. $GABA_A$ RECEPTOR FUNCTION

The changes in $GABA_A$-receptor gene expression induced in cerebellar granule cells by long-term exposure to PROG were accompanied by changes in receptor function. These functional changes were characterized with the voltage-clamp technique and receptors transplanted from cultured granule cells into Xenopus oocytes. Receptor transplantation was accomplished by injecting crude membrane vesicles prepared from granule cells into the oocytes. The transplanted receptors are efficiently inserted into the oocyte plasma membrane, likely as a result of fusion of the membrane vesicles with the plasma membrane, where they form clusters that retain their native properties (Morales et al., 1995; Sanna et al., 1996, 1998). Twelve to 18 h after injection of oocytes with granule cell membrane vesicles, GABA induces an inward Cl^- current with a peak amplitude that is dependent on the

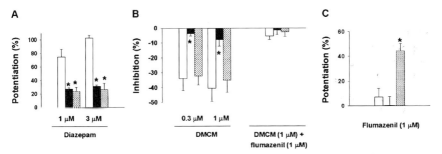

FIG. 4. Effects of long-term treatment with and subsequent withdrawal of PROG on the modulation of GABA$_A$-receptor function by various benzodiazepine receptor ligands. *Xenopus* oocytes were injected with crude membranes prepared either from control (vehicle-treated) cerebellar granule cells (open bars), from granule cells treated with 1 µM progesterone for 5 days (black bars), or from granule cells incubated with 1 µM progesterone for 5 days and then in the absence of the steroid for 6 h (shaded bars). (**A**) The effects of the indicated concentrations of diazepam; (**B**) DMCM in the absence or presence of flumazenil or (**C**) flumazenil on the amplitude of the Cl$^-$ current induced by GABA (at a concentration yielding 10% of the maximal response) were examined. Data are expressed as percentage potentiation or percentage inhibition of the response to GABA and are means ± SEM of values obtained from three to seven different oocytes. (*$P < 0.05$ vs receptors from control cells.) [From Follesa *et al.* (2000). Allopregnanolone synthesis in cerebellar granule cells: Roles in regulation of GABA$_A$ receptor expression and function during progesterone treatment and withdrawal. *Mol. Pharmacol.* 57, 1262–1270; adapted with permission.]

neurotransmitter concentration; maximal current amplitudes, induced by 10 mM GABA, usually range from 100 to 200 nA.

The benzodiazepine diazepam markedly potentiated GABA-evoked Cl$^-$ currents in oocytes expressing GABA$_A$ receptors from control granule cells (Fig. 4A). This effect was concentration dependent, with potentiation values of $74.4 \pm 12\%$ and $102.6 \pm 4\%$ at 1 and 3 µM diazepam, respectively. In oocytes injected with membrane vesicles prepared from granule cells after exposure to 1 µM PROG for 5 days, the potentiating effect of diazepam was much less pronounced ($27.5 \pm 2\%$ and $31.6 \pm 2\%$ at 1 and 3 µM, respectively). The anxiogenic and convulsant β-carboline derivative DMCM, a benzodiazepine receptor inverse agonist (Braestrup *et al.*, 1982, 1983), induced a marked inhibition ($34.1 \pm 8\%$ and $40.5 \pm 9\%$ at 0.3 and 1 µM, respectively) of GABA-evoked Cl$^-$ currents in oocytes injected with membranes from control granule cells (Fig. 4B). Consistent with previous results (Whittemore *et al.*, 1996), higher concentrations (10–30 µM) of this drug enhanced the response of control GABA$_A$ receptors to GABA (data not shown). In oocytes expressing GABA$_A$ receptors from PROG-treated granule cells, DMCM at 0.3 or 1 µM had no effect on GABA-evoked Cl$^-$ currents. The inhibitory effect of DMCM on GABA-evoked Cl$^-$ currents in oocytes injected with membrane vesicles from control granule cells was completely blocked by 1 μM flumazenil (Fig. 4B). Moreover, this benzodiazepine receptor antagonist alone had no significant effect on GABA-evoked Cl$^-$ currents in oocytes injected with membranes from either control granule cells or those subjected to long-term treatment with PROG (Fig. 4C). Finally, the changes in diazepam and DMCM sensitivity apparent with GABA$_A$ receptors from PROG-treated granule cells were prevented by the inclusion of finasteride in the 5-day incubation of cells with this steroid (data not shown).

These data show that long-term exposure of cultured cerebellar granule cells to PROG mimics the effect of chronic treatment of cultured cortical neurons with 3α,5α-TH PROG (Yu *et al.*, 1996a,b). The decrease in the abilities of diazepam and DMCM to potentiate and inhibit, respectively, GABA-evoked Cl$^-$ currents after long-term exposure of granule cells to PROG is consistent with the decreases in the abundance of α_1-, α_3-, α_5-, and γ_2-subunit mRNAs induced by such treatment. Thus, both α and γ_2 subunits are required for GABA$_A$ receptors to show maximal sensitivity to benzodiazepines as well as to benzodiazepine receptor inverse agonists (Pritchett *et al.*, 1989; Barnard *et al.*, 1998). Although it is likely that such changes in the abundance of receptor subunit mRNAs result in corresponding changes in the synthesis of the encoded proteins, the relations between the amount of receptor subunit mRNAs and the amount of subunit proteins expressed on the cell surface remain to be determined.

The ability of granule cells to metabolize PROG to 3α,5α-TH PROG reinforces the notion that neuronal metabolism of PROG produced by peripheral organs likely contributes to the amount of 3α,5α-TH PROG in the brain and to the physiological modulation of GABAergic synapses. Physiological or pharmacologically induced fluctuations in PROG production by the gonads or the adrenal glands therefore likely affect the expression of specific GABA$_A$-receptor subunit genes and GABA$_A$ receptor activity in specific regions of the brain.

III. Effects of PROG Withdrawal on GABA$_A$-Receptor Gene Expression and Function *in Vitro*

The discontinuation of long-term exposure of cultured granule cells to PROG and the consequent sudden decrease in the production of 3α,5α-TH PROG by these cells resulted in a selective increase in the abundance of the GABA$_A$-receptor α_4-subunit mRNA (Fig. 5). The decreases in the amounts of α_1- and γ_2L-subunit mRNAs elicited by persistent exposure to PROG remained apparent after PROG withdrawal.

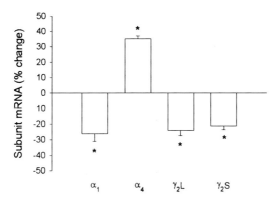

FIG. 5. Effects of PROG withdrawal on the abundance of GABA$_A$-receptor-subunit mRNAs in cerebellar granule cells. Cultured cerebellar granule cells were incubated for 5 days in the presence of 1 μM progesterone and then for 6 h in the absence of this steroid, after which total RNA was extracted and the amounts of the indicated subunit mRNAs were measured by RNase protection assay. Data are expressed as the percentage change relative to values for control (vehicle-treated) cells and are means ± SEM of values from six independent experiments. ($*P < 0.01$ vs control values.) [From Follesa et al. (2000). Allopregnanolone synthesis in cerebellar granule cells: Roles in regulation of GABA$_A$ receptor expression and function during progesterone treatment and withdrawal. Mol. Pharmacol. 57, 1262–1270; adapted with permission.]

The presence of the α_4 subunit in recombinant GABA$_A$ receptors is associated with a reduced sensitivity to classical benzodiazepine agonists and zolpidem as well as with distinct patterns of regulation by flumazenil, DMCM, and other positive and negative modulators (Barnard et al., 1998). Electrophysiological recording of the pharmacological responses of GABA$_A$ receptors introduced into Xenopus oocytes revealed that receptors derived from cells subjected to PROG withdrawal were both markedly less sensitive to the potentiating effect of diazepam than were those derived from control cells as well as positively modulated by the benzodiazepine receptor antagonist flumazenil. Thus, in oocytes expressing GABA$_A$ receptors derived from granule cells 6 h after PROG withdrawal, diazepam potentiated GABA-evoked Cl$^-$ currents by only $23.9 \pm 6\%$ and $27.0 \pm 9\%$ at 1 and 3 μM, respectively (Fig. 4A), whereas flumazenil potentiated GABA-evoked Cl$^-$ currents by $44.4 \pm 6\%$ (Fig. 4C). These characteristics are consistent with those previously determined for GABA$_A$ receptors containing the α_4 subunit (Whittemore et al., 1996; Barnard et al., 1998). Withdrawal from chronic PROG treatment restored the sensitivity of GABA$_A$ receptors to the inhibitory action of the benzodiazepine receptor inverse agonist DMCM, sensitivity that was markedly reduced during PROG exposure (Fig. 4B).

Finally, the changes in diazepam, flumazenil, and DMCM sensitivity apparent for $GABA_A$ receptors derived from granule cells subjected to PROG withdrawal were prevented by inclusion of finasteride in the 5-day incubation of the neurons with PROG (data not shown). Given that recombinant α_4-subunit-containing receptors, like α_1-subunit-containing receptors, are negatively modulated by DMCM (Whittemore et al., 1996), our data suggest that the increased sensitivity of $GABA_A$ receptors to DMCM induced by PROG withdrawal is attributable to the increase in abundance of the α_4 subunit mRNA (Follesa et al., 2000). Such increased sensitivity to endogenous inverse agonists conferred by an increase in the expression of the α_4 subunit may contribute to the pathogenesis of PROG withdrawal syndrome. Consistent with this notion, the increase in the abundance of the α_4-subunit mRNA apparent during withdrawal from PROG in a rat pseudo-pregnancy model is associated with changes in the kinetics of hippocampal $GABA_A$-receptor-mediated currents, with experimental anxiety, and with an increased susceptibility to seizures (Smith et al., 1998b; Reddy and Rogawski, 2000). Our observation that the effects of PROG withdrawal in vitro were prevented by inclusion of finasteride in the incubation with PROG further suggests that the withdrawal-induced increase in the abundance of the α_4-subunit mRNA is triggered by the associated decrease in $3\alpha,5\alpha$-TH PROG accumulation. This conclusion is consistent with our previous demonstration that, by preventing the pregnancy-induced increase in the brain concentration of $3\alpha,5\alpha$-TH PROG, finasteride antagonized the decrease in $GABA_A$-receptor function and gene expression normally observed during pregnancy in rats (Concas et al., 1998) as well as with the anticonvulsant effect elicited by this PROG metabolite both in pseudo-pregnant rats and in mice treated with pentylenetetrazol (Kokate et al., 1999; Reddy and Rogawski, 2000).

IV. Effects of Long-Term Exposure to and Subsequent Withdrawal of PROG in Pseudo-pregnancy

The effects of PROG withdrawal in vivo have been investigated with the rat pseudo-pregnancy paradigm (Smith, Gong, Li et al., 1998; Reddy and Rogawski, 2000). Pseudo-pregnancy is thought to be a more physiological experimental model than are multiple-injection or chronic-implant paradigms, in that PROG is produced both from an endogenous source (the ovaries) and according to a time course resembling that apparent during the luteal phase of the menstrual cycle. Pseudo-pregnancy is induced in rats by gonadotropin treatment and is associated with long-term increases in the serum concentrations of PROG and $3\alpha,5\alpha$-TH PROG; these increases are followed by abrupt PROG and neurosteroid withdrawal induced by ovariectomy or administration of 5α-reductase inhibitors, such as MK-906 or finasteride (Rittmaster, 1994; Azzolina *et al.*, 1997).

Neurosteroid withdrawal in pseudo-pregnant rats is associated with increases both in anxiety-like behavior, as revealed in the plus-maze paradigm (Smith *et al.*, 1998b), and in susceptibility to pentylentetrazol-induced seizures (Reddy and Rogawski, 2000) compared with control animals. Electrophysiological recording of GABA-gated Cl^- currents in acutely dissociated CA1 hippocampal pyramidal neurons from pseudo-pregnant rats undergoing neurosteroid withdrawal revealed a marked decrease in the decay time for recovery of currents (Smith *et al.*, 1998b), indicative of a pronounced decrease in the total GABA-gated current. Such a change in the kinetic properties of GABA-gated Cl^- channels likely results in a decrease in inhibitory tone and a consequent increase in neuronal excitability that may contribute to withdrawal symptoms, such as anxiety and seizure susceptibility.

Further evaluation of the allosteric modulation of GABA-gated Cl^- channels revealed that neurosteroid withdrawal in pseudo-pregnant rats induces marked alterations in the pharmacology of $GABA_A$ receptors (Smith *et al.*, 1998b). Neurosteroid withdrawal is thus associated with a marked decrease both in the ability of benzodiazepines to enhance and in that of β-carboline inverse agonists to inhibit GABA-gated currents. In addition, the benzodiazepine receptor competitive antagonist flumazenil, which lacks intrinsic activity in neurons from control or pseudo-pregnant rats, acts as a positive modulator of GABA-evoked Cl^- currents in neurons isolated from animals undergoing neurosteroid withdrawal. These results are consistent with the behavioral observations of reduced sedative (Smith *et al.*, 1998b) and anticonvulsant (Reddy and Rogawski, 2000) actions of benzodiazepines during neurosteroid withdrawal in pseudo-pregnant rats. Furthermore, they suggest that neurosteroid withdrawal in pseudo-pregnant rats is associated with the development of cross-tolerance to benzodiazepines, an effect similar to the benzodiazepine insensitivity observed in women with premenstrual syndrome (Sundstrom *et al.*, 1997). Neurosteroid withdrawal was also associated with an inability of 3α,5α-TH PROG to potentiate GABA-gated Cl^- currents (Smith *et al.*, 1998b). This latter observation, however, contrasts with data obtained by Reddy and Rogawski (2000) showing that neurosteroid withdrawal in pseudo-pregnant rats enhanced the anticonvulsant activity of ganaxolone, a synthetic 3β-methyl analog of 3α,5α-TH PROG.

Semiquantitative reverse transcription–polymerase chain reaction analysis and immunoblot analyses revealed that neurosteroid withdrawal induced a marked and selective increase in the abundance of both $GABA_A$-receptor $α_4$-subunit mRNA and protein in the hippopcampus of pseudo-pregnant rats (Smith *et al.*, 1998b). These data are consistent with the effects of neurosteroid withdrawal on α_4-subunit expression observed by the same group with a model of chronic intermittent administration of PROG (Smith *et al.*, 1998a) as well as by us with cultured cerebellar granule neurons (Follesa *et al.*, 2000). In addition, administration of α_4-subunit antisense RNA, but not of vehicle or of a control oligoribonucleotide, not only blocked the increase in the expression of the α_4 subunit but also prevented the development of insensitivity to benzodiazepines, the change in Cl^- current kinetics, and the increase in sensitivity to convulsant agents induced by neurosteroid withdrawal in pseudo-pregnant rats (Smith *et al.*, 1998a). These data indicate that the enhancement of α_4-subunit gene expression underlies most of the electrophysiological, pharmacological, and behavioral effects induced by neurosteroid withdrawal.

Together, these studies have shown that abrupt discontinuation of long-term treatment with PROG in rats is associated with increased anxiety-like behavior (Gallo and Smith, 1993; Smith *et al.*, 1998b), increased sensitivity to convulsants, and an increased expression of the $GABA_A$-receptor α_4-subunit (Moran *et al.*, 1998; Smith *et al.*, 1998a; Reddy and Rogawski, 2000). The behavioral symptoms and the molecular changes induced by PROG withdrawal are similar to those elicited by the withdrawal of other positive modulators of $GABA_A$ receptors, including benzodiazepines, barbiturates, and ethanol (Majchrowicz, 1975; Tseng *et al.*, 1993; Devaud *et al.*, 1997; Mahmoudi *et al.*, 1997; Schweizer and Rickels, 1998; Follesa *et al.*, unpublished observation). The anxiogenic, proconvulsant, and neurochemical effects of PROG withdrawal appear to result from termination of the persistent interaction of $3\alpha,5\alpha$-TH PROG with $GABA_A$ receptors, given that they are prevented by prior administration of indomethacin, a 3αHSD blocker (Gallo and Smith, 1993).

V. Effects of Long-Term Exposure to and Subsequent Withdrawal of PROG in Pregnancy

A. NEUROACTIVE STEROID CONCENTRATIONS IN THE BRAIN

Pregnancy is a physiological condition that is associated with marked changes in the hormonal milieu. PROG concentrations are maximal during pregnancy, and the concentrations of other steroids are also greatly increased. The concentrations of $3\alpha,5\alpha$-TH PROG and $3\alpha,5\alpha$-tetrahyrodeoxycorticosterone ($3\alpha,5\alpha$-TH DOC) are increased in plasma as a result of the high concentrations of their precursors, PROG and deoxycorticosterone. In addition, the activities of the enzymes responsible for the synthesis of $3\alpha,5\alpha$-TH PROG and $3\alpha,5\alpha$-TH DOC are increased in both maternal (especially the placenta) and fetal tissue (Milewich *et al.,* 1979; Buster, 1983). We have also shown in rats that the plasma concentrations of PROG, $3\alpha,5\alpha$-TH PROG, and $3\alpha,5\alpha$-TH DOC peak on Days 15 or 19 of pregnancy, return to control values immediately before delivery (Day 21), and remain unchanged for 2 days after delivery (Concas *et al.,* 1998, 1999; Biggio *et al.,* 2000). Whereas PROG concentrations in both the brain and plasma are maximal on Day 15 (approximately 12 and 10 times the estrus values, respectively) and remain substantially increased on Day 19, the cerebrocortical concentrations of $3\alpha,5\alpha$-TH PROG and $3\alpha,5\alpha$-TH DOC do not peak until Day 19 of pregnancy (+208% and +90%, respectively), suggesting that the synthesis of $3\alpha,5\alpha$-TH PROG and $3\alpha,5\alpha$-TH DOC in the brain is not simply a function of the plasma concentration of PROG. One possible explanation for this difference in kinetics is that steroid hormones such as estrogens or other agents may regulate the activity or expression of either 5α-reductase or 3αHSD in the brain during pregnancy. Indeed, estradiol has been shown to regulate 3αHSD activity in the rat brain (Penning *et al.,* 1985).

Given that $3\alpha,5\alpha$-TH PROG and $3\alpha,5\alpha$-TH DOC each positively modulates $GABA_A$ receptors (Majewska, 1992; Lambert *et al.,* 1995), several studies have been undertaken to determine whether the physiological fluctuations in the concentrations of these neuroactive steroids that occur during pregnancy and after delivery affect the plasticity and function of $GABA_A$ receptors in various regions of the brain.

B. $GABA_A$-Receptor Density and Function in the Cortex and Hippocampus

Majewska *et al.* (1989) demonstrated that pregnancy induces changes in $GABA_A$-receptor sensitivity in the maternal brain. Thus, the affinity of $GABA_A$ receptors in the rat forebrain for [^{3}H]muscimol was increased on Days 15 and 19 of gestation. A further increase in receptor affinity associated with a decrease in receptor density was also apparent in the postpartum period. In addition, Weizman *et al.* (1997) showed that the density of central benzodiazepine binding sites was increased in the hippocampus and decreased in the hypothalamus on Day 19 of pregnancy in the rat; these researchers did not detect changes in [^{3}H]muscimol binding in the cerebral cortex. These changes in the density of $GABA_A$ receptors were proposed to result from an action of $3\alpha,5\alpha$-TH PROG and $3\alpha,5\alpha$-TH DOC, the concentrations of which increase markedly in the placenta and the adrenal glands during pregnancy.

The idea that a functional relationship exists between neuroactive steroids and $GABA_A$ receptors during pregnancy and lactation is supported by results from our laboratory (Concas et al., 1998, 1999; Follesa et al., 1998; Biggio et al., 2000). We have thus shown that the marked increases in the concentrations of neuroactive steroids in plasma and the brain during pregnancy exert a tonic modulatory action on the activity and expression of cerebrocortical $GABA_A$ receptors, whereas the rapid and substantial decreases in the concentrations of these compounds apparent immediately before delivery and their low levels during lactation may represent a withdrawal-like phenomenon.

We showed that GABA binding sites in the brain are markedly modified during pregnancy and lactation in rats. Increases in both the density of low-affinity [^{3}H]GABA binding sites (B_{max}) and in the corresponding K_d value were apparent in the cerebral cortex on Day 19 of pregnancy (Fig. 6A). The density of these binding sites had returned to control (estrus) values immediately before delivery (Day 21), was significantly decreased 2 days after delivery, and had returned to control values again by 7 days after delivery. Benzodiazepine receptors labeled by [^{3}H]flunitrazepam in the cerebral cortex showed a similar pattern of changes (Fig. 6B).

Pregnancy also affected the $GABA_A$-receptor-coupled Cl^- channel. The density of [^{35}S]TBPS binding sites thus showed a temporal pattern of changes similar to those observed for GABA and benzodiazepine binding sites (Fig. 6C). In addition, measurement of muscimol-stimulated $^{36}Cl^-$ uptake by cerebrocortical membrane vesicles revealed that the sensitivity of the Cl^- channel to the action of this GABA agonist was reduced during the last week of pregnancy (Fig. 7). In contrast, the effect of muscimol on $^{36}Cl^-$ uptake was markedly potentiated 2 days after delivery. These data may reflect physiological changes in the activity of the $GABA_A$-receptor-associated Cl^- channel during pregnancy and after delivery. The reduced sensitivity of the Cl^- channel to the action of muscimol during pregnancy is thus suggestive of a reduced activity of this channel *in vivo*. Consistent with these results, the abilities of diazepam and $3\alpha,5\alpha$-TH PROG to potentiate the effect of muscimol on $^{36}Cl^-$ uptake were decreased during pregnancy and increased after delivery (Follesa et al., 1998).

The temporal changes in the density and function of $GABA_A$ receptors during pregnancy and after delivery are similar to those in the concentrations of $3\alpha,5\alpha$-TH PROG and $3\alpha,5\alpha$-TH DOC. The cortical concentrations of these neurosteroids peak on Day 19 of pregnancy, coincident with the maximal increase in the density of GABA, benzodiazepine, and TBPS recognition sites and the maximal reduction in Cl^- channel function. Furthermore, the normalization of the plasma and cortical concentrations of neurosteroids that precedes delivery (Day 21 of pregnancy) is paralleled by a return of the density and function of $GABA_A$ receptors to control values.

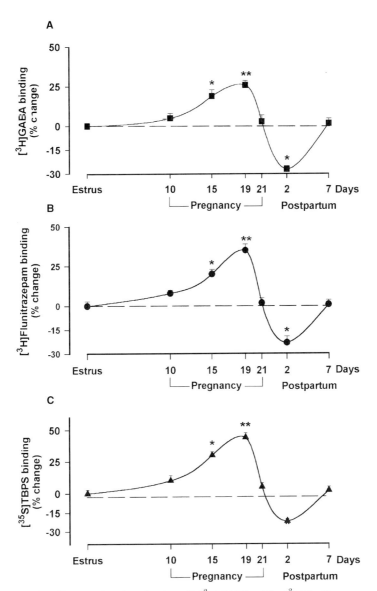

FIG. 6. (**A**) Changes in the density of [^{3}H]GABA, (**B**) [^{3}H]flunitrazepam, and (**C**) [^{35}S]TBPS binding sites in the cerebral cortex during pregnancy and after delivery in the rat. Saturation binding analysis was performed with 12 concentrations (10–1800 nM) of [^{3}H]GABA, eight concentrations (0.125–16 nM) of [^{3}H]flunitrazepam, and seven concentrations (2.5–500 nM) of [^{35}S]TBPS. Data are expressed as the percentage change with respect to the estrus value and are means ± SEM of values from four to six experiments, each performed in triplicate. (*$P < 0.05$, **$P < 0.01$ vs estrus value.)

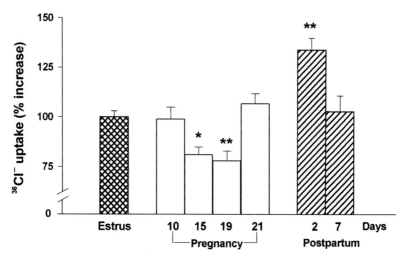

FIG. 7. Effect of muscimol on ^{36}Cl$^-$ uptake by membrane vesicles prepared from the cerebral cortex of rats during pregnancy and after delivery. Membrane vesicles prepared from rats on the indicated days were incubated for 10 min at 30°C before the addition of ^{36}Cl$^-$ in the absence (basal value) or presence of 5 μM muscimol. Uptake of ^{36}Cl$^-$ was terminated after 5 s. Data are expressed as the percentage increase in ^{36}Cl$^-$ uptake induced by 5 μM muscimol relative to the basal value and are means ± SEM of values from six experiments, each performed with two rats per group. (*$P < 0.05$, **$P < 0.01$ vs estrus.) [From Concas et al. (1998). Role of brain allopregnanolone in the plasticity of γ-aminobutyric acid type A receptor in rate brain during pregnancy and after delivery. Proc. Natl. Acad. Sci. USA 95, 13284–13289; adapted with pemission.]

These observations suggested that cortical neuroactive steroids might play a role in the modulation of GABA$_A$ receptor density and activity during pregnancy and after delivery. To investigate this possibility further, we administered the specific 5α-reductase inhibitor finasteride to pregnant rats to inhibit the synthesis of 3α,5α-TH PROG and 3α,5α-TH DOC. Administration of finasteride from Day 12 to Day 18 of pregnancy markedly reduced the increases in the plasma and cerebrocortical concentrations of 3α,5α-TH PROG and 3α,5α-TH DOC normally apparent on Day 19 of pregnancy (Concas et al., 1998, 1999; Biggio et al., 2000). Finasteride also induced an apparent normalization of GABAergic transmission, as reflected by its prevention both of the increases in the density and K_d of [^{3}H]flunitrazepam and [^{35}S]TBPS binding sites and of the decrease in the stimulatory effect of muscimol on ^{36}Cl$^-$ uptake normally observed on Day 19 of pregnancy (Fig. 8).

These results demonstrated that the pregnancy-induced changes in the density and sensitivity of GABA$_A$ receptors are functionally related to the increases in the concentrations of 3α,5α-TH PROG and 3α,5α-TH DOC apparent during the last week of pregnancy. Consistent with this conclusion, long-term treatment with PROG induces an upregulation of GABA and

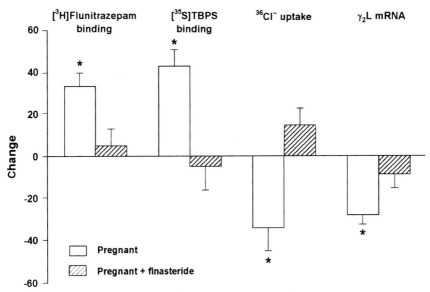

FIG. 8. Effects of finasteride on [^{3}H]flunitrazepam and [^{35}S]TBPS binding, muscimol-stimulated ^{36}Cl$^-$ uptake, and the abundance of the γ_2L subunit mRNA in the cerebral cortex of pregnant rats. Finasteride (25 mg/kg subcutaneously) or vehicle was injected daily from Day 12 to Day 18 of pregnancy, and the rats were killed on Day 19. Data are expressed as the percentage change relative to the estrus value and are means ± SEM of values from three to four experiments, each performed in triplicate. (*$P < 0.05$ vs estrus value.) [From Concas et al. (1998). Role of brain allopregnanolone in the plasticity of γ-aminobutyric acid type A receptor in rate brain during pregnancy and after delivery. Proc. Natl. Acad. Sci. USA 95, 13284–13289; adapted with permission.]

benzodiazepine binding sites in specific brain regions (Gavish et al., 1987; Canonaco et al., 1989). Furthermore, steroid hormone deprivation elicited by ovariectomy and adrenalectomy, which can be considered comparable to the sudden decrease in the concentrations of neuroactive steroids that occurs immediately before delivery, results in a decrease in GABA$_A$-receptor density in rat brain (Jussofie et al., 1995). In addition, long-term exposure of cultured mammalian cortical neurons to neurosteroids reduces both the ability of GABA to stimulate ^{36}Cl$^-$ uptake and the efficacy of benzodiazepines and neurosteroids with regard to enhancement of this effect (Yu and Ticku, 1995a,b).

C. GABA$_A$-Receptor Gene Expression in the Cortex and Hippocampus

We have shown that the pregnancy-induced changes in the density and function of GABA$_A$-receptors are associated with changes in the expression

TABLE II
ABUNDANCE OF GABA$_A$ RECEPTOR α_4-, α_5-, AND γ_2L-SUBUNIT mRNAs IN THE RAT CEREBRAL CORTEX AND HIPPOCAMPUS DURING PREGNANCY AND AFTER DELIVERY

	Cerebral cortex			Hippocampus	
	α_4	α_5	γ_2L	α_4	γ_2L
Estrus	100 ± 2.0	100 ± 1.85	100 ± 1.50	100 ± 1.30	100 ± 2.30
Pregnancy Day 10	98 ± 5.0	87.5 ± 4.03	77.5 ± 4.79*	103 ± 5.15	80.9 ± 5.75
Pregnancy Day 15	97 ± 2.0	87.2 ± 4.49	76.4 ± 4.01*	99 ± 3.58	70.7 ± 3.45*
Pregnancy Day 19	100 ± 2.0	87.7 ± 2.85	74.7 ± 3.40*	99 ± 2.73	77 ± 3.00*
Pregnancy Day 21	90 ± 5.0	76.9 ± 4.07*	103 ± 6.30	95 ± 3.10	90.9 ± 2.29
Postpartum Day 2	99 ± 3.0	104 ± 4.89	98.5 ± 6.17	106 ± 3.87	111 ± 5.55
Postpartum Day 7	105 ± 4.0	109 ± 5.22	87.1 ± 4.40	130 ± 6.13*	85 ± 4.23

From Follesa *et al.* (1998). Molecular and functional adaptation of GABA$_A$ receptor complex during pregnancy and after delivery in the rat brain. *Eur. J. Neurosci.* 10, 2905–2912; adapted with permission.

Note. Data are expressed as a percentage of the values for rats in estrus and are means ± SEM of values from 9 (three experiments) to 30 (10 experiments) animals.

*$P < 0.05$ vs estrus.

of GABA$_A$-receptor subunit genes (Follesa *et al.*, 1998). The amount of the γ_2L-subunit mRNA, measured by RNase protection assay, decreased progressively during pregnancy in both the rat cerebral cortex and the hippocampus. The decrease was apparent on Day 10 and maximal (–25 to –30%) on Day 15 or Day 19 of pregnancy (Table II). The abundance of this subunit mRNA in both brain regions had returned to control (estrus) values on Day 21 of pregnancy and did not differ significantly from control values during the postpartum period.

A similar pattern of changes was observed for the abundance of the γ_2 protein. Immunoblot analysis thus revealed that the amount of this protein in the cerebral cortex was reduced (–30%) on Days 15 and 19 of pregnancy and had returned to control values 2 days after delivery (Follesa *et al.*, 1998). The decrease in the abundance of the γ_2L-subunit mRNA thus appears to result in a reduction in the production of the γ_2 protein and possibly in a reduction in the number of newly assembled GABA$_A$ receptors that contain this subunit. Characterization of recombinant GABA$_A$ receptors expressed in mammalian cells has revealed that the γ_2 subunit is required for benzodiazepine sensitivity (Pritchett *et al.*, 1989). The reduced ability of diazepam to potentiate GABA-induced ^{36}Cl$^-$ uptake that is apparent on Day 19 of pregnancy may thus result from a reduction in the number of GABA$_A$ receptors that contain the γ_2-subunit. Diazepam also fails to induce sedation or loss of the righting reflex in mice with a targeted disruption of the γ_2-subunit gene (Günther *et al.*, 1995). Moreover, the absence of the γ_2 subunit in such mice results in a substantial impairment in the ability of $GABA_A$ receptors to form postsynaptic clusters and a consequent marked reduction in $GABA_A$-receptor function (Essrich *et al.*, 1998); such a deficit might explain the reduction in the activity of the $GABA_A$-receptor-associated Cl^- channel observed during pregnancy.

The abundance of mRNAs encoding the α_5 and α_4 subunits of the $GABA_A$ receptor was also affected by pregnancy and lactation, respectively (see Table II). The amount of the α_5-subunit mRNA in the cerebral cortex decreased during pregnancy, with the maximal reduction (–25%) apparent on Day 21, and returned to control values after delivery. In contrast, the abundance of the α_4-subunit mRNA was increased (+30%) in the hippocampus (but not in the cortex) only on Day 7 after delivery. Given that $GABA_A$ receptors containing the α_5-subunit exhibit the highest affinity for GABA (Sigel *et al.*, 1990), the decrease in the abundance of the α_5-subunit mRNA might be related to the low affinity of GABA binding sites observed during pregnancy. These results, together with the observation that the amounts of α_1-, α_2-, α_3-, β_1-, β_2-, β_3-, and γ_2S-subunit mRNAs in the cortex or hippocampus did not change during pregnancy or after delivery, suggest that the time- and region-dependent changes in the abundance of the γ_2L-, α_5-, and α_4-subunit mRNAs are specific. The abundance of the α_1-subunit mRNA was previously shown to increase during pregnancy specifically in hypothalamic magnocellular neurons, whereas an increase in γ_2 gene expression was apparent in the same brain region only after delivery (Fenelon and Herbison, 1996; Brussaard *et al.*, 1997). Thus, perhaps only specific neurons containing specific populations of $GABA_A$ receptors contribute to the changes in $GABA_A$-receptor function and expression observed during pregnancy and after delivery.

Our studies have thus shown that the abundance of the γ_2L-subunit mRNA, the concentrations of $3\alpha,5\alpha$-TH PROG and $3\alpha,5\alpha$-TH DOC, and $GABA_A$-receptor density and function in the cerebral cortex change with similar time courses during pregnancy and after delivery. In contrast, the time courses of the changes in the abundance of α_5- and α_4-subunit mRNAs do not match those of the changes in neuroactive steroid concentrations. The maximal decrease in the amount of the α_5-subunit mRNA in the cortex was observed on Day 21 of pregnancy, and the increase in the amount of the α_4-subunit mRNA in the hippocampus was significant only on Day 7 after delivery, times at which the concentrations of PROG and its metabolites have returned to control values. The increase in the amount of the α_4-mRNA in the hippocampus apparent 7 days after delivery appears to be consistent with the observations that withdrawal from $3\alpha,5\alpha$-TH PROG, either after intermittent administration of PROG (Smith *et al.*, 1998a) or in a model of pseudo-pregnancy (Smith et al., 1998b), results in an increase in the abundance of the α_4-subunit mRNA and peptide in the rat hippocampus. However, the long period between the normalization of neurosteroid concentrations on Day 21 of pregnancy and the increase in the amount of the α_4-subunit mRNA apparent 7 days after parturition also suggests the possibility that the latter effect is not the result of sudden neurosteroid withdrawal. In contrast, the notion that the changes in the amount of the γ_2L-subunit mRNA during pregnancy are attributable to the increased concentrations of neurosteroids is supported by our observation that treatment of dams from Day 12 to Day 18 of pregnancy with finasteride prevented the decrease in the abundance of the γ_2L-subunit mRNA that is normally apparent in both the cerebral cortex (see Fig. 8) and hippocampus on Day 19 of pregnancy.

D. GABA$_A$-Receptor Gene Expression in Hypothalamic Magnocellular Neurons

GABA$_A$ receptors play an important role in regulating the activity of the magnocellular neurons within the supraoptic nucleus of the rat hypothalamus (Moos, 1995; Voisin et al., 1995). These neurons release oxytocin in an episodic manner directly into the general circulation at the time of birth to facilitate both delivery and lactation. The GABA$_A$ receptors in these neurons are composed of combinations of only α_1, α_2, β_2, β_3, and γ_2 subunits (Fenelon et al., 1995), and the abundance of these subunits undergoes substantial changes during pregnancy and lactation (Fenelon and Herbison, 1996). These hypothalamic oxytocin neurons thus provide a model system with which to study GABA$_A$-receptor plasticity in the CNS.

Fenelon and Herbison (1996) showed that the amount of the α_1-subunit mRNA in both the supraoptic and the posterior paraventricular nuclei increases during pregnancy until Day 19 and then decreases on the day of parturition (Fig. 9). The abundance of α_2-, β_4-, and γ_2-subunit mRNAs in the supraoptic nucleus did not change during pregnancy or at delivery (Fig. 9), although the amount of α_2- (Fig. 9) and of γ_2-subunit (Fenelon and Herbison, 1996) mRNAs was increased in the paraventricular nucleus on Day 10 of pregnancy and on Day 14 after delivery, respectively. These results were interpreted to reflect a temporal switch in the synthesis of α_1-containing receptor subtypes to the synthesis of α_2-containing receptors around delivery.

Electrophysiological recording from neurons of the supraoptic nucleus during late pregnancy revealed a relatively fast-decaying spontaneous inhibitory current (sIPSC) that was prolonged by $3\alpha,5\alpha$-TH PROG (Fig. 10).

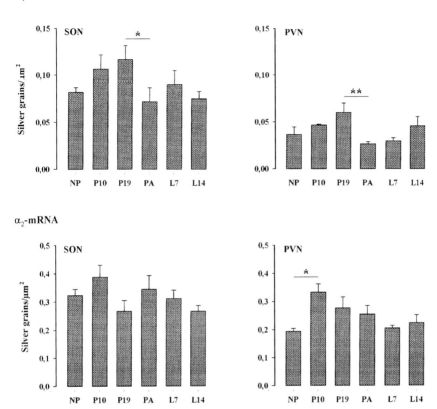

FIG. 9. Quantitative (Silver-Grain Density) analysis of GABA$_A$ receptor α_1- (top panels) and α_2-(bottom panels) subunit mRNAs in the supraoptic nucleus (SON) and posterior paraventricular nucleus (PVN) of nonpregnant ovariectomized (NP) rats, rats on Day 10 (P10) and Day 19 (P19) or pregnancy, parturient (PA) rats, and rats on Day 7 (L7) and Day 14 (L14) of lactation. Data are expressed as relative numbers of silver grains per square micrometer and are means ± SEM of values from five animals. (*$P < 0.05$, **$P < 0.01$.) [From Fenelon and Herbison (1996). Plasticity in GABA$_A$ receptor subunit in mRNA expression by hypothalamic magnocellular neurons in the adult rat. *J. Neurosci.* 16, 4872–4880; adapted with permission.]

In contrast, at the time of parturition, the sIPSC decay was slower and less sensitive to the action of 3α,5α-TH PROG. These results suggested that the increase during late pregnancy in the proportion of GABA$_A$ receptors containing the α_1 subunit is related to the relatively fast decay in the sIPSC. By the time of delivery, the proportion of α_1-subunit-containing receptors has decreased and that of α_2-subunit-containing receptors has increased, effects

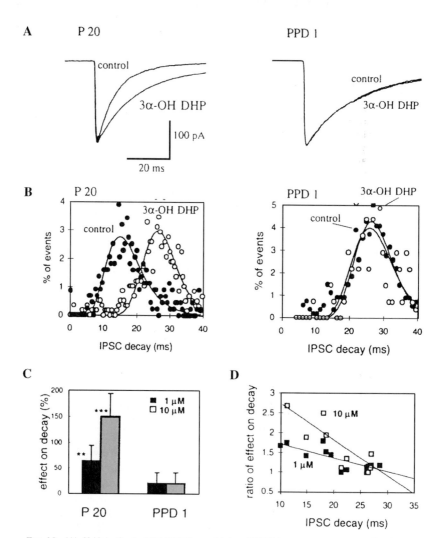

FIG. 10. (A) Shift in $3\alpha,5\alpha$-TH PROG sensitivity of $GABA_A$ receptors as a result of parturition. (B) Superimposed averages of 50 spontaneous inhibitory postsynaptic currents (IPSCs) recorded from dorsal supraoptic nucleus neurons, both before (Day 20 of pregnancy, or P20) and after (postpartum Day 1, or PPD1) parturition, in the absence or presence of 1 μM 3α-OH DHP ($3\alpha,5\alpha$,-TH PROG). (C) Lognormal curves fitted to the IPSC decay time-constant histogram before (control) and after 3α-OH DHP application (1 μM) at P20 and PPD1. The overall effect of 3α-OH DHP at 1 μM and 10 μM on spontaneous IPSC decay in P20 ($n=5$) and PPD1 ($n=7$) animals. ANOVA, $F(5, 23) = 4.271; P < 0.007$. (D) Post-hoc comparison indicated $**P < 0.05, ***P < 0.001$ compared with the control decay. Scatter plots of the decay time constants (each point represents the average spontaneous IPSC decay time constant of one neuron in the absence of neurosteroid) vs the relative effect at 1 μM or 10 μM 3α-OH DHP on the spontaneous IPSC decay. [From Brussaard et al. (1997). Plasticity fast synaptic inhibition of adult oxytocin neurons caused by switch in $GABA_A$ receptor subunit expression. Neuron 19, 1103–1114; reprinted with permission.]

that likely underlie the slower decay in the sIPSC and its relative insensitivity to 3α,5α-TH PROG that are apparent at this time (Brussaard et al., 1997). Consistent with this conclusion, antisense-RNA-induced depletion of the $α_2$ subunit in neurons expressing both $α_1$ and $α_2$ subunits resulted in the loss of slowly decaying sIPSCs (Brussaard et al., 1997).

Given that $GABA_A$ receptors containing the $α_1$ subunit are more sensitive to the action of 3α,5α-TH PROG than are $α_2$-subunit-containing receptors (Puia et al., 1993; Belelli et al., 1996), the increasing levels of circulating 3α,5α-TH PROG during late pregnancy strongly potentiate the $GABA_A$-receptor-mediated transmission by increasing the sIPSC decay time constants. The marked increase in synaptic inhibition elicited by 3α,5α-TH PROG observed in hypothalamic nuclei during late pregnancy may therefore reflect a tonic GABAergic input to these neurons to prevent premature release of oxytocin. This inhibitory influence may be relaxed at the time of delivery to enable the release of oxytocin and thus to facilitate delivery and lactation.

The evidence that 3α,5α-TH PROG concentrations are highest during late pregnancy, that $GABA_A$ receptors containing the $α_1$ subunit are more sensitive to the action of 3α,5α-TH PROG than are those containing the $α_2$ subunit (Puia et al., 1993; Belelli et al., 1996), and that 3α,5α-TH PROG potentiates an inhibitory input to oxytocin neurons only during pregnancy (Brussaard et al., 1997) suggests that fluctuations in the concentrations of this neurosteroid during the reproductive cycle contribute to $GABA_A$-receptor plasticity in the rat hypothalamus.

VI. Oral Contraceptives and $GABA_A$-Receptor Plasticity

The observations that long-term fluctuations in the cerebrocortical concentrations of neuroactive steroids, either those induced by pharmacological treatments (PROG or 3α,5α-TH PROG) or those associated with physiological conditions such as pregnancy, delivery, and lactation, affect $GABA_A$-receptor plasticity and function prompted us to evaluate whether a pharmacologically induced persistent reduction in the plasma and brain concentrations of these steroids might also affect $GABA_A$-receptor gene expression. Oral contraceptives prevent ovulation by suppressing pulsatile secretion of gonadotropins (luteinizing and follicle-stimulating hormones), which results in a long-lasting decrease in the synthesis of endogenous steroids (estrogens and PROG) (Goodman et al., 1981; Kuhl et al., 1984). Given that 3α, 5α-TH PROG and 3α,5α-TH DOC are PROG metabolites, animals treated with oral contraceptives might represent a pharmacological

TABLE III
EFFECTS OF LONG-TERM TREATMENT WITH ORAL CONTRACEPTIVES ON
CEREBROCORTICAL AND PLASMA CONCENTRATIONS OF NEUROACTIVE STEROIDS

	Cerebral cortex		Plasma	
	Control	Treated	Control	Treated
Pregnenolone	117.9 ± 7.9	69.3 ± 7.18*	5.38 ± 0.27	3.48 ± 0.22*
Progesterone	98.0 ± 17	20.9 ± 5.70*	6.03 ± 0.64	3.41 ± 0.40†
3α,5α-TH PROG	40.4 ± 2.8	6.26 ± 1.43*	18.5 ± 0.75	11.15 ± 0.54*
3α,5α-TH DOC	9.68 ± 1.27	4.41 ± 0.50†	7.78 ± 0.41	5.21 ± 0.32*

Note. Female rats were injected daily for 6 weeks with ethynylestradiol (0.030 mg, subcutaneously) and levonorgestrel (0.125 mg subcutaneously), or with vehicle (control), and were killed 24 h after the last treatment. Data are means ± SEM of values obtained from 10 to 15 rats.
*$P < 0.001$
†$P < 0.005$ vs the control group.

model with which to obtain greater insight into the physiological role of these neuroactive steroids in the modulation of GABAergic transmission.

Daily administration of a combination of ethynylestradiol (0.030 mg subcutaneously) and levonorgestrel (0.125 mg subcutaneously) to female rats for 6 weeks resulted in marked decreases in the cerebrocortical concentrations of both PROG (–79%) and 3α,5α-TH PROG (–85%), as well as in a smaller decrease in that of 3α,5α-TH DOC (–54%) (Table III). The same treatment also reduced, albeit to a lesser extent than in the cerebral cortex, the concentrations of these steroids in plasma. These results suggested that oral contraceptives might directly affect the synthesis and accumulation of neurosteroids in the brain.

The decreases in the concentrations of neuroactive steroids induced by ethynylestradiol and levonorgestrel were paralleled by increases in the abundance of mRNAs encoding the $\gamma_2 S$ (+22%) and $\gamma_2 L$ (+32%) subunits of the $GABA_A$ receptor in the cerebral cortex (Table IV). The amounts of the mRNAs encoding α_1, α_4, β_1, β_2, and β_3 subunits were not affected by this treatment (data not shown).

The persistent reduction in the brain concentrations of the $GABA_A$-receptor-active PROG metabolites elicited by chronic administration of oral contraceptives was thus associated with a plastic adaptation of $GABA_A$-receptor gene expression in the rat cerebral cortex. This observation is consistent with evidence that the long-term exposure of $GABA_A$ receptors to neuroactive steroids during PROG treatment or pregnancy induces a marked decrease in γ_2-subunit mRNA abundance in the same brain region (Concas *et al.*, 1998; Follesa *et al.*, 1998) and in cerebellar culture

TABLE IV

EFFECTS OF LONG-TERM TREATMENT WITH ORAL CONTRACEPTIVES ON THE ABUNDANCE OF mRNAs ENCODING THE γ_2L AND γ_2S SUBUNITS OF THE GABA$_A$ RECEPTOR IN THE RAT CEREBRAL CORTEX

	γ_2L mRNA (% of control)	γ_2S mRNA (% of control)
Control	100 ± 3.38	100 ± 2.87
Treated	$132 \pm 3.73^*$	$122 \pm 2.94^*$

Note. Female rats were injected daily for 6 weeks with ethynylestradiol (0.030 mg subcutaneously) and levonorgestrel (0.125 subcutaneously), or with vehicle (control) and were killed 24 h after the last treatment. Data are means ± SEM of values obtained from 15 rats. ($^*P < 0.01$ vs control group.)

(Follesa *et al.*, 2000), an effect that is reversed by treatment with finasteride or by the decrease in the concentrations of PROG and 3α,5α-TH PROG that precedes parturition. A substantial increase in the abundance of γ_2-subunit mRNA has also been observed in the posterior paraventricular nucleus in association with low brain concentrations of neurosteroids during lactation in rats (Fenelon and Herbison, 1996). Together, these data further support a role for 3α,5α-TH PROG and 3α,5α-TH DOC in the physiological modulation of central GABAergic transmission.

The functional relevance of the changes in γ_2-subunit expression in the cerebral cortex after oral contraceptive treatment, as well as of those apparent during pregnancy, remains to be elucidated. The γ_2 subunit forms intersubunit contacts with α_1 and β_3 subunits (Klausberger *et al.*, 2000); contributes to most GABA$_A$-receptor subtypes; and is essential for benzodiazepine action, normal channel gating, and clustering of GABA$_A$ receptors and gephyrin at postsynaptic sites (Günther *et al.*, 1995; Sieghart, 1995; Essrich *et al.*, 1998). Moreover, mice heterozygous for deletion of the γ_2-subunit gene represent a genetically defined model of trait anxiety (Crestani *et al.*, 1999). These mice also exhibit enhanced behavioral inhibition to natural aversive stimuli as well as enhanced emotional memory for negative associations.

The γ_2 subunit of GABA$_A$ receptors has also been shown to play an active role in maintenance of godanotropin-releasing hormone (GnRH) biosynthesis in GnRH neurons in the mouse (Simonian *et al.*, 2000). This observation is consistent with evidence suggesting that GABA, acting at GABA$_A$ receptors, exerts an inhibitory action on GnRH release from the hypothalamus, which results in an immediate reduction in the secretion of luteinizing hormone (Leonhardt *et al.*, 1995; Seong *et al.*, 1995). Given that luteinizing hormone release is important for fertility and is suppressed by contraceptives (Kuhl *et al.*, 1984), which increase the abundance of γ_2-subunit mRNA in the cerebral cortex, an increase in γ_2-subunit expression in the hypothalamus and the consequent increase in $GABA_A$-receptor function (Essrich *et al.*, 1998) may contribute to the prevention by these drugs of the luteinizing hormone surge that is normally apparent at the time of ovulation.

VII. Mechanism of the Effect of Long-Term PROG Exposure on $GABA_A$-Receptor Plasticity

It remains to be determined whether changes in $GABA_A$-receptor plasticity induced by long-term exposure to PROG or its metabolites are mediated by allosteric modulation of $GABA_A$ receptors or by an indirect genomic action. The PROG metabolites $3\alpha,5\alpha$-TH PROG and $3\alpha,5\alpha$-TH DOC do not influence gene expression by direct interaction with the intracellular PROG receptor but do modulate neuronal excitability by acting at membrane-bound $GABA_A$ receptors. However, it remains possible that they influence gene expression indirectly by their conversion to 5α-pregnan-3,20-dione (5α-dehydroprogesterone, or 5α-DH PROG) and 21-hydroxy-5α-pregnan-3,20-dione (5α-dehydrocorticosterone, or 5α-DH DOC), both of which bind to and activate the intracellular PROG receptor. Exposure of neuroblastoma cells to $3\alpha,5\alpha$-TH PROG or $3\alpha,5\alpha$-TH DOC, in the absence of PROG, thus increases the expression of a luciferase reporter gene construct containing a PROG response element through the conversion of these neurosteroids to 5α-DH PROG and 5α-DH DOC (Rupprecht *et al.*, 1993, 1996). However, the changes in $GABA_A$-receptor gene expression induced by PROG treatment or pregnancy are unlikely to be mediated by the interaction of 5α-DH PROG and 5α-DH DOC with intracellular PROG receptors. The high concentrations of PROG present both in PROG-treated cells (Follesa *et al.*, 2000) and in the brain of pregnant rats (Concas *et al.*, 1998) would be expected to preclude an action at the intracellular PROG receptor of 5α-DH PROG and 5α-DH DOC, given that the concentrations of these metabolites and their affinities for this receptor are markedly smaller than those of PROG. Although a genomic action of PROG mediated by other transcriptional regulatory proteins that contribute to the expression of $GABA_A$-receptor subunit genes cannot be excluded, the observations that finasteride prevents the increase in $3\alpha,5\alpha$-TH PROG concentrations both in cultured cerebellar granule cells exposed to PROG and in the cerebral cortex of pregnant rats, as well as inhibits the associated changes in $GABA_A$-receptor function and gene expression, suggest that these latter changes are the consequence of $3\alpha,5\alpha$-TH PROG action at the steroid recognition site of the $GABA_A$ receptor. This conclusion is also supported by data showing that fluctuations in the concentration of $3\alpha,5\alpha$-TH PROG are responsible for modifying $GABA_A$-receptor gene expression in the hippocampus (Weiland and Orchinik, 1995; Smith *et al.*, 1998a).

Fenelon and Herbison (2000) suggested that PROG, rather than $3\alpha,5\alpha$-TH PROG, influences $GABA_A$-receptor plasticity in the supraoptic nucleus. Thus, whereas treatment of rats for 7 days with PROG reduced the abundance of the α_1-subunit mRNA in the supraoptic nucleus, no change in the amount of this subunit mRNA was detected after similar treatment with $3\alpha,5\alpha$-TH PROG. In contrast, treatment with $3\alpha,5\alpha$-TH PROG reduced the abundance of the α_1-subunit mRNA in the hippocampus. Although these results suggest that neurosteroids may modulate α_1-subunit expression in the supraoptic nucleus and the hippocampus by different mechanisms, the effect of PROG in combination with finasteride on the abundance of the α_1-subunit mRNA in the supraoptic nucleus was not described.

VIII. Conclusions

The studies described in this chapter have demonstrated that fluctuations in the brain concentrations of PROG and its metabolite $3\alpha,5\alpha$-TH PROG play a major role in the temporal pattern of expression of various subunits of the $GABA_A$ receptor. Thus, rapid and long-lasting increases or decreases in the concentrations of these steroid derivatives associated with pregnancy, pseudo-pregnancy, or pharmacological treatments, including the administration of contraceptives, elicit selective changes in $GABA_A$-receptor gene expression and function in specific neuronal populations in various regions of the brain. Given both the importance of $GABA_A$ receptors in the regulation of neuronal excitability and the large fluctuations in the plasma and brain concentrations of neuroactive steroids associated with physiological conditions and the response to environmental stimuli, these compounds are likely among the most relevant endogenous modulators of emotional and affective behaviors.

References

Azzolina, B., Ellsworth, K., Andersson, S., Geissler, W., Bull, H. G., and Harris, G. S. (1997). Inhibition of rat α-reductases by finasteride: Evidence for isozyme differences in the mechanism of inhibition. *J. Steroid Biochem. Mol. Biol.* **61,** 55–64.

Barnard, E. A., Skolnick, P., Olsen, R. W., Mohler, H., Sieghart, W., Biggio, G., Braestrup, C., Bateson, A. N., and Langer, S. Z. (1998). International Union of Pharmacology: XV. Subtypes of γ-aminobutyric acid$_A$ receptors: Classification on the basis of subunit structure and receptor function. *Pharmacol. Rev.* **50**, 291–310.

Belelli, D., Lambert, J. J., Peters, J. A., Gee, K. W., and Lan, N. C. (1996). Modulation of human recombinant GABA(A) receptors by pregnanediols. *Neuropharmacology* **35**, 1223–1231.

Bicikova, M., Dibbelt, L., Hill, M., Hampl, R., and Starka, L. (1998). Allopregnanolone in women with premenstrual syndrome. *Horm. Metab. Res.* **30**, 227–230.

Biggio, G., Barbaccia, M. L., Follesa, P., Serra, M., Purdy, R. H., and Concas, A. (2000). Neurosteroids and GABA$_A$ receptor plasticity. *In* "GABA in the Nervous System: The View at Fifty Years" (D. L. Martin and R. W. Olsen, Eds.), pp. 207–232. Lippincott Williams & Wilkins, New York.

Bitran, D., Purdy, R. H., and Kellog, C. K. (1993). Anxiolytic effect of progesterone is associated with increases in cortical allopregnanolone and GABA$_A$ receptor function. *Pharmacol. Biochem. Behav.* **45**, 423–428.

Bitran, D., Shiekh, M., and McLeod, M. (1995). Anxiolytic effect of progesterone is mediated by neurosteroid allopregnanolone at brain GABA$_A$ receptors. *J. Neuroendocrinol.* **7**, 171–177.

Braestrup, C., Schmiechen, R., Neef, G., Nielsen, M., and Petersen, E. N. (1982). Interaction of convulsive ligands with benzodiazepine receptors. *Science* **216**, 1241–1243.

Braestrup, C., Nielsen, M., and Honoré, T. (1983). Binding of [^{3}H]DMCM, a convulsive benzodiazepine ligand, to rat brain membranes: Preliminary studies. *J. Neurochem.* **41**, 454–465.

Brussaard, A. B., Kits, K. S., Baker, R. E., Willems, W. P. A., Leyting-Vermeulen, J. W., Voorn, P., Smith, A. B., Bicknell, R. J., and Herbison, A. E. (1997). Plasticity in fast synaptic inhibition of adult oxytocin neurons caused by switch in GABA$_A$ receptor subunit expression. *Neuron* **19**, 1103–1114.

Buster, J. E. (1983). Gestational changes in steroid hormone biosynthesis, secretion, metabolism, and action. *Clin. Perinatol.* **10**, 527–552.

Canonaco, M., O'Connor, L. H., Pfaff, D. W., and McEwen, B. S. (1989). Longer term progesterone treatment induces changes of GABA$_A$ receptor levels in forebrain sites in the female hamster: Quantitative autoradiography study. *Exp. Brain Res.* **77**, 407–411.

Concas, A., Mostallino, M. C., Porcu, P., Follesa, P., Barbaccia, M. L., Trabucchi, M., Purdy, R. H., Grisenti, P., and Biggio, G. (1998). Role of brain allopregnanolone in the plasticity of γ-aminobutyric acid type A receptor in rat brain during pregnancy and after delivery. *Proc. Natl. Acad. Sci. USA* **95**, 13284–13289.

Concas, A., Follesa, P., Barbaccia, M. L., Purdy, R. H., and Biggio, G. (1999). Physiological modulation of GABA$_A$ receptor plasticity by progesterone metabolites. *Eur. J. Pharmacol.* **375**, 225–235.

Crestani, F., Lorez, M., Baer, K., Essrich, C., Benke, D., Laurent, J. P., Belzung, C., Fritschy, J.-M., Lüscher, B., and Mohler, H. (1999). Decreased GABA$_A$-receptor clustering results in enhanced anxiety and a bias for threat cues. *Nature Neurosci.* **2**, 833–839.

Devaud, L. L., Fritschy, J. M., Sieghart, W., and Morrow, A. L. (1997). Bidirectional alterations of GABA(A) receptor subunit peptide levels in rat cortex during chronic ethanol consumption and withdrawal. *J. Neurochem.* **69**, 126–130.

Essrich, C., Lorez, M., Benson, J. A., Fritschy, J.-M., and Lüscher, B. (1998). Postsynaptic clustering of major GABA$_A$ receptor subtypes requires the γ_2 subunit and gephyrin. *Nature Neurosci.* **1**, 563–571.

Fenelon, V. S., and Herbison, A. E. (1996). Plasticity in GABA$_A$ receptor subunit mRNA expression by hypothalamic magnocellular neurons in the adult rat. *J. Neurosci.* **16**, 4872–4880.

Fenelon, V. S., and Herbison, A. E. (2000). Progesterone regulation of GABA$_A$ receptor plasticity in adult rat supraoptic nucleus. *Eur. J. Neurosci.* **12,** 1617–1623.

Fenelon, V. S., Sieghart, W., and Herbison, A. E. (1995). Cellular localization and differential distribution of GABA$_A$ receptor subunit proteins and messenger RNAs within hypothalamic magnocellular neurons. *Neuroscience* **64,** 1129–1143.

Follesa, P., Floris, S., Tuligi, G., Mostallino, M. C., Concas, A., and Biggio, G. (1998). Molecular and functional adaptation of the GABA$_A$ receptor complex during pregnancy and after delivery in the rat brain. *Eur. J. Neurosci.* **10,** 2905–2912.

Follesa, P., Serra, M., Cagetti, E., Pisu, M. G., Porta, S., Floris, S., Massa, F., Sanna, E., and Biggio, G. (2000). Allopregnanolone synthesis in cerebellar granule cells: Roles in regulation of GABA$_A$ receptor expression and function during progesterone treatment and withdrawal. *Mol. Pharmacol.* **57,** 1262–1270.

Freeman, E., Purdy, R. H., Coutifaris, K. R., and Paul, S. M. (1993). Anxiolytic metabolites of progesterone: Correlation with mood and performance measures following oral progesterone administration to healthy female volunteers. *Neuroendocrinology* **58,** 478–484.

Friedman, L., Gibbs, T. T., and Farb, D. H. (1993). γ-Aminobutyric acid$_A$ receptor regulation: chronic treatment with pregnenolone uncouples allosteric interactions between steroid and benzodiazepine recognition sites. *Mol. Pharmacol.* **44,** 191–197.

Friedman, L., Gibbs, T. T., and Farb, D. H. (1996). γ-Aminobutyric acid$_A$ receptor regulation: heterologous uncoupling of modulatory site interactions induced by chronic steroid, barbiturate, benzodiazepine, or GABA treatment in culture. *Brain Res.* **707,** 100–109.

Gallo, M. A., and Smith, S. S. (1993). Progesterone withdrawal decreases latency to and increases duration of electrified prod burial: A possible rat model of PMS anxiety. *Pharmacol. Biochem. Behav.* **46,** 897–904.

Gavish, M., Weizman, A., Youdim, M. B. H., and Okun, F. (1987). Regulation of central and peripheral benzodiazepine receptors in progesterone-treated rats. *Brain Res.* **409,** 386–390.

Genazzani, A. R., Petraglia, F., Bernardi, F., Casarosa, E., Salvestroni, C., Tonetti, A., Nappi, R. E., Luisi, S., Palumbo, M., Purdy, R. H., and Luisi, M. (1998). Circulating levels of allopregnanolone in humans: Gender, age, and endocrine influences. *Clin. Endocrinol. Metab.* **83,** 2099–2103.

Goodman, R. L., Bittman, E. L., Foster, D. L., and Karsch, F. J. (1981). The endocrine basis of the synergistic suppression of luteinizing hormone by estradiol and progesterone. *Endocrinology* **109,** 1414–1417.

Günther, U., Benson, J. A., Benke, D., Fritschy, J.-M., Reyes, G., Knoflach, F., Crestani, F., Aguzzi, A., Arigoni, M., Lang, Y., Bluethmann, H., Mohler, H., and Lüscher, B. (1995). Benzodiazepine-insensitive mice generated by targeted disruption of the γ-aminobutyric acid type A receptors. *Proc. Natl. Acad. Sci. USA* **92,** 7749–7753.

Harrison, N. L., and Simmonds, M. A. (1984). Modulation of GABA receptor complex by a steroid anesthetic. *Brain Res.* **323,** 284–293.

Holt, R. A., Bateson, A. N., and Martin, I. L. (1996). Chronic treatment with diazepam or abecarnil differently affects the expression of GABA$_A$ receptor subunit mRNAs in the rat cortex. *Neuropharmacology* **35,** 1457–1463.

Holt, R. A., Martin, I. L., and Bateson, A. N. (1997). Chronic diazepam exposure decreases transcription of the rat GABA$_A$ receptor γ_2-subunit gene. *Mol. Brain Res.* **48,** 164–166.

Hu, Z. Y., Borreau, E., Jung-Testa, I., Robel, P., and Baulieu, E.-E. (1987). Neurosteroids: Oligodendrocyte mitochondria convert cholesterol to pregnenolone. *Proc. Natl. Acad. Sci. USA* **84,** 8215–8219.

Impagnatiello, F., Pesold, C., Longone, P., Caruncho, H., Fritschy, J. M., Costa, E., and Guidotti, A. (1996). Modifications of γ-aminobutyric acid$_A$ receptor subunit expression in rat neocortex during tolerance to diazepam. *Mol. Pharmacol.* **49,** 822–831.

Jussofie, A., Körner, I., Schell, C., and Hiemke, C. (1995). Time course of the effects of steroid hormone deprivation elicited by ovariectomy or ovariectomy plus adrenalectomy on the affinity and density of GABA binding sites in distinct rat brain areas. *Exp. Clin. Endocrinol.* **103**, 196–204.

Klausberger, T., Fuchs, K., Mayer, B., Ehya, N., and Sieghart, W. (2000). $GABA_A$ receptor assembly. Identification and structure of γ_2 sequences forming the intersubunit contacts with α_1 and β_3 subunits. *J. Biol. Chem.* **275**, 8921–8928.

Kokate, T. G., Banks, M. K., Tmagee, T., Yamaguchi, S., and Rogawski, M. A. (1999). Finasteride, a 5α-reductase inhibitor, blocks the anticonvulsant activity of progesterone in mice. *J. Pharmacol. Exp. Ther.* **288**, 679–684.

Kuhl, H., Weber, W., Mehlis, W., Sandow, J., and Taubert, H. D. (1984). Time- and dose-dependent alterations of basal and LH-RH-stimulated LH-release during treatment with various hormonal contraceptives. *Contraception* **30**, 467–482.

Lambert, J. J., Belelli, D., Hill-Venning, C., and Peters, J. A. (1995). Neurosteroids and $GABA_A$ receptor function. *Trends Pharmacol. Sci.* **16**, 295–303.

Le Goascogne, C., Robel, P., Gouezou, M., Sananes, N., Baulieu, E.-E., and Waterman, M. (1987). Neurosteroids: Cytochrome P-450_{500} in rat brain. *Science* **237**, 1212–1214.

Leonhardt, S., Seong, J. Y., Kim, K., Thorun, Y., Wuttke, W., and Jarry, H. (1995). Activation of central $GABA_A$- but not of $GABA_B$-receptors rapidly reduces pituitary LH release and GnRH gene expression in the preoptic/anterior hypothalamic area of ovariectomized rats. *Neuroendocrinology* **61**, 655–662.

Löw, K., Crestani, F., Keist, R., Benke, D., Brünig, I., Benson, J. A., Fritschy, J.-M., Rülicke, T., Bluethmann, H., Möhler, H., and Rudolph, U. (2000). Molecular and neuronal substrate for the selective attenuation of anxiety. *Science* **290**, 131–134.

Mahmoudi, M., Kang, M. H., Tillakaratne N., Tobin A. J., and Olsen, R. W. (1997). Chronic intermittent ethanol treatment in rats increases $GABA_A$ receptor α_4 subunit expression—Possible relevance to alcohol dependence. *J. Neurochem.* **68**, 2485–2492.

Majchrowicz, E. (1975). Induction of physical dependence upon ethanol and the associated behavioral changes in rats. *Psychopharmacologia* **43**, 245–254.

Majewska, M. D. (1992). Neurosteroids: Endogenous bimodal modulators of the $GABA_A$ receptor. Mechanism of action and physiological significance. *Prog. Neurobiol.* **38**, 379–395.

Majewska, M. D., Harrison, N. L., Schwartz, R. D., Barker, J. L., and Paul, S. M. (1986). Steroid hormone metabolites are barbiturate-like modulators of the GABA receptor. *Science* **232**, 1004–1007.

Majewska, M. D., Ford-Rice, F., and Falkay, G. (1989). Pregnancy-induced alterations of $GABA_A$ receptor sensitivity in maternal brain: An antecedent of post-partum "blues"? *Brain Res.* **482**, 397–401.

Mathur, C., Prasad, V. V. K., Raju, V. S., Welch, M., and Lieberman, S. (1993). Steroids and their conjugates in the mammalian brain. *Proc. Natl. Acad. Sci. USA* **90**, 85–88.

McKernan, R. M., Rosahl, T. W., Reynolds, D. S., Sur, C., Wafford, K. A., Atack, J. R., Farrar, S., Myers, J., Cook, G., Ferris, P., Garrett, L., Bristow, L., Marshall, G., Macaulay, A., Brown, N., Howell, O., Moore, K. W., Carling, R. W., Street, L. J., Castro, J. L., Ragan, C. I., Dawson, G. R., and Whiting, P. J. (2000). Sedative but not anxiolytic properties of benzodiazepines are mediated by the $GABA_A$ receptor α_1 subtype. *Nature Neurosci.* **3**, 587–592.

Mellon, S. H., and Deshepper, C. F. (1993). Neurosteroid biosynthesis: Genes for adrenal steroidogenic enzymes are expressed in the brain. *Brain Res.* **629**, 283–292.

Mhatre, M. C., and Ticku, M. K. (1992). Chronic ethanol administration alters γ-aminobutyric $acid_A$ receptor gene expression. *Mol. Pharmacol.* **42**, 415–422.

Mhatre, M. C., Pena, G., Sieghart, W., and Ticku, M. K. (1993). Antibodies specific for GABA$_A$ receptor α subunits reveal that chronic alcohol treatment down-regulates α-subunit expression in rat brain regions. *J. Neurochem.* **61,** 1620–1625.

Milewich, L., Gant, N. F., Schwarz, B. E., Chen, G. T., and MacDonald, P. C. (1979). 5α-Reductase activity in human placenta. *Am. J. Obstetr. Gynecol.* **133,** 611–617.

Montpied, P., Ginns, E. I., Martin, B. M., Roca, D., Farb, D. H., and Paul, S. M. (1991). γ-Aminobutyric acid (GABA) induces a receptor-mediated reduction in GABA$_A$ receptor α subunit messenger RNAs in embryonic chick neurons in culture. *J. Biol. Chem.* **266,** 6011–6014.

Moos, F. C. (1995). GABA-induced facilitation of the periodic bursting activity of oxytocin neurones in suckled rats. *J. Physiol. (London)* **488,** 103–114.

Morales, A., Aleu, J., Ivorra, I., Ferragut, J. A., Gonzales-Ros, J. M., and Miledi, R. (1995). Incorporation of reconstituted acetylcholine receptors from Torpedo into *Xenopus* oocytes membrane. *Proc. Natl. Acad. Sci. USA* **92,** 8468–8472.

Moran, M. H., Goldberg, M., and Smith, S. S. (1998). Progesterone withdrawal. II: Insensitivity to the sedative effects of benzodiazepine. *Brain Res.* **807,** 91–100.

Morrow, A. L., Montpied, P., Lingford-Hughes, A., and Paul, S. M. (1990). Chronic ethanol and pentobarbital administration in the rat: Effects on GABA$_A$ receptor function and expression in brain. *Alcohol* **7,** 237–244.

Penning, T. M., Sharp, R. B., and Krieger, N. R. (1985). Purification and properties of 3 alpha-hydroxysteroid dehydrogenase from rat brain cytosol. Inhibition by non-steroidal anti-inflammatory drugs and progestins. *J. Biol. Chem.* **260,** 15266–15272.

Picazo, O., and Fernández-Guasti, A. (1995). Anti-anxiety effects of progesterone and some of its reduced metabolites: An evaluation using the burying behavior test. *Brain Res.* **680,** 135–141.

Prasad, V. V., Vegesna, S. R., Welch, M., and Lieberman, S. (1994). Precursor of the neurosteroids. *Proc. Natl. Acad. Sci. USA* **91,** 3220–3223.

Pritchett, D. B., Sontheimer, H., Shivers, B. D., Ymer, S., Kettenmann, H., Schofield, P. R., and Seeburg, P. H. (1989). Importance of a novel GABA$_A$ receptor subunit for benzodiazepine pharmacology. *Nature* **338,** 582–585.

Puia, G., Ducic, I., Vicini, S., and Costa, E. (1993). Does neurosteroid modulatory efficacy depend on GABA$_A$ receptor subunit composition? *Receptors Channels* **1,** 135–142.

Rapkin, A. J., Morgan, M., Goldman, L., Brann, D. W., Simone, D., and Mahesh, V. B. (1997). Progesterone metabolite allopregnanolone in women with premenstrual syndrome. *Obstetr. Gynecol.* **90,** 709–714.

Reddy, D. S., and Rogawski, M. A. (2000). Enhanced anticonvulsant activity of ganaxolone after neurosteroid withdrawal in a rat model of catamenial epilepsy. *J. Pharmacol. Exp. Ther.* **294,** 909–915.

Rittmaster, R. S. (1994). Finasteride. *N. Engl. J. Med.* **330,** 120–125.

Roca, D. J., Rozenberg, I., Farrant, M., and Farb, D. H. (1989). Chronic agonist exposure induces down-regulation and allosteric uncoupling of the γ-aminobutyric acid/benzodiazepine receptor complex. *Mol. Pharmacol.* **37,** 37–43.

Roca, D. J., Shiller, G. D., Friedman, L., Rozenberg, I., Gibbs, T. T., and Farb, D. H. (1990). γ-Aminobutyric acidA receptor regulation in culture: Altered allosteric interaction following prolonged exposure to benzodiazepines, barbiturates, and methylxanthines. *Mol. Pharmacol.* **37,** 710–719.

Romeo, E., Ströhle, S., Spalletta, G., di Michele, F., Hermann, B., Holsboer, F., Pasini, A., and Rupprecht, R. (1998). Effects of antidepressant treatment on neuroactive steroids in major depression. *Am. J. Psychiatry* **155,** 910–913.

Rudolph, U., Crestani, F., Benje, D., Brünig, I., Benson, J. A., Fritschy, J., Martin, J. R., Bluethmann, H., and Möhler, H. (1999). Benzodiazepine actions mediated by specific γ-aminobutyric acid$_A$ receptor subtypes. *Nature* **401,** 796–800.
Rupprecht, R., Reul, J. M. H. M., Trapp, T., van Steensel, B., Wetzel, C., Damm, K., Zieglgansberger, W., and Holsboer, F. (1993). Progesterone receptor-mediated effects of neuroactive steroids. *Neuron* **11,** 523–530.
Rupprecht, R., Hauser, C. A. E., Trapp, T., and Holsboer, F. (1996). Neurosteroids: Molecular mechanism of action and psychopharmacological significance. *J. Steroid Biochem. Mol. Biol.* **56,** 163–168.
Sanna, E., Motzo, C., Murgia, A., Amato, F., Deserra, T., and Biggio, G. (1996). Expression of native GABA$_A$ receptors in *Xenopus* oocytes injected with rat brain synaptosomes. *J. Neurochem.* **67,** 2212–2214.
Sanna, E., Motzo, C., Usala, M., Pau, D., Cagetti, E., and Biggio, G. (1998). Functional changes in rat nigral GABA$_A$ receptors induced by degeneration of the striatonigral GABAergic pathway: An electrophysiological study of receptors incorporated into *Xenopus* oocytes. *J. Neurochem.* **70,** 2539–2544.
Schmidt, P. J., Purdy, R. H., Moore, Jr., P. H., Paul, S. M., and Rubinow, D. R. (1994). Circulating levels of anxiolytic steroids in the luteal phase in women with premenstrual syndrome and in control subjects. *J. Clin. Endocrinol. Metab.* **79,** 1256–1260.
Schweizer, E., and Rickels, K. (1998). Benzodiazepine dependence and withdrawal: A review of the syndrome and its clinical management. *Acta Psychiatr. Scand.* **98,** 95–101.
Seong, J. Y., Jarry, H., Kuhnemuth, S., Leonhardt, S., Wuttke, W., and Kim, K. (1995). Effect of GABAergic compounds on gonadotropin-releasing hormone receptor gene expression in the rat. *Endocrinology* **136,** 2587–2593.
Sieghart, W. (1995). Structure and pharmacology of γ-aminobutyric acid$_A$ receptor subtypes. *Pharmacol. Rev.* **47,** 181–234.
Sigel, E., Baur, R., Trube, G., Mohler, H., and Malherbe, P. (1990). The effect of subunit composition of rat brain GABA$_A$-receptors on channel function. *Neuron* **5,** 703–711.
Simonian, S. X., Skynner, M. J., Sieghart, W., Essrich, C., Luscher, B., and Herbison, A. E. (2000). Role of the GABA$_A$ receptor γ$_2$ subunit in the development of gonadotropin-releasing hormone neurons *in vivo. Eur. J. Pharmacol.* **12,** 3488–3496.
Smith, S. S., Gong, Q. H., Hsu, F.-C., Markowitz, R. S., ffrench-Mullen, J. M. H., and Li, X. (1998a). GABA$_A$ receptor α$_4$ subunit suppression prevents withdrawal properties of an endogenous steroid. *Nature* **392,** 926–930.
Smith, S. S., Gong, Q. H., Li, X., Moran, M. H., Bitran, D., Frye, C. A., and Hsu, F.-C. (1998b). Withdrawal from 3α-OH-5α-pregnan-20-one using a pseudopregnancy model alters the kinetics of hippocampal GABA$_A$-gated current and increases the GABA$_A$ receptor α$_4$ subunit in association with increased anxiety. *J. Neurosci.* **18,** 5275–5284.
Ströhle, A., Romeo, E., Hermann, B., Pasini, A., Spalletta, G., Di Michele, F., Holsboer, F., and Rupprecht, R. (1999). Concentrations of 3α-reduced neuroactive steroids and their precursors in plasma of patients with major depression and after clinical recovery. *Biol. Psychiatry* **45,** 274–277.
Sundstrom, I., Ashbrook, D., and Backstrom, T. (1997). Reduced benzodiazepine sensitivity in patients with premenstrual syndrome: A pilot study. *Psychoneuroendocrinology* **22,** 25–38.
Tseng, Y. T., Miyaoka, T., and Ho, I. K. (1993). Region-specific changes of GABA$_A$ receptors by tolerance to and dependence upon pentobarbital. *Eur. J. Pharmacol.* **236,** 23–30.
Uzunova, V., Sheline, Y., Davis, J. M., Rasmusson, A., Uzunov, D. P., Costa, E., and Guidotti, A. (1998). Increase in the cerebrospinal fluid content of neurosteroids in patients with unipolar major depression who are receiving fluoxetine or fluovoxamine. *Proc. Natl. Acad. Sci. USA* **95,** 3239–3244.

Voisin, D. L., Herbison, A. E., and Poulain, D. A. (1995). Central inhibitory effects of muscimol and bicuculline on the milk ejection reflex in the anaesthetized rat. *J. Physiol. (London)* **481,** 211–224.

Wang, M., Seippel, L., Purdy, R. H., and Backstrom, J. (1996). Relationship between symptom severity and steroid variation in women with premenstrual syndrome: Study on serum pregnenolone, pregnenolone sulfate, 5α-pregnane-3,20-dione, and 3α-hydroxy-5α-pregnan-20-one. *J. Clin. Endocrinol. Metab.* **81,** 1076–1082.

Weiland, N. G., and Orchinik, M. (1995). Specific subunit mRNAs of the GABA$_A$ receptor are regulated by progesterone in subfields of the hippocampus. *Mol. Brain Res.* **32,** 271–278.

Weizman, R., Dagan, E., Snyder, S. H., and Gavish, M. (1997). Impact of pregnancy and lactation on GABA$_A$ receptor and central-type and peripheral-type benzodiazepine receptors. *Brain Res.* **752,** 307–314.

Whittemore, E. R., Yang, W., Drewe, J. A., and Woodward, R. M. (1996). Pharmacology of the human γ-aminobutyric acid$_A$ receptor α$_4$ subunit expressed in *Xenopus laevis* oocytes. *Mol. Pharmacol.* **50,** 1364–1375.

Wu, F. S., Gibbs, T. T., and Farb, D. H. (1990). Inverse modulation of gamma-aminobutyric acid- and glycine-induced currents by progesterone. *Mol. Pharmacol.* **37,** 597–602.

Yu, R., and Ticku, M. K. (1995a). Chronic neurosteroid treatment produces functional heterologous uncoupling at the γ-aminobutyric acid type A/benzodiazepine receptor complex in mammalian cortical neurons. *Mol. Pharmacol.* **47,** 603–610.

Yu, R., and Ticku, M. K. (1995b). Chronic neurosteroid treatment decreases the efficacy of benzodiazepine ligands and neurosteroids at the γ-aminobutyric acid$_A$ receptor complex in mammalian cortical neurons. *J. Pharmacol. Exp. Ther.* **275,** 784–789.

Yu, R., Follesa, P., and Ticku, M. K. (1996a). Down-regulation of the GABA receptor subunit mRNA levels in mammalian cultured cortical neurons following chronic neurosteroid treatment. *Mol. Brain Res.* **41,** 163–168.

Yu, R., Hay, M., and Ticku, M. K. (1996b). Chronic neurosteroid treatment attenuates single cell GABA$_A$ response and its potentiation by modulators in cortical neurons. *Brain Res.* **706,** 160–162.

STRESS AND NEUROACTIVE STEROIDS

Maria Luisa Barbaccia,*,[1] Mariangela Serra,[†] Robert H. Purdy,[‡] and Giovanni Biggio[†]

*Department of Neuroscience, University of Rome "Tor Vergata"
00133 Rome, Italy
[†]Department of Experimental Biology "Bernardo Loddo," University of Cagliari
09100 Cagliari, Italy
[‡]Department of Psychiatry, Veterans Affairs Medical Center Research Service
San Diego, California, and Department of Neuropharmacology, The Scripps
Research Institute, La Jolla, California 92037

I. Stress and $GABA_A$ Receptors
II. Effect of Stress on Brain Concentrations of Neuroactive Steroids
   A. Acute Stress
   B. Chronic Stress
III. Possible Mechanisms Underlying the Stress-Induced Changes in Brain Neurosteroid Concentrations
IV. Conclusions
   References

The discovery that the endogenous steroid derivatives 3α-hydroxy-5α-pregnan-20-one (allopregnanolone, or $3\alpha,5\alpha$-TH PROG) and $3\alpha,21$-dihydroxy-5α-pregnan-20-one (allotetrahydrodeoxycorticosterone, or $3\alpha, 5\alpha$-TH DOC) elicit marked anxiolytic and anti-stress effects and selectively facilitate γ-aminobutyric acid (GABA)-mediated neurotransmission in the central nervous system (see Chapter 3) has provided new perspectives for our understanding of the physiology and neurobiology of stress and anxiety. Evidence indicating that various stressful conditions that downregulate GABAergic transmission and induce anxiety-like states (Biggio *et al.*, 1990) also induce marked increases in the plasma and brain concentrations of these neuroactive steroids (Biggio *et al.*, 1996, 2000) has led to the view that stress, neurosteroids, and the function of $GABA_A$ receptors are intimately related. Changes in the brain concentrations of neurosteroids may play an important role in the modulation of emotional state as well as in the homeostatic mechanisms that counteract the neuronal overexcitation elicited by acute stress. Indeed, neurosteroids not only interact directly with $GABA_A$ receptors but also regulate the expression of genes that encode subunits of this receptor complex. This chapter summarizes observations from our laboratories and others, suggesting that neurosteroids and GABAergic

[1]Author to whom correspondence should be sent.

transmission are important contributors to the changes in emotional state induced by environmental stress. © 2001 Academic Press.

I. Stress and GABA$_A$ Receptors

Central γ-aminobutyric acid (GABA)-ergic transmission plays a key role both in the regulation of an individual's reactivity to rapid changes in environmental conditions that may lead to anxiety and in the control of emotional state (Fig. 1). Benzodiazepines and their congeners that act as positive allosteric modulators of the GABA$_A$-receptor complex relieve anxiety in humans (Haefely, 1994). Conversely, a reduction in GABAergic transmission, such as that induced by negative allosteric modulators of the GABA$_A$ receptor, induces in humans psychic and somatic symptomatology reminiscent of the manifestations of anxiety and panic attacks (Dorow et al., 1983).

We showed many years ago that, in animals, GABA$_A$ receptors in the brain are affected by emotional state, thus, stress caused by handling was shown to reduce the density of low-affinity GABA$_A$ receptors in rat cerebral cortex (Biggio et al., 1980, 1981; Biggio, 1983). These data were obtained by comparing naive animals that were handled only before killing (stressed) with rats that were habituated for 4 or 5 days before their actual demise to the manipulations that precede killing (nonstressed). We further showed

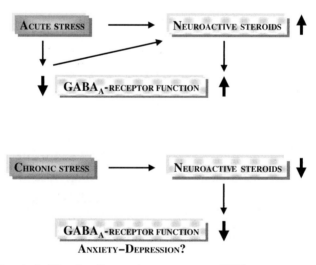

Fig. 1. Hypothetical functional interactions among stress, GABA$_A$ receptors, and emotional behavior.

that the administration of foot shock immediately before killing resulted in a marked decrease in [^{3}H]GABA binding to brain membranes isolated from handling-habituated rats but failed to further reduce [^{3}H]GABA binding in those isolated from naive animals (Biggio, 1983; Concas et al., 1985; Corda and Biggio, 1986; Biggio et al., 1987).

The notion that GABA$_A$ receptors are affected by stress was further supported by the observations that various acute stress paradigms, including forced inhalation of carbon dioxide (CO_2), forced swimming, and exposure to a new environment (Biggio et al., 1981; Concas et al., 1983, 1987; Medina et al., 1983; Drugan et al., 1989; Serra et al., 1991; Andrews et al., 1992; File et al., 1993), all of which also elicit anxiety-related behavior, induced a rapid and reversible downregulation of GABA$_A$ receptor function. The latter was assessed by measuring GABA-stimulated Cl$^-$ flux in synaptoneurosomes or the binding of [^{3}H]benzodiazepine ligands or of t-[^{35}S]butylbicyclophosphorothionate (TBPS), which interacts with the GABA$_A$-receptor-associated Cl$^-$ channel, to brain membranes.

Moreover, certain β-carboline derivatives that bind selectively to the benzodiazepine recognition site of the GABA$_A$ receptor and act as negative allosteric modulators at these receptors (Biggio et al., 1984; Concas et al., 1984, 1988a,b) were also shown to elicit anxiety-like behavior in rodents and primates as well as anxiety in humans (Ninan et al., 1982; Corda et al., 1983; Dorow et al., 1983). The biochemical and behavioral effects of acute stress and of negative allosteric modulators of the GABA$_A$ receptor are prevented, in a flumazenil-sensitive manner, by the systemic administration of benzodiazepine ligands that act as positive allosteric modulators of the GABA$_A$ receptor (Biggio et al., 1990).

More recently, impairment of GABAergic transmission either by inactivating the gene for one of the two isoforms of glutamic acid decarboxylase (GAD 65, which catalyzes the synthesis of GABA from glutamate) (Kash et al., 1999) or by reducing the expression of the γ_2 subunit of the GABA$_A$ receptor (which results in reduced clustering of GABA$_A$ receptors at synaptic sites) (Crestani et al., 1999) in mice has been shown to induce anxiety-related behavior in various behavioral tests. Together, these observations support the view that GABAergic transmission, particularly that mediated by the GABA$_A$ receptor complex, is highly responsive to sudden changes in environmental conditions that may require a rapid increase in the state of arousal and attention.

Several mechanisms may contribute to the downregulation of GABA$_A$ receptor function induced by acute stress: changes in the phosphorylation state of receptor subunits (Moss et al., 1992; Gyenes et al., 1994; Wan et al., 1997; McDonald et al., 1998; Nusser et al., 1999), changes in the extent of GABA synthesis or release (File et al., 1993), and changes in the local concentrations of allosteric modulators (neuroactive steroids).

Whereas changes in the phosphorylation state of $GABA_A$ receptor subunits in response to acute stress have not been described, a decrease in GABA release (File et al., 1993) and changes in the abundance of steroids active at the $GABA_A$ receptor (Purdy et al., 1991; Barbaccia et al., 1994, 1996a,b,c, 1997; Biggio et al., 1996, 2000) have been demonstrated in stressed animals.

II. Effect of Stress on Brain Concentrations of Neuroactive Steroids

A. ACUTE STRESS

The term neuroactive steroid was coined to describe the ability of certain steroids to interfere rapidly with synaptic activity by modulating the function of ligand-gated ionotropic receptors (Paul and Purdy, 1992). Thus, neuroactive steroids have been shown to potentiate [3α-hydroxylated, 5α-reduced metabolites of progesterone (PROG), deoxycorticosterone, and dehydroepiandrosterone (DHEA)] or to inhibit [DHEA sulfate (DHEAS) and pregnenolone sulfate (PREGS)] GABAergic transmission mediated by $GABA_A$ receptors, to facilitate glutamatergic transmission through N-methyl-D-aspartate-sensitive receptors (DHEA, DHEAS, and PREGS), or to inhibit cholinergic transmission mediated by nicotinic receptors (PROG). Whereas the potentiation of $GABA_A$ receptor function by 3α-hydroxylated, 5α-reduced metabolites of PROG occurs at submicromolar concentrations, consistent with the concentrations of these compounds in the brain, the inhibition of these receptors by DHEAS or PROGS requires concentrations of ≥ 10 μM, suggesting that such inhibition may be of only pharmacological relevance.

The first evidence that a synthetic 3α-hydroxylated, 5α-reduced steroid, alphaxalone, could interact in a stereospecific manner with the $GABA_A$ receptor complex, and thereby potentiate GABA-induced Cl^- currents in brain slices and dissociated neurons in culture, was obtained in 1984 (Harrison and Simmonds, 1984). Majewska et al. (1986) subsequently showed that $3\alpha,5\alpha$-TH PROG (epiallopregnanolone) and $3\alpha,5\alpha$-TH DOC (tetrahydrodeoxycorticosterone), two endogenous 3α-hydroxylated, 5α-reduced metabolites of PROG and deoxycorticosterone, shared the ability of alphaxalone to act, at submicromolar concentrations, as positive allosteric modulators at $GABA_A$ receptors. Evidence now indicates that $3\alpha,5\alpha$-TH PROG and $3\alpha,5\alpha$-TH DOC are likely the most potent and efficacious endogenous positive allosteric modulators of the $GABA_A$ receptor, exerting anxiolytic, anticonvulsant, and sedative-hypnotic actions when administered *in vivo* (see Chapter 3). It was therefore of interest to determine whether the changes in GABAergic transmission induced by acute stress are associated with fluctuations in the plasma and brain concentrations of $3\alpha,5\alpha$-TH PROG and $3\alpha,5\alpha$-TH DOC.

1. *Brain and Plasma Concentrations of Neuroactive Steroids*

The existence of a link between acute stress and changes in the concentrations of neurosteroids was first detected by Purdy *et al.* (1991), who showed that forced swimming induced a time-dependent increase in the amounts of PROG and its metabolites, $3\alpha,5\alpha$-TH PROG and $3\alpha,5\alpha$-TH DOC, in rat brain and plasma. We subsequently showed that the concentrations of these neurosteroids were much lower in the brains of handling-habituated rats than in those of naive animals or of handling-habituated rats that had been subjected to foot shock (Barbaccia *et al.*, 1994, 1996a,b,c,1997; Biggio *et al.*, 1996). Handling-habituated rats, which exhibit a greater number of $GABA_A$ receptors in the brain than do naive animals, represent an experimental model of relatively nonstressed animals. Handling-habituated adult male rats exposed to foot shock (0.2-mA, 500-ms pulses delivered each second for 5 min) showed a time-dependent increase in the plasma and brain concentrations of $3\alpha,5\alpha$-TH PROG and $3\alpha,5\alpha$-TH DOC as well as in those of their precursors PREG and PROG (Barbaccia *et al.*, 1996b, 1997). The brain and plasma concentrations of $3\alpha,5\alpha$-TH PROG increased in parallel, reaching a peak (+390% and +300%, respectively) 30 min after foot shock. Similarly, rats forced to inhale a mixture of 65% oxygen (O_2) and 35% CO_2 for 1 min showed a time-dependent increase in the cerebral cortical concentration of $3\alpha,5\alpha$-TH PROG which was maximal (+218%) 30 min after CO_2 inhalation and still apparent (+121%) at 60 min (Barbaccia *et al.*, 1996). Although CO_2 inhalation also increased the plasma concentration of $3\alpha,5\alpha$-TH PROG, this effect was no longer apparent at 60 min. A dissociation between the time courses of changes in $3\alpha,5\alpha$-TH PROG concentration in the brain and plasma was also apparent in female rats during pregnancy (Concas *et al.*, 1998) and may reflect differences in the synthesis or metabolism of neuroactive steroids between the brain and plasma. Indeed, as discussed in previous chapters, $3\alpha,5\alpha$-TH PROG,$3\alpha,5\alpha$-TH DOC, and other neuroactive steroids are produced by neural cells, either from cholesterol or through metabolism of blood-borne precursors (Baulieu and Robel, 1990). The regional distribution of steroidogenic enzymes (especially 3α-hydroxysteroid oxidoreductase and 5α-reductase) (Li *et al.*, 1997) may explain the uneven concentrations of neuroactive steroids detected in different regions either of human postmortem brain (Bixo *et al.*, 1997) or of the brains of adrenalectomized-castrated rats treated with selective agonists of the mitochondrial benzodiazepine receptor (MBR) (Romeo *et al.*, 1992; Serra *et al.*, 1999) or with PROG (Cheney *et al.*, 1995).

Our demonstration that acute stress increases the abundance of neurosteroids in the brain of rats in which GABAergic transmission is reduced therefore suggested that (1) downregulation of $GABA_A$-mediated neurotransmission may facilitate the increase in the plasma and brain concentrations of neurosteroids, and (2) neurosteroids may function to counteract the inhibitory effect of stress on $GABA_A$ receptor function. We thus investigated the role of $GABA_A$-receptor-mediated neurotransmission in the modulation of plasma and brain concentrations of neurosteroids as well as the role of these steroids as physiological modulators of both $GABA_A$-receptor function and stress-induced behavioral changes.

2. GABAergic Transmission and Neuroactive Steroids

Further insight into the putative functional role of GABAergic transmission in the stress-induced increase in plasma and brain concentrations of neurosteroids was provided by the observations that isoniazid (an inhibitor of GABA synthesis) and β-carboline derivatives that act as negative allosteric modulators of $GABA_A$-receptor function mimic this effect of stress in rats. Isoniazid thus increased the cerebrocortical (Fig. 2) and plasma concentrations of $3\alpha,5\alpha$-TH PROG and $3\alpha,5\alpha$-TH DOC.

The plasma concentration of corticosterone was also increased 40 min after isoniazid administration but had returned to control values with 80 min. Isoniazid induced tonic-clonic seizures in about 80% of rats within 50 min. The possibility that the isoniazid-induced increase in the brain concentrations of $3\alpha,5\alpha$-TH PROG and $3\alpha,5\alpha$-TH DOC might result from the generalized increase in neuronal activity associated with seizures, rather than being a direct consequence of the reduction in GABAergic transmission, was evaluated by treating animals with the β-carboline derivative FG 7142, an anxiogenic but not convulsant drug that acts as a selective negative allosteric modulator of $GABA_A$ receptors.

The administration of FG 7142 also increased the cerebrocortical and plasma concentrations of $3\alpha,5\alpha$-TH PROG and $3\alpha,5\alpha$-TH DOC with similar time courses (Fig. 2), which were also consistent with that of the $GABA_A$-receptor-mediated anxiogenic and proconvulsant action of this drug (Ninan *et al.*, 1982; Concas *et al.*, 1983, 1984). The contribution of a decrease in $GABA_A$ receptor function to the increase in the brain and plasma concentrations of $3\alpha,5\alpha$-TH PROG and $3\alpha,5\alpha$-TH DOC induced by isoniazid or FG 7142 was further demonstrated by the observation that abecarnil (0.3 mg/kg, intraperitoneally), a β-carboline derivative that acts as a positive allosteric modulator at $GABA_A$ receptors and is a potent anxiolytic and anticonvulsant (Stephens *et al.*, 1990), prevented these effects. The intraperitoneal injection of abecarnil alone, at a dose previously shown to induce anxiolytic and anticonvulsant effects, failed to modify the basal concentrations

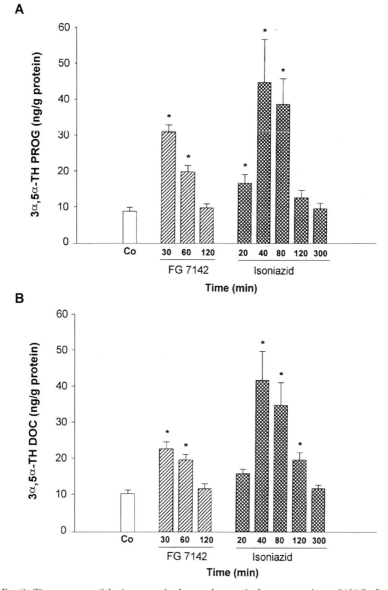

FIG. 2. Time courses of the increases in the cerebrocortical concentrations of (**A**) 3α,5α-TH PROG and (**B**) 3α,5α-TH DOC (**B**) induced by isoniazid and FG 7142. Handling-habituated male rats were treated with vehicle (Co), FG 7142 (20 mg/kg intraperitoneally), or isoniazid (375 mg/kg, subcutaneously) and killed by focused microwave irradiation to the head at the indicated times thereafter. Data are means ± SEM of values obtained from at least 10 rats. (*$P < 0.05$ vs control.) (Adapted with permission from M. L. Barbaccia, G. Roscetti, M. Trabbucci, R. H. Purdy, M. C. Mostallino, A. Concas, A., and G. Biggio. The effects of inhibitors of GABAergic transmission and stress on brain and plasma allopregnanolone concentrations. *The British Journal of Pharmacology*, 1997; 120:1582–1588.)

of $3\alpha,5\alpha$-TH PROG and $3\alpha,5\alpha$-TH DOC in the brain of handling-habituated rats (Barbaccia *et al.*, 1996a, 1997).

3. *Role of the Increase in Neuroactive Steroid Concentrations Induced by Acute Stress*

We showed that the time courses of changes in proconflict behavior (measured by Vogel's test) and in $GABA_A$ receptor function (monitored as the binding of [^{35}S]TBPS to fresh, unwashed cerebrocortical membranes, which is inversely proportional to receptor occupancy by GABA) induced by forced inhalation of CO_2 for 1 min were similar (but opposite) magnitude and differed from that of the associated increase in the cerebrocortical concentration of $3\alpha,5\alpha$-TH PROG (Fig. 3). The anxiety-related behavior and $GABA_A$ receptor function were maximally increased and decreased, respectively, 5 min after stress delivery and had returned to control values by 60 min. In contrast, the cerebrocortical concentration of $3\alpha,5\alpha$-TH PROG was maximally increased after 30 min, remained significantly increased at 60 min, and had returned to control values by 90 min after CO_2 inhalation. Such acute stress increases the concentrations of $3\alpha,5\alpha$-TH PROG and $3\alpha,5\alpha$-TH DOC in rat cerebral cortex from 2–3 nM to 8–12 nM, the latter values being within the range of those shown to potentiate GABA action at $GABA_A$ receptors (Matsumoto *et al.*, 1999; Pinna *et al.*, 2000). These observations suggested that the increase in the brain concentrations of $3\alpha,5\alpha$-TH PROG, which positively modulates $GABA_A$ receptor function even more efficiently when the GABA concentration is reduced (Concas *et al.*, 1996), and $3\alpha,5\alpha$-TH DOC in response to acute stress may serve to limit the extent and duration of the stress-induced impairment in GABAergic transmission and the associated behavioral consequences.

4. *Acute Stress and the Hypothalamic-Pituitary-Adrenal Axis*

Our observations that foot shock and CO_2 inhalation also increased the plasma concentration of corticosterone (with maximal effects of $+320\%$ and $+58\%$ apparent 10 and 30 min after foot shock and CO_2 inhalation, respectively) suggested that the increases in neuroactive steroid concentrations induced by these treatments might be related to the stress-induced activation of the hypothalamic-pituitary-adrenal (HPA) axis. Given that GABA, acting at $GABA_A$ receptors, inhibits the release of corticotropin-releasing factor (CRF) from the hypothalamus (Calogero *et al.*, 1988), the acute-stress-induced downregulation of central $GABA_A$ receptor function may contribute to the activation of the HPA axis in response to such stress. This notion is also consistent with our observations that the decrease in $GABA_A$ receptor function elicited by systemic administration of isoniazid or FG 7142 mimicked the effect of stress on brain and plasma concentrations of $3\alpha,5\alpha$-TH PROG and $3\alpha,5\alpha$-TH DOC and that abecarnil prevented the effects of

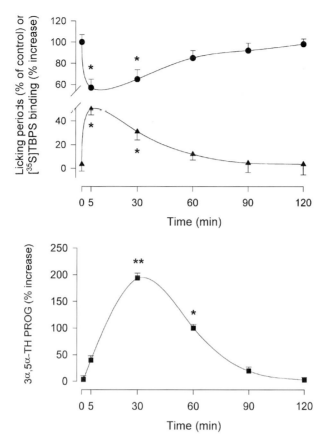

FIG. 3. Time courses of the effects of CO_2 inhalation on Vogel's conflict test (●), [^{35}S]TBPS binding to unwashed cerebrocortical membranes (▲), and the cortical concentration of $3\alpha,5\alpha$-TH PROG (■). Rats were exposed to CO_2 for 1 min. Vogel's conflict test was performed 10 min after CO_2 inhalation, and the number of licking periods was measured during the punished session at a current of 0.4 mA. For the biochemical studies, rats were killed by focused microwave irradiation to the head (for $3\alpha,5\alpha$-TH PROG assay) or by guillotine ([^{35}S]TBPS binding) at the indicated times after stress. Data are expressed as a percentage of control values or as a percentage increase and are means ± SEM of values obtained from at least 10 rats. (*$P < 0.05$, **$P < 0.01$ vs control.) (Adapted with permission of S. Karger AG, Basel, from M. L. Barbaccia, G. Roscetti, M. Trabucchi, M. C. Mostallino, A. Concas, R. H. Purdy, and G. Biggio. Time dependent changes in rat brain neuroactive steroid concentrations and $GABA_A$ receptor function after acute stress. *Neuroendocrinology* 1996;63:166–172.)

both of these agents as well as acute stress on neuroactive steroid abundance (Barbaccia et al., 1996a, b, c, 1997; Biggio et al., 1996). Acute stress failed to affect the brain and plasma concentrations of DHEA, which is synthesized from PREG by the action of the enzyme cytochrome 17α-hydroxylase/17,20-lyase($P450_{c17}$); this enzyme is present only at low levels in adrenal glands and has not yet been detected in the brains of adult rodents (Mellon and Deschepper, 1993).

The role of the HPA axis in the effects of acute stress on neuroactive steroid concentrations was further supported by the observations that neither foot shock, CO_2 inhalation, nor the pharmacologically induced downregulation of GABAergic transmission increased 3α,5α-TH PROG or 3α,5α-TH DOC abundance in the brains of rats whose major peripheral steroidogenic organs (adrenals and gonads) had been removed (Fig. 4) (Barbaccia et al., 1997). Adrenalectomy had previously been shown only to reduce the increase in the brain concentration of 3α,5α-TH PROG induced by forced swimming for 20 min (Purdy et al., 1991). This apparent discrepancy might be attributable to (1) the difference in the type of surgery, adrenalectomy-orchiectomy vs adrenalectomy, given that GABA also appears to modulate the hypothalamic-pituitary-gonadal axis (Masotto et al., 1989) in addition to the HPA axis, (2) the difference in the method of killing (focused microwave

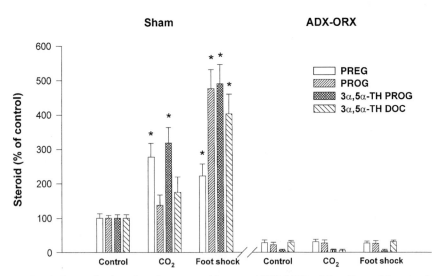

FIG. 4. Prevention by adrenalectomy-orchiectomy (ADX-ORX) of the effects of foot shock or CO_2 inhalation on the cerebrocortical concentrations of PREG, PROG, 3α,5α-TH PROG, and 3α,5α-TH DOC. Three weeks after ADX-ORX or sham surgery, male rats were subjected to CO_2 inhalation for 1 min or foot shock (0.2 mA for 500 ms every second) for 5 min. The animals were killed 30 min later by focused microwave irradiation to the head. Data are means ± SEM of 10 animals. (*$P < 0.05$ vs respective control value.)

irradiation to the head vs guillotine), or (3) the difference in the type of stressor (foot shock or CO_2 inhalation vs forced swimming), which might result in a difference in the processing or integration of afferent stimuli.

With regard to this latter point, the acute stress associated with the brief handling of animals before killing, which downregulates central $GABA_A$ receptor function (Biggio *et al.*, 1990) and reduces the extent of [^{3}H]flunitrazepam binding to brain membranes (Andrews *et al.*, 1992), yielded higher concentrations of $3\alpha,5\alpha$-TH PROG and $3\alpha,5\alpha$-TH DOC in the cerebral cortex, but not in the plasma, of naive rats compared with those in handling-habituated rats (Barbaccia *et al.*, 1997). This observation, together with the fact that neither foot shock nor CO_2 inhalation increased brain $3\alpha,5\alpha$-TH PROG and $3\alpha,5\alpha$-TH DOC concentrations in adrenalectomized-orchiectomized rats, whose plasma concentrations of CRF and adrenocorticotropic hormone (ACTH) are presumably increased because of the lack of negative feedback control by circulating glucocorticoids, suggests that brain neurosteroidogenesis is not regulated by pituitary factors that control peripheral steroidogenesis (Le Goascogne *et al.*, 1987; Corpéchot *et al.*, 1993; Cheney *et al.*, 1995; Roscetti *et al.*, 1998).

B. CHRONIC STRESS

The only experimental model in which the effects of chronic stress on the brain and plasma concentrations of neuroactive steroids have been investigated is social isolation of rats or mice that have been subjected to prolonged periods of individual housing. Although the extent of behavioral changes induced by social isolation appears to depend on the age at the time of isolation (prepubertal vs adult), on the type of cage used for the individual housing, and on the strain of animal (Holson *et al.*, 1991; Kim and Kirkpatrick, 1996; Paulus *et al.*, 2000), individual housing for prolonged periods (1–10 weeks) results in increased aggression and neophobia, increased anxiety-related behavior in the elevated plus-maze, and increased locomotor activity (Parker and Morinan, 1986; Hilakivi *et al.*, 1989; Wong-witdecha and Marsden, 1996). These effects of individual housing are thus consistent with the notion (Hatch *et al.*, 1963) that isolation is stressful for these normally gregarious animals and that the abnormal behavior is the product of prolonged stress.

1. *Behavioral Effects of Social Isolation*

Consistent with previous data (Parker and Morinan, 1986), we observed that male rats isolated for 30 days immediately after weaning (25 days of age) exhibited anxiety-related behavior in the elevated plus-maze (Serra *et al.*, 2000). The time that isolated animals spent (14 ± 0.2%) in the open arms of the maze was thus markedly reduced compared with that spent by group-housed rats ($31 \pm 0.7\%$; means ± SEM of values for 10–15 rats, $P < 0.01$). The number of entries into the closed arms of the maze did not differ significantly between the two groups of animals. Moreover, social isolation significantly reduced the punished consumption of water in Vogel's conflict test; the number of licking periods in 3 min during punishment for isolated rats was 4 ± 0.1, compared with a value of 10 ± 0.1 (means ± SEM for 10–15 rats, $P < 0.01$) for group-housed animals.

2. Brain and Plasma Concentrations of Neuroactive Steroids

Social isolation for 30 days without any additional stressor significantly reduced the cerebrocortical and hippocampal concentrations of PREG, PROG, $3\alpha,5\alpha$-TH PROG, and $3\alpha,5\alpha$-TH DOC compared with the corresponding values for group-housed animals (Fig. 5). The concentrations of DHEA in the same two brain regions were not affected by social isolation. Whereas social isolation also reduced the plasma concentrations of all neuroactive steroids measured, it significantly increased the plasma corticosterone concentration (184 ± 21 vs 139 ± 16 ng/ml; $P < 0.05$). This observation, although consistent with previous data showing that chronic or intermittent exposure to stressful conditions, including social isolation, resulted in a persistent increase in the plasma concentration of corticosterone (Kant et al., 1983; Rivier and Vale, 1987; Greco et al., 1989; Hauger et al., 1988), is at variance with the results of other studies showing that isolation either reduced (Miachon et al., 1993; Sanchez et al., 1998) or failed to modify (Haller and Halasz, 1999) plasma corticosterone concentration. This discrepancy may be due to a difference in the proportion of dominant (lower plasma corticosterone) or subordinate (higher plasma corticosterone) rats in the nonisolated groups (Pohorecky et al., 1999). However, the relatively small size of the increase (+32%) in the plasma corticosterone concentration that we detected in isolated animals indicates that individual housing represents a relatively mild stressor, and this finding is consistent with the anxiety-like behavior exhibited by these animals in Vogel's conflict test and the elevated plus-maze test.

The effects of social isolation on the cerebrocortical concentrations of neuroactive steroids were shown to be time dependent (Fig. 6). Whereas the concentrations of PREG, PROG, $3\alpha,5\alpha$-TH PROG, and $3\alpha,5\alpha$-TH DOC remained unchanged during the first 48 h of isolation, they were significantly decreased after 7 days and remained at similar reduced levels at 30 days. The effects of social isolation for 30 days on the cortical concentrations of these molecules were prevented by handling of the animals twice daily; the steroid concentrations thus did not differ significantly between handled, isolated rats and group-housed animals (Serra et al., 2000).

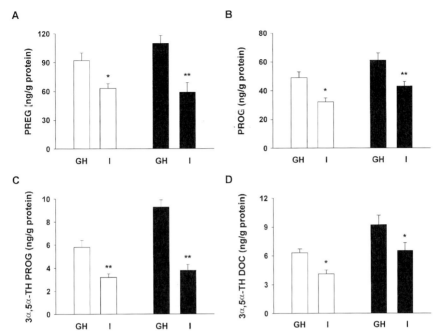

FIG. 5. Effects of social isolation on the concentrations of neuroactive steroids in the cerebral cortex (white columns) and hippocampus (black columns) of rats. Animals were either group-housed (GH) or isolated (I) for 30 days, after which they were killed by focused microwave irradiation to the head and the concentrations of (**A**) PREG (**B**) PROG, (**C**) 3α,5α-TH PROG, and (**D**) 3α,5α-TH DOC were determined. Data are means ± SEM of values from at least 12 rats. ($^*P < 0.05$, $^{**}P < 0.01$ vs group-housed animals.) (Adapted with permission from M. Serra, M. G. Pisu, M. Littera, G. Papi, E. Sanna, F. Tuveri, L. Usala, R. H. Purdy, and G. Biggio. Social isolation-induced decrease in both the abundance of neuroactive steroids and GABA$_A$ receptor function in rat brain. *Journal of Neurochemistry* 2000; 75:732–740.)

The molecular mechanisms that underlie the persistent decrease in the plasma and brain concentrations of neuroactive steroids apparent in individually housed rats remain unclear. Given that the concentrations of all neurosteroids examined, with the exception of that of DHEA, were reduced in both the brain and the plasma of socially isolated animals, an effect of social isolation on the activity or of expression of enzymes that contribute to the synthesis of specific neuroactive steroids, as suggested by Matsumoto *et al.* (1999), appears to be unlikely.

Various types of chronic stress, including social isolation, activate hypothalamic neurons that release CRF (Rivier and Vale, 1987; deGoeij *et al.*, 1991; Ojima *et al.*, 1995). Moreover, various chronic stress paradigms increase the amount of CRF mRNA in rat brain (Makino *et al.*, 1995; Albeck

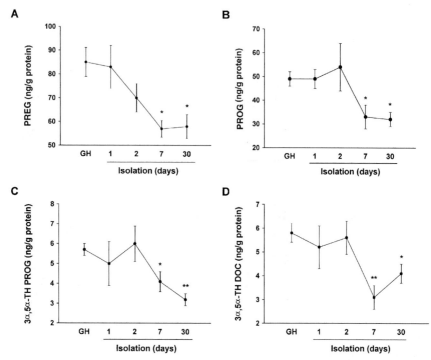

FIG. 6. Time courses of the effects of social isolation on the concentrations of neuroactive steroids in the cerebral cortex of rats. Animals were either group-housed for 30 days or isolated for the indicated times, after which they were killed by focused microwave irradiation to the head and the cortical concentrations of (**A**) PREG (**B**) PROG (**C**) $3\alpha,5\alpha$-TH PROG, and (**D**) $3\alpha,5\alpha$-TH DOC were determined. Data are means ± SEM of values from at least eight rats. (*$P < 0.05$, **$P < 0.01$ vs group-housed animals.) (Adapted with permission from M. Serra, M. G. Pisu, M. Littera, G. Papi, E. Sanna, F. Tuveri, L. Usala, R. H. Purdy, and G. Biggio. Social isolation-induced decrease in both the abundance of neuroactive steroids and $GABA_A$ receptor function in rat brain. *Journal of Neurochemistry* 2000; 75:732–740.)

et al., 1997), and the intracerebroventricular administration of CRF in rats induces anxiety-like behavior similar to the conflict behavior detected in socially isolated animals (Britton *et al.,* 1986; Dunn and File, 1987). Given that long-term exposure of cultured pituitary cells to CRF induces desensitization to the effect of this factor on ACTH release (Hoffman *et al.,* 1985), a reduction in pituitary responsiveness to the high concentrations of CRF associated with chronic stress might result in a decrease in the extent of ACTH secretion and, in turn, a reduced stimulation of the synthesis of neuroactive steroids in the adrenal glands. Consistent with this scenario, the ACTH response to chronic foot shock or immobilization stress in rats is reduced,

TABLE I

Effects of Acute Foot Shock Stress on the Cerebrocortical Concentrations of Neuroactive Steroids in Group-Housed and Isolated Rats

Group	PREG	PROG	$3\alpha,5\alpha$-TH PROG	$3\alpha,5\alpha$-TH DOC
Group housed				
Control	104 ± 13	52 ± 7.4	8.3 ± 0.9	6.9 ± 0.8
Foot shock	$169 \pm 4.5^*$	$94 \pm 4.9^*$	$14.8 \pm 1.4^*$	$14.3 \pm 1.5^*$
Isolated				
Control	$66 \pm 4.6^{**}$	$36 \pm 3.7^{**}$	$4.3 \pm 0.3^{**}$	$4.8 \pm 0.4^{**}$
Foot shock	$195 \pm 17^*$	$150 \pm 22^*$	$21.3 \pm 1.9^*$	$18.9 \pm 1.3^*$

Note. Animals were housed in groups or isolation for 30 days, after which they were exposed or not (control) to foot shock stress for 5 min. The animals were killed 30 min thereafter, and cortical concentrations of steroids were measured. Data are means ± SEM of values from 8 to 10 rats.

$^*P < 0.01$ vs respective control animals.
$^{**}P < 0.05$ vs control group-housed animals.

compared with that to acute stress (Rivier and Vale, 1987; Hauger *et al.,* 1988). Acute foot shock markedly increased the cerebral cortical (Table I) and plasma (data not shown) concentrations of neuroactive steroids in both group-housed and isolated rats; however, the percentage increases in the cortical and plasma concentrations were markedly greater in isolated rats than in group-housed rats. These observations are consistent with the notion that a "facilitatory trace," characterized by hyper-responsiveness of the HPA axis to new stimuli, may develop during chronic stress (Akana *et al.,* 1992). Thus, the reduced basal concentrations of neuroactive steroids associated with social isolation immediately after weaning may result from altered regulation of the HPA axis, rather than from a change in adrenal secretory capability per se.

3. *GABA$_A$ Receptor Function*

The persistent decrease in the concentrations of neuroactive steroids in the brain of socially isolated rats was shown to be associated with a decrease in the function of brain GABA$_A$ receptors. Prolonged social isolation affected the functional coupling between the recognition site for GABA and those for allosteric modulators, such as benzodiazepines and anxiogenic β-carbolines. Thus, in *Xenopus* oocytes injected with synaptosomes derived from isolated rats, the potentiating effect of diazepam on GABA-evoked Cl$^-$ current was less pronounced than that observed in oocytes injected with synaptosomes purified from group-housed animals; the potency of diazepam did not differ between the two preparations (Fig. 7). Similar results were obtained with oocytes expressing hippocampal GABA$_A$ receptors from

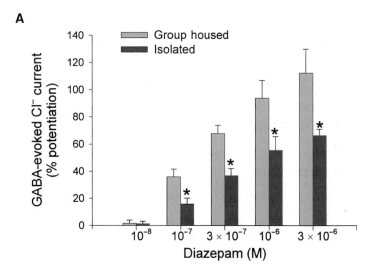
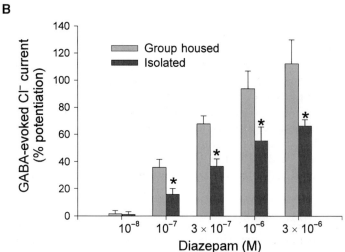

FIG. 7. Effects of social isolation on the modulation of GABA$_A$-receptor function by diazepam. *Xenopus* oocytes were injected with synaptosomes purified from the (**A**) cerebral cortex or (**B**) hippocampus of rats that were housed in groups or in isolation for 30 days. Oocytes were voltage-clamped at −90 mV, and the effect of various concentrations of diazepam on the amplitude of GABA-evoked Cl$^-$ current was measured. Data are expressed as percentage potentiation of the GABA response and are means ± SEM of values from six to eight oocytes injected with synaptosomes from five rats. (*$P < 0.05$ vs respective value for group-housed animals.) (Adapted with permission from M. Serra, M. G. Pisu, M. Littera, G. Papi, E. Sanna, F. Tuveri, L. Usala, R. H. Purdy, and G. Biggio. Social isolation-induced decrease in both the abundance of neuroactive steroids and GABA$_A$ receptor function in rat brain. *Journal of Neurochemistry* 2000; 75:732–740.)

TABLE II
EFFECT OF SOCIAL ISOLATION ON THE INHIBITORY ACTION OF β-CCE
ON GABA-EVOKED CL$^-$ CURRENT

β-CCE (μM)	GABA-evoked Cl$^-$ current (% inhibition)	
	Group housed	Isolated
Cerebral cortex		
1	15.3 ± 2.8	33.7 ± 3.1*
3	7.2 ± 2.8	20.5 ± 4.2*
Hippocampus		
1	20.9 ± 3.7	15.3 ± 3.0
3	10.0 ± 3.1	10.7 ± 3.4

Note. Xenopus oocytes were injected with synaptosomes purified from the cerebral cortex or hippocampus of rats that were housed in groups or in isolation for 30 days. Oocytes were voltage-clamped at -90 mV. Data are expressed as percentage inhibition of the GABA response and are means ± SEM of values from six to eight oocytes injected with synaptosomes from five rats.

*$P < 0.05$ vs respective value for group-housed animals.

socially isolated and group-housed rats. In contrast, social isolation potentiated the inhibitory action of the benzodiazepine receptor inverse agonist β-carboline-3 carboxylic acid ethyl ester (β-CCE) at GABA$_A$ receptors derived from the cerebral cortex, although it had no effect on the action of β-CCE at receptors prepared from the hippocampus (Table II).

The binding of [^{35}S]TBPS to cortical or hippocampal membranes derived from rats individually housed for 30 days was significantly increased (+24 and +22%, respectively), compared with the corresponding values for group-housed rats; social isolation for 7 days had no effect on this parameter (Serra *et al.*, 2000). The effect of social isolation for 30 days on [^{35}S]TBPS binding to cortical or hippocampal membranes was prevented by handling the animals twice daily. Given that acute stress or administration of negative modulators of GABAergic transmission increases [^{35}S]TBPS binding to brain membranes and induces anxiety in animals or humans (Ninan *et al.*, 1982; Corda *et al.*, 1983; Dorow *et al.*, 1983; Concas *et al.*, 1988a,b; Serra *et al.*, 1989; Biggio *et al.*, 1990), the increase in [^{35}S]TBPS binding to both cortical and hippocampal membranes of socially isolated animals further supports the notion that social isolation results in a decrease in GABAergic transmission mediated by GABA$_A$ receptors. These data are thus, also consistent with our results showing that social isolation increases the level of anxiety in rats as well as with the observations of other researchers showing that isolation reduces the extent of radioligand binding to brain benzodiazepine receptors (Insel, 1989; Miachon *et al.*, 1990), impairs the anxiolytic effect of diazepam in rats subjected to a social interaction test (Wongwitdecha and Marsden, 1996), and reduces the ability of GABA to stimulate $^{36}Cl^-$ uptake into mouse synaptoneurosomes (Ojima et al., 1997).

With regard to the mechanism of the downregulation of GABAergic transmission in individually housed rats, it is possible that the decrease in the brain concentrations of neuroactive steroids affects either the conformation of $GABA_A$ receptors or the expression of specific receptor subunits. The latter proposal is consistent with observations showing that physiological or pharmacologically induced fluctuations in the brain concentrations of neuroactive steroids, such as those that occur during pregnancy or pseudopregnancy, induce a selective modulation of the function and expression of $GABA_A$ receptor subunits in the cerebral cortex and hippocampus of rats (Fenelon and Herbison, 1996; Concas et al., 1998, 1999; Follesa et al., 1998; Smith et al., 1998a, b).

Social isolation of adult mice for 6–10 weeks was also shown to reduce both the concentration of $3\alpha,5\alpha$-TH PROG in the cerebral cortex and the duration of the pentobarbital-induced loss of the righting reflex (Matsumoto et al., 1999). The effect of social isolation in this experimental model was selective for $3\alpha,5\alpha$-TH PROG, given that the cortical concentrations of PREG and PROG were not affected. Moreover, the administration to isolated mice of fluoxetine, which in addition to being a selective inhibitor of serotonin reuptake is also a potent activator of 3α-hydroxysteroid oxidoreductase (which catalyzes the synthesis of $3\alpha,5\alpha$-TH PROG from 5α-dihydroprogesterone) (Griffin and Mellon, 1999), restored the cortical concentration of $3\alpha,5\alpha$-TH PROG and the duration of the pentobarbital-induced loss of the righting reflex to values observed in nonisolated mice. Matsumoto et al. (1999) thus concluded that $3\alpha,5\alpha$-TH PROG in the brain plays a permissive role in the pentobarbital-induced potentiation of GABA action at $GABA_A$ receptors.

Individual housing for 15 days selectively reduced the cerebrocortical and hippocampal, but not plasma, concentrations of $3\alpha,5\alpha$-TH PROG in 3- to 4-month-old Sardinian alcohol-preferring rats, which are characterized by a high level of innate anxiety-like behavior compared with age-matched Sardinian non-alcohol-preferring rats (Barbaccia, et al., 1999). Individually housed, alcohol-preferring rats also exhibited a reduced density of MBRs, which contribute to the synthesis of PREG from cholesterol, in the cerebral cortex and hippocampus compared with that apparent in group-housed animals.

The effects of social isolation on brain and plasma concentrations of neuroactive steroids thus appear to differ somewhat among the various studied versions of this experimental model. Factors likely to be important in these differences include the age at which the animals are exposed to social isolation as well as their basal anxiety or GABAergic tone. Nevertheless, the association between changes in brain neurosteroid concentrations and those in $GABA_A$ receptor function suggests that these steroids play a physiological role in modulation of central GABAergic transmission. Whereas the acute-stress-induced increase in the concentrations of $3\alpha,5\alpha$-TH PROG and $3\alpha,5\alpha$-TH DOC is thought to represent a short-term homeostatic mechanism to restore rapidly the function of $GABA_A$ receptors, the decrease in the concentrations of neuroactive steroids induced by a prolonged period of social isolation may, through impairment of $GABA_A$-receptor function, confer vulnerability, especially in juvenile animals, to mood- and anxiety-related psychopathology (Fig. 1).

Few studies in humans have addressed this issue. Mood alterations and anxiety symptomatology associated with either the menstrual cycle or menopause have been shown to be related to imbalances in the circulating concentrations of neuroactive steroids (Wang *et al.*, 1996; Barbaccia *et al.*, 2000; Monteleone *et al.*, 2000). Moreover, major depression has been associated with low concentrations of $3\alpha,5\alpha$-TH PROG in cerebrospinal fluid (Uzunova *et al.*, 1998) and plasma (Romeo *et al.*, 1998), and successful treatment with antidepressants restored $3\alpha,5\alpha$-TH PROG concentrations to those typical of normal controls. However, the plasma concentration of $3\alpha,5\alpha$-TH DOC has been shown to be increased in depressed individuals, and successful treatment with fluoxetine reduced this increase (Strohle *et al.*, 2000). Such a dissociation between the plasma concentrations of $3\alpha,5\alpha$-TH PROG and $3\alpha,5\alpha$-TH DOC in depressed individuals contrasts with the situation in animal models of either acute or chronic stress, in which the plasma concentrations of these steroids usually change in the same direction.

III. Possible Mechanisms Underlying the Stress-Induced Changes in Brain Neurosteroid Concentrations

Various mechanisms may contribute to the changes in the brain concentrations of neurosteroids induced by exposure of animals to either acute or chronic stress. The failure of acute foot shock and forced CO_2 inhalation to affect the brain concentrations of $3\alpha,5\alpha$-TH PROG and $3\alpha,5\alpha$-TH DOC in adrenalectomized-orchiectomized rats is consistent with a prominent role for the HPA axis in the increases in the concentrations of these steroids induced by acute stress in intact animals. The activation of the HPA axis is likely mediated by the stress-induced downregulation of GABAergic transmission. It should be noted, however, that adrenalectomy-orchiectomy is itself a stressful procedure. Even though the rats were allowed 2 weeks to recover from surgery, the extent of [^{35}S]TBPS binding to brain membranes was greater for adrenalectomized-orchiectomized animals at this time than for intact controls, suggesting that $GABA_A$ receptor function may be reduced in the adrenalectomized-orchiectomized rats. Thus, the basal concentrations of neuroactive steroids in the brain of these rats may reflect, at least to a certain extent, their intrinsic stressed condition and may mask an increase in brain neurosteroid content in response to additional stressors.

Acute stress, acting through the associated increase in ACTH output, may induce rapid changes in the density of MBRs or in the expression of endogenous ligands for these receptors in the adrenal glands and the brain (Gavish *et al.*, 1999). The MBR shows high affinity for various classes of drugs, including benzodiazepines, imidazopyridines, isoquinolines, and 2-aryl-indole-acetamide derivatives (Guarneri *et al.*, 1992; Korneyev *et al.*, 1993; Gavish *et al.*, 1999; Serra *et al.*, 1999; Trapani *et al.*, 1999), as well as for the endogenous peptide diazepam-binding inhibitor (DBI) and its processing product triakontatetraneuropeptide (TTN) (Costa and Guidotti, 1991). The binding of these peptides and synthetic ligands to the MBR increases the efficiency of cholesterol transport to the mitochondrial inner membrane and thereby promotes the synthesis of PREG (Papadopoulos, 1993; Gavish *et al.*, 1999). The number of MBRs and the expression of DBI in the periphery and the brain were shown to be increased in rats subjected to an acute stress (Novas *et al.*, 1987; Ferrarese *et al.*, 1991).

Acute and chronic stress appear to have opposite effects on MBR density and DBI expression in the brain and peripheral tissues of rats. Whereas acute stress increases both parameters in the adrenal glands and the hippocampus (Ferrarese *et al.*, 1991), chronic (social isolation) stress reduces DBI expression in the hypothalamus (Dong *et al.*, 1999). A reduced density of MBRs, associated with reduced synthesis of PREG by cortical minces, has also been detected in the cerebral cortex of Sardinian alcohol-preferring rats after social isolation for 2 weeks compared with values for group-housed animals (Barbaccia *et al.*, 1999).

The parallel decreases in the plasma and brain concentrations of neuroactive steroids apparent in rats socially isolated for 1–4 weeks after weaning suggest that the reduced brain concentrations of $3\alpha,5\alpha$-TH PROG and $3\alpha,5\alpha$-TH DOC are a consequence of reduced adrenal output (Serra *et al.*, 2000). Such a reduced output, which does not appear to be a result of a decrease in adrenal secretory capability per se (given that challenge with a novel acute stress increases the brain and plasma concentrations of these steroids in isolated rats to a greater extent than it does in group-housed animals) (Serra *et al.*, 2000), has been suggested to result from a hyporesponsiveness of the HPA axis. Whether this hyporesponsiveness also involves a decrease in the number of MBRs in the adrenal glands or in DBI expression, which is controlled by ACTH (Besman *et al.*, 1989; Massotti *et al.*, 1991), is not known.

Apparent dissociations between changes in $3\alpha,5\alpha$-TH PROG and $3\alpha,5\alpha$-TH DOC concentrations in the brain and plasma in various animal models of stress may reflect effects on neurosteroidogenesis in brain tissue. As discussed extensively in other chapters of this book, $3\alpha,5\alpha$-TH PROG, $3\alpha,5\alpha$-TH DOC, and other neuroactive steroids are produced by brain cells. Thus, glial cells, like the steroidogenic cells of the gonads and the adrenal glands, express both cytochrome P450side-chain cleavage, the mitochondrial enzyme that cleaves the side chain of cholesterol to yield PREG (Hu *et al.*, 1989; Le Goascogne *et al.*, 1987), and MBRs (Mukhin *et al.*, 1989). Neurons and glia also express 3β-hydroxysteroid dehydrogenase, 5α-reductase, and 3α-hydroxysteroid oxidoreductase, the enzymes that convert PREG and PROG to their active metabolites (Robel *et al.*, 1991; Celotti *et al.*, 1992; Mellon and Deschepper, 1993; Griffin and Mellon, 1999). Thus, brain cells may produce neuroactive steroids *in situ* either through metabolism of circulating precursors (Follesa *et al.*, 2000) or through *de novo* synthesis (Baulieu and Robel, 1990). An important distinction between *de novo* neurosteroidogenesis in the brain and peripheral steroidogenesis is that the former does not appear to be regulated by the pituitary factors that control adrenal and gonadal steroid hormone production (Corpéchot *et al.*, 1993; Mathur *et al.*, 1993; Cheney *et al.*, 1995; Roscetti *et al.*, 1998). Exposure of rat brain tissue *in vitro* to dibutyryl cyclic adenosine monophospate (cAMP), to forskolin (which increases the intracellular concentration of cAMP), or to L-ascorbic acid (which also appears to act through adenylate cyclase) results in an increase in neurosteroid production that varies in a brain-region-dependent manner (Barbaccia *et al.*, 1992; Roscetti *et al.*, 1994, 1998). This cAMP-induced stimulation of neurosteroidogenesis is inhibited by PK 11195, a partial agonist of the MBR (Roscetti *et al.*, 1994). These data suggest that acute stress or prolonged social isolation may selectively alter MBR density, DBI expression, and neurosteroid production in the brain by a cAMP-mediated mechanism. Indeed, acute and chronic stress increase or decrease, respectively, the brain concentration of cAMP (Stone and John, 1992; Stone *et al.*, 1984).

IV. Conclusions

Evidence accumulated since the 1980s has revealed a new mechanism of action of steroids. In addition to the classical mechanism of steroid action, in which these compounds interact with specific intracellular receptors to regulate gene expression directly, steroids have been shown to modulate intercellular communication in the brain by affecting synaptic transmission.

Persistent exposure of specific neurotransmitter receptors to neuroactive steroids may also affect the expression of receptor subunits. The demonstration that 3α,5α-TH PROG derived from neuronal metabolism of PROG is able to modulate the expression of $GABA_A$ receptor subunit genes in primary cultures of cerebellar granule cells (Follesa et al., 2000) suggests that changes in the peripheral secretion of PROG and the metabolism of this steroid in the brain might influence both $GABA_A$-receptor function and plasticity, thereby affecting functional connectivity of central neurons. Given the high potencies of 3α,5α-TH PROG and 3α,5α-TH DOC, these two steroids have received the most attention with regard to regulation of the $GABA_A$-receptor complex. However, other derivatives, such as certain 3α-hydroxylated, 5α-reduced metabolites of DHEA (androsterone and 5α-androstane-3α,17β-diol), share with 3α,5α-TH PROG and 3α,5α-TH DOC the ability to potentiate GABAergic transmission (Gee et al., 1988; Lambert et al., 1995; Frye et al., 1996a, b) and may be relevant, because of the high-circulating concentrations of DHEA and DHEA sulfate, in humans. Indeed, reduced concentrations of DHEA and 5α-androstane-3α,17β-diol in plasma correlate with higher anxiety scores in menopausal women (Barbaccia et al., 2000).

The brain concentrations of 3α,5α-TH PROG, 3α,5α-TH DOC, and other neuroactive steroids may be affected by regulatory mechanisms intrinsic to the brain and by changes in the activity of the HPA or hypothalamic-pituitary-gonadal axis and consequent variation in the supply of either neuroactive steroids or their precursors to the brain. Current evidence suggests that, depending on the type of acute or chronic stress or on experimental conditions, both mechanisms are operative in the stress-induced changes in the brain concentrations of neuroactive steroids and that such changes may have important effects on central GABAergic transmission, thereby modulating behavior and mood.

References

Akana, S. F., Scribner, K. A., Bradbury, M. J., Strack, A. M., Walker, C.-D., and Dallman, M. F. (1992). Feedback sensitivity of the rat hypothalamo-pituitary-adrenal axis and its capacity to adjust to endogenous corticosterone. *Endocrinology* **131**, 585–594.

Albeck, D. S., McKittrick, C. R., Blanchard, D. C., Blanchard, R. J., Nikulina, J., McEwen, B. S., and Sakai, R. R. (1997). Chronic social stress alters levels of corticotropin-releasing factor and arginine vasopressin mRNA in rat brain. *J. Neurosci.* **17**, 4895–4903.

Andrews, N., Zharkowsky, A., and File, S. E. (1992). Acute stress down regulates benzodiazepine receptors: Reversal by diazepam. *Eur. J. Pharmacol.* **210**, 247–251.

Barbaccia, M. L., Roscetti, G., Trabucchi, M., Ambrosio, C., and Massotti, M. (1992). Cyclic-AMP-dependent increase of steroidogenesis in brain cortical minces. *Eur. J. Pharmacol.* **219,** 485–486.

Barbaccia, M. L., Roscetti, G., Trabucchi, M., Cuccheddu, T., Concas, A., and Biggio, G. (1994). Neurosteroids in the brain of handling habituated and naive rats: Effect of CO_2 inhalation. *Eur. J. Pharmacol.* **261,** 317–320.

Barbaccia, M. L., Roscetti, G., Bolacchi, F., Concas, A., Mostallino, M. C., Purdy, R. H., and Biggio, G. (1996a). Stress-induced increase in brain neuroactive steroids: Antagonism by abecarnil. *Pharmacol. Biochem. Behav.* **54,** 205–210.

Barbaccia, M. L., Roscetti, G., Trabucchi, M., Mostallino, M. C., Concas, A., Purdy, R. H., and Biggio, G. (1996b). Time dependent changes in rat brain neuroactive steroid concentrations and $GABA_A$ receptor function after acute stress. *Neuroendocrinology* **63,** 166–172.

Barbaccia, M. L., Roscetti, G., Trabucchi, M., Purdy, R. H., Mostallino, M. C., Perra, C., Concas, A., and Biggio, G. (1996c). Isoniazid-induced inhibition of GABAergic transmission enhances neurosteroid content in rat brain. *Neuropharmacology* **35,** 1299–1305.

Barbaccia, M. L., Roscetti, G., Trabucchi, M., Purdy, R. H., Mostallino, M. C., Concas, A., and Biggio, G. (1997). The effects of inhibitors of GABAergic transmission and stress on brain and plasma allopregnanolone concentrations. *Br. J. Pharmacol.* **120,** 1582–1588.

Barbaccia, M. L., Affricano, D., Trabucchi, M., Purdy, R.., Colombo, G., Agabio, R., and Gessa, G. L. (1999). Allopregnanolone and THDOC brain concentrations in Sardinian-alcohol-preferring (sP) rats. *Amer. Soc. Neurosci.* (Abstract n. 816.3, vol 2, p. 2049).

Barbaccia, M. L., Lello, S., Sidiropoulou, T., Cocco, T., Sorge, R. P., Cocchiarale, A., Piermarini, V., Sabato, A. F., Trabucchi, M., and Romanini, C. (2000). Plasma 5α-androstane-3α, 17β-diol, an endogenous steroid that positively modulates $GABA_A$ receptor function, and anxiety: A study in menopausal women. *Psychoneuroendocrinology* **25,** 659–675.

Baulieu, E.-E., and Robel, P. (1990). Neurosteroids: A new brain function? *J. Steroid Biochem. Mol. Biol.* **37,** 395–403.

Besman, M. J., Yanagibashi, K., Lee, T. D., Kawamura, M., Hall, P. F., and Shively, J. E. (1989). Identification of des-(Gly-Ile)-endozepine as an effector of corticotropin-dependent adrenal steroidogenesis: Stimulation of cholesterol delivery is mediated by peripheral benzodiazepine receptor. *Proc. Natl. Acad. Sci. USA* **86,** 4897–4901.

Biggio, G. (1983). The action of stress, β-carbolines, diazepam and Ro 15-1788 on GABA receptors in the rat brain. *In* "Benzodiazepine Recognition Site Ligands: Biochemistry and Pharmacology" (G. Biggio and E. Costa, Eds.), pp.105–117, Raven Press, New York.

Biggio, G., Corda, M. G., Demontis, G., Concas, A., and Gessa, G. L. (1980). Sudden decrease in cerebellar GABA binding induced by stress. *Pharmacol. Res. Commun.* **12,** 489–493.

Biggio, G., Corda, M. G, Concas, A., Dermontis, G., Rossetti, Z., and Gessa, G. L. (1981). Rapid changes in GABA binding induced by stress in different areas of the rat brain. *Brain Res.* **229,** 363–369.

Biggio, G., Concas, A., Serra, M., Salis, M., Corda, M. G., Nurchi, V., Crisponi, C., and Gessa, G. L. (1984). Stress and β-carbolines decrease the density of low affinity GABA binding sites: An effect reversed by diazepam. *Brain Res.* **305,** 13–18.

Biggio, G., Concas, A., Mele, S., and Corda, M. G. (1987). Changes in GABAergic transmission by zolpidem, an imidazopyridine with preferential affinity for type I benzodiazepine receptors. *Eur. J. Pharmacol.* **161,** 173–180.

Biggio, G., Concas, A., Corda, M. G., Giorgi, O., Sanna, E., and Serra, M. (1990). GABAergic and dopaminergic transmission in the rat cerebral cortex: Effect of stress, and anxiolytic and anxiogenic drugs. *Pharmacol. Ther.* **48,** 121–142.

Biggio, G., Concas, A., Mostallino, M. C., Purdy, R. H., Trabucchi, M., and Barbaccia, M. L. (1996). Inhibition of GABAergic transmission enhances neurosteroid concentrations in the rat brain. In "The Brain: Source and Target for Sex Steroid Hormones" (A. R. Genazzani, F. Petraglia, and R. H. Purdy, Eds.), pp. 43–62, Parthenon, New York.

Biggio, G., Barbaccia, M. L., Follesa, P., Serra, M., Purdy, R. H., and Concas, A. (2000). Neurosteroids and GABA$_A$ receptor plasticity. In "GABA in the Nervous System. The View at Fifty Years" (R. W. Olsen and D. L. Martin, Eds.), pp. 207–232, Lippincott, New York.

Bixo, M., Andersson, A., Winblad, B., Purdy, R. H., and Backstrom, T. (1997). Progesterone, 5α-pregnane-3-20-dione and 3α-hydroxy-5α-pregnane-20-one in specific regions of the human brain in different endocrine states. Brain Res. **764**, 173–178.

Britton, K. T., Lee, G., Dana, R., Risch, S. C., and Koob, G. F. (1986). Activating and "anxiogenic" effects of corticotropin-releasing factor are not inhibited by blockade of the pituitary-adrenal system with dexamethasone. Life Sci. **39**, 1281–1286.

Calogero, A. E., Gallucci, W. T., Chousos, G. P., and Gold, P. W. (1988). Interaction between GABAergic neurotransmission and rat hypothalamic corticotropin releasing hormone secretion in vitro. Brain Res. **463**, 223–228.

Celotti, F., Melcangi, R. C., and Martini, L. (1992). The 5α-reductase in the brain: Molecular aspects and relation to brain function. Front. Neuroendocrinol. **13**, 163–215.

Cheney, D. L., Uzunov, D. P., Costa, E., and Guidotti, A. (1995). Gas chromatographic-mass fragmentographic quantitation of 3α- hydroxy-5α-pregnan-20-one (allopregnanolone) and its precursors in blood and brain of adrenalectomized and castrated rats. J. Neurosci. **15**, 4641–4650.

Concas, A., Salis, M., Serra, M., Corda, M. G., and Biggio, G. (1983). Ethyl β-carboline-3-carboxylate decreases ^{3}H-GABA binding in membrane preparations of rat cerebral cortex. Eur. J. Pharmacol. **89**, 179–181.

Concas, A., Serra, M., Salis, M., Nurchi, V., Crisponi, G., and Biggio, G. (1984). Evidence for an involvement of GABA receptors in the mediation of the proconvulsant action of ethyl-β-carboline-3-carboxylate. Neuropharmacology **23**, 323–326.

Concas, A., Corda, M. G., and Biggio, G. (1985). Involvment of benzodiazepine recognition sites in the foot shock-induced decrease of low affinity GABA receptors in the rat cerebral cortex. Brain Res. **341**, 50–56.

Concas, A., Mele, S., and Biggio, G. (1987). Foot shock stress decreases chloride efflux from rat brain neurosynaptosomes. Eur. J. Pharmacol. **135**, 423–427.

Concas, A., Serra, M., Atsoggiu, T., and Biggio, G. (1988a). Foot shock stress and anxiogenic β-carbolines increase t-[^{35}S]butylbicyclophosphorothionate binding in the rat cerebral cortex, an effect opposite to anxiolytics and γ-aminobutyric acid mimetics. J. Neurochem. **51**, 1868–1876.

Concas, A., Serra, M., Corda, M. G., and Biggio, G. (1988b). Changes of ^{36}Cl$^-$ flux and ^{35}S-TBPS binding induced by stress and GABAergic drugs. In "Chloride Channels and Their Modulation by Neurotransmitters and Drugs" (G. Biggio and E. Costa, Eds.), pp. 227–246, Raven Press, New York.

Concas, A., Mostallino, M. C., Perra, C., Lener, R., Roscetti, G., Barbaccia, M. L., Purdy, R. H., and Biggio, G. (1996). Functional correlation between allopregnanolone and [^{35}S]TBPS binding in the brain of rats exposed to isoniazid, pentylenetetrazol or stress. Br. J. Pharmacol. **118**, 839–846.

Concas, A., Mostallino, M. C., Porcu, P., Follesa, P., Barbaccia, M. L., Trabucchi, M., Purdy, R. H., Grisenti, P., and Biggio, G. (1998). Role of brain allopregnanolone in the plasticity of γ-aminobutyric acid type A receptor in rat brain during pregnancy and after delivery. Proc. Natl. Acad. Sci. USA **95**, 13284–13289.

Concas, A., Follesa, P., Barbaccia, M. L., Purdy, R. H., and Biggio, G. (1999). Physiological modulation of GABA$_A$ receptor plasticity by progesterone metabolites. *Eur. J. Pharmacol.* **375,** 225–235.

Corda, M. G., and Biggio, G. (1986). Proconflict effect of GABA receptor complex antagonists. Reversal by diazepam. *Neuropharmacology* **25,** 541–544.

Corda, M. G., Blaker, W. D., Mendelson, W. B., Guidotti, A., and Costa, E. (1983). β-Carbolines enhance shock-induced suppression of drinking in rats. *Proc. Natl. Acad. Sci. USA* **80,** 2072–2076

Corpéchot, C., Young, J., Calvel, M., Wehrey, C., Veltz, J. N., Touyer, G., Mouren, M., Prasad, V. V. K., Banner, C., Sjovall, J., Baulieu, E.-E., and Robel, P. (1993). Neurosteroids: 3α-Hydroxy-5α-pregnan-20-one and its precursors in the brain, plasma and steroidogenic glands of male and female rats. *Endocrinology* **133,** 1003–1009.

Costa, E., and Guidotti, A. (1991). Diazepam binding inhibitor (DBI): A pepetide with multiple biological functions. *Life Sci.* **49,** 325–341.

Crestani, F., Lorez, M., Baer, K., Essrich, C., Benke, D., Laurent, J. P., Belzung, C., Fritschy, J.-M., Luscher, B., and Mohler, H. (1999). Decreased GABA$_A$- receptor clustering results in enhanced anxiety and a bias for threat cues. *Nature Neurosci.* **2,** 833–839.

deGoeij, D. C., Kvetnansky, R., Whitnall, M. H., Jezova, D., Berkenbosch, F., and Tilders, F. J. (1991). Repeated stress-induced activation of corticotropin-releasing factor neurons enhances vasopressin stores and colocalization with corticotropin-releasing factor in the median eminence of rats. *Neuroendocrinology* **53,** 150–159.

Dong, E., Matsumoto, K., Tohda, M., Kaneko, Y., and Watanabe, H. (1999). Diazepam binding inhibitor (DBI) gene expression in the brains of socially isolated and group-housed rats. *Neurosci. Res.* **33,** 171–177.

Dorow, R., Horowsky, R., Paschelke, G., Amin, M., and Braestrup, C. (1983). Severe anxiety induced by FG-7142, a β-carboline ligand for benzodiazepine receptors. *Lancet* **ii,** 98–99.

Drugan, R. C., Morrow, A. L., Weizman, R., Weizman, A., Deutsch, S. I., Crawley, J. N., and Paul, S. M. (1989). Stress-induced behavioural depression in the rat is associated with a decrease in GABA receptor-mediated chloride ion flux and brain benzodiazepine receptor occupancy. *Brain Res.* **487,** 45–51.

Dunn, A. J., and File, S. E. (1987). Corticotropin-releasing factor has anxiogenic action in the social interaction test. *Horm. Behav.* **21,** 193–202.

Fenelon, V. S., and Herbison, A. E. (1996). Plasticity in GABA$_A$ receptor subunit mRNA expression by hypothalamic magnocellular neurons in the adult rat. *J. Neurosci.* **16,** 4872–4880.

Ferrarese, C., Mennini, T., Pecora, N., Pierpaoli, C., Frigo, M., Marzorati, C., Gobbi, M., Bizzi, A., Codegoni, A., Garattini, S., and Frattola, L. (1991). Diazepam binding inhibitor (DBI) increases after acute stress in rat. *Neuropharmacology* **30,** 1445–1452.

File, S. E., Zangrossi, H., Jr., and Andrews, N. (1993). Novel environment and cat odor change GABA and 5HT release and uptake in the rat. *Pharmacol. Biochem. Behav.* **45,** 931–934.

Follesa, P., Floris, S., Tuligi, G., Mostallino, M. C., Concas, A., and Biggio, G. (1998). Molecular and functional adaptation of the GABA$_A$ receptor complex during pregnancy and after delivery in the rat brain. *Eur. J. Neurosci.* **10,** 2905–2912.

Follesa, P., Serra, M., Cagetti, E., Pisu, M. G., Porta, S., Floris, S., Massa, F., Sanna, E., and Biggio, G. (2000). Allopregnanolone synthesis in cerebellar granule cells: Roles in regulation of GABA$_A$ receptor expression and function during progesterone treatment and withdrawal. *Mol. Pharmacol.* **57,** 1262–1270.

Frye, C. A., Duncan, J. E., Basham, M., and Erskine, M. S. (1996a). Behavioral effects of 3α-androstanediol. II: Hypothalamic and preoptic area actions via a GABAergic mechanism. *Behav. Brain Res.* **79,** 119–130.

Frye, C. A., Van Keuren, K. R., and Erskine, M. S. (1996b). Behavioral effects of 3α-androstanediol. I: Modulation of sexual receptivity and promotion of GABA-stimulated chloride flux. *Behav. Brain Res.* **79,** 109–118.

Gavish, M., Bachman, I., Shoukrun, R., Katz, Y., Veenman, L., Weisinger, G., and Izman, A. (1999). Enigma of the peripheral benzodiazepine receptor. *Pharmacol. Rev.* **51,** 629–650.

Gee, K. W., Bolger, M. B., Brinton, R. E., Coirini, H., and McEwen, B. S. (1988). Steroid modulation of the chloride ionophore in rat brain: Structure–activity requirements, regional dependence and mechanism of action. *J. Pharmacol. Exp. Ther.* **246,** 803–812.

Greco, A. M., Gambardella, P., Sticchi, R., D'Aponte, D., Di Rienzo, G., and De Franciscis, P. (1989). Effect of individual housing on circadian rhythms of adult rats. *Physiol. Behav.* **45,** 363–366.

Griffin, L. D., and Mellon, S. H. (1999). Selective serotonin reuptake inhibitors directly alter activity of neurosteroidogenic enzymes. *Proc. Natl. Acad. Sci. USA* **96,** 13512–13517.

Guarneri, P., Papadopoulos, V., Pan, B., and Costa, E. (1992). Regulation of pregnenolone synthesis in C62B glioma cells by 4′-chlorodiazepam. *Proc. Natl. Acad. Sci. USA* **89,** 5118–5122.

Gyenes, M., Wang, Q., Gibbs, T. T., and Farb, D. H. (1994). Phosphorylation factors control neurotransmitter and neuromodulator actions at the γ-aminobutyric acid type A receptor. *Mol. Pharmacol.* **46,** 542–549.

Haefely, W. E. (1994). Allosteric modulation of the $GABA_A$ receptor channel: A mechanism for interaction with a multitude of central nervous system functions. *In* "The Challenge of Neuropharmacology" (H. C. Mohler and M. Da Prada, Eds.), pp. 15–39, Editions Roche, Basel.

Haller, J., and Halasz, J. (1999). Mild social stress abolished the effects of isolation on anxiety and chlordiazepoxide reactivity. *Psychopharmacology (Berlin)* **144,** 311–315.

Harrison, N. L., and Simmonds, M. A. (1984). Modulation of GABA receptor complex by a steroid anesthetic. *Brain Res.* **323,** 284–293.

Hatch, A. M., Balasz, T., Wiberg, G. S., and Grice, H. C. (1963). Long term isolation in rats. *Science* **142,** 507–508.

Hauger, R. L., Lorang, M., Irwin, M., and Aguilera, G. (1988). CRF receptor regulation and sensitization of ACTH response to acute ether stress during chronic intermittent immobilization stress. *Brain Res.* **532,** 34–40.

Hilakivi, L. A., Ota, M., and Lister, R. G. (1989). Effect of isolation on brain monoamines and the behavior of mice in tests of exploration, locomotion, anxiety and behavioral "despair." *Pharmacol. Biochem. Behav.* **33,** 371–374.

Hoffman, A. R., Ceda, G., and Reisine, T. D. (1985). Corticotropin-releasing factor desensitization of adrenocorticotropic hormone release is augmented by arginine vasopressin. *J. Neurosci.* **5,** 234–242.

Holson, R. R., Scallet, A., Ali, S. F., and Turner, B. B. (1991). "Isolation stress" revisited: Isolation-rearing effects depend on animal care methods. *Physiol. Behav.* **49,** 1107–1118.

Hu, Z. Y., Bourreau, E., Jung-Testas, I., Robel, P., and Baulieu, E.-E. (1989). Neurosteroids: Biosynthesis of pregnenolone and progesterone in primary cultures of rat glial cells. *Endocrinology* **93,** 1157–1162.

Insel, T. R. (1989). Decreased in vivo binding to brain benzodiazepine receptors during social isolation. *Psychopharmacology (Berlin)* **97,** 142–144.

Kant, G. J., Bunnel, B. N., Mougey, E. H., Pennington, L. L., and Meyerhoff, J. L. (1983). Effects of repeated stress on pituitary cyclic AMP and plasma prolactin, corticosterone and growth hormone in male rats. *Pharmacol. Biochem. Behav.* **18,** 967–971.

Kash, S. F., Tecott, L. H., Hodge, C., and Baekkeshov, S. (1999). Increased anxiety and altered response to anxiolytics in mice deficient in the 65 kDa isoform of glutamic acid decarboxylase. *Proc. Natl. Acad. Sci. USA* **96,** 1698–1703.

Kim, J. W., and Kirkpatrick, B. (1996). Social isolation in animal models of relevance to neuropsychiatric disorders. *Biol. Psychiatry* **40,** 918–922.

Korneyev, A., Guidotti, A., and Costa, E. (1993). Regional and interspecies differences in brain progesterone metabolism. *J. Neurochem.* **61,** 2041–2047.

Lambert, J. J., Belelli, D., Hill Venning, C., and Peters, J. A. (1995). Neurosteroids and $GABA_A$ receptor function. *Trends Pharmacol. Sci.* **16,** 295–303.

Le Goascogne, C., Robel, P., Gouezou, M., Sananes, N., Baulieu, E.-E., and Waterman, M. (1987). Neurosteroids: Cytochrome P-450$_{scc}$ in rat brain. *Science* **237,** 1212–1214.

Li, X., Bertics, P. J., and Karavolas, H. J. (1997). Regional distribution of cytosolic and particulate 5α-dihydroprogesterone and 3α-hydroxysteroid oxidoreductase in female rat brain. *J. Steroid Biochem. Mol. Biol.* **60,** 311–331.

Majewska, M. D., Harrison, N. L., Schwartz, R. D., Barker, J. L., and Paul, S. M. (1986). Steroid hormone metabolites are barbiturate-like modulators of the GABA receptor. *Science* **232,** 1004–1007.

Makino, S., Smith, M. A., and Gold, P. W. (1995). Increased expression of corticotropin-releasing hormone and vasopressin messenger ribonucleic acid (mRNA) in hypothalamic paraventricular nucleus during repeated stress: Association with reduction in glucocorticoid receptor mRNA levels. *Endocrinology* **136,** 3299–3309.

Masotto, C., Wisniewski, G., and Negro-Vilar, A. (1989). Different gamma γ-aminobutyric acid receptor subtypes are involved in the regulation of opiate-dependent and independent luteinizing hormone-releasing hormone secretion. *Endocrinology* **125,** 548–553.

Massotti, M., Slobodiansky, E., Konkel, D., Costa, E., and Guidotti, A. (1991). Regulation of diazepam binding inhibitor in rat adrenal gland by adrenocorticotropin. *Endocrinology* **129,** 591–596.

Mathur, C., Prasad, V. V. K., Raju, V. S., Welch, M., and Lieberman, S. (1993). Steroids and their conjugates in the mammalian brain. *Proc. Natl. Acad. Sci. USA* **90,** 85–88.

Matsumoto, K., Uzunova, V., Pinna, G., Taki, K., Uzunov, D. P., Watanabe, H., Mienville, J.-M., Guidotti, A., and Costa, E. (1999). Permissive role of brain allopregnanolone content in the regulation of pentobarbital-induced righting reflex loss. *Neuropharmacology* **38,** 955–963.

McDonald, B. J., Amato, A., Connolly, C. N., Benke, D., Moss, S. J., and Smart, T. (1998). Adjacent phosphorylation sites on $GABA_A$ receptor beta subunits determine regulation by cAMP-dependent protein kinase. *Nature Neurosci.* **1,** 23–28.

Medina, J. H., Novas, M. L., Wolfman, C. N., Levi de Stein, M., and De Robertis, E. (1983). Benzodiazepine receptors in rat cerebral cortex and hippocampus undergo rapid and reversible changes after acute stress. *Neuroscience* **9,** 331–335.

Mellon, S. H., and Deschepper, C. F. (1993). Neurosteroid biosynthesis: Genes for adrenal steroidogenic enzymes are expressed in rat brain. *Brain Res.* **629,** 283–292.

Miachon, S., Manchon, M., Fromentin, J. R., and Buda, M. (1990). Isolation-induced changes in radioligand binding to benzodiazepine binding sites. *Neurosci. Lett.* **111,** 246–251.

Miachon, S., Rochet, T., Mathian, B., Barbagli, B., and Claustrat, B. (1993). Long-term isolation of Wistar rats alters brain monoamine turnover, blood corticosterone and ACTH. *Brain Res. Bull.* **32,** 611–614.

Monteleone, P., Luisi, S., Tonetti, A., Bernardi, F., Gennazzani, A. D., Luisi, M., Petraglia, F., and Gennazzani, A. (2000). Allopregnanolone concentrations and premenstrual syndrome. *Eur. J. Endocrinol.* **142,** 269–273.

Moss, S. J., Smart, T. J., Blackstone, C. D., and Huganir, R. L. (1992). Functional modulation of GABA$_A$ receptors by cAMP-dependent protein phosphorylation. *Science* **257,** 661–665.

Mukhin, A. G., Papadopoulos, V., Costa, E., and Krueger, K. E. (1989). Mitochondrial benzodiazepine receptors regulate steroid biosynthesis. *Proc. Natl. Acad. Sci. USA* **86,** 9813–9816.

Ninan, P. T., Insel, T. M., Cohen, R. M., Cook, J. M., Skolnick, P., and Paul, S. M. (1982). Benzodiazepine receptor-mediated experimental "anxiety" in primates. *Science* **218,** 1332–1334.

Novas, M. L., Medina, J. H., Calvo, D., and De Robertis, E. (1987). Increase of peripheral benzodiazepine binding sites in kidney and olfactory bulb in acutely stressed rats. *Eur. J. Pharmacol.* **135,** 243–246.

Nusser, Z., Sieghart, W., and Mody, I. (1999). Differential regulation of synaptic GABA$_A$ receptors by cAMP-dependent protein kinase in mouse cerebellar and olfactory bulb neurons. *J. Physiol. (London)* **521,** 421–435.

Ojima, K., Matsumoto, K., and Watanabe, H. (1995). Hyperactivity of central noradrenergic CRF system is involved in social isolation-induced decrease in pentobarbital sleep. *Brain Res.* **684,** 87–94.

Ojima, K., Matsumoto, K., and Watanabe, H. (1997). Flumazenil reverses the decrease in the hypnotic activity of pentobarbital by social isolation stress: Are endogenous benzodiazepine receptor ligands involved? *Brain Res.* **745,** 127–133.

Papadopoulos, V. (1993). Peripheral type benzodiazepine/diazepam binding inhibitor receptor: Biological role in steroidogenic cell function. *Endocrine Rev.* **14,** 222–240.

Parker, V., and Morinan, A. (1986). The socially isolated rat as a model for anxiety. *Neuropharmacology* **25,** 663–664.

Paul, S. M., and Purdy, R. H. (1992). Neuroactive steroids. *FASEB J.* **6,** 2311–2322.

Paulus, M. P., Varty, G. B., and Geyer, M. A. (2000). The genetic liability to stress and postweaning isolation have a competitive influence on behavioral organization in rats. *Physiol. Behav.* **68,** 389–394.

Pinna, G., Uzunova, V., Matsumoto, K., Puia, G., Mienville, J.-M., Costa, E., and Guidotti, A. (2000). Brain allopregnanolone regulates the potency of the GABA$_A$ receptor agonist muscimol. *Neuropharmacology* **39,** 440–448.

Pohorecky, L. A., Skiandos, A., Zhang, X., Rice, K., and Benjamin, D. (1999). Effect of social chronic stress on δ-opioid receptor function in the rat. *J. Pharmacol. Exp. Ther.* **290,** 196–206.

Purdy, R. H., Morrow, A. L., Moore, P. H. Jr., and Paul, S. M. (1991). Stress-induced elevations of γ-aminobutyric acid type A receptor- active steroids in the rat brain. *Proc. Natl. Acad. Sci. USA* **88,** 4553–4557.

Rivier, C., and Vale, W. (1987). Diminished responsiveness of the hypothalamic-pituitaryadrenal axis of the rat during exposure to prolonged stress: A pituitary-mediated mechanism. *Endocrinology* **121,** 1320–1328.

Robel, P., Jung-Testas, I., Hu, Z. Y., Akwa, Y., Sananes, N., Kabbadji, K., Eychenne, B., Sancho, M. J., Kang, K. I., Zucman, D., Morfin, R., and Baulieu, E.-E. (1991). Neurosteroids: Biosynthesis and metabolism in cultured rodent glia and neurons. *In* "Neurosteroids and Brain Function" (E. Costa and S. M. Paul, Eds.), pp. 147–154, Fidia Research Foundation Symposium Series, Vol. 8, Thieme, New York.

Romeo, E., Auta, J., Kozikowski, A. P., Ma, D., Papadopoulos, V., Puia, G., Costa, E., and Guidotti, A. (1992). 2-Aryl-3-indoleacetamides (FGIN-1): A new class of potent and specific ligands for the mitochondrial DBI receptor (MDR). *J. Pharmacol. Exp. Ther.* **262,** 971–978.

Romeo, E., Strohle, A., Spalletta, G., Di Michele, F., Hermann, B., Holsboer, F., Pasini, A., and Rupprecht, R. (1998). Effects of antidepressant treatment on neuroactive steroids in major depression. *Am. J. Psychiatry* **155,** 910–913.

Roscetti, G., Ambrosio, C., Trabucchi, M., Massotti, M., and Barbaccia, M. L. (1994). Modulatory mechanisms of cyclic AMP-stimulated steroid content in rat brain cortex. *Eur. J. Pharmacol.* **269,** 17–24.

Roscetti, G., Del Carmine, R., Trabucchi, M., Massotti, M., Purdy, R. H., and Barbaccia, M. L. (1998). Modulation of neurosteroid synthesis/accumulation by L-ascorbic acid in rat brain tissue: Inhibition by selected serotonin antagonists. *J. Neurochem.* **71,** 1108–1117.

Sanchez, M. M., Aguado, F., Sanchez-Toscano, F., and Saphier, D. (1998). Neuroendocrine and immunocytochemical demonstrations of decreased hypothalamo-pituitary-adrenal axis responsiveness to restraint stress after long-term social isolation. *Endocrinology* **139,** 579–587.

Serra, M., Sanna, E., and Biggio, G. (1989). Isoniazid, an inhibitor of GABAergic transmission, enhances [^{35}S]TBPS binding in the rat cerebral cortex. *Eur. J. Pharmacol.* **164,** 385–388.

Serra, M., Sanna, E., Concas, A., Foddi, C., and Biggio, G. (1991). Foot shock stress enhanced the increase of [^{35}S]TBPS binding in the rat cerebral cortex and the convulsions induced by isoniazid. *Neurochem. Res.* **16,** 17–22.

Serra, M., Madau, P., Chessa, M. F., Caddeo, M., Sanna, E., Trapani, G., Franco, M., Liso, G., Purdy, R. H., Barbaccia, M. L., and Biggio, G. (1999). 2-Phenyl-imidazo[1,2-a]pyridine derivatives as ligands for peripheral benzodiazepine receptors: Stimulation of neurosteroid synthesis and anticonflict action in rats. *Br. J. Pharmacol.* **127,** 177–187.

Serra, M., Pisu, M. G., Littera, M., Papi, G., Sanna, E., Tuveri, F., Usala, L., Purdy, R. H., and Biggio, G. (2000). Social isolation-induced decrease in both the abundance of neuroactive steroids and GABA$_A$ receptor function in rat brain. *J. Neurochem.* **75,** 732–740.

Smith, S. S., Gong, Q. H., Hsu, F. C., Markowitz, R. S., French-Mullen, J. M. H., and Li, X. (1998). GABA$_A$ receptor α_4 subunit suppression prevents withdrawal properties of an endogenous steroid. *Nature* **392,** 926–930.

Smith, S. S., Gong, Q. H., Li, X., Moran, M. H., Bitran, D., Frye, C. A., and Hsu, F. C. (1998). Withdrawal from 3α-OH-5α-pregnan-20-one using a pseudopregnancy model alters the kinetics of hippocampal GABA$_A$-gated current and increases the GABA$_A$ receptor α_4 subunit in association with increased anxiety. *J. Neurosci.* **18,** 5275–5284.

Stephens, D. N., Schneider, H. H., Kehz, W., Andrews, J. S., Rettig, K. J., Turski, L., Schmiechen, R., Turner, J. D., Jensen, L. H., Petersen, E. N., Honoré, T., and Bondo-Hansen, J. (1990). Abecarnil, a metabolically stable, anxioselective β-carboline acting at benzodiazepine receptors. *J. Pharmacol. Exp. Ther.* **253,** 334–343.

Stone, E. A., and John, S. M. (1992). Stress-induced increase of extracellular levels of cyclic AMP in rat cortex. *Brain Res.* **597,** 144–147.

Stone, E. A., Platt, J. E., Trullas, R., and Slucky, A. V. (1984). Reduction of the cAMP response to norepinephrine in rat cerebral cortex following repeated restraint stress. *Psychopharmacology (Berlin)* **82,** 403–405.

Strohle, A., Pasini, A., Romeo, E., Hermann, B., Spalletta, G., Di Michele, F., Holsboer, F., and Rupprecht, R. (2000). Fluoxetine decreases concentrations of 3alpha, 5alpha-tetrahydrodeoxycorticosterone (THDOC) in major depression. *J. Psychiatr. Res.* **34,** 183–186.

Trapani, G., Franco, M., Latrofa, A., Ricciardi, L., Carotti, A., Serra, M., Sanna, E., Biggio, G., and Liso, G. (1999). Novel 2-phenylimidazo [1,2-a] pyridine derivatives as potent and selective ligands for peripheral benzodiazepine receptors: Synthesis, binding affinity, and in vivo studies. *J. Med. Chem.* **42,** 3934–3941.

Uzunova, V., Sheline, Y., Davis, J. M., Rasmusson, A., Uzunov, D. P., Costa, E., and Guidotti, A. (1998). Increase in the cerebrospinal fluid content of neurosteroids in patients with unipolar major depression who are receiving fluoxetine or fluvoxamine. *Proc. Natl. Acad. Sci. USA* **95,** 3239–3244.

Wan, Q., Man, H. Y., Braunton, J., Wang, W., Salter, M. W., Becker, L., and Wang, Y. T. (1997). Modulation of $GABA_A$ receptor function by tyrosine phosphorylation of β subunits. *J. Neurosci.* **17,** 5062–5069.

Wang, M., Seippel, L., Purdy, R. H., and Backstrom, T. (1996). Relationship between symptom severity and steroid variation in women with premenstrual syndrome: Study on serum pregnenolone sulfate, 5α-pregnan-3,20-dione and 3α-hydroxy-5α-pregnan-20-one. *J. Clin. Endocrinol. Metab.* **81,** 1076–1082.

Wongwitdecha, N., and Marsden, C. A. (1996). Social isolation stress increases aggressive behavior and alters the effects of diazepam in the rat social interaction test. *Behav. Brain Res.* **75,** 27–32.

NEUROSTEROIDS IN LEARNING AND MEMORY PROCESSES

Monique Vallée,* Willy Mayo,* George F. Koob,[†] and Michel Le Moal*

*INSERM U.259, Institut François Magendie, Domaine de Carreire, 33077 Bordeaux, France, and [†]Department of Neuropharmacology, The Scripps Research Institute, La Jolla, California 92037

I. Introduction
  A. Neurosteroids of Principal Interest
  B. Role of Neurosteroids in Learning and Memory Processes
II. Learning and Memory Processes and Animal Models
  A. Different Types of Learning and Memory Processes
  B. Animal Models and Behavioral Tasks
III. Pharmacological Effects of Neurosteroids
  A. Introduction
  B. Excitatory Neurosteroids: PREG, DHEA, PREGS, and DHEAS
  C. Inhibitory Neurosteroids: PROG, 5α-DH PROG, 3α,5α-TH PROG
  D. Conclusions
IV. Mechanisms of Action
  A. Behavioral Approach
  B. Cellular Approach (Brain Plasticity)
  C. Conclusions
V. Physiological Significance
  A. Animal Studies: Correlation Studies between PREGS Levels and Learning and Memory Performances
  B. Human Studies
VI. Conclusions and Future Perspectives
  A. Which Neurosteroids for Which Behavioral Effects?
  B. Stable Analogs of Neurosteroids
  C. New Method of Quantification of Neurosteroid Levels
  D. Application for Clinical Studies
  References

The discovery that neurosteroids could be synthesized *de novo* in the brain independent from the periphery and display neuronal actions led to great enthusiasm for the study of their physiological role. Pharmacological studies suggest that neurosteroids may be involved in several physiological processes, such as learning and memory. This chapter summarizes the effects of the administration of neurosteroids on learning and memory capabilities in rodents and in models of amnesia. We address the central mechanisms involved in mediating the modulation of learning and memory processes by neurosteroids. In this regard, the neurosteroid-modulated neurotransmitter systems, such as γ-aminobutyric acid type A, N-methyl-D-aspartate, and cholinergic and σ opioid systems, appear to be potential targets for the rapid memory alteration actions of neurosteroids. Moreover, given that some neurosteroids affect neuronal plasticity, this neuronal change could be involved in the long-term modulation of learning and memory processes.

To understand the role of endogenous neurosteroids in learning and memory processes, we present some physiological studies in rodents and humans. However, the latter do not successfully prove a role of endogenous neurosteroids in age-related memory impairments. Finally, we discuss the relative implication of a given neurosteroid vs its metabolites. For this question, a new approach using the quantitative determination of traces of neurosteroids by mass spectrometry seems to have potential for examining the role of each neurosteroid in discrete brain areas in learning and memory alterations, as observed during aging. © 2001 Academic Press.

I. Introduction

A. Neurosteroids of Principal Interest

Neurosteroids have been described as steroids present in the brain at higher concentrations than at the periphery. Their capacity to remain present within the brain 15 days after removing the peripheral glands of steroidogenesis suggests a synthesis of these steroids in the brain rather than an accumulation from the periphery (Baulieu, 1981; Corpéchot et al., 1993). The biosynthesis pathways of neurosteroids in the brain also have been demonstrated by several studies (for review, see Mellon and Compagnone, 1999). Neurosteroids, thus defined, include pregnenolone (PREG); dehydroepiandrosterone (DHEA); their sulfate derivatives, PREGS and DHEAS; and progesterone (PROG), which is metabolized to 5α-dihydroprogesterone (5α-DH PROG) and $3\alpha,5\alpha$-tetrahydroprogesterone ($3\alpha,5\alpha$-TH PROG), also named allopregnanolone. Distinct neurotransmitter-mediated effects have been reported, mainly through γ-aminobutyric acid type A ($GABA_A$) receptors, N-methyl-D-aspartate (NMDA)-type glutamatergic receptors, and σ opioid receptors (Wu et al., 1991; Majewska, 1992; Maurice et al, 1999; Park-Chung et al., 1999). Based on their interactions with these brain neurotransmitters, PREG, DHEA, and their sulfate derivatives are hypothesized to display excitatory cellular actions, while PROG and its metabolites, 5α-DH PROG and $3\alpha,5\alpha$-TH PROG, have inhibitory cellular properties. As a consequence, neurosteroids have been hypothesized to exhibit a broad spectrum of biological actions, from interactions with development to such complex processes as learning and memory.

In the 1990s, neurosteroids became of great interest for scientists because an extensive number of studies had examined their effects on neuronal functions, with the aim of understanding their physiological role. This chapter presents data suggesting a potential role of neurosteroids in learning and memory processes.

B. Role of Neurosteroids in Learning and Memory Processes

To analyze the role of neurosteroids in learning and memory processes, this chapter includes five sections. The first section briefly describes the different types of learning and memory processes as well as the animal models and behavioral tasks that have been used in the neurosteroid research. The second section summarizes the pharmacological effects of neurosteroids that have been reported in these models, and the third section explores possible mechanisms of actions for these effects. In the fourth section, we discuss the physiological significance of the effects of neurosteroids on learning and memory abilities in animals and describe the controversies encountered in human studies. Finally, in conclusion, we present some future perspectives that we believe are necessary to fully understand the functional role of neurosteroids in cerebral processes, such as learning and memory.

II. Learning and Memory Processes and Animal Models

A. Different Types of Learning and Memory Processes

1. *Definitions*

Living organisms acquire new information from their environment by learning and subsequent retention of the information or skills and abilities. Learning is defined as the acquisition of information and refers to processes whereby we acquire new knowledge about events in the environment. Memory refers to the processes through which we retain that knowledge and involves consolidation and retrieval of the new information (for review, see Squire, 1987; Morris *et al.*, 1988). Consolidation is defined as the process by which short-term memories are converted into long-term memories. Memory retrieval is intimately related to such parameters as perception, attention, motivation, stimulus selection, environmental cues, arousal, and vigilance. It is generally accepted that the storage of information causes physiochemical changes in synapses or synaptic plasticity, which participate in information processing.

2. The Nature of Learning

Learning is subdivided into several types, such as simple learning (perceptual learning), associative learning (conditioned learning or stimulus–response learning), and more complex learning, such as spatial learning (for review, see Carlson, 1999; Crawley, 1999). Perceptual learning is the ability to learn to recognize a particular stimulus that has been perceived previously. Stimulus–response learning involves the ability to perform an automatic response when a particular stimulus is present. It includes two categories of learning, classical conditioning and operant conditioning. Classical conditioning is a learning procedure that occurs when a stimulus that initially produces no particular response is followed several times by an unconditioned stimulus that produces an aversive or appetitive response (unconditioned response); the first stimulus is then called the conditioned stimulus, which itself evokes a conditioned response. Operant learning refers to a learning procedure whereby the effects of a particular situation increase (reinforce) or decrease (punish) the probability of the behavior. Spatial learning is more complex, where learning the relationship among stimuli is required.

3. Types of Memory Processes

Separate memory systems or processes have been proposed (Fig. 1).

a. Declarative and Nondeclarative (Procedural) Memory. One memory process affords humans the ability to store information explicitly so that it is available later as a conscious recollection. This ability is lost in case of amnesia. Squire and Zolan-Morgan (1988) used the term "declarative" to describe this kind of memory ability, which can be declared (i.e., it can be brought to mind as a proposition or an image). *Declarative memories* have been defined as those that are "explicitly available to conscious recollections as facts, events, or specific stimuli" (Squire *et al.*, 1989).

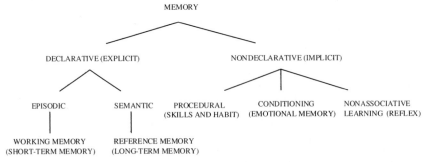

FIG. 1. Different types of memory processes. (Adapted from Squire and Zolan-Morgan, 1988, 1996.)

Nondeclarative memory comprises a heterogeneous collection of learning and memory abilities, all of which afford the capacity to acquire information implicitly. Squire and Zolan-Morgan (1988) used the term "procedural" to describe this skill-learning ability. Procedural memory seems to operate automatically, without conscious awareness.

b. Episodic and Semantic Memory. Declarative memory can be subdivided into episodic and semantic memory (Tulving, 1972). Episodic memory refers to memory of the events of an individual's past (i.e., accumulating as an individual's autobiography). Semantic memory refers to one's knowledge about the world. The content of semantic memory is explicitly known and can be recalled. Episodic and semantic memories include working and reference memories, respectively (Tulving, 1987; Baddeley, 1992).

B. ANIMAL MODELS AND BEHAVIORAL TASKS

Based on the observation in humans that hippocampal damage causes anterograde amnesia (i.e., loss of the ability to establish new declarative memories), researchers began making lesions in the hippocampal formation of animals and testing their learning ability. However, given that most of the learning tasks that the animals were given tested simple associative learning, animals remained capable of learning these tasks. Other tasks have been developed that required more complex learning. On such tasks, laboratory animals with hippocampal lesions showed learning and memory deficits, such as were observed in humans. We describe here several tasks requiring simple associative learning, conditioning learning, and complex learning that have been most commonly used for studying the role of neurosteroids in learning and memory processes.

1. *Simple Learning Tasks*

Simple learning can involve learning to recognize new stimuli or changes and variations in familiar stimuli. It often requires short-term memory (e.g., memory for a stimulus that has just been received). Most studies of short-term memory employ a delayed-matching-to-sample task that requires the subject to indicate which of several stimuli has just been received. Adapted tasks for rodents are based on the natural tendency of rodents to prefer exploration of new environments vs familiar ones. In the commonly used Y-maze or T-maze tasks, learning can be based on the ability of the animal to perform spontaneous choice exploration, resulting in alternating exploration of the three arms and thus assessing working memory. In these tasks, one arm can be reinforced with food. Learning abilities have also been evaluated in mice and rats by measuring the transfer latency from the open arms to the closed arms of an elevated plus-maze task (Itoh *et al.*, 1990; Sharma and Kulkarni, 1992).

2. *Conditioning Learning (Aversive/Reinforcing Learning) Tasks*

In rodents, aversive learning paradigms are frequently used because of ease of control of the unconditioned stimulus. The animals avoid (i.e., do not enter or else leave quickly) a location where they previously received an aversive stimulus (e.g., foot shock). An avoidance task can require either an active (i.e., moving) or passive (i.e., resting) position. Passive and active avoidance tasks measure memory of an aversive experience through simple avoidance of a location in which the aversive experience occurred. When the aversive stimulus is made predictable (i.e., conditioned stimulus), the animal can actively respond (i.e., active avoidance). The conditioning test consists of two phases, the training and the retention sessions, usually separated by a 24 h delay, but longer delays (a week) have also been used. The learning performance is measured by the ability of the animal to avoid (in a passive or active way) the compartment where a shock was received during the training phase. One-way step-through or step-down avoidance procedures have been used with rats and mice (for review, see Crawley, 1999). In step-through procedures, the apparatus consists of two connected compartments, one lit and one dark. Rodents tend to prefer dark environments and will immediately enter the darkened compartment. In passive avoidance paradigms, during the training session, a foot shock is delivered as soon as the animal enters the dark compartment. The animal remains in the dark compartment for 10–20 sec after the shock to allow formation of the association between the dark and the foot shock. Learning has occurred when the latency is significantly higher during the retention session than in the training session. For active avoidance paradigms, using the same compartment and the same parameters, the animal must move into the opposite compartment to avoid receiving a foot shock. Latency to enter the nonshocked compartment is the measure of learning. In step-down procedures, latency to jump onto a platform to avoid the shock is the measure of learning.

Cue and contextual conditioning is a fear-conditioning task that also measures memory of an aversive experience and the stimuli present during the aversive experience (Fanselow, 1990). Depending on whether cue and contextual conditioning are tested separately or in the same test, two or three sessions are conducted, respectively. The following procedure is used for measuring both cue and contextual conditioning at the same time. During the training session, the animal is placed in a chamber illuminated by normal house lighting. An auditory stimulus (80 dB white noise) is presented for 30 sec, and a foot shock (0.5 mA, 2 sec) is then delivered. The animal is left in the cage for 30 sec and then returned to its home cage. On the 24-h-delay retention session, the animal is returned to the same testing chamber. The time spent in freezing behavior is measured over a 5-min period. One hour later, the animal is placed in the same testing chamber, but the context has been changed. Freezing is quantitated for 3 min. Then the auditory cue is presented, and freezing is measured over the next 3 min in the presence of the sound. In this procedure, memory is assessed under three conditions: context, altered context, and auditory cue (Crawley, 1999).

In reinforcement paradigms, animals can learn an operant lever-press to gain access to an appetitive stimulus (water or food). Food- or water-restricted animals are habituated to the reinforced chamber and trained to press the lever for the food or water reinforcer. The number of trials required for the lever-press task is the measure of learning. In a more complex version of the task, called the discriminative go–no go task, the animal must press the lever only when the cue light is illuminated to obtain the reinforcement or press the lever only during the period in which a tone is sounding.

3. Complex Learning (Spatial Learning) Tasks

Complex learning task are behavioral tasks that require an animal to use multiple sources of information simultaneously to learn (Crawley *et al.*, 1997).

One of the most-used tasks that examines spatial learning is the Morris water maze (Morris, 1984). The task requires the animals, usually rats or mice, to find a particular location in space (a platform) using visual distal room cues. The maze consists of a circular pool (1.3–1.8 m in diameter), filled with a mixture of water and powdered milk. The milk mixture hides the location of the platform, which is situated just beneath the surface of the liquid. The researcher puts the rat into the water and lets it swim until it encounters the hidden platform and climbs onto it. The rat is released from a new position during each trial. The Morris water maze requires spatial learning because, to navigate around the maze, the animals get their bearings from the relative locations of stimuli located outside of the maze. The test involves reference memory because the position of the platform remains constant throughout all of the trials of acquisition. The latency and distance traveled to find the platform on each trial provide the measure of acquisition.

Often, spatial selectivity is measured on a probe trial, in which the platform is removed. In a probe trial, an animal that has learned the position of the platform will spend more time in the area of the pool that previously contained the platform vs other areas of the pool. The maze can be also used for simple (stimulus–response) learning when the platform is elevated just above the surface so that the animal can see it. This version of the task also enables examination of the visual acuity of the animals, which is a critical parameter in spatial learning tasks. Even rats with hippocampal lesions quickly learn to swim directly toward the visible platform. Similarly, if the animals are always released from the same place in the hidden platform procedure, then they learn to move in a particular direction (Eichenbaum *et al.*, 1990).

Another test of a rodent's spatial/reference abilities is the Barnes holeboard maze (Barnes, 1979). Acquisition of the location of a reinforcer over repeated trials provides the measure of learning and memory. The Barnes maze is circular (1.3 m in diameter), with numerous holes equally spaced around the perimeter of the circle. One of the holes leads to a dark enclosed box, located just below the circular platform. The hole can be reinforced by the presence of food. The time to reach the "goal" hole, distance traveled, and number of errors on each training day (i.e., visits of nonreinforced holes) provide the measure of acquisition. As in the Morris water maze, the learning requires that the animal "build" a spatial map by using external cues.

A two-trials spatial learning task in a Y-maze has been validated in rats (Dellu *et al.*, 1992, 1997a; Conrad *et al.*, 1996, 1997) and mice (Dellu *et al.*, 2000). The spatial learning requires recognizing new stimuli vs familiar stimuli that have been previously received with a long delay (30 min–6 h). This task is based on a free-choice exploration paradigm, which involves working memory. During the first trial (training/acquisition), the animal is allowed to visit two arms of the Y-maze, and the third is blocked by a door. During the second trial (retention/retrieval), the door is opened, and the animal has access to all three arms. Because the three arms are identical, discrimination of novelty vs familiarity requires distinction among their respective spatial representations that the animal perceives by building a spatial map of the environment using visual room cues. The number of visits to the novel arm and time spent in the novel arm vs the two familiar arms measure learning performance.

In the previously described spatial learning tasks, the animal's behavior can be driven by an aversive stimulus, where the goal is to find a refuge (e.g., platform in the water-maze task, escape tunnel in the hole maze), by an appetitive stimulus (e.g., search for food in the Barnes maze), or by spontaneous choice exploration (e.g., novel arm vs familiar arms in the Y-maze task).

III. Pharmacological Effects of Neurosteroids

A. INTRODUCTION

The effects of the administration of various neurosteroids on learning and memory processes have been studied in animals in several learning tasks described in the previous section. Here, we summarize the effects of neurosteroids *per se* (i.e., without other pharmacological treatments), according to their excitatory or inhibitory neuronal actions. Although the majority of pharmacological studies have been conducted in adult animals, some groups have also studied the effects of neurosteroids on learning and memory processes in rat pups and in aged rats and mice. These studies are of particular interest, given the previously demonstrated change in steroid (glucocorticoids) levels early in the lives of animals that interfere with the learning and memory abilities throughout life (Vallée *et al.*, 1999a). Thus, steroids can alter the decline in learning and memory abilities that has been commonly described in aged rodents (Dellu *et al.*, 1997b; Vallée *et al.*, 1999a) and humans (for review, see Grady and Craik, 2000).

B. Excitatory Neurosteroids: PREG, DHEA, PREGS, and DHEAS

1. *Development*

During development, effects of steroids on the central nervous system are classified as organizational effects. Subsequent exposure to steroids activates, modulates, or inhibits the function of the existing neural circuits (Whiting *et al.*, 1998).

Fleshner *et al.* (1997) studied the effects of DHEAS (0.1 mg/ml) given in drinking water for 5 days in 30-day-old rat pups in contextual-fear conditioning and auditory-cue-fear conditioning tasks. The authors used two procedures. In the first procedure, the retention occurred immediately after training, which measured short-term memory. The second procedure used a 24-h-delay retention and assessed long-term memory. In the contextual-fear conditioning task, chronic oral administration of DHEAS impaired (by decreasing the percentage of freezing) the long-term memory performance, whereas short-term memory was not altered. However, DHEAS failed to impair short- and long-term memory in the auditory-cue-fear conditioning. These results suggest that DHEAS impairs learning performance in contextual-fear conditioning; however, no results were shown in control animals (without shock).

2. *Adulthood*

The effects of neurosteroids on learning and memory processes in adult animals have usually been examined in two-session (training/retention) paradigms. Several groups have attempted to evaluate the effects of neurosteroids on the different stages, acquisition, consolidation, and retrieval of learning and memory processes by administering the neurosteroids before training, immediately following training, or before retention.

a. Pretraining Administration of Neurosteroids. Melchior and Ritzmann (1996) tested the effects of acute pretraining administration of neurosteroids in a food-appetitive T-maze task in mice. The authors used a win-shift foraging paradigm to assess working memory (Melchior *et al.*, 1993; Ritzmann *et al.*, 1993). Neurosteroids were injected intraperitoneally 30 min before training, and retention was performed 180 sec later. The administration (0.001–1 mg/kg) of PREG, DHEA, or their sulfates enhanced memory by producing bell-shaped dose-response curves. The maximal effect was observed with the 0.001 mg/kg dose for PREGS, 0.01 mg/kg dose for DHEA and PREG, and the 0.1 mg/kg dose for DHEAS.

Reddy and Kulkarni (1999) also evaluated the effects of pretraining injection of DHEAS using a spatial long-term memory task in female and male rats. The decreased transfer latency between the training and 24-h-delay retention sessions was used as an index of learning and memory in the elevated plus-maze task. DHEAS (5 mg/kg), injected subcutaneously 30 min before training improved memory in male rats but not in female rats. Because the elevated plus-maze is usually used as a model of anxiety-type task and because DHEAS displays anxiolytic properties in this task (Melchior and Ritzmann, 1994), the previous results could be related to changes in anxiety. However, it has been shown that the transfer latency was not affected by anxiolytics and anxiogenics administration in mice (Miyazaki *et al.*, 1995), suggesting that the results found by Reddy and Kulkarni (1999) were independent of the stress response.

The effects of chronic pretraining treatment with PREGS and DHEAS have also been evaluated. For example, Ladurelle *et al.* (2000) studied the effects of prolonged intracerebroventricular injections of PREGS before the training session of a two-trials Y-maze discrimination arm test in mice. PREGS was given for 3 days or 6 days through an osmotic minipump delivering a constant flow rate of 0.5 μl/h. Administration of PREGS for 3 days improved the 2-h-delay retention performance with 10, 50, and 100 ng/h doses. For the 6-days treatment, a memory-enhancing effect was observed only for the 10 ng/h dose, suggesting a dose-dependent effect of PREGS on memory performance.

b. Post-training Administrations of Neurosteroids. The majority of studies tested the effects of neurosteroids using post-training administration. Roberts *et al.* (1987) was one of the first group to examine these effects. They showed that injection of DHEA and DHEAS immediately after training could enhance retention performance a week later in a T-maze active avoidance task in mice. Intracerebroventricular injections of the neurosteroids increased the retention score during the retention session. The memory-enhancing effect was maximal with a $5.4\ 10^{-10}$ mol/mouse ($\sim 4.4\ 10^{-3}$ mg/kg) dose for DHEA and $6.2\ 10^{-10}$ mol/mouse ($\sim 6.5\ 10^{-3}$ mg/kg) dose for DHEAS. Given that DHEA was dissolved in dimethyl sulfoxide (DMSO), which by itself induces memory impairment, the authors concluded that DHEAS and DHEA could enhance memory and alleviate amnesia, respectively, by their actions in the central nervous system, either directly or via their metabolites. Flood *et al.* (1988) have also reported memory-enhancing effects of DHEA and DHEAS in mice in 1-week memory retention in foot shock active avoidance learning. Post-training intracerebroventricular injection of DHEA (1 ng/2 μl–120 ng/2 μl) reversed the amnesia induced by the vehicle DMSO, while lower (0.5 ng/2μl) or higher doses (160 ng/2μl) had no effect. DHEAS differently enhanced retention whether injected intracerebroventricularly or subcutaneously immediately post-training. DHEAS enhanced retention but had no effect on acquisition performance. The maximum effective doses were: 162 ng/mouse intracerebroventricularly; 700 μg/mouse subcutaneously; and 1.45 mg/mouse/day orally. The memory-retention effects of post-training injection of DHEAS were time dependent, as a significant enhancement of retention was observed when it was given either immediately (2 min) or at 30 and 60 min after training, but not at 90 and 120 min. DHEAS given intracerebroventricularly (54–216 ng/mouse) also improved retention for step-down passive avoidance.

In other experiments, Flood *et al.* (1992, 1995) compared the potency of PREG, PREGS, DHEA, and DHEAS in improving retention performance in a 1-week retention foot shock active avoidance task in mice. The neurosteroids were first given intracerebroventricularly immediately post-training (Flood *et al.*, 1992). Dose–response curves were obtained for PREG, PREGS, and DHEA; PREG and PREGS were the most potent, with PREGS showing significant effects at $3.5\ 10^{-15}$ mol/mouse ($\sim 4.2\ 10^{-8}$ mg/kg). However, in the protocol used in this study, it is difficult to conclude that there was a memory-enhancing effect of the neurosteroids *per se*, as the significant memory effects observed were compared with the vehicle/DMSO-induced memory deficit. Moreover, retention performance of neurosteroid-treated animals was not improved compared with the retention performance of saline-treated animals. On the basis that PREGS was the most potent memory-enhancing neurosteroid, Flood *et al.* (1995) then tested the effects of injections of PREGS into different brain areas. The authors found a memory improvement of post-training administration of PREGS in 1-week retention foot-shock-buzzer, sound-paired active avoidance tasks in mice. The retention was enhanced when PREGS was injected into the hippocampus, amygdala, septum, and mammillary bodies but not into the caudate nucleus. Intra-amygdala injection was the most potent in a molar basis for producing memory enhancement, and the authors calculated that in 150 molecules of PREGS were able to enhance post-training memory processes. The authors concluded that the amygdala was the most sensitive brain region for memory enhancement in the active avoidance task, which is in accord with data reported that the amygdala could be the critical mediator of fear learning. Moreover, the authors confirmed that PREGS was the most potent memory enhancer, as intra-hippocampal injection of DHEAS significantly enhanced retention performance at higher doses than did PREGS. The same group showed that DHEAS also enhanced 1-week retention in foot shock active avoidance task when injected bilaterally into the hippocampus immediately following training in mice (Flood *et al.*, 1999).

PREGS also displays memory-enhancing properties in a two-trial spatial memory task in rats (Darnaudéry *et al.*, 2000). Indeed, PREGS injected intracerebroventricularly immediately after training dose-dependently altered retention performance tested 4 h later. An enhancement of retention was observed for the injection of 12 nmol/20μl ($\sim$5 μg/20 μl) dose, while the 192 nmol/20 μl ($\sim$80 μg/20 μl) dose failed to have an effect.

c. Pretraining vs Post-training Administrations of Neurosteroids. Isaacson *et al.* (1994) examined the effects of PREGS and PREG in a holeboard food-search task in rats. PREGS was injected (100 μg/day, subcutaneously), and PREG was given through subcutaneous pellet (2.5 mg, 100 μg/day). Training was given on an every-other-day schedule for 5 days, and then a retention session consisting of eight trials occurred 10 days after training. In the first experiment, PREG was given 8 days before training, in the second experiment, PREGS was administered before and during the training in association with PREG pellets implanted between the training and retention sessions. In the two experiments, none of the treatments altered training performance, but significant enhancement of retention was found. However, the enhancement was observed only during the fifth and sixth trials and when the treated and control groups of the two experiments were combined, thus limiting conclusions regarding the time specificity of the memory-enhancing effects of PREG and PREGS in this study.

Mayo *et al.* (1993) compared also the effects of pre- and post-training injections of PREGS using a two-trial Y-maze discrimination arm task in rats. PREGS (5 ng/0.5μl) was injected into the nucleus basalis magnocellularis (NBM). Injection 15 min before training had no effect on the 4-h-retention performance, while injection immediately post-training improved retention. These results suggest that PREGS injection into the NBM affects only the consolidation and/or retention processes but not the acquisition processes.

d. *Post-training vs Preretention Administrations of Neurosteroids.* Isaacson *et al.* (1995) studied the effects of post-training or preretention injection of PREGS in a passive avoidance task in rats. Post-training subcutaneous injection of PREGS facilitated either 24-h-delay retention with 1, 10, and 100 ng doses, or 48-h-delay retention with a 100-ng dose. For preretention administration of PREGS, only the dose of 10 ng significantly enhanced 24-h-delay retention performances. These results suggest that postacquisition administration of PREGS is effective in enhancing retention of passive avoidance behavior, whereas preretention administration is not.

e. *Pretraining vs Post-training vs Preretention Administration of Neurosteroids.* Reddy and Kulkarni (1998a) tested the effects of using pretraining and different post-training time intervals for injections of PREGS and DHEAS in a step-down passive avoidance task in mice. PREGS and DHEAS were given subcutaneously in doses from 0.125 to 10 mg/kg. Injections of PREGS or DHEAS 60 min before training or immediately post-training improved 24-h-delay retention performance, while injections 60 min before retention failed to enhance memory. The effects of PREGS and DHEAS were dose dependent. The maximal effective dose was 0.5 mg/kg for PREGS and 1 mg/kg for DHEAS, while no effect was observed at the lowest and highest doses. Moreover, pretraining administration of PREGS and DHEAS decreased the number of errors during training with 0.5 and 1 mg/kg doses for PREGS and with 1 and 5 mg/kg doses for DHEAS. First, these results confirm that the memory-enhancing effects of post-training administration of neurosteroids are time dependent, as previously reported (Isaacson *et al.*, 1995). Second, PREGS again appears to be the more potent memory enhancer among excitatory neurosteroids, as previously described by Flood *et al.* (1992, 1995).

All of these results were obtained in two-session (training–retention) memory tasks. The effects of neurosteroids have been also examined in other tasks measuring the learning abilities in successive learning sessions.

f. *Administration of Neurosteroids during Learning Sessions.* Frye and Sturgis (1995) studied the effects of the administration of PREGS and DHEAS in two learning tasks in ovariectomized female rats. First, finding the location of a hidden platform in a water-maze task assessed reference/spatial memory. Second, working memory was measured by finding the food-reinforced arm of a Y-maze using a delayed, nonmatching sample procedure. The water-maze trial consisted of 2 successive days (six trials/day). The injection of PREGS or DHEAS (3.2, 6.4 mg/kg subcutaneously) 30 min before the first trial on the second testing day decreased the distance traveled to reach the hidden platform in the water-maze task. The Y-maze task consisted of 2 successive days of habituation (Days 1 and 2) and 2 days of testing (Days 3 and 5). Pretesting PREGS injection (3.2, 6.4 mg/kg, subcutaneously) did not change memory performance, while DHEAS dose-dependently altered the performance. The subcutaneous injection of DHEAS at the 3.2 mg/kg dose increased the latency to the goal box but had no effect on the percentage of correct responses, suggesting a slight impairment of working memory. In contrast, DHEAS at the 6.4 mg/kg dose increased the percentage of correct responses, revealing a better working memory. Moreover, DHEAS injected intracerebroventricularly (1–2 μg) improved the learning performance in both the water-maze and Y-maze tasks. Together, these results suggest that both PREGS and DHEAS facilitate reference/spatial memory, while they have distinct effects on working/spatial memory processes. PREGS did not affect performance, while the performance was altered in a biphasic manner by DHEAS. However, an extended range of doses should confirm this biphasic effect.

3. *Aging*

Flood and Roberts (1988) were also the first to examine the effects of neurosteroids on learning and memory processes during aging. They studied the effects of DHEAS in a foot-shock active avoidance task in middle-age (18 months) and old (24 months) mice. The injection of DHEAS (20 mg/kg subcutaneously) immediately after training improved, 1 week later, the retention performance in middle-age and old mice to the high level observed in 2-month-old mice.

The ability of PREGS and DHEAS to modulate the age-induced learning impairment was also tested in 16-month-old mice using two different behavioral models of long-term memory, the step-down type of passive avoidance and elevated plus-maze paradigms (Reddy and Kulkarni, 1998b). Decreased step-down latency and increased transfer latency were observed in the passive avoidance and elevated plus-maze tasks, respectively, in 16-month-old mice compared with 3-month-old mice, revealing retention deficits in old mice. Pretraining injections of PREGS or DHEAS (1–20 mg/kg subcutaneously) dose-dependently improved the 24-h-delay retention performances in both tasks in old mice. The maximal effects were obtained with 5 mg/kg dose for PREGS and with 10 mg/kg dose for DHEAS in both tasks.

Moreover, we investigated the effects of systemic PREGS administration in 24-month-old rats in a Y-maze arm discrimination task. Pretraining administration of PREGS (47.5 mg/kg, intraperitoneally) was able to restore 6-h-delay retention deficits in old rats and, interestingly, this beneficial effect lasted for 10 days (Vallée *et al.*, 1997).

One study reported a lack of memory-enhancing effect of neurosteroids in old mice. Indeed, Shi *et al.* (2000) found that DHEA (20 mg/kg subcutaneously) did not affect the retention deficits observed in aged mice in the water-maze task. The procedure used by the authors was not the one commonly used in the water-maze task, as the test consisted of six trials of training and one retention session performed 1, 2, 4, or 5 weeks later. Thus, the lack of effect of post-training injection of DHEA in improving the performance could be the result of the long delay used between training and retention. Moreover, because DHEA was injected after training, no conclusion can be drawn regarding the effect of DHEA on the learning impairment usually observed in aged rodents in the water-maze task. The authors also tested the effect of 7-oxo-DHEA, an endogenous DHEA metabolite found in human urine and rabbit liver slices (Schneider and Mason, 1948; Fukushima *et al.*, 1954). Given that 7-oxo-DHEA was able to increase the retention performance at delays of 1, 2, and 4 weeks, the authors suggested that 7-oxo-DHEA was more effective than its parent steroid DHEA in increasing memory performance in old mice.

C. Inhibitory Neurosteroids: PROG, 5α-DH PROG, 3α,5α-TH PROG

Among the inhibitory neurosteroids, the effects of 3α,5α-TH PROG (allopregnanolone) on learning and memory processes were the most studied.

1. *Development*

Zimmerberg *et al.* (1995) reported that injection of 3α,5α-TH PROG in the lateral ventricle of 6-day-old rat pups could alter their retention performance in an odor-conditioning test. During the training, a novel odor (conditioning stimulus) was paired with an appetitive reward (milk), and the retention response was tested 1 h later in a two-odor choice preference task. Post-training injections of 3α,5α-TH PROG dose-dependently impaired learning and memory, while injections 20 min before the training did not disrupt retention.

2. *Adulthood*

a. 3α,5α-TH PROG and Learning and Memory Processes. The effects of 3α,5α-TH PROG on learning and memory processes were first tested in a two-trial Y-maze discrimination arm task in rats (Mayo *et al.*, 1993). Pretraining administration of 3α,5α-TH PROG (0.2 and 2 ng/5 μl) in the NBM decreased the one-h delay retention performance, while no effect was observed when 3α,5α-TH PROG was injected immediately following training. These results suggest that 3α,5α-TH PROG impaired recognition by acting on acquisition and not on consolidation and/or retrieval phases of learning and memory processes. Using the same spatial memory task, Ladurelle *et al.* (2000) tested the effects of a prolonged administration of 3α,5α-TH PROG in mice. Pretraining intracerebroventricular injection of 3α,5α-TH PROG for 3 days or 6 days impaired the 2-h-delay retention performance. For 3 days treatment, amnesia effects were observed for 0.1, 0.5, and 1 ng/h doses; for 6 days treatment, impairment occurred with 0.5 and 1 ng/h doses. Impairment of learning has been also reported following pretraining administration of 3α,5α-TH PROG in female and male rats in an elevated plus-maze task (Reddy and Kulkarni, 1999). Injection of 3α,5α-TH PROG (0.25 mg/kg subcutaneously) 30 min before training significantly decreased transfer latency during the 24-h-delay retention session. However, the learning abilities of the animals were not completely abolished.

In contrast, administration of 3α,5α-TH PROG has been reported to enhance learning performance in ovariectomized female rats (Frye and Sturgis, 1995). The authors first measured the reference/spatial memory in a water-maze task in which the animal had to find the location of a hidden platform. Second, working memory was assessed using a delayed nonmatching-to-sample task in a Y-maze in which one arm was reinforced with food. The injection of 3α,5α-TH PROG (3.2, 6.4 mg/kg subcutaneously) improved learning performance in the water-maze task, as the animals treated with 3α,5α-TH PROG were quicker to find the hidden platform than vehicle-treated animals. Injection of 3α,5α-TH PROG (3.2 mg/kg subcutaneously) also increased the percentage of correct responses in the Y-maze, revealing a learning improvement. At a higher dose, however, 3α,5α-TH PROG (6.4 mg/kg subcutaneously) increased latencies to reach the goal box without affecting the percentage of correct responses, which suggests a learning impairment. Similarly, 3α,5α-TH PROG injected intracerebroventricularly (1–2 μg) increased latencies and distance traveled to the hidden platform in the water-maze task and increased latencies and decreased percentage correct in the Y-maze, indicating an impairment of learning performance. The authors concluded that low levels of 3α,5α-TH PROG—similar to those during diestrus—produced memory enhancement, while high levels of 3α,5α-TH PROG were associated with memory decrements. However, measurement of 3α,5α-TH PROG levels following its administration is required to confirm this conclusion.

b. PROG and Learning and Memory Processes. Frye and Sturgis (1995) tested the effects of PROG on learning performances in the paradigms described for 3α,5α-TH PROG in ovariectomized female rats. The administration of PROG (3.2, 6.4 mg/kg subcutaneously) increased learning performance in the water-maze task but did not affect performance in the Y-maze. The effect of PROG has also been investigated on acquisition performance in an active avoidance-conditioning task in female rats (Díaz-Véliz, 1994). The injection of PROG (5 mg/kg subcutaneously) 6 h before testing increased performance in female rats at estrus when estrogen levels were low. However, PROG was not able to modify the acquisition of conditioned avoidance responses in intact rats, neither at diestrus nor in ovariectomized female rats. The authors concluded that the effects of PROG are dependent on the presence of the ovary.

It has also been shown that PROG could antagonize the reverse effect of DHEAS and PREGS on dizolcipine-induced working memory impairment in rats (Zou et al., 2000). The authors explained this result by an antagonist action of PROG on σ_1 receptors.

D. CONCLUSIONS

1. *Excitatory Neurosteroids and Learning and Memory Processes*

The effects of the administration of excitatory neurosteroids have been studied in distinct learning paradigms, including simple associative learning, conditioned learning, and spatial learning (Table I). Emerging from the findings described here, two major concepts can be highlighted regarding the effects of excitatory neurosteroids on learning and memory processes. The first concept is dose dependency. Indeed, a majority of the studies described biphasic effects of excitatory neurosteroids: very low and very high concentrations are usually not effective in altering learning and memory processes. Thus, the effects of neurosteroids follow a bell-shape or an inverted U-shape dose–response curve, which is typically reported for memory-enhancing drugs. However, given that the dose effective for a memory-enhancing effect varied from one study to another, it is difficult to elaborate on the optimal range concentration.

The second concept concerns the time-dependent effects of excitatory neurosteroids. Neurosteroids act differently during the different stages of learning and memory processes (i.e., acquisition, consolidation, and retrieval stages). In two-session (training/retention) tasks, the memory-enhancing effects of pretraining administrations of neurosteroids can be the result of alterations in the three stages of learning and memory processes, while post-training injections allow one to see their effects on consolidation and/or retrieval processes. Depending on the task, pretraining administration of neurosteroids either improved or did not affect retention performance. Short-delay post-training administration of neurosteroids improved performance, while long-delay post-training or preretention administration failed to have an effect. This time dependency of postacquisition interventions in altering memory storage or retrieval processes is a classic phenomenon described for memory-enhancing drugs (McGaugh, 1966, 1983). In conclusion, excitatory neurosteroids can either affect or not affect the acquisition of information, depending on the experimental procedure

TABLE I
MEMORY-ENHANCING EFFECTS OF NEUROSTEROIDS ON LEARNING AND MEMORY ABILITIES IN RODENTS

Neurosteroids	Administration route	Administration time	Behavioral paradigms	Dose-, time-, and localization-dependent effects	References
PREG, PREGS, DHEA, DHEAS	i.p.	Pretraining	Food appetitive T-maze (working memory)	Inverted-U curve (0.001–1 mg/kg), PREGS is the more potent	Reddy and Kulkarni (1999)
DHEAS	s.c.	Pretraining	Plus-maze (transfer learning)	5 mg/kg	Reddy and Kulkarni (1999)
PREGS	Minipump i.c.v. infusion	3-days pretraining	Two-trial Y-maze discrimination arm (working memory)	5, 10, 50 ng/h	Ladurelle et al. (2000)
		6-days pretraining		Dose-dependent effect (5, 10, 50 ng/h)	
DHEA, DHEAS	i.c.v.	Post training	Foot shock active avoidance	Dose-dependent effect for DHEA (10^{-12}–10^{-9} mol) and DHEAS (10^{-10}–10^{-9} mol)	Roberts et al. (1987)
DHEA	i.c.v.	Post-training	Foot shock active avoidance	Inverted-U curve (0.5–160 ng)	Flood et al. (1988)
DHEAS	i.c.v., s.c., or oral	Post-training	Foot shock active avoidance	Time-dependent effects (2, 30, 90, and 120 min post-training delays)	
DHEAS	i.c.v.	Post-training	Step-down passive avoidance		
PREG, PREGS, DHEA, DHEAS	i.c.v.	Post-training	Foot shock active avoidance	Inverted-U curve ($3.5\ 10^{-14}$–$3.5\ 10^{-9}$ mol) PREGS is the most potent	Flood et al. (1992)
PREGS	Hippocampus, amygdala, septum	Post-training	Foot shock-buzzer sound paired active avoidance	Most potent effect in amygdala	Flood et al. (1995)

Compound	Site/Route	Timing	Task	Effects	Reference
DHEAS	Hippocampus				Flood et al. (1999)
DHEAS	Hippocampus			PREGS more potent than DHEAS	
PREGS	i.c.v.	Post-training	Foot shock active avoidance	Dose-dependent effect (12, 192 nmol)	Darnaudéry et al. (2000)
PREG, PREGS	Chronic s.c.	Post-training	Two-trial spatial memory	Time-dependent effects (post-training effects, no pretraining effects)	Isaacson et al. (1994)
PREGS	NBM	Pretraining, post-training	Holeboard food search (reference memory)	Time-dependent effects (post-training effects, no pretraining effects)	Mayo et al. (1993)
PREGS	s.c.	Pretraining, post-training	Two-trial Y-maze discrimination arm (working memory)	Time-dependent effects (post-training effects, no pretretention effects)	Isaacson et al. (1995)
PREGS, DHEAS	s.c.	Post-training Preretention	Passive avoidance	Inverted U-curve (0.125–10 mg/kg); PREGS more potent than DHEAS; Time-dependent effects (pre- and post-training effects, no preretention effects)	Reddy and Kulkarni (1998a)
PREG, DHEAS, $3\alpha,5\alpha$-TH PROG	s.c.	Pretraining, post-training, preretention	Step-down passive avoidance		
PREGS	s.c.	Before the first trial (second day)	Water-maze (reference memory)	3.2, 6.4 mg/kg	Frye and Sturgis (1995)
DHEAS, $3\alpha,5\alpha$-TH PROG	s.c.	Before testing	Y-maze food research task (working memory)	3.2, 6.4 mg/kg	
	s.c.	Before testing	Y-maze food research task (working memory)	Dose-dependent effects (3.2, 6.4 mg/kg)	

Note. i.p. = intraperitoneal; s.c. = subcutaneous; i.c.v. = intracerebroventricular.

used. Moreover, they consistently facilitate the consolidation processes while they fail to affect the retrieval processes.

2. *Inhibitory Neurosteroids and Learning and Memory Processes*

The effects of inhibitory neurosteroids on learning and memory processes have been studied to a lesser extent than those of excitatory neurosteroids (see Table I). Controversies have been reported regarding the effects of $3\alpha,5\alpha$-TH PROG, the most frequently studied inhibitory neurosteroid. A majority of studies described an impairment of learning and memory performances. $3\alpha,5\alpha$-TH PROG impaired the acquisition of performance in young and adult rodents (Zimmerberg *et al.*, 1995; Reddy and Kulkarni, 1999; Ladurelle *et al.*, 2000). Moreover, $3\alpha,5\alpha$-TH PROG impaired acquisition without altering consolidation and/or retrieval phases of learning and memory processes (Mayo *et al.*, 1993). In contrast, Frye and Sturgis (1995) reported a biphasic effect of $3\alpha,5\alpha$-TH PROG: At low concentrations, it improved performance, while, at high concentrations, it impaired performance.

Given that anxiolytic-like actions have been described for $3\alpha,5\alpha$-TH PROG (Wieland *et al.*, 1991; Brot *et al.*, 1997), it is likely that its behavioral profile displays analogies with anxiolytic drugs, like diazepam. In conclusion, the role of $3\alpha,5\alpha$-TH PROG in learning and memory processes is not yet fully understood, and additional data are needed to confirm which memory process is altered by $3\alpha,5\alpha$-TH PROG and whether $3\alpha,5\alpha$-TH PROG displays a biphasic effect.

IV. Mechanisms of Action

A. BEHAVIORAL APPROACH

1. *Neurotransmitter Systems and Learning and Memory Processes*

Among the numerous neurotransmitter systems that play a role in learning and memory processes, we briefly present here the systems that, first, are known to play a role in learning and memory processes (for reviews, see McGaugh, 1989; McGaugh and Cahill, 1997) and, second, have been described as possible targets for the action of neurosteroids on learning and memory processes.

a. Cholinergic Systems. Basic and clinical studies have long recognized the importance of cholinergic mechanisms in cognitive functioning. For instance, considerable psychopharmacological evidence shows that antagonists of the muscarinic cholinergic receptors interfere with memory function in humans, nonhuman primates, and rodents (Hasselmo and Bower, 1993). It is also well established in animal studies that post-training administration of muscarinic cholinergic agonists and antagonists (e.g., scopolamine) enhance and impair, respectively, retention of a variety of tasks (for review, see McGaugh and Cahill, 1997). Moreover, drugs that increase synaptic acetylcholine (ACh) levels are the most frequently used for the treatment of cognitive deficits associated with central nervous system disorders, such as Alzheimer's disease, albeit with limited effectiveness (Benzi and Moretti, 1998).

The forebrain cholinergic pathways have been reported to play a role in learning and memory processes, and degeneration of the cholinergic neurons of the basal forebrain has been shown to occur in Alzheimer's disease (Fibiger, 1991; Sarter et al., 1996). The basal forebrain cholinergic neurons are divided into two classes. One innervates the neocortex from the NBM and the septo-hippocampal pathway, which originates in the medial septum/diagonal band (MSDB) (Lehmann et al., 1980; Amaral and Kurz, 1985; Frotcher and Leranth, 1985). It was hypothesized that improvements in septo-hippocampal pathway-related learning and memory tasks, which can be induced by intraseptal infusion of muscarinic receptor agonists, occur as a result of an increase in hippocampal ACh release. However, Wu et al. (2000) showed that muscarinic receptor agonists do not activate septo-hippocampal cholinergic neurons, but rather excite noncholinergic, septo-hippocampal GABA-type MSDB neurons, which appear to be another neuronal group involved in mediating learning and memory.

b. GABAergic Systems. In addition to the previous findings, it is commonly reported that $GABA_A$ receptors downregulate memory consolidation processes. For instance, it has been shown in animal studies that post-training injections of GABAergic antagonists (e.g., bicuculline, picrotoxine) and agonists (e.g., muscimol, baclofen) enhanced and impaired, respectively, retention of several types of training (for review, see McGaugh and Cahill, 1997). Moreover, it has been suggested that GABAergic interneurons in the hippocampus are important for setting the conditions for synaptic changes in hippocampal neurons during learning (Paulsen and Moser, 1998).

c. Glutamatergic Systems. Considerable evidence indicates that glutamatergic neurotransmission is involved in biochemical events underlying learning and memory processes (for review, see Ungerer et al., 1998). Supporting this, modulations of the NMDA type of glutamate receptor induced learning and memory alterations. For example, competitive NMDA-receptor antagonists, such as D-2-amino-5-phosphovalerate (D-AP5), 3-(($\pm$)-2-carboxypiperazine-4-yl)-propyl-1-phosphonic acid (CPP), and D(−)-(E)-4-(3-phosphonoprop-2-enyl)piperazine-2-carboxylic acid (D-CPPene), and the noncompetitive NMDA-receptor antagonist(+)-5-methyl-10,11-dihydro-5H-dibenzocyclohepten-5,10-imine maleate (MK-801, or dizolcipine), disturb acquisition and retention in various learning tasks. Moreover, it has been shown that animals deficient in hippocampal NMDA receptors (NMDAR1 knockout mice) display impaired hippocampal representation and spatial memory (McHugh *et al.*, 1996; Tsien *et al.*, 1996).

d. σ Systems. σ_1 Receptors have been described as mediating the neuromodulatory role on the cholinergic transmission and the NMDA system. Regarding learning and memory processes, σ_1-receptor agonists do not affect memory capacities by themselves but exert anti-amnesic effects (for review, see Maurice and Lockhart, 1997). For example, they block learning impairments induced by the muscarinic receptor antagonist, scopolamine, or the NMDA-receptor antagonist, dizolcipine.

2. *Interactions between Neurosteroids and Pharmacological Compounds Acting on Different Neurotransmitter Systems*

In this section, we describe the capacity of neurosteroids to alleviate several pharmacological models of amnesia induced by cholinergic, GABAergic, or glutamate drugs. Moreover, we report experimental data where σ_1-receptor ligands interfere with the memory-enhancing or anti-amnesic properties of neurosteroids (see Table II).

a. Cholinergic Systems. A number of studies used scopolamine as a model of amnesia in several learning paradigms, and they described reversal effects of neurosteroids on scopolamine-induced deficits.

DHEAS (162 ng/2 μl intracerebroventricularly) has been shown to block the effects of scopolamine (1 mg/kg subcutaneously) when both are injected after training in a foot shock active avoidance task in adult mice (Flood *et al.*, 1988). Pretraining administration of DHEAS (20 mg/kg intraperitoneally) also reversed the amnesia induced by pretraining injection of scopolamine (1 mg/kg intraperitoneally) in a passive avoidance task in adult rats (Li *et al.*, 1995). Similarly, PREGS displays anti-amnesic effects. PREGS (0.1–10 nmol intracerebroventricularly) co-administered with scopolamine (3 mg/kg subcutaneously) before training dose-dependently blocks the scopolamine-induced learning deficits in a go–no go visual discrimination task in adult mice (Meziane *et al.*, 1996). We also observed that PREGS (0.5; 1 nmol intracerebroventricularly) co-administered with scopolamine (1 mg/kg subcutaneously) 30 min before training dose-dependently altered the 24-h-delay retention deficit induced by scopolamine in a passive avoidance test in rats (Vallée *et al.*, 1999b).

Moreover, Urani *et al.* (1998) showed in adult mice that pretraining administration of PREGS or DHEAS (5-20 mg/kg doses subcutaneously) prevented alternation deficits induced by scopolamine (2 mg/kg subcutaneously) in a Y-maze task, and that administration of DHEAS (20 mg/kg subcutaneously) attenuated scopolamine-induced learning deficits in a water-maze task. In the same study, the authors showed that PROG (5–20 mg/kg subcutaneously) failed to affect scopolamine-induced deficits in the Y-maze task. Finally, Shi *et al.* (2000) reported recently that DHEA (20 mg/kg subcutaneously) blocked the retention deficit induced by scopolamine (1 mg/kg subcutaneously) in a water-maze task in adult mice. The drugs were injected after the last trial of the training session, and the retention session was assessed 6 days later.

Together, these studies suggest a role for central cholinergic systems in the memory-enhancing action of neurosteroids. This idea is supported by the findings that the memory-enhancing effect of PREGS has been related to an increase of extracellular levels of acetylcholine in the hippocampus measured by *in vivo* microdialysis in rats (Vallée *et al.*, 1997; Darnaudéry *et al.*, 2000).

b. GABAergic Systems. The interactions between neurosteroids and GABAergic drugs have not been directly tested. However, given that ethanol has $GABA_A$- agonist properties, ethanol-induced amnesia is often used as the model of amnesia related to GABAergic systems. This is supported by the fact that $GABA_A$ agonists and antagonists enhance and decrease, respectively, memory impairment induced by ethanol (Castellano and Pavone, 1990).

By using ethanol-induced amnesia, Melchior and Ritzmann (1996) showed that, in adult mice, PREG, DHEA, and their sulfates (0.05 mg/kg intraperitoneally) injected 20 min before pretraining administration of a low concentration of ethanol (0.5 g/kg intraperitoneally) could block ethanol-induced retention impairment. The authors showed these effects by testing working memory in a food-appetitive T-maze task. The authors concluded that the observed memory-enhancing effect of neurosteroids was consistent with their GABAergic modulatory effects; however, because ethanol can modulate neurosteroid levels (Baulieu and Robel, 1996; Vallée, unpublished data), the interaction between neurosteroids and GABAergic system concerning their memory effects cannot be directly examined in this study.

c. Glutamatergic Systems. Several studies have stated that neurosteroids can block the amnesia induced by NMDA-receptor antagonists. Romeo *et al.* (1994) first reported that pretraining administration of PREGS (48 μmol/kg intraperitoneally) and $3\alpha,5\alpha$-TH PROG (15 μmol/kg intravenously) reduced the passive avoidance retention deficit elicited by the pretraining injection of the noncompetitive NMDA receptor antagonist dizolcipine (0.3 μmol/kg intraperitoneally) in adult rats. The effect of pretraining injection of dizolcipine (0.15 mg/kg intraperitoneally) is also

TABLE II
ANTI-AMNESIC EFFECTS OF NEUROSTEROIDS IN AGE-RELATED AND PHARMACOLOGICAL MODELS OF AMNESIA IN RODENTS

Amnesia models	Related drugs	Neurosteroids	Administration route	Administration time	Behavioral paradigms	Effective doses	References
Age-related deficits		DHEAS	s.c.	Post-training	Foot shock active avoidance	20 mg/kg	Flood and Roberts (1988)
Age-related deficits		PREGS, DHEAS	s.c.	Pretraining	Step-down passive avoidance; plus-maze (transfer learning)	Inverted-U curve (1–20 mg/kg); PREGS more potent than DHEAS	Reddy and Kulkarni (1998b)
Age-related deficits		PREGS	i.p.	Pretraining	Y-maze discrimination arm (working memory)	47.5 mg/kg	Vallée et al. (1997)
Age-related deficits		DHEA	Hippocampus s.c.	Post-training Post-training	Water maze	5 ng/0.5 µl Lack of effect (20 mg/kg)	Shi et al. (2000)
Scopolamine		DHEAS	i.c.v.	Post-training	Foot shock active avoidance	162 ng/2 µl	Flood et al. (1988)
Scopolamine		DHEAS	i.p.	Pretraining	Passive avoidance	20 mg/kg	Li et al. (1995)
Scopolamine		PREGS	i.c.v.	Pretraining	Go-no go visual discrimination	Inverted-U curve (0.1–10 nmol)	Meziane et al. (1996)
Scopolamine		PREGS	i.c.v.	Pretraining	Passive avoidance	Dose-dependent effect (0.5, 1 nmol)	Vallée et al. (1999b)
Scopolamine		PREGS, DHEAS DHEAS PROG	s.c. s.c. s.c.	Pretraining Pretraining Pretraining	Alternance Y-maze Water maze Alternance Y-maze	5–20 mg/kg 20 mg/kg Lack of effect (5–20 mg/kg)	Urani et al. (1998)
NE-100		PREGS, DHEAS	s.c.	Pretraining	Alternance Y-maze	NE-100 blocks the anti-amnesic effects of PREGS and DHEAS (20 mg/kg)	
NE-100		DHEAS	s.c.	Pretraining	Water maze		

Drug	Neurosteroid	Route	Timing	Task	Dose	Reference
Scopolamine	DHEA	s.c.	Post-training	Water-maze task	20 mg/kg	Shi et al. (2000)
Ethanol	PREG, PREGS, DHEA, DHEAS	i.p.	Pretraining	Food appetitive T-maze task (working memory)	0.05 mg/lkg	Melchior and Ritzmann (1996)
Dizolcipine	PREGS 3α,5α-TH PROG	i.p. i.v.	Pretraining	Passive avoidance	48 μmol/kg 15 μmol/kg	Romeo et al. (1994)
Dizolcipine	PREGS	i.v.	Pretraining	Passive avoidance	5–20 mg/kg	Cheney et al. (1995)
Dizolcipine	DHEAS	s.c.	Pretraining	Alternance Y-maze step-down passive avoidance	10–20 mg/kg	Maurice et al. (1997)
	BMY-14802	s.c.			σ1 antagonists block the anti-amnesic effects of DHEAS (20 mg/kg)	
	Haloperidol					
Dizolcipine	DHEAS	s.c.	Pretraining	Plus-maze (transfer learning)	Inverted-U curve (1–20 mg/kg)	Reddy and Kulkarni (1998b)
	PREGS, DHEAS	s.c.	Pretraining	Passive avoidance	5–10 mg/kg	
D-CPPene	PREGS	i.v.	Pretraining	Passive avoidance	0.84–1680 pmol	Cheney et al. (1995)
CPP	PREGS	i.c.v.	Pretraining	Passive avoidance	0.001–0.1 nmol	Mathis et al. (1994)
D-AP5	PRECS	i.c.v.	Post-training	Y-maze active avoidance		Mathis et al. (1996)
β25-35 amyloid peptide	Haloperidol PREG, DHEA PREGS, DHEAS	s.c. s.c.	Pretesting Pretraining	Alternance Y-maze Passive avoidance	Haloperidol blocks the anti-amnesic effects of neurosteroids (20 mg/kg)	Maurice et al. (1998)

Note. s.c. = subcutaneous; i.p. = intraperitoneal; i.c.v. = intracerebroventricular; i.v. = intraveneous.

reversed by PREGS (5–20 mg/kg intraveneously) in the passive avoidance task in female rats (Cheney *et al.*, 1995). Moreover, learning deficits induced by dizolcipine (0.15 mg/kg intraperitoneally) are attenuated by DHEAS (10–20 mg/kg subcutaneously) in the alternance Y-maze and step-down passive avoidance tasks in mice (Maurice *et al.*, 1997). Reddy and Kulkarni (1998b) confirmed these results, showing that pretraining administration of PREGS (1–20 mg/kg subcutaneously) and DHEAS (1–20 mg/kg subcutaneously) reversed the impairment of the passive avoidance retention and plus-maze learning transfer induced by pretraining injection of dizolcipine (0.1 mg/kg intraperitoneally) in mice. The reverse effects of PREGS and DHEAS on the dizolcipine-induced impairment of learning were partial, leading to 60–70% antagonism, and showed an inverted-U shaped retention function.

Similarly, PREGS alters the amnesia induced by the competitive NMDA-receptor antagonists, such as D-CPPene (Cheney *et al.*, 1995), CPP (Mathis *et al.*, 1994), or D-AP5 (Mathis *et al.*, 1996). Pretraining co-administration of PREGS (5–10 mg/kg intraventricularly) and D-CPPene (2.5 mg/kg intraperitoneally), or PREGS (0.84–1680 pmol intracerebroventricularly) and CPP (1.2, 1.6 nmol intracerebroventricularly) reversed the impairment of passive avoidance retention in adult adrenalectomized/castrated (Adx/CX) female or intact male rats, respectively (Mathis *et al.*, 1994; Cheney *et al.*, 1995). Post-training administration of PREGS also blocked the deficits induced by D-AP5 (0.02 nmol/mouse) on retention performance in a Y-maze active avoidance task and in an appetitively reinforced lever- press task in mice (Mathis *et al.*, 1996). In the experimental procedures used in the previous studies, PREGS alone has limited effects on learning and memory abilities, suggesting that the anti-amnesia effects of PREGS is not due to intrinsic memory-enhancing properties, but rather to its action on NMDA receptor.

d. σ *Systems.* Maurice *et al.* (1997) tested in several experiments the relationships between neurosteroids and σ systems. The first study showed that σ_1-receptor antagonists (BMY-14802 and haloperidol) were able to antagonize the anti-amnesic effect of DHEAS in the spontaneous choice Y-maze and step-down passive avoidance tasks in mice. Indeed, the attenuation of dizolcipine-induced learning impairment by DHEAS (20 mg/kg subcutaneously) was suppressed by co-administration of BMY-14802 (5 mg/kg intraperitoneally) or by a chronic treatment with haloperidol (4 mg/kg/day subcutaneously) for 7 days. The authors suggested that DHEAS attenuates learning impairment induced by the NMDA-receptor antagonist dizolcipine via an interaction with σ_1 receptors. Other experiments by the same group confirmed the interaction between σ_1 receptor and the anti-amnesic effects of neurosteroids. Maurice *et al.* (1998) showed that administration of haloperidol (0.1 mg/kg intraperitoneally) blocked the attenuating effects of PREG and DHEA (20 mg/kg subcutaneously). They also blocked the effects of PREGS and DHEAS (20 mg/kg subcutaneously) on spontaneous alternation and passive avoidance retention deficits, respectively. In this study, the amnesia was induced by central administration of the β_{25-35} amyloid peptide (3 nmol/mouse). In the same study, PROG (20 mg/kg subcutaneously) was able to block the beneficial effects of σ_1-receptor agonists [(+)-pentazocine, PRE-084, and SA4503] on the spontaneous alternation deficits induced by the amyloid peptide. Moreover, Urani *et al.* (1998) found that another σ_1 antagonist, NE-100 (1 mg/kg intraperitoneally), blocked the attenuating effects of DHEAS and PREGS (20 mg/kg subcutaneously) on scopolamine-induced learning deficits in the Y-maze. It similarly blocked the effects of DHEAS (20 mg/kg subcutaneously) in the water maze in mice. NE-100 did not affect either the learning performance by itself or the scopolamine-induced amnesia in both tasks. Reddy and Kulkarni (1998a) also reported that the memory-enhancing effects induced by post-training administration of PREGS (0.5 mg/kg subcutaneously) and DHEAS (1 mg/kg subcutaneously) were completely blocked by the co-administration of haloperidol (0.25 mg/kg subcutaneously) in the passive avoidance task in mice, while, by itself, haloperidol did not affect the retention performance.

Overall, the data for σ systems suggest that alterations of these systems can affect the anti-amnesic and memory-enhancing effects of neurosteroids.

B. CELLULAR APPROACH (BRAIN PLASTICITY)

1. *Brain Plasticity and Memory Processes*

Storage of information as long-term memory is commonly assumed to involve modifications of relevant synapses. In the mammalian brain, long-term potentiation (LTP) is an enduring form of synaptic plasticity that can be detected at every excitatory synapse in the hippocampus. It has been hypothesized as a cellular basis for certain forms of learning and memory (for review, see Hölscher, 1999; Malenka and Nicoll, 1999; Miller and Mayford, 1999). However, although some studies indicate that hippocampal LTP can be modulated by various cognitive enhancers and neurotransmitter-agents affecting memory functions, other studies have found no correlation between the inducibility of LTP in the hippocampus and the ability of animals to learn hippocampal-dependent tasks. It is usually proposed that NMDA-receptor-dependent LTP in the hippocampus corresponds to a synaptic mechanism of memory. In addition to glutamate, the neurotransmitters ACh and GABA have been proposed as potentials factors involved in mediating the regulation of LTP. Finally, new evidence showed that LTP also occurred in the amygdala (for review, see Maren, 1999), a brain structure that is essential for simple forms of emotional learning and for memory.

Thus, on the basis that LTP may at least form a synaptic model for the plasticity involved in learning and memory in mammals, and because of the known interactions between neurosteroids and neurotransmitters related LTP regulation, we discuss the effects of neurosteroids on brain plasticity and LTP as possible mechanisms by which neurosteroids could modulate learning and memory processes.

2. *Neurosteroids and Brain Plasticity*

Little is known about the effects of neurosteroids on LTP; however, several studies reported an effect for DHEAS on hippocampal plasticity.

The first study suggesting a modulation of neuronal plasticity by neurosteroids is the study of Carette and Poulain (1984), showing that *in situ* application of DHEAS and PREGS produced excitatory effects on single neurons in the septo-preoptic area of the guinea pig. Other studies confirm this finding in rats. Steffensen (1995) investigated the effects of *in situ* microelectrophoretic application of DHEAS on evoked response and cellular activity in the hippocampus in anesthetized rats. Application of DHEAS in the CA1 hippocampal subfield increased population excitatory postsynaptic potential (pEPSP) and population spike (PS) amplitudes. Similarly, Meyer *et al.* (1999) found that application of DHEAS (10 μM) alters excitatory synaptic transmission at the Shaffer collateral-CA1 synapse by using an *in vitro* rat hippocampal slice preparation. This effect of DHEAS involved GABA$_A$-receptor-mediated inhibitory synaptic responses. Moreover, Yoo *et al.* (1996) reported that DHEAS (10–30 mg/kg intraveneously) enhances LTP development in a dose-dependent manner in the dentate gyrus in anesthetized rats. Injections of 20 and 30 mg of DHEAS produced significant increases of the EPSP and PS components of the evoked response in relation to 10-mg doses. However, the intensity of the DHEAS response (i.e., the peak after a single bolus injection) showed a similar pattern for the three doses.

In conclusion, the reported *in vitro* and *in vivo* synaptic action of DHEAS is consistent with its effects on learning and memory processes. However, to draw conclusions about a possible mechanism of action for neurosteroids, the effect of the other enhancing-memory neurosteroids should be investigated.

C. CONCLUSIONS

The behavioral studies described in this section demonstrate that the memory-enhancing effects of neurosteroids are related to cholinergic, glutamate, and/or GABAergic systems in simple and complex learning paradigms. The anti-amnesic effects of neurosteroids, especially on the amnesia induced by cholinergic and glutamate compounds, have been related to σ systems in spontaneous choice and passive avoidance paradigms. The interactions between neurosteroids and neurotransmitter systems are consistent with the rapid memory-enhancing or anti-amnesic effects described for neurosteroids. However, the long-term effects also reported for neurosteroids might involve other mechanisms. In this regard, the electrophysiological data are in favor of a modulation of hippocampal plasticity by the neurosteroids. Still, additional studies are needed to understand better the neurochemical and cellular mechanisms of actions involved in the action of endogenous neurosteroids on learning and memory processes.

V. Physiological Significance

A. Animal Studies: Correlation Studies between PREGS Levels and Learning and Memory Performances

Given that the animal studies described in the previous sections used a pharmacological approach, concerns can be pointed out regarding the physiological relevance of such studies (Warner and Gustafsson, 1995).

To assess the role of endogenous neurosteroids in learning and memory processes, we measured PREGS levels in different brain areas in aged rats previously tested for their learning and memory abilities (Vallée et al., 1997, 1998). We decided to study aged rats because it has been well documented that learning and memory abilities decrease with age and, interestingly, display a higher variability, which we believed was a good model for studying a correlation phenomenon. We tested PREGS because it was the most potent memory-enhancing neurosteroid described in pharmacological studies in rodents (see Section III). Rats that were 24 months old were tested in a Y-maze discrimination arm task and 1 week later in the spatial learning water-maze task. Then, the brain areas and blood samples were collected, and PREGS concentrations were assessed using a radioimmunoassay technique. First, as expected, we observed a great variability within the learning performance in the water-maze test in the aged rats. The distinct individual performance abilities were confirmed in the Y-maze task. Indeed, the distance traveled to reach the hidden platform in the water-maze correlated with the arm discrimination performance in the Y-maze (Spearman's, $\rho = -0.67$, $P < 0.001$). These results suggest that the performance ability of the population of aged rats was consistent in two spatial memory paradigms. On

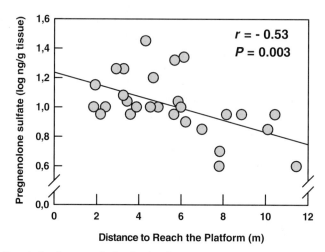

FIG. 2. Correlation between the levels of endogenous PREGS in the hippocampus and the learning performance in the water-maze task of individual 24-month-old rats. PREGS concentrations are expressed in log (ng/g). The performance was assessed by measuring the distance to reach the hidden platform during the last 3 days of learning in the water maze. Animals that swam for the longest distance (thus exhibiting worse performance) had the lowest level of PREGS in the hippocampus ($y = [-7.01 \pm 2.18] \times +12.61$).

this basis, we were confident in studying the relationship between individual performance and PREGS levels. And, interestingly, we found that the learning performance in the water-maze was correlated with PREGS levels in the hippocampus (Fig. 2). This correlation seemed specific to the hippocampus, as no correlation was found with PREGS levels either in other brain areas, such as the amygdala, frontal cortex, cortex, and striatum, or in plasma. Moreover, we found that PREGS levels in the hippocampus were decreased and displayed a higher variability in 24-month-old rats compared with 2-month-old rats (Fig. 3), which was consistent with the age-related memory decline usually described in aged rodents.

These data demonstrate a potential physiological role of hippocampal PREGS in the age-related learning and memory alterations. Nevertheless, some drawbacks can be outlined from the previous correlation study. First, the correlation found does not necessarily mean a cause–effect relationship. We attempted to demonstrate this relationship by injecting PREGS in memory-deficient, aged rats. Figure 4 shows that PREGS (5 ng/0.5 μl) administered in the hippocampus immediately after training was able to restore the retention performance in a Y-maze arm discrimination task at levels similar to those of young rats. The second drawback concerns the specific role of PREGS regarding the other neurosteroids (discussed in detail in

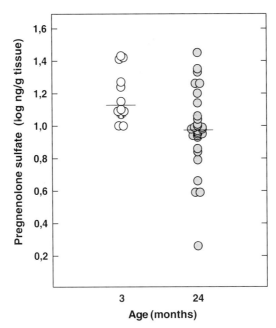

FIG. 3. PREGS levels in the hippocampus of 3-month-old and 24-month-old rats. Concentrations of PREGS were decreased but expanded in aged animals compared with young animals. The lines represent the median of each population.

Section VI). Indeed, the radioimmunoassay dosage allowed us to measure only PREGS in one biological sample but not the other memory-related neurosteroids, which might be physiologically involved in the regulation of learning and memory processes. Finally, it would be interesting to see whether hippocampal PREGS is involved in other learning and memory processes.

B. HUMAN STUDIES

1. *DHEA and DHEAS as Biomarkers of Healthy Aging*

The literature concerning the role of neurosteroids in human learning and memory processes is contradictory. Most studies have focused on DHEA and DHEAS, which are the neurosteroids found most abundantly in plasma and cerebrospinal fluid in humans (Orentreich *et al.*, 1984; Guazzo *et al.*, 1996).

These two neurosteroids appeared to be of great interest for elderly populations for two major reasons. The first is that circulating levels of

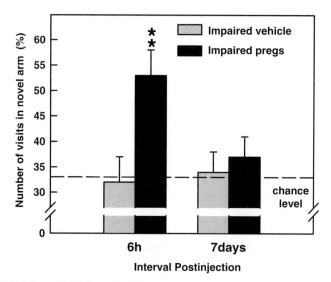

FIG. 4. The bilateral injection of PREGS (5ng/0.5 μl) into the dorsal hippocampus restored memory deficits of 22-month-old rats in the Y-maze discrimination arm task. The retention performance is expressed as the percentage of the number of visits to the novel arm. The animals that were not previously able to discriminate between the novel and the familiar arms of the maze could perform this discrimination after the injection of PREGS, while vehicle-treated animals still displayed impaired performance. [$^{**}P < 0.01$, compared to chance level. The dotted line expresses the level of equivalent exploration of the three arms (chance level, 33%)].

these steroids decline progressively and markedly with age, reaching levels at age 80 that are about 20% of those at age 20 (Orentreich et al., 1984, 1992; Vermeulen, 1995). Second, numerous animal studies have convincingly demonstrated the beneficial effect of DHEA and DHEAS in preventing age-related health deficits. Collectively, these observations have led investigators to speculate that some of the degenerative changes associated with human aging may be related to a progressive deficit in circulating DHEA or DHEAS. The idea of DHEA as a possible "fountain of youth" (Baulieu, 1996) derived from these speculations. However, given that most of the experimental studies in animals have been conducted in rodents, which have little, if any, circulating DHEA, the extrapolation from animal models to humans is not immediately obvious.

We describe here studies that investigated, first, whether DHEA and/or DHEAS levels in plasma are associated with cognitive performance, and, second, whether DHEA and/or DHEAS administration can improve performance. These studies were conducted in different elderly populations, either healthy or in residential care, or in Alzheimer's disease patients, who present cognitive deficits.

2. *Association between DHEA or DHEAS Plasma Levels and Cognitive Performance*

Studies attempting to relate DHEA and DHEAS plasma levels and cognitive functioning in elderly humans in a cross-sectional or longitudinal design are summarized in Table III.

The findings of these studies are contradictory (for review, see Wolf and Kirschbaum, 1999a). Cognitive dysfunctions have been associated with low DHEAS levels (Yanase *et al.*, 1996), high DHEAS levels (Morrison *et al.*, 1998, 2000), or high DHEA levels (Miller *et al.*, 1998; Morrison *et al.*, 2000). In other studies, no relationship was found (Barrett-Connor and Edelstein, 1994; Birkenhager-Gillesse *et al.*, 1994; Legrain *et al.*, 1995; Ravaglia *et al.*, 1998; Carlson and Sherwin, 1998, 1999; Moffat *et al.*, 2000; Morrison *et al.*, 2000).

a. Healthy Elderly Population. Prospective studies have measured DHEAS plasma levels and the risk of cognitive decline in elderly, but DHEAS levels, measured many years before the cognitive testing, were not associated with cognitive function (Barrett-Connor and Edelstein, 1994). Similarly, a 4-year longitudinal study in a population of elderly women (65–80 years) showed that change in cognitive performance over time was not associated with plasma DHEAS levels (Yaffe *et al.*, 1998). Moreover, a cross-sectional study and a longitudinal study for an 18-month period in healthy aged men and women (mean: 72 years) reported no association between plasma DHEAS levels and cognitive performance (Carlson and Sherwin, 1998, 1999). Accordingly, a 12-year longitudinal study reported that decline in endogenous DHEAS concentration was independent of cognitive status and cognitive decline in healthy aging men (Moffat *et al.*, 2000). Indeed, neither the rate of decline in serum DHEAS concentrations in men nor the mean DHEAS concentrations within individuals were related to memory status or memory decline. A comparison between the highest and the lower DHEAS quartiles revealed no memory performance differences, despite the fact that these groups differed in endogenous DHEAS concentrations by more than a factor of 4 for a mean duration of 12 years (Moffat *et al.*, 2000). A tendency for an inverse association between plasma DHEAS levels and cognitive impairment has also been reported in a 2-year prospective study in 189 healthy participants ages 55–80 years (Kalmijn *et al.*, 1998).

b. Residential Care Population. In a frail elderly residential care population (78–95 years), Morrison *et al.* (2000) found that DHEA and DHEAS

TABLE III
ASSOCIATIONS BETWEEN PLASMA DHEA, DHEAS LEVELS AND COGNITIVE PERFORMANCE IN HUMANS

Population type	Study design	Main findings	References
Healthy elderly men and women	Prospective study (16-year delay)	DHEAS levels did not predict the cognitive change over time.	Barrett-Connor and Edelstein (1994)
Elderly women population (65–80 years)	Prospective study (4-year delay)	Baseline DHEAS levels were not associated with cognitive decline	Yaffe et al. (1998)
Healthy men and women (72 years)	Cross sectional and longitudinal studies	DHEAS levels were not related to cognitive decline	Carlson and Sherwin (1998, 1999)
Healthy men (50–91 years)	12-year longitudinal study	Decline in DHEAS levels independent from cognitive decline	Moffat et al. (2000)
Healthy men and women (67 years)	Cross-sectional and prospective (2-year delay) studies	Inverse but not significant association between DHEAS and cognitive impairment	Kalmijn et al. (1998)
Frail elderly residential care women (78–95 years)	Cross-sectional study	Inverse correlation between DHEA or DHEAS and cognitive impairment	Morrison et al. (1998, 2000)
Aged men and women (91–104 years) in long-stay facilities	Cross-sectional study	No association between DHEAS levels and cognitive performance	Ravaglia et al. (1998)
Alzheimer's patients	Cross-sectional study (comparison to age- and gender-comparable elderly control individuals)	Decrease in DHEAS levels; no change of DHEA level	Yanase et al. (1996)
		Decrease in DHEAS levels	Leblhuber et al. (1990); Näsman et al. (1991); Scheider et al. (1992); Sunderland et al. (1998)
Alzheimer's patients (70–104 years)	3-year longitudinal study	Decrease in DHEAS levels	Hillen et al. (2000)
Alzheimer's men and women patients (moderate cognitive impairments)	6-month longitudinal study	DHEA levels did not predict cognitive impairment	Miller et al. (1998)

plasma levels increased with the level of cognitive dysfunction in women, while no association was found in men. The finding in women corresponds to an unexpected inverse correlation between DHEA or DHEAS and cognitive impairment, which the authors believe could be the result of the unusual population tested. These data confirm previous findings of Morrison's group, showing that cognitive dysfunction in another sample of women residents from the same nursing home correlated with increased rather than decreased plasma DHEAS levels (Morrison *et al.*, 1998). Another study conducted in men and women (91–104 years) in long-term care facilities for the elderly reported different results because no association between plasma DHEAS levels and cognitive testing scores was found either in men or in women (Ravaglia *et al.*, 1998).

c. Alzheimer's Disease Patients. It has been shown that plasma DHEAS levels in men and women patients (mean: 73 years) with Alzheimer's disease exhibited a decrease of 54% in men and 50% in women, compared with age- and gender-comparable elderly control individuals (Yanase *et al.*, 1996). However, no change was found in plasma DHEA levels. The decreased plasma DHEAS levels in Alzheimer's disease patients has also been found in several other studies (Leblhuber *et al.*, 1990; Näsman *et al.*, 1991; Schneider *et al.*, 1992; Sunderland *et al.*, 1989) and has been confirmed in a 3-year longitudinal study (70–104 years) (Hillen *et al.*, 2000). These findings suggest that low DHEAS plasma levels might be associated with the cognitive impairment observed in patients with Alzheimer's disease.

In contrast, a 6-month longitudinal study, conducted in men and women patients with Alzheimer's disease who show moderate cognitive impairment, unexpectedly found that lower plasma DHEA levels were associated with better memory performance at the beginning of the study (Miller *et al.*, 1998). However, the initial DHEA plasma levels did not predict decline in cognitive function over time.

3. Effects of DHEA Treatment on Cognitive Performance

Human studies have reported improvement of learning and memory dysfunction after DHEA administration to individuals with low DHEAS levels (Bonnet and Brown, 1990; Wolkowitz *et al.*, 1995), but other studies have failed to detect significant cognitive effects of DHEA administration (Wolf *et al.*, 1997a,b, 1998a,b; Bloch *et al.*, 1999). Table IV summarizes the findings concerning DHEA replacement studies in humans.

A first, single-case study reported that a 47-year-old woman showing a lifelong history of specific learning disabilities and deficient circulating DHEA and DHEAS levels could recover from some memory dysfunction following

TABLE IV
EFFECTS OF DHEA TREATMENT ON COGNITIVE PERFORMANCE IN HUMANS

Population type	Study design	Main findings	References
47-year-old woman (long history of learning disabilities, low DHEA and DHEAS levels)	Single case study	Chronic DHEA treatment normalized DHEA and DHEAS levels and reversed some memory dysfunctions	Bonnet and Brown (1990)
Elderly patients with low DHEA levels	Clinical trial ($n=6$)	Chronic DHEA treatment improved memory	Wolkowitz et al. (1995)
Young healthy subjects (25-year old)		Single administration of DHEA (300 mg) had no effect on memory performance	Wolf et al. (1997a)
Elderly healthy men and women (70 years old)	2 weeks of DHEA substitution (50 mg/day) or placebo	No improvement of attention and declarative memory	Wolf et al. (1997b, 1998a,b)
Old men and women (45–63 years) with midlife-onset dysthymia	Double-blind cross-over treatment (3 weeks on 90 mg DHEA, 3 weeks on 450 mg DHEA, and 6 weeks on placebo)	No specific effect on cognitive performance	Bloch et al. (1999)

chronic DHEA treatment, which normalized the plasma levels of DHEA and DHEAS (Bonnet and Brown, 1990). Similarly, a clinical trial in six depressed elderly patients with low DHEA levels reported an improvement in memory after subchronic DHEA treatment (Wolkowitz et al., 1995).

Studies carried out by Wolf's et al. (1997a) did not support the previous findings. Indeed, a single administration of DHEA (300 mg) in young healthy subjects (25 years old) had no effect on performance in several tests covering different aspects of memory, such as visual, verbal, or declarative memory. Moreover, 2 weeks of DHEA substitution (50 mg/day) in healthy elderly men and women (mean age: 70 years) failed to improve cognitive abilities, such as attention and declarative memory (Wolf et al., 1997b, 1998a,b). Similarly, a double-blind, crossover treatment with DHEA did not specifically alter the cognitive performance in men and women ages 45–63 years with midlife-onset dysthymia (Bloch et al., 1999). In this study, the treatment consisted of 3 weeks on 90 mg DHEA, 3 weeks on 450 mg DHEA, or 6 weeks on placebo.

4. Conclusions

Human experimental studies regarding, first, the relationships between DHEA or DHEAS plasma levels and learning and memory abilities, and, second, the effects of DHEA replacement on cognition have not led to a clear picture. Thus, human cross-sectional and longitudinal studies suggest that plasma DHEAS might be associated with global measures of well-being and functioning; however, a consistent relationship with cognition has not been detected. The lack of relationship between the plasma levels of neurosteroids and cognitive function corroborates with our findings observed in aged rodents (Vallée *et al.*, 1997, 1998). These data support a central action of neurosteroids synthesized in the brain, independent from the peripheral steroidogenesis.

Second, given that nonbeneficial effects of DHEA treatment have been reported, the hypothesis that positive long-term effects occur after DHEA replacement awaits experimental demonstration (Wolf and Kirschbaum, 1998, 1999b). This is in sharp contrast to the media "hype" of DHEA in certain countries. Thus, the therapeutic use of neurosteroids for cognitive dysfunctions is uncertain. Indeed, the beneficial effects of neurosteroids on cognitive functions have to be demonstrated; moreover, the possible negative side effects of long-term DHEA replacement have been poorly explored. To this end, expectations come from long-term experimental studies, such as the recent 1-year longitudinal study of Baulieu *et al.* (2000), which should further understanding of the role of neurosteroids in improving age-related neurodegenerative changes, such as learning and memory process alterations.

VI. Conclusions and Future Perspectives

A. Which Neurosteroids for Which Behavioral Effects?

Given that all neurosteroids are derived from PREG and given the existence of some reverse metabolic pathways, the possibility that several neurosteroids contribute to the effect observed following the administration of a single neurosteroid cannot be ruled out.

In this regard, several studies suggest that the effect of PREG or PREGS could be attributed to its conversion to allopregnanolone. This suggestion emerged from observations that administration of PREG and/or PREGS induced an increase of $3\alpha,5\alpha$-TH PROG levels and that the inhibition of $3\alpha,5\alpha$-TH PROG synthesis abolishes the effects of PREG or PREGS. For example, the administration of PREGS (20 mg/kg intraveneously), which reversed the amnesia induced by the NMDA-receptor antagonist dizolcipine in a passive avoidance test, increased the whole brain content of PREG by 50-fold and PROG, 5α-DH PROG, and 3α,5α-TH PROG by five- to sixfold in Adx/CX rats (Cheney *et al.*, 1995). The increase of 5α-DH PROG and 3α,5α-TH PROG levels and the antagonism of dizolcipine amnesia observed after the injection of PREGS were reversed by inhibiting the conversion of PROG to 5α-DH PROG with the 5α-reductase blocker SKF 105111 (Cheney *et al.*, 1995). Similarly, Romeo *et al.* (1994) reported that PREGS administration (48 μmol/kg intraperitoneally) increased the whole brain content of PREGS by 10-fold, of PREG and 3α,5α-TH PROG by sevenfold, and PROG by twofold. As in the study by Cheney *et al.* (1995), the administration of SKF 105111 eliminated the protective action of PREGS, but not 3α,5α-TH PROG, on the passive avoidance retention disruption elicited by dizolcipine. Together, these results suggest that PREGS prevents dizolcipine-induced memory deficit via an increase of brain 5α-DH PROG and/or 3α,5α-TH PROG content. However, given that PREGS and 3α,5α-TH PROG are negative and positive modulators of $GABA_A$ receptors, respectively, a common GABA-related memory-enhancing property is unexpected and suggests distinct mechanisms of action for the two neurosteroids.

It has been further suggested that the sulfated form of DHEA might be the active agent for memory enhancement. This hypothesis has been assumed when some authors reported that steroid sulfatase inhibitors could potentiate the memory-enhancing effect of DHEAS (Li *et al.*, 1995, 1997; Rhodes *et al.*, 1997; Flood *et al.*, 1999; Johnson *et al.*, 2000). The two sulfatase inhibitors used, the estrone-3-*O*sulfamate (EMATE) and (p-*O*-sulfamoyl)-*N*-tetracanoyl tyramine (DU-14), blocked the conversion of DHEAS to DHEA, resulting in an increase of the endogenous levels of DHEAS in the blood and brain of rats and a decrease of DHEA levels (Johnson *et al.*, 1997; Rhodes *et al.*, 1997). The inhibitor EMATE has been shown to potentiate the enhancement of performance of mice induced by DHEAS, but not PREGS, in an active avoidance T-maze memory paradigm (Flood *et al.*, 1999). EMATE alone could also block the scopolamine-induced amnesia in a passive avoidance task in rats (Li *et al.*, 1995). However, given that EMATE had estrogenic properties, the memory-enhancing effects observed could be mediated through estrogenic effects. To control for the possible estrogenic effects, the nonestrogenic steroid sulfatase inhibitor DU-14 has been used. It has been shown that DU-14 could reverse scopolamine-induced amnesia in a passive avoidance task in rats (Li *et al.*, 1995; Rhodes *et al.*, 1997). Similarly, DU-14 reversed the scopolamine-induced impairment and by itself increased performance in learning and spatial memory in the Morris water maze in rats (Johnson *et al.*, 2000). However, because no comparison between the effects of DHEA alone and DHEA+sulfate inhibitor has been examined and because DHEAS displays dose-dependently memory-enhancing properties, the conclusion that the memory-enhancing effect of DHEA is due to its conversion into DHEAS needs further confirmation.

B. STABLE ANALOGS OF NEUROSTEROIDS

In addition of the use of steroidogenesis enzyme inhibitors, the synthesis of stable analogs of neurosteroids could be helpful to determine the relative effect of one neurosteroid vs its metabolites.

For example, Chu *et al.* (1998) synthesized the stable analog of DHEAS, 17-oxoandrosta-3,5-dien-3-methyl sulfonate, which is not metabolized in DHEA. Thus, this compound could be used to examine the specific effect of DHEAS vs DHEA on learning and memory processes. Accordingly, the stable analog of PREGS, such as its (−) enantiomer, has been synthesized by Covey's group, and identical modulatory properties on $GABA_A$ receptors have been reported for both enantiomers (Nilsson *et al.*, 1998).

C. NEW METHOD OF QUANTIFICATION OF NEUROSTEROID LEVELS

One of the best approaches to determine which neurosteroids are involved in learning and memory processes might be to study the relationship between the memory-related drug effect and the content of the entire spectrum of neurosteroids in brain areas involved in these processes. To this end, given that the commonly used radioimmunoassay method allows measurement of only one neurosteroid at one time, a new method of quantification of neurosteroids levels is needed. Several groups have proposed new methods for the simultaneous quantification of traces of neurosteroids using the mass spectrometry technique (Uzunov *et al.*, 1996; Kim *et al.*, 2000; Lierre *et al.*, 2000; Shimada and Yago, 2000; Vallée *et al.*, 2000). Thus, these methods should allow exploration of the variations of neurosteroid levels in specific brain areas during alterations of learning and memory processes, such as those that occur during aging.

D. APPLICATION FOR CLINICAL STUDIES

Although clinical studies have not yet been successful in finding a beneficial effect of DHEA replacement on age-related memory impairments, the new method of quantification of neurosteroids should provide a useful tool for understanding the role of neurosteroids in altering learning and memory impairments associated with normal aging and/or with pathological aging, such as Alzheimer's disease.

Acknowledgments

This is Publication Number 13678-NP from The Scripps Research Institute. The authors gratefully thank Robert H. Purdy for critically reading the manuscript.

References

Amaral, D. G., and Kurz, J. (1985). An analysis of the origins of the cholinergic and non-cholinergic septal projections to the hippocampal formation of the rats. *J. Comp. Neurol.* **240**, 37–59.

Baddeley, A. (1992). Working memory. *Science* **255**, 556–559.

Barnes, C. A. (1979). Memory deficits associated with senescence: A neurophysiological and behavioral study in the rat. *J. Comp. Physiol. Psychol.* **93**, 74–194.

Barrett-Connor, E., and Edelstein, S. L. (1994). A prospective study of dehydroepiandrosterone sulfate and cognitive function in an older population: The Rancho Bernardo Study. *J. Am. Geriatr. Soc.* **42**, 420–423.

Baulieu, E.-E. (1981). Steroid hormones in the brain: Several mechanisms. *In* "Steroid Hormone Regulation of the Brain" (K. Fuxe, J. A. Gustafsson, and L. Weterberg, Eds.), pp. 3–14. Pergamon Press, Oxford.

Baulieu, E.-E. (1996). Dehydroepiandrosterone (DHEA): A fountain of youth? *J. Clin. Endocrinol. Metab.* **81**, 3147–3151.

Baulieu, E.-E., and Robel, P. (1996). Dehydroepiandrosterone and dehydroepiandrosterone sulfate as neuroactive neurosteroids. *J. Endocrinol.* **150**, S221–S239.

Baulieu, E.-E., Thomas, G., Legrain, S., Lahlou, N., Roger, M., Debuire, B., Faucounau, V., Girard, L., Hervy, M.-P., Latour, F., Leaud, M.-C., Mokrane, A., Pitti-Ferrandi, H., Trivalle, C., de Lacharrière, O., Nouveau, S., Rakoto-Arison, B., Souberbielle, J.-C., Raison, J., Le Bouc, Y., Raynaud, A., Girerd, X., and Forrette, F. (2000). Dehydrepiandrosterone (DHEA), DHEA sulfate, and aging: Contribution of the DHEAge study to a sociobiomedical issue. *Proc. Natl. Acad. Sci. USA* **97**, 4279–4284.

Benzi, G., and Moretti, A. (1998). Is there a rationale for the use of acetylcholinesterase inhibitors in the therapy of Alzheimer's disease? *J. Pharmacol.* **346**, 1–13.

Birkenhager-Gillesse, E. G., Derksen, J., and Lagaay, A. M. (1994). Dehydroepiandrosterone sulphate (DHEAS) in the oldest old, aged 85 and over. *Ann. NY Acad. Sci.* **719**, 543–552.

Bloch, M., Schmidt, P. J., Danaceau, M. A., Adams, L. F., and Rubinow, D. R. (1999). Dehydroepiandrosterone treatment of midlife dysthymia. *Biol. Psychiatry* **45**, 1533–1541.

Bonnet, K. A., and Brown, R. P. (1990). Cognitive effects of DHEA replacement therapy. *In* "The Biologic Role of Dehydroepiandrosterone" (M. Kalimi, and W. Regelson, Eds.), pp. 65–79. Walter de Gruyter & Co., Berlin.

Brot, M. D., Akwa, Y., Purdy, R. H., Koob, G. F., and Britton, K. T. (1997). The anxiolytic-like effects of the neurosteroid allopregnanolone: Interactions with GABA$_A$ receptors. *Eur. J. Pharmacol.* **325**, 1–7.

Carette, B., and Poulain, P. (1984). Excitatory effect of dehydroepiandrosterone, its sulphate ester and pregnenolone sulphate, applied by iontophoresis and pressure, on single neurones in the septo- preoptic area of the guinea pig. *Neurosci Lett.* **45,** 205–210.

Carlson, L. E., and Sherwin, B. B. (1998). Steroid hormones, memory and mood in a healthy elderly population. *Psychoneuroendocrinology* **23,** 583–603.

Carlson, L. E., and Sherwin, B. B. (1999). Relationships among cortisol (CRT), dehydroepiandrosterone-sulfate (DHEAS), and memory in a longitudinal study of healthy elderly men and women. *Neurobiol. Aging* **20,** 315–324.

Carlson, N. R. (1999). Learning and memory. *In* "Foundations of Physiological Psychology," 4th ed. (N. R. Carlson, Ed.), pp. 339–381. Allyn and Bacon, Needham Heights, MA.

Castellano, C., and Pavone, F. (1990). Effects of ethanol on passive avoidance behavior in the mouse: Involvement of GABAergic mechanisms. *Pharmacol. Biochem. Behav.* **29,** 321–324.

Cheney, D. L., Uzunov, D., and Guidotti, A. (1995). Pregnenolone sulfate antagonize dizolcipine amnesia: Role for allopregnanolone. *Neuroreport* **6,** 1697–1700.

Chu, G. H., Jagannathan, S., and Li, P. K. (1998). Synthesis of 17-oxoandrosta-3,5-dien-3-methyl sulfonate as stable analog of dehydroepiandrosterone sulfate. *Steroids* **63,** 214–217.

Conrad, C. D., Galea, L. A., Kuroda, Y., and McEwen, B. S. (1996). Chronic stress impairs rat spatial memory on the Y maze, and this effect is blocked by tianeptine pretreatment. *Behav. Neurosci.* **110,** 1321–1334.

Conrad, C. D., Lupien, S. J., Thanasoulis, L. C., and McEwen, B. S. (1997). The effects of type I and type II corticosteroid receptor agonists on exploratory behavior and spatial memory in the Y-maze. *Brain Res.* **759,** 76–83.

Corpéchot, C., Young, J., Calvel, M., Wehrey, C., Veltz, J. N., Trouyer, G., Mouren, M., Prasad, V. V. K., Banner, C., Sjövall, S., Baulieu, E.-E., and Robel, P. (1993). Neurosteroids. 3α-Hydroxy-5α-pregnan-20-one and its precursors in the brain, plasma, and steroidogenic glands of male and female rats. *Endocrinology* **133,** 1003–1009.

Crawley, J. N. (1999). Behavioral phenotyping of transgenic and knockout mice: Experimental design and evaluation of general health, sensory functions, motor abilities, and specific behavioral tests. *Brain Res.* **835,** 18–26.

Crawley, J. N., Belknap, J. K., Collins, A., Crabbe, J. C., Frankel, W., Henderson, N., Hitzemann, R. J., Maxson, S. C., Miner, L. L., Silva, A. J., Wehner, J. M., Wynshaw-Boris, A., and Paylor, R. (1997). Behavioral phenotypes of inbred mouse strains: Implications and recommendations for molecular studies. *Psychopharmacology* (*Berlin*) **132,** 107–124.

Darnaudéry, M., Koehl, M., Piazza, P.-V., Le Moal, M., and Mayo, W. (2000). Pregnenolone sulfate increases hippocampal acetylcholine release and spatial recognition. *Brain Res.* **852,** 173–179.

Dellu, F., Mayo, W., Cherkaoui, J., Le Moal, M., and Simon, H. (1992). A two-trial memory task with automated recording: Study in young and aged rats. *Brain Res.* **588,** 132–139.

Dellu, F., Fauchey, V., Le Moal, M., and Simon, H. (1997a). Extension of a new two-trial memory task in the rat: Influence of environmental context on recognition processes. *Neurobiol. Learn. Mem.* **67,** 112–120.

Dellu, F., Mayo, W., Vallée, M., Le Moal, M., and Simon, H. (1997b). Facilitation of cognitive performance in aged rats by past experience depends on the type of information processing involved: A combined cross-sectional and longitudinal study. *Neurobiol. Learn. Mem.* **67,** 121–128.

Dellu, F., Contarino, A., Simon, H., Koob, G. F., and Gold, L. H. (2000). Genetic differences in response to novelty and spatial memory using a two- trial recognition task in mice. *Neurobiol. Learn. Mem.* **73,** 31–48.

Díaz-Véliz, G., Urresta, F., Dussaubat, N., and Mora, S. (1994). Progesterone effects on the acquisition of conditioned avoidance responses and other motoric behaviors in intact and ovariectomized rats. *Psychoneuroendocrinology* **19,** 387–394.

Eichenbaum, H., Stewart, C., and Morris, R. G. (1990). Hippocampal representation in place learning. *J. Neurosci.* **10,** 3531–3542.
Fanselow, M. S. (1990). Factors governing one-trial contextual conditioning. *Animal Learn Behav.* **18,** 264–270.
Fibiger, H. C. (1991). Cholinergic mechanisms in learning, memory and dementia: A review of recent evidence. *Trends Neurosci.* **14,** 220–223.
Fleshner, M., Pugh, C. R., Trembaly, D., and Rudy, J. W. (1997). DHEA-S selectively impairs contextual-fear conditioning: Support for the antiglucocorticoid hypothesis. *Behav. Neurosci.* **111,** 512–517.
Flood, J. F., and Roberts, E. (1988). Dehydroepiandrosterone sulfate improves memory in aging mice. *Brain Res.* **448,** 178–181.
Flood, J. F., Smith, G. E., and Roberts, E. (1988). Dehydroepiandrosterone and its sulfate enhance memory retention in mice. *Brain Res.* **447,** 269–278.
Flood, J. F., Morley, J. E., and Roberts, E. (1992). Memory-enhancing effects in male mice of pregnenolone and steroids metabolically derived from it. *Proc. Natl. Acad. Sci. USA* **89,** 1567–1571.
Flood, J. F., Morley, J. E., and Roberts, E. (1995). Pregnenolone sulfate enhances post-training memory processes when injected in very low doses into limbic system structures: The amygdala is by far the most sensitive. *Proc. Natl. Acad. Sci. USA* **92,** 10806–10810.
Flood, J. F., Farr, S. A., Johnson, D. A., Li, P.-K., and Morley, J. E. (1999). Peripheral steroid sulfatase inhibition potentiates improvement of memory retention for hippocampally administered dehydroepiandrosterone sulfate but not pregnenolone sulfate. *Psychoneuroendocrinology* **24,** 799–811.
Frotcher, M., and Leranth, C. (1985). Cholinergic innervation of the rat hippocampus as revealed by choline acetyltransferase immunochemistry: A combined light and electron microscopic study. *J. Comp. Neurol.* **239,** 237–246.
Frye, C. A., and Sturgis, J. D. (1995). Neurosteroids affect spatial/reference, working, and long-term memory in female rats. *Neurobiol. Learn. Mem.* **64,** 83–96.
Fukushima, D., Kemp, A. D., Schneider, R., Stokem, M., and Gallagher, T. F. (1954). Studies in steroid metabolism XXV. Isolation and characterisation of new urinary steroids. *J. Biol. Chem.* **210,** 129–137.
Grady, C. L., and Craik, F. I. N. (2000). Changes in memory processing with age. *Curr. Opin. Neurobiol.* **10,** 224–231.
Guazzo, E. P., Kirkpatrick, P. J., Goodyer, I. M., Shiers, H. M., and Herbert, J. (1996). Cortisol, dehydroepiandrosterone (DHEA), and DHEA sulfate in the cerebrospinal fluid in man: Relation to blood levels and the effects of age. *J. Clin. Endocrinol. Metab.* **81,** 3951–3960.
Hasselmo, M. E., and Bower, J. M. (1993). Acetylcholine and memory. *Trends Sci.* **16,** 218–222.
Hillen, T., Lun, A., Reischies, F. M., Borchelt, M., Steinhagen-Thienssen, E., and Schaub, R. T. (2000). DHEA-S plasma levels and incidence of Alzheimer's disease. *Biol. Psychiatry* **47,** 161–163.
Hölscher, C. (1999). Synaptic plasticity and learning and memory: LTP and beyond. *J. Neurosci. Res.* **58,** 62–75.
Isaacson, R. L., Yoder, P. E., and Varner, J. (1994). The effects of pregnenolone on acquisition and retention of a food search task. *Behav. Neur. Biol.* **61,** 170–176.
Isaacson, R. L., Varner, J. A., Baars, J.-M., and de Wied, D. (1995). The effects of pregnenolone sulfate and ethylestrenol on retention of a passive avoidance task. *Brain Res.* **689,** 79–84.
Itoh, J., Nabeshima, T., and Kameyama, T. (1990). Utility of an elevated plus-maze for the evaluation of memory in mice: Effects of nootropics, scopolamine and electroconvulsive shock. *Psychopharmacology (Berlin)* **101,** 27–33.

Johnson, D. A., Rhodes, M. E., Boni, R. L., and Li, P. K. (1997). Chronic steroid sulfatase inhibition by (p-O-sulfamoyl)-N-tetradecanoyl tyramine increases dehydroepiandrosterone sulfate in whole brain. *Life Sci.* **61,** 355–359.

Johnson, D. A., Wu, T., Li, P., and Maher, T. J. (2000). The effect of steroid sulfatase inhibition on learning and spatial memory. *Brain Res.* **865,** 286–290.

Kalmijn, S., Launer, L. J., Stolk, R. P., de Jong, F. H., Pols, H. A., Hofman, A., Breteler, M. M., and Lamberts, S. W. (1998). A prospective study on cortisol, dehydroepiandrosterone sulfate, and cognitive function in the elderly. *J. Clin. Endocrinol. Metab.* **83,** 3487–3492.

Kim, Y.-S., Zhang, H., and Kim, H.-Y. (2000). Profiling neurosteroids in cerebrospinal fluids and plasma by gas chromatography/electron capture negative chemical ionization mass spectrometry. *Anal. Biochem.* **277,** 187–195.

Ladurelle, N., Eychenne, B., Denton, D., Blair-West, J., Schumacher, M., Robel, P., and Baulieu, E.-E. (2000). Prolonged intracerebroventricular infusion of neurosteroids affects cognitive performances in the mouse. *Brain Res.* **858,** 371–379.

Leblhuber, F., Windhager, E., Reisecker, F., Steinparz, F. X., and Dienstl, E. (1990). Dehydroepiandrosterone sulphate in Alzheimer's disease. *Lancet* **336,** 449.

Legrain, S., Berr, C., Frenoy, N., Gourlet, V., Debuire, B., and Baulieu, E.-E. (1995). Dehydroepiandrosterone sulfate in a long-term care aged population. *Gerontology* **41,** 343–351.

Lehmann, J., Nagy, J. I., Atmadja, S., and Fibiger, H. C. (1980). The nucleus basalis magnocellularis: The origin of a cholinergic projection to the neorcortex of the rat. *Neuroscience* **5,** 1161–1174.

Li, P. K., Rhodes, M. E., Jagannathan, S., and Johnson, D. A. (1995). Reversal of scopolamine induced amnesia in rats by the steroid sulfatase inhibitor estrone-3-O-sulfamate. *Brain Res. Cogn. Brain Res.* **2,** 251–254.

Li, P. K., Rhodes, M. E., Burke, A. M., and Johnson, D. A. (1997). Memory enhancement mediated by the steroid sulfatase inhibitor (p-O-sulfamoyl)-N-tetradecanoyl tyramine. *Life Sci.* **60,** 45–51.

Lierre, P., Akwa, Y., Weill-Engerer., S., Eychenne, B., Pianos, A., Robel, P., Sjövall, J., Schumacher, M., and Baulieu, E.-E. (2000). Validation of an analytical procedure to measure trace amounts of neurosteroids in brain tissue by gas chromatography–mass spectrometry. *J. Chromatogr. B.* **739,** 301–312.

Majewska, M. D. (1992). Neurosteroids: Endogenous bimodal modulators of the $GABA_A$ receptor. Mechanism of action and physiological significance. *Prog. Neurobiol.* **28,** 379–395

Malenka, R. C., and Nicoll, R. A. (1999). Long-term potentiation-a decade of progress? *Science* **285,** 1870–1874.

Maren, S. (1999). Long-term potentiation in the amygdala: A mechanism for emotional learning and memory. *Trends Neurosci.* **22,** 561–567.

Mathis, C., Paul, S. M., and Crawley, J. N. (1994). The neurosteroid pregnenolone sulfate blocks NMDA antagonist-induced deficits in a passive avoidance memory task. *Psychopharmacology* **116,** 201–206.

Mathis, C., Vogel, E., Cagniard, B., Criscuolo, F., and Ungerer, A. (1996). The neurosteroid pregnenolone sulfate blocks deficits induced by a competitive NMDA antagonist in active avoidance and lever-press learning tasks in mice. *Neuropharmacology* **35,** 1057–1064.

Maurice, T., and Lockhart, B. P. (1997). Neuroprotective and anti-amnesic potential of sigma receptors ligands. *Prog. Neuropsychopharmacol. Biol. Psychiatry* **21,** 69–102.

Maurice, T., Junien, J.-L., and Privat, A. (1997). Dehydroepiandrosterone sulfate attenuates dizolcipine-induced learning impairment in mice via σ_1-receptors. *Behav. Brain Res.* **83,** 159–164.

Maurice, T., Su, T.-P., and Privat, A. (1998). Sigma$_1$ (σ_1) receptor agonists and neurosteroids attenuate B25-35-amyloid peptide-induced amnesia in mice through a common mechanism. *Neuroscience* **83,** 413–428.

Maurice, T., Phan, V. L., Urani, A., Kamei, H., Noda, Y., and Nabeshima, T. (1999). Neuroactive neurosteroids as endogenous effectors for the sigma$_1$ (sigma$_1$) receptor: Pharmacological evidence and therapeutic opportunities. *Jpn. J. Pharmacol.* **81,** 125–155.

Mayo, W., Dellu, F., Robel, P., Cherkaoui, J., Le Moal, M., Baulieu, E.-E., and Simon, H. (1993). Infusion of neurosteroids into the nucleus basalis magnocellularis affects cognitive processes in the rat. *Brain Res.* **607,** 324–328.

McGaugh, J. L. (1966). Time-dependent processes in memory storage. *Science* **153,** 1351–1358.

McGaugh, J. L. (1983). Preserving the presence of the past: Hormonal influences on memory storage. *Am. Psychologist* **38,** 161–174.

McGaugh, J. L. (1989). Involvement of hormonal and neuromodulatory systems in the regulation of memory storage. *Ann. Rev. Neurosci.* **12,** 255–287.

McGaugh, J. L., and Cahill, L. (1997). Interaction of neuromodulatory systems in modulating memory storage. *Behav. Brain Res.* **83,** 31–38.

McIIugh, T. J., Blum, K. I., Tsien, J. Z., Tonegawa, S., and Wilson, M. A. (1996). Impaired hippocampal representation of space in CA1-specific NMDAR1 knockout mice. *Cell* **87,** 1339–1349.

Melchior, C. L., Glasky, A. J., and Ritzmann, R. F. (1993). A low dose of ethanol impairs working memory in mice in a win-shift foraging paradigm. *Alcohol* **10,** 491–493.

Melchior, C. L., and Ritzmann, R. F. (1994). Dehydroepiandrosterone is an anxiolytic in mice on the plus maze. *Pharmacol. Biochem. Behav.* **47,** 437–441.

Melchior, C. L., and Ritzmann, R. F. (1996). Neurosteroids block the memory-impairing effects of ethanol in mice. *Pharm. Biochem. Behav.* **53,** 51–56.

Mellon, S. H., and Compagnone, N. A. (1999). Molecular biology and developmental regulation of the enzymes involved in the biosynthesis and metabolism of neurosteroids. *In* "Neurosteroids: A New Regulatory Function in the Nervous System" (E.-E. Baulieu, M. Shumacher, and P. Robel, Eds.), pp. 27–50. Humana Press, Totowa, NJ.

Meyer, J. H., Lee, S., Wittenberg, G. F., Randall, R. D., and Gruol, D. L. (1999). Neurosteroid regulation of inhibitory synaptic transmission in the rat hippocampus in vitro. *Neuroscience* **90,** 1177–1183.

Meziane, H., Mathis, C., Paul, S. M., and Ungerer, A. (1996). The neurosteroid pregnenolone sulfate reduces learning deficits induced by scopolamine and has promnestic effects in mice performing an appetitive learning task. *Psychopharmacology* **126,** 323–330.

Miller, S., and Mayford, M. (1999). Cellular and molecular mechanisms of memory: The LTP connection. *Curr. Opin. Genet. Dev.* **9,** 333–337.

Miller, T. P., Taylor, J., Rogerson, S., Mauricio, M., Kennedy, Q., Schatzberg, A., Tinklenberg, J., and Yesavage, J. (1998). Cognitive and non cognitive symptoms in dementia patients: Relationship to cortisol and dehydroepiandrosterone. *Int. Psychogeriatr* **10,** 85–96.

Miyazaki, S., Imaizumi, M., and Machida, H. (1995). The effects of anxiolytics and anxiogenics on evaluation of learning and memory in an elevated plus-maze test in mice. *Methods Find Exp Clin. Pharmacol.* **17,** 121–127.

Moffat, S. D., Zonderman, A. B., Harman, S. M., Blackman, M. R., Kawas, C., and Resnick, S. M. (2000). The relationship between longitudinal declines in dehydroepiandrosterone sulfate concentrations and cognitive performance in older men. *Arch Intern Med.* **160,** 2193–2198.

Morris, R. G. (1984). Developments of a water-maze procedure for studying spatial learning in the rat. *J. Neurosci. Methods* **11,** 47–60.

Morris, R. G., Kandel, E. R., and Squire, L. R. (1988). The neuroscience of learning and memory: Cells, neural circuits and behavior. *Trends Neurosci.* **11,** 125–127.

Morrison, M. F., Katz, I. R., Parmelee, P., Boyce, A. A., and TenHave, T. (1998). Dehydroepiandrosterone sulfate (DHEA-S) and psychiatric and laboratory measures of frailty in a residential care population. *Am. J. Geriatr. Psychiatry* **6,** 277–284.

Morrison, M. F., Redei, E., TenHave, T., Parmelee, P., Boyce, A. A., Sinha, P. S., and Katz, I. R. (2000). Dehydroepiandrosterone sulfate and psychiatric measures in a frail, elderly residential care population. *Biol. Psychiatry* **47,** 144–150.

Näsman, B., Olsson, T., Bäckström, T., Eriksson, S., Grankvist, K., Viitanen, M., and Bucht, G. (1991). Serum dehydroepiandrosterone sulfate in Alzheimer's disease and in multi-infarct dementia. *Biol. Psychiatry* **30,** 684–690.

Nilsson, K. R., Zorumski, C. F, and Covey, D. F. (1998). Neurosteroid analogues. 6. The synthesis and GABA$_A$ receptor pharmacology of enantiomers of dehydroepiandrosterone sulfate, pregnenolone sulfate, and (3α,5β)-3-hydroxypregnan-20-one sulfate. *J. Med. Chem.* **41,** 2604–2613.

Orentreich, N., Brind, J. L., Rizer, R. L., and Vogelman, J. H. (1984). Age changes and sex differences in serum dehydroepiandrosterone sulfate concentrations throughout adulthood. *J. Clin. Endocrinol. Metab.* **59,** 551–555.

Orentreich, N., Brind, J. L., Vogelman, J. H., Andres, R., and Baldwin, H. (1992). Long-term longitudinal measurement of plasma dehydroepiandrosterone sulfate in normal men. *J. Clin. Endocrinol. Metab.* **75,** 1002–1004.

Park-Chung, M., Malayev, A., Purdy, R. H., Gibbs, T. T., and Farb, D. H. (1999). Sulfated and unsulfated steroids modulate gamma-aminobutyric acid$_A$ receptor function through distinct sites. *Brain Res.* **830,** 72–87.

Paulsen, O., and Moser, E. I. (1998). A model of hippocampal memory encoding and retrieval: GABAergic control of synaptic plasticity. *Trends Neurosci.* **21,** 271–278.

Ravaglia, G., Forti, P., Maioli, F., Boschi, F., De Ronchi, D., Bernardi, M., Pratelli, L., Pizzoferrato, A., and Cavalli, G. (1998). Dehydroepiandrosterone sulphate and dementia. *Arch. Gerontol. Geriatr. Suppl.* **6,** 423–426.

Reddy, D. S., and Kulkarni, S. K. (1998a). The effects of neurosteroids on acquisition and retention of a modified passive-avoidance learning task in mice. *Brain Res.* **791,** 108–116.

Reddy, D. S., and Kulkarni, S. K. (1998b). Possible role of nitric oxide in the nootropic and antiamnesic effects of neurosteroids on aging- and dizolcipine-induced learning impairment. *Brain Res.* **799,** 215–229.

Reddy, D. S., and Kulkarni, S. K. (1999). Sex and estrous cycle-dependent changes in neurosteroid and benzodiazepine effects on food consumption and plus-maze learning behaviors in rats. *Pharmacol. Biochem. Behav.* **62,** 53–60.

Rhodes, M. E., Li, P.-K., Burke, A. M., and Johnson, D. A. (1997). Enhanced plasma DHEAS, brain acetylcholine and memory mediated by steroid sulfatase inhibition. *Brain Res.* **773,** 28–32.

Ritzmann, R. F., Kling, A., Melchior, C. L., and Glasky, A. J. (1993). Effect of age and strain on working memory in mice as measured by win-shift paradigm. *Pharmacol. Biochem. Behav.* **44,** 805–807.

Roberts, E., Bologa, L., Flood, J. F., and Smith, G. E. (1987). Effects of dehydroepiandrosterone and its sulfate on brain tissue in culture and on memory in mice. *Brain Res.* **406,** 357–362.

Romeo, E., Cheney, D. L., Zivkovic, I., Costa, E., and Guidotti, A. (1994). Mitochondrial diazepam-binding inhibitor receptor complex agonists antagonize dizolcipine amnesia: Putative role for allopregnanolone. *J. Pharm. Exp. Ther.* **270,** 89–96.

Sarter, M., Bruno, J. P., Givens, B., Moore, H., McGaugh, J., and McMahon, K. (1996). Neuronal mechanisms mediating drug-induced cognition enhancement: Cognitive activity as a necessary intervening variable. *Cogn. Brain Res.* **3,** 329–343.

Schneider, J. J., and Mason, H. L. (1948). Studies on intermediary steroid metabolism. Isolation of Δ^5-androstene-3 (β), 17 (α)-diol, and Δ^5-androstene-3(β), 16 (β), 17 (α) triol following the incubation of dehydroisoandrosterone with surviving rabbit liver slices. *J. Biol. Chem.* **172,** 771–782.

Schneider, L. S., Hinsey, M., and Lyness, S. (1992). Plasma dehydroepiandrosterone sulfate in Alzheimer's disease. *Biol. Psychiatry* **31,** 205–208.

Sharma, A. C., and Kulkarni, S. K. (1992). Evaluation of learning and memory mechanisms employing elevated plus-maze in rats and mice. *Prog. Neuropsychopharmacol. Biol. Psychiatry* **16,** 117–125.

Shi, J., Schulze, S., and Lardy, H. A. (2000). The effect of 7-oxo-DHEA acetate on memory in young and old C57BL/6 mice. *Steroids* **65,** 124–129.

Shimada, K, and Yago, K. J. (2000). Studies on neurosteroids. Determination of pregnenolone and dehydroepiandrosterone in rat brains using gas chromatography–mass spectrometry-mass spectrometry. *Chromatogr. Sci.* **38,** 6–10.

Squire, L. R. (1987). Memory: Neural organization and behavior. *In* "The Nervous System" (V. N. Mountcastle, F. Plum, and S. R. Geiger, Eds.), pp. 295–371. American Physiological Society, Bethesda, MD.

Squire, L. R., and Zola-Morgan, S. (1988). Structure and function of declarative and nondeclarative memory system. *Proc. Natl. Acad. Sci. USA* **93,** 13515–13522.

Squire, L. R., and Zola-Morgan, S. (1996). Memory: Brain systems and behavior. *Trends Neurosci.* **11,** 170–175.

Squire, L. R., Shimamura, A. P., and Amaral, D. G. (1989). Memory and the hippocampus. *In* "Neural Models of Plasticity: Experimental and Theoretical Approaches" (J. H. Byrne and W. O. Berry, Eds.), Academic Press, San Diego.

Steffensen, S. C. (1995). Dehydroepiandrosterone sulfate suppresses hippocampal recurrent inhibition and synchronizes neuronal activity to theta rhythm. *Hippocampus* **5,** 320–328.

Sunderland, T. S., Merril, C. R., Harrington, M. G., Lawlor, M. G., Molchan, S. E., Martinez, R., and Murphy, D. L. (1989). Reduced plasma dehydroepiandrosterone concentrations in Alzheimer's disease. *Lancet* **2,** 570.

Tsien, J. Z., Huerta, P. T., and Tonegawa, S. (1996). The essential role of hippocampal CA1 NMDA receptor-dependent synaptic plasticity in spatial memory. *Cell* **87,** 1327–1338.

Tulving, E. (1972). Episodic and semantic memory. *In* "Organization of Memory" (E. Tulving and W. Donaldson Eds.), pp. 381–403. Academic Press, New York.

Tulving, E. (1987). Multiple memory systems and consciousness. *Hum. Neurobiol.* **6,** 67–80.

Ungerer, A., Mathis, C., and Melan, C. (1998). Are glutamate receptors specifically implicated in some forms of memory processes? *Exp. Brain Res.* **123,** 45–51.

Urani, A., Privat, A., and Maurice, T. (1998). The modulation by neurosteroids of the scopolamine-induced learning impairment in mice involves an interaction with sigma$_1$ (σ_1) receptors. *Brain Res.* **799,** 64–77.

Uzunov, D. P., Cooper, T. B., Costa, E., and Guidotti, A. (1996). Fluoxetine-elicited changes in brain neurosteroid content measured by negative mass fragmentation. *Proc. Natl. Acad. Sci. USA.* **93,** 12599–12604.

Vallée, M., Mayo, W., Corpéchot, C., Young, J., Le Moal, M., Baulieu, E.-E., Robel, P., and Simon, H. (1997). Neurosteroids: Cognitive performance in deficient aged rats depends on low pregnenolone sulfate levels in the hippocampus. *Proc. Natl. Acad. Sci. USA* **94,** 14865–14870.

Vallée, M., Robel, P., Le Moal, M., Baulieu, E.-E., and Mayo, W. (1998). Cognitive deficits in aged rats: Implication of neurosteroids. *Alzheimer's Report* **1,** 49–54.

Vallée, M., Maccari, S., Dellu, F., Simon, H., Le Moal, M., and Mayo, W. (1999a). Long-term effects of prenatal stress and postnatal handling on age-related glucocorticoid secretion and cognitive performance. A longitudinal study in rats. *Eur. J. Neurosci.* **11,** 2906–2916.

Vallée, M., Shen, W., Heinrichs, S. C., Zorumski, C. F., Covey, D. F., Koob, G. F., and Purdy, R. H. (1999b). Differential effects of the neurosteroid pregnenolone sulfate and its synthetic analogs in the passive avoidance memory task. *Soc. Abs. Neurosci.* **25**, 630.

Vallée, M., Rivera, J. D., Koob, G. F., Purdy, R. H., and Fitzgerald, R. (2000). Quantification of neurosteroids by negative chemical ionization gas chromatography–mass spectrometry. Study in plasma and brain tissues. *Anal. Biochem.* **287**, 153–166.

Vermeulen, A. (1995). Dehydroepiandrosterone sulfate and aging. *Ann. NY Acad. Sci.* **774**, 121–127.

Warner, M., and Gustafsson, J. A. (1995). Cytochrome P450 in the brain: Neuroendocrine functions. *Front Neuroendocrinol* **16**, 224–236.

Whiting, K. P., Restall, C. J., and Brain, P. F. (1998). Changes in the neuronal membranes of mice related to steroid hormone influences. *Pharmacol. Biochem. Behav.* **59**, 829–833.

Wieland, S., Lan, N. C., Mirasedeghi, S., and Gee, K. W. (1991). Anxiolytic activity of the progesterone metabolite 5 alpha-pregnan-3 alpha-o1-20-one. *Brain Res.* **565**, 263–268.

Wolf, O. T., and Kirschbaum, C. (1998). Wishing a dream came true: DHEA as a rejuvenating treatment? *J. Endocrinol. Invest.* **21**, 133–135.

Wolf, O. T., and Kirschbaum, C. (1999a). Actions of dehydroepiandrosterone and its sulfate in the central nervous system: Effects on cognition and emotion in animals and humans. *Brain Res. Rev.* **30**, 264–288.

Wolf, O. T., and Kirschbaum, C. (1999b). Dehydroepiandrosterone replacement in elderly individuals: Still waiting for the proof of beneficial effects on mood or memory. *J. Endocrinol. Invest.* **22**, 316.

Wolf, O. T., Köster, B., Kirschbaum, C., Pietrowsky, R., Kern, W., Hellhammer, D. H., Born, J., and Fehm, H. L. (1997a). A single administration of dehydroepiandrosterone does not enhance memory performance in young healthy adults, but immediately reduces cortisol levels. *Biol. Psychiatry* **42**, 845–848.

Wolf, O. T., Neuman, O., Hellhammer, D. H., Geiben, A. C., Straburger, C. J., Dressendörfer, R. A., Pirke, K.-M., and Kirschbaum, C. (1997b). Effects of a two week physiological dehydroepiandrosterone (DHEA) substitution on cognitive performance and well being in healthy elderly women and men. *J. Clin. Endocrinol. Metab.* **82**, 2363–2367.

Wolf, O. T., Kudielka, B. M., Hellhammer, D. H., Hellhammer, J., and Kirschbaum, C. (1998a). Opposing effects of DHEA replacement in elderly subjects on declarative memory and attention after exposure to a laboratory stressor. *Psychoneuroendocrinology* **23**, 617–629.

Wolf, O. T., Naumann, E., Hellhammer, D. H., and Kirschbaum, C. (1998b). Effects of dehydroepiandrosterone replacement in elderly men on event-related potentials, memory, and well-being. *J. Gerontol.* **53A**, M385–M390.

Wolkowitz, O. M., Reus, V. I., Roberts, E., Manfredi, F., Chan, T., Ormiston, S., Johnson, R., Canick, J., Brizendine, L., and Weingartner, H. (1995). Antidepressant and cognition-enhancing effects of DHEA in major depression. *Ann. NY. Acad. Sci.* **774**, 337–339.

Wu, F. S., Gibbs, T. T., and Farb, D. H. (1991). Pregnenolone sulfate: A positive allosteric modulator at the N-methyl-D-aspartate receptor. *Molec. Pharmacol.* **40**, 33–36.

Wu, M., Shanabrough, M., Leranth, C., and Alreja, M. (2000). Cholinergic excitation of septo-hippocampal GABA but not cholinergic neurons: Implications for learning and memory. *J. Neurosci.* **20**, 3900–3908.

Yaffe, K., Ettinger, B., Pressman, A., Seeley, D., Whooley, M., Schaefer, C., and Cummings, S. (1998). Neuropsychiatric function and dehydroepiandrosterone sulfate in elderly women: A prospective study. *Biol. Psychiatry* **43**, 694–700.

Yanase, T., Fukahori, M., Taniguchi, S., Nishi, Y., Sakai, Y., Takayanagi, R., Haji, M., and Nawata, H. (1996). Serum dehydroepiandrosterone (DHEA) and DHEA-sulfate (DHEA-S) in Alzheimer's disease and in cerebrovascular dementia. *Endocr. J.* **43**, 119–123.

Yoo, A., Harris, J., and Dubrovsky, B. (1996). Dose–response study of dehydroepiandrosterone sulfate on dentate gyrus long-term potentiation. *Exp Neurol.* **137,** 151–156.

Zimmerberg, B., Drucker, P. C., and Weider, J. M. (1995). Differential behavioral effects of the neuroactive steroid allopregnanolone on neonatal rats prenatally exposed to alcohol. *Pharmacol. Biochem. Behav.* **51,** 463–468.

Zou, L., Yamada, K., Sasa, M., Nakata, Y., and Nabeshima, T. (2000). Effects of sigma(1) receptor agonist SA4503 and neuroactive steroids on performance in a radial arm maze task in rats. *Neuropharmacology* **39,** 1617–1627.

NEUROSTEROIDS AND BEHAVIOR

Sharon R. Engel and Kathleen A. Grant

Department of Physiology and Pharmacology, Wake Forest University School of Medicine,
Winston-Salem, North Carolina 27157-1083

I. Introduction
II. Anxiety and Stress
III. Cognition
IV. Aggression
V. Sleep, Feeding, and Reinforcement
VI. Discriminative Stimulus Effects
References

Neurosteroid production may be a mechanism to counteract the negative effects of stress and return organisms toward homeostasis. Stress induces an increase in neurosteroid production. Neurosteroids affect two of the most widely distributed neurotransmitter and receptor systems in the central nervous system (CNS): γ-aminobutyric acid (GABA) and glutamate. This ability of this class of compounds to affect both the primary excitatory and the inhibitory systems in the CNS allows the modulation of a wide array of behaviors. For example, neurosteroids modulate anxiety, cognition, sleep, ingestion, aggression, and reinforcement. In general, neurosteroids that are positive modulators of N-methyl-D-aspartate receptors enhance cognitive performance and decrease appetite. Neurosteroids that are positive modulators of $GABA_A$ receptors decrease anxiety, increase feeding and sleeping, and exhibit a bimodal effect on aggression that may be secondary to effects on anxiety and cognition. Some data suggest that neurosteroids have reinforcing effects, which could affect their clinical utility. Drug discrimination studies are helping scientists to dissect more closely the receptor systems affected by neurosteroids at the behavioral level. © 2001 Academic Press.

I. Introduction

Neurosteroids appear to have evolved in mammals to mediate complex, conditioned behavioral processes that are tightly regulated by physiological processes, such as response to stressful events, cognition, anxiety, and aggression. The purpose of this chapter is to review the behavioral data implicating neurosteroids in each of these behavioral responses. A fundamental premise of this review is that neurosteroids are involved in behaviors that are important for the maintenance of homeostasis. Thus, a major emphasis is placed on behaviors associated with stressful conditions—for example, anxiolytic responses. In addition, the role of neurosteroids in other behavioral outcomes closely associated with stress is reviewed, including cognitive processes, drug reinforcement, aggression, and sleeping patterns. Finally, data are reviewed from drug discrimination procedures that examine the receptor-mediated activity of neurosteroids on a behavioral level and that help to verify that behavioral effects are mediated through direct modulation of ionotropic γ-aminobutyric acid type A ($GABA_A$) and N-methyl-D-aspartate (NMDA) glutamate receptor systems.

II. Anxiety and Stress

Stress is a term used rather freely to describe a reaction to events that results in a wide variety of measures showing deviation from normal values within mechanical, biological, and ecological systems. Stress is found in biological systems when there is disruption in homeostasis, a concept introduced by the French physiologist Claude Bernard (1865) to reflect the general principle of maintaining a constancy of the internal milieu. The primary physiological system that has evolved to maintain internal consistency is the endocrine system and in particular the hypothalamic-pituitary-adrenal (HPA) axis in mammals. The neurosteroids were first classified and identified by Hans Selye (1941) as representing a complementary endogenous steroid mechanism that rapidly responds to elements of stressful events and acts in concert with gonadal and adrenal steroids in the modulation of genetic expression in response to an activated HPA axis. There is now ample evidence that neurosteroids exert their effects through specific membrane-bound receptor mechanisms, most prominently $GABA_A$ and NMDA ionophores. The rapid ability to maintain a balance between excitation and inhibition of brain activity is clearly important in situations when fast and accurate decisions could mean the difference between an individual's survival or its death (i.e., flight or fight).

Although the brain possesses the enzymatic pathways necessary to synthesize a vast array of neurosteroids from cholesterol, in response to acute stressful events, endogenous production of neurosteroids fluctuates with adrenal function. In male rats exposed to brief ambient-temperature swim stress, there was a rapid (<5 min) and robust (4- to 20-fold) increase in the plasma and brain levels of allopregnanolone ($3\alpha,5\alpha$-TH PROG) (Purdy et al., 1991).

Levels of $3\alpha,5\alpha$-TH PROG in the brain (10–30 mM) were within the range of concentrations demonstrated to augment GABA-activated chloride currents in electrophysiological studies (Puia *et al.*, 1990). There was no significant level of $3\alpha,5\alpha$-TH PROG in the plasma of adrenalectomized male rats before or after swim stress, but this neurosteroid was detectable in the cerebral cortex of both groups. In addition, acute stress increased brain levels of $3\alpha,5\alpha$-TH PROG before its elevation in plasma. These results are consistent with some biosynthesis of $3\alpha,5\alpha$-TH PROG in the brain (Holzbauer *et al.*, 1985; Purdy *et al.*, 1991). Pregnenolone (PREG), progesterone (PROG), $3\alpha,5\alpha$-TH PROG, and deoxycorticosterone (DOC) also show a time-dependent increase in the cerebral cortex of rats following the acute stress of carbon dioxide inhalation (Barbaccia *et al.*, 1996b). Similarly, foot shock leads to a time-dependent increase of PREG, PROG, and $3\alpha,5\alpha$-TH DOC in the cortex (Barbaccia *et al.*, 1996a). These stress-induced increases in neurosteroids peak between 10 and 30 min. Thus, it appears that a consistent physiological response to stress is an increase in circulating and brain neurosteroid production.

The functional outcome of an increase in circulating and brain levels of neurosteroids is to counteract the short- and long-term effects of stress. Long-term effects of stress are usually associated with extreme, unavoidable stressors. In humans, postraumatic stress disorder can be considered a long-term effect. In animals, only one study has been directed at preventing the long-term effects of stress. This study used maternal deprivation, which result in heightened HPA responses to stress in adulthood (Patchev *et al.*, 1997). This effect on the HPA is blocked when $3\alpha,5\alpha$-TH DOC is administered concurrently with maternal deprivation, suggesting a neuroprotective effect of neurosteroids (Patchev *et al.*, 1997). Two other studies addressed the effect of neurosteroids on stress-induced neurochemical changes. Physical restraint acts as a stressor in rodents and leads to an increase in cortical and striatal dopamine. Pretreatment with $3\alpha,5\alpha$-TH DOC blocked this stress-induced dopamine accumulation in the prefrontal cortex but not in the striatum (Grobin *et al.*, 1992). Using foot shock as a stressor, $3\alpha,5\alpha$-TH DOC also blocked the stress-induced increase in acetylcholine in the hippocampus in a manner similar to the benzodiazepine midazolam (Dazzi *et al.*, 1996). Thus, neurosteroids are increased in response to stress and can prevent neurochemical changes associated with stress responses. It is hypothesized that neurosteroids may act as endogenous agents to counteract the effects of stress and may play a role in neuroprotection by modulating excessive responses to stress.

The effects of acute stress on neurosteroid production and the ability of neurosteroids to alter responses to acute stress have been the subject of many investigations. These studies have used a wide variety of procedures to measure stress and anxiety. A widely used procedure is the elevated plus-maze, which is based on the natural tendency of rodents to avoid open, exposed places. In this test, the rodent has the choice to hide in the walled arms of the maze or to explore the exposed, open arms of the maze. More time spent exploring the open arms and/or an increase in the number of times the rodent enters the open arms is believed to correspond with a lack of anxiety. The light–dark shuttle box also uses rodents' natural tendency to avoid open, brightly lit space. In this test, the animal has the choice of choosing a small, dark box or a large, lighted compartment. Decreased latency toward entering the dark compartment, increased time spent in the darkened side, and a decrease in the number of transitions between the two compartments all can be used as measures of anxiety. Rodents will also go through the motions of burying an object when they experience anxiety. Thus, an increase in burying behavior is also considered to be a measure of anxiety. Simiand *et al.* (1984) described a "staircase" procedure for measuring anxiety. In this procedure, the animal is placed on a small "staircase." An increase in the number of times the animal rears up on its hind legs is considered to be an increase in anxiety, whereas climbing is the activity component. Another test uses the anxiety produced by being placed in an environment where a mild electric shock is delivered. Under such circumstances, ongoing behaviors, such as responding on a lever, will be suppressed. The suppression of behavior is defined as punished responding and is often measured in the Geller–Seifter conflict test. Specifically, in a conflict procedure, there are two components with different environmental stimuli, such as differently colored lights. In the presence of one set of stimuli, the animal can respond on the lever, and receive food without any shock (the unpunished component). In the presence of another set of stimuli, when the animal responds on the lever, food is delivered but so is a shock under a random schedule (the punished component). An increase in punished responding is considered to be reflective of a decrease in anxiety or anxiolysis.

$3\alpha,5\alpha$-TH PROG has been reported to have anxiolytic properties in the mirrored chamber test, the Geller–Seifter test, and burying behavior (Fernandez-Guasti and Picazo, 1995; Carboni *et al.*, 1996; Brot *et al.*, 1997; Reddy and Kulkarni, 1997a). In the Geller–Seifter conflict test, $3\alpha,5\alpha$-TH PROG increased punished responding in a dose-dependent manner (Carboni *et al.*, 1996; Brot *et al.*, 1997). In the mirrored chamber test of anxiolysis, $3\alpha,5\alpha$-TH PROG decreased the latency to enter the chamber and increased the number of chamber entries as well as the total time spent in the chamber (Reddy and Kulkarni, 1997a). Consistent with an anxiolytic effect, treatment with $3\alpha,5\alpha$-TH PROG decreases burying behavior in response to a stressful event in rats (Fernandez-Guasti and Picazo, 1995; Picazo and Fernandez-Guasti, 1995). In addition, $3\alpha,5\alpha$-TH PROG (Zimmerberg *et al.*, 1994; Vivian *et al.*, 1997) and $3\alpha,5\alpha$-TH DOC (Patchev *et al.*, 1997) decreased ultrasonic vocalizations of rat pups that are separated from their dam. These behavioral changes are interpreted as decreases in anxiety. Genetic background can alter responses to neurosteroids. Although $3\alpha,5\alpha$-TH PROG was an effective anxiolytic in C57BL/6 and DBA/2 strains of mice, the C57BL/6 mice were more sensitive to the anxiolytic effects than were the DBA/2 mice (Finn *et al.*, 1997b). Either endogenous production or receptor modulation by neurosteroids may be a basis for individual differences in anxiety.

The endogenous neurosteroids, $3\alpha,5\alpha$-TH DOC and PROG, also exhibit anxiolytic properties in a number of behavioral tests. $3\alpha,5\alpha$-TH DOC decreased the number of rearing events in the staircase procedure (Pick *et al.*, 1997), increased the number of transitions in the light–dark shuttle box, and increased punished responding (Crawley *et al.*, 1986). PROG showed anxiolytic effects in the burying behavior and the elevated plus-maze (Bitran *et al.*, 1995; Picazo and Fernandez-Guasti, 1995). Other metabolites of PROG are also anxiolytic; 5α-Pregnane-$3\alpha,20\alpha$-diol, 5β-pregnane-$3\alpha,20\alpha$-diol, 5β-pregnane-$3\alpha,20\beta$-diol, and 5α-pregnane-$3\alpha,20\beta$-diol increased punished responding (Carboni *et al.*, 1996). Bitran *et al.* (1995) showed the anxiolytic effect of PROG was mediated by its metabolite at $GABA_A$ receptors. PROG pretreatment showed a direct correlation with the increase in blood $3\alpha,5\alpha$-TH PROG and the ability of GABA to increase ^{36}Cl flux through the $GABA_A$ channel in synaptosomes prepared from the brains of PROG-treated rats. Blockade of the enzymatic conversion of PROG to DH PROG and $3\alpha,5\alpha$-TH PROG blocked the anxiolytic effect of PROG. Furthermore, the intracellular progestin receptor antagonist, RU 38486, had no effect on anxiolysis, but anxiolysis was dependent on the functioning of $GABA_A$ receptors. Together, these data suggest that the mechanism of PROG-mediated anxiolysis is by the metabolism of PROG to neurosteroids, such as $3\alpha,5\alpha$-TH PROG, which then act as positive modulators of $GABA_A$ receptors. Furthermore, the efficacy of the neurosteroids as anxiolytics is similar to that of the prototypic anxiolytics, the benzodiazepines (Pick *et al.*, 1996; Reddy and Kulkarni, 1997a,b).

Synthetic neuroactive steroids, such as alphaxalone and Co 3-0593, that maintain the 3α-hydroxy-reduced pregnane configuration are also anxiolytic. Alphaxalone decreased the number of rearing events in the staircase procedure (Pick *et al.*, 1997), and Co 3-0593 increased punished responding (Wieland *et al.*, 1997). Furthermore, consistent with electrophysiological data of $GABA_A$-positive modulation by 3α-hydroxy-reduced pregnane neurosteroids, the 3β-substitutes of PROG were devoid of anxiolytic activity in the burying behavior test (Picazo and Fernandez-Guasti, 1995).

The receptor site of $3\alpha,5\alpha$-TH PROG to elicit anxiolysis has been investigated with various compounds that alter $GABA_A$-receptor activity. In the Geller-Seifter conflict test, Ro 15-4513, a partial inverse agonist at $GABA_A$

receptors, blocked the 3α,5α-TH PROG-induced anxiolysis (Brot et al., 1997). Picrotoxin and isopropylbicyclophosphate, which binds the picrotoxin site on the GABA$_A$ receptor, blocked the 3α,5α-TH PROG-induced anxiolysis in the Geller–Seifter conflict test and the mirrored chamber test (Brot et al., 1997; Reddy and Kulkarni, 1997a). Picrotoxin (Bitran et al., 1995; Reddy and Kulkarni, 1996) and bicuculline (Reddy and Kulkarni, 1996) also blocked the PROG-induced anxiolysis in the elevated plus-maze. However, bicuculline and picrotoxin had no effect on the suppression of ultrasonic vocalizations produced by 3α,5α-TH PROG in maternally deprived rat pups (Vivian et al., 1997). Together, these data suggest that positive modulation of GABA$_A$ receptors affects the anxiolytic activity of 3α,5α-TH PROG. However, the ability of GABA$_A$ inverse agonists or convulsants to attenuate the effects of 3α,5α-TH PROG is dependent on the behavioral procedure.

There have been mixed results with the ability of the benzodiazepine antagonist flumazenil to block 3α,5α-TH PROG-induced anxiolysis. Flumazenil pretreatment did not block the effects of 3α,5α-TH PROG in the mirrored chamber test, the Geller–Seifter conflict test (Brot et al., 1997; Reddy and Kulkarni, 1997a), or the suppression of ultrasonic vocalizations produced by 3α,5α-TH PROG in maternally deprived rat pups (Vivian et al., 1997). In addition, flumazenil did not block the PROG-induced anxiolysis in the elevated plus-maze (Reddy and Kulkarni, 1996). In contrast, flumazenil blocked the anxiolysis induced by 3α,5α-TH DOC, but not that by alphaxalone, in the staircase test (Pick et al., 1996, 1997), and 3α,5α-TH PROG induced a decrease in stress-induced burying (Fernandez-Guasti and Picazo, 1995).

One reason for the inconsistent effects of flumazenil in these studies could be related to dose. Flumazenil was not effective in blocking the anxiolytic effect of 3α,5α-TH PROG in a procedure that used a high dose of 3α,5α-TH PROG (8 mg/kg) relative to flumazenil (up to 6 mg/kg) (Brot et al., 1997) or a low dose of 3α,5α-TH PROG (0.5 mg/kg) with a low dose of flumazenil (2 mg/kg) Reddy and Kulkarni(1997a) Whereas when a low dose of 3α,5α-TH PROG (0.5 mg/kg) was used with higher doses of flumazenil (5 and 10 mg/kg), flumazenil was able to block the anxiolytic effect of 3α,5α-TH PROG (Fernandez-Guasti and Picazo, 1995). Thus, when flumazenil was able to block the 3α,5α-TH PROG-induced anxiolysis, the dose of flumazenil was 10–20 times that of the neurosteroid.

An alternative interaction of anxiolytic neurosteroids with the benzodiazepines is through the diazepam-binding inhibitor (DBI), which binds to its receptor on the mitochondria and leads to an increases in availability of cholesterol to the neurosteroid biosynthetic pathways (Papadopoulos et al., 1992; Korneyev et al., 1993). The DBI agonist, 4′chlordiazepam, also leads to an increase in neurosteroid synthesis (Papadopoulos *et al.*, 1992). 4′Chlordiazepam decreased the ability of immobilization stress to increase nocioception in the tail-flick test (Reddy and Kulkarni, 1996) and the DBI receptor antagonist, PK 11195; bicuculline, but not flumazenil, blocked this effect. These data suggest that the anxiolytic effect was a result of, the production of neurosteroids and their subsequent action at nonbenzodiazepine sites on $GABA_A$ receptors

In summary, the complete $GABA_A$-channel blockade with picrotoxin consistently blocks the neurosteroid-induced anxiolysis, suggesting that the function of the $GABA_A$ receptors is required for the action of the neurosteroid. Alternatively, picrotoxin is a highly efficacious agent, and the loss of anxiolytic action in its presence may be the result of the additive effects of the two agents (picrotoxin and the neurosteroid) at different sites. Flumazenil binds to the benzodiazepine site on $GABA_A$ receptors and consistently blocks the anxiolytic effect of benzodiazepines. However, the ability of flumazenil to block the anxiolytic effects of neurosteroids is inconsistent. Electrophysiological data suggest that flumazenil is unable to alter the neurosteroid-induced increase in $GABA_A$ current (Le Foll *et al.*, 1997). Thus, when flumazenil and neurosteroids are effective in anxiolytic measures, the effects may also be the result of additive effects at separate sites.

The sulfated forms of the 3α-reduced neurosteroids are generally negative modulators of $GABA_A$ receptors and, as one would predict, they are anxiogenic 3B-Hydroxyandrost-5-en-17-one sulfate (DHEAS) increased anxiety measures in the mirrored chamber test (Reddy and Kulkarni, 1997a,b), and PREG was anxiogenic in the elevated plus maze (Melchior and Ritzmann, 1994). 3B-Hydroxypregn-5-en-20-one sulfate (PREGS) has a biphasic effect on anxiety-producing anxiolysis at low doses (0.1 μg/kg), becoming anxiogenic at higher doses (≥ 1 μg/kg) (Melchior and Ritzmann, 1994). This is in agreement with electrophysiological data showing PS-enhanced $GABA_A$ currents at low doses (10 nM) and a decrease in these currents at higher doses (10 μM) (Puia *et al.*, 1993). In another test, PS (0.5 and 2 mg/kg) was anxiolytic in the mirrored chamber test, and this effect was insensitive to picrotoxin and flumazenil pretreatment, suggesting that the effect was via a site other than the benzodiazepine and $GABA_A$ receptor (Reddy and Kulkarni, 1997a).

Overall, these data show that the reduced-pregnane neurosteroids are anxiolytic and that their mechanism of action is by modulation of the $GABA_A$ receptors. Antagonism studies have yet to elucidate the site of action of neurosteroid modulation of $GABA_A$ receptors. Multiple factors can affect neurosteroid action on anxiety and, therefore, it is important to remember that anxiety measures can vary widely among laboratories. Stress and anxiety are intimately intertwined, and data suggest that neurosteroids that are positive modulators of GABA$_A$ receptors may be endogenous agents to combat the negative effects of stress and anxiety. Future studies that investigate the effects of chronic stress on neurosteroid production and/or efficacy will be helpful in addressing the question of the role of neurosteroids and stress.

III. Cognition

Cognition is the operation by which an organism is aware of objects or conditions, and it includes the processes of learning and memory. One common method of testing learning and memory is to pair an aversive stimulus with an environmental stimulus. The aversive stimulus is generally a mild foot shock. The animal learns to associate the foot shock with a particular environment. These tests also use the animals', often rodents', natural tendency to explore and to hide. In the step-down, passive avoidance task, the animal is placed on a platform above a grid floor through which a mild electric current can be passed. When the animal steps down from the platform onto the grid floor, it receives a mild electric shock. Later, usually 24–48 h later, the animal is again placed on the platform, and an increase in the time before the animal steps down from the platform onto the grid floor is used as a measure of learning or memory retention. The step-through, passive avoidance task is similar to that of the step-down avoidance task in that it is usually used as a model of single trial learning. The animal is placed in a lighted area of a box with a darkened chamber available for the animal to enter. Upon entering the darkened chamber, the animal receives a mild foot shock. The following day, the animal is again placed in the chamber, and the latency to enter the darkened compartment is used as a measure of memory. In contrast to the passive avoidance procedures, active avoidance of a mild foot shock is performed in the T- or Y-maze test. In the T- or Y-maze, shock avoidance task, the animal must learn to run into one of the arms of the maze avoid the shock. The impending shock can be signaled by the opening of a door that allows access to the rest of the maze or a sound. Consecutive training trials are usually given on the same day or on consecutive days until the animals are consistently able to avoid the shock. Testing occurs at a later time, and the ability of the animal to recall how to avoid the shock is measured as memory retention. The number of trials required to avoid the shock consistently in the training phase can be used as a measure of memory acquisition. The Morris water maze uses the animal's natural aversion to water. Although rodents are efficient swimmers, they will avoid water when possible. In the Morris water maze, the animal is placed in a tub of water with a hidden platform submerged just under the surface of the water. The animal swims until it finds the platform and is able to get out of the water. The time required or distance traveled to find the platform on subsequent trials is used as a measure of memory function.

Another Y-maze uses the natural tendency of a rodent to explore a novel area. The animals are allowed to explore two arms of a Y-maze and, at a later time, are returned to the maze and allowed to explore all three arms. The more time the animal spends exploring the newly exposed arm of the maze over that of the other two familiar arms is used as a measure of recall. Other tasks use food reinforcement in mildly food-restricted animals. In the Go–No Go visual discrimination task with food reinforcement, animals are trained to run down an alley to receive food reinforcement within a short (e.g., 30 sec) time period. Two separate alleys are trained, but only one is reinforced. An increase in completion of the task on the reinforced side is used as a measure of learning. In the lever-press task, food-restricted mice are trained to press a lever for food reinforcement, and retention of the task is measured as the ability of the animal to get the reinforcers at a later time.

Neurosteroids enhance learning and memory in these tasks. In particular, the sulfated forms of the neurosteroids are able to enhance cognitive performance. This action is thought to be mediated primarily by positive modulation of NMDA receptors and/or negative modulation at $GABA_A$ receptors. The sulfated forms DHEAS and PREGS, enhance cognitive performance. DHEAS enhanced the acquisition of learning in the T-maze avoidance and step-down avoidance tasks (Flood et al., 1992). In this study, the neurosteroids were administered immediately after the acquisition phase of the task, and testing commenced 1 week after training. When administered immediately or at 30 or 60 min after training, DHEAS significantly improved retention of the tasks after a 1-week interval. Route of administration had no effect on the ability of DHEAS to improve memory retention (Flood et al., 1988). DHEAS given subcutaneously, intraventricularly, or chronically (1 week) in the drinking water was able to improve memory retention. Although mice receiving DHEAS in the drinking water continuously for 1 week showed an improved memory retention, the time required to learn the avoidance tasks was unchanged. These data suggest a facilitory role of DHEAS in memory retention but not acquisition. PREGS enhanced performance in the cognitive Go–No Go visual discrimination task (Meziane et al., 1996) and an active avoidance task (Isaacson et al., 1995). The memory-enhancing effects of the PREGS and other neurosteroids may depend on the timing of their availability during memory processing. PREGS and the synthetic anabolic neurosteroid ethylestrenol improved memory retention in an active avoidance task when administered immediately after training but not before testing (24 and 48 h after training) (Isaacson et al., 1995). Other studies have not been able to demonstrate this memory-enhancing effect of PREGS (Mathis *et al.*, 1994, 1996; Li *et al.*, 1995; Maurice *et al.*, 1997). The reason for this discrepancy is unclear, as there was overlap in the timing, dose, and route of neurosteroid administration as well as in the type of testing employed.

Other neurosteroids and steroids have memory-enhancing effects (Flood *et al.*, 1988, 1992). Specifically, PREG, DHEA, aldosterone, testosterone, and dihydrotestosterone enhanced memory retention in a foot shock avoidance task in mice (Flood *et al.*, 1992). However, in this study, the steroids were injected intraventrically in dimethyl sulfoxide (DMSO), which alone had a slight impairment on retention of the shock avoidance task (Flood *et al.*, 1988). Thus, these steroids may have been blocking the effects of DMSO-induced memory deficits as opposed to having a direct action on memory enhancement. However, the ability of a substance to block memory deficits induced by other substances does allow investigation of possible mechanisms of action and may suggest treatment strategies for the reversal or blockade of memory deficits.

Scopolamine and related anticholinergic agents impair performance in a wide variety of learning tasks and are often used as amnestics to test the promnestic effects of other agents, in a shock-avoidance task and in a Go–No Go visual discrimination test, DHEAS (Li *et al.*, 1997) and PREGS (Meziane *et al.*, 1996) blocked the amnesic effects of scopolamine. Blockade of the enzymatic breakdown of DHEAS further enhanced the ability of DHEAS to block scopolamine-induced amnesia (Li *et al.*, 1995, 1997). Although the steroid sulfatase inhibitor p-*O*-(sulfamonyl)-*N*-tetradecanoyl tyramine, results in a significant inhibition of sulfatase activity within 24 h, there was no improvement in the learning task when the sulfatase was administered without DHEAS (Li *et al.*, 1997). However, when the enzyme inhibitor estrone-3-*O*-sulfamate was given chronically (10 days), this treatment alone was sufficient to block scopolamine-induced amnesia (Li *et al.*, 1995). This suggests that the prolonged inhibition of DHEAS breakdown allowed this (or other) neurosteroid(s) to accumulate to promnestic concentrations. PREGS causes a dose-dependent release of acetylcholine that correlates with doses of PREGS that are promnestic (Darnaudery *et al.*, 2000). Thus, the mechanism by which neurosteroids are able to enhance cognitive function via NMDA receptor activation may be through the release of acetylcholine.

PREGS also blocked learning deficits induced by competitive NMDA antagonists (Mathis *et al.*, 1994, 1996; Maurice *et al.*, 1997). The competitive NMDA antagonist 3-(($\pm$)-2-carboxypiperazin-4-yl)-propyl-1-phosphonic acid (CPP) induced memory deficit in a step-through, passive avoidance task (Mathis *et al.*, 1994). CPP administered 15 min before the training session resulted in a dose-dependent decrease in the latency to enter a darkened chamber, and this effect was completely blocked by the co-administration of PREGS. Similar results were obtained with the competitive NMDA antagonist D-2-amino-5-phosphonovalerate (D-AP5) (Mathis *et al.*, 1996), which increased the number of errors in a Y-maze shock avoidance task and in a lever-press task for food reinforcement. PREGS completely reversed the learning deficits induced by D-AP5. PREGS was able to block the NMDA-antagonist-induced memory deficits whether given before or after the training sessions (Mathis *et al.*, 1994, 1996).

Other studies suggest that neurosteroid modulation of σ_1 receptors is responsible for the cognition-enhancing effects of neurosteroids. DHEAS blocked learning impairment induced by the NMDA antagonist dizocilpine in the Y-maze and passive avoidance tasks (Maurice *et al.*, 1997). The σ-receptor antagonist BMY-14802 blocked the ability of DHEAS to improve dizocilpine-induced memory impairment; however, BMY-14802 alone was unable to effect dizocilpine-induced impairment (Maurice *et al.*, 1997). Haloperidol downregulates σ-receptor expression (Maurice *et al.*, 1997), and chronic treatment with haloperidol (4 mg/kg for 7 days) attenuated the ability of DHEAS to block the dizocilpine-induced deficit but alone had no effect on memory performance (Maurice *et al.*, 1997). These data suggest that DHEAS may be acting at σ-receptors or that σ-receptors are required for the action of DHEAS at NMDA receptors.

The σ_1-receptor agonists enhance cognitive performance, and the alteration of endogenous neurosteroids alters the number of σ_1-receptor binding sites in the cortex and the hippocampus (Phan *et al.*, 1999). Blockade of the conversion of PREG to PROG with the 3β-hydroxysteroid-dehydrogenase inhibitor trilostane leads to an accumulation of PREG and an increase in σ_1 binding sites, whereas blockade of the metabolism of PROG with the 5α-reductase inhibitor finasteride leads to an accumulation of PROG and a decrease in σ_1 binding sites (Phan *et al.*, 1999). Neither treatment alone had any effect on dizocilpine-induced learning impairment, suggesting that the increase in cognition was due to the σ_1-receptor. In addition, the anti-amnesic effects of a σ_1-agonist, PRE-084, were further enhanced in the trilostane-treated animals but ineffective in the finasteride-treated group. Thus, neurosteroids modify the expression of σ_1-receptors in the cortex and the hippocampus, and activation of σ_1-receptors allows the memory-impairing effects of NMDA antagonism to be circumvented.

The decline in neurosteroids with age and the subsequent decrease in acetylcholine release may explain some age-related cognitive dysfunction. PREGS is decreased in the hippocampus of aged rats, and this decrease is positively correlated with memory impairment in learning tasks (Vallée *et al.*, 1997). The ability of aged rats (24 months) to find the platform in the Morris water maze was positively correlated with a higher PREGS content in the hippocampus. PREGS either intraperitoneally or intraventricularly, significantly increased exploration of the novel arm of the maze 6 and 7 h post-injection, and this effect was lost at Day 10 (Vallée *et al.*, 1997). DHEAS was also able to improve memory performance in aged (18 and 24 months) mice on a shock avoidance task (Flood and Roberts, 1988). Mice received DHEAS immediately after avoidance training and were tested 1 week later. The aged mice took longer to learn the shock avoidance task than young (2-month-old) mice. Upon testing 1 week later, the aged mice that had received DHEAS performed significantly better than aged mice that had received saline. The performance on retention testing of aged mice that had received DHEAS was equal to that of young (2-month-old) mice. Thus, DHEAS did not improve acquisition of the task in aged mice; however, memory retention was enhanced.

A common feature of the cognition enhancing effects of the neurosteroids was the U-shaped dose–response curve, with the low to medium doses being more efficacious than the higher doses (Flood *et al.*, 1988, 1992; Mathis *et al.*, 1994; Meziane *et al.*, 1996). The loss of effect at higher doses may reflect the conversion of the neurosteroids to other compounds. At a high enough concentration, enzymes with high K_m values may be able to divert the neurosteroid into alternate pathways whose products may have opposite effects on cognition. There may be an optimal "tone" at which cognition is enhanced, and this tone may be a reflection of the glutamate and GABAergic levels of activation modulated by these neurosteroids.

Several electrophysiological measures in the hippocampus also support the role of sulfated neurosteroids in the facilitation of learning and memory. DHEAS increases primed-burst potentiation (Diamond *et al.*, 1996) and long-term potentiation (LTP) (Yoo *et al.*, 1996). The most widely studied neural correlate of learning and memory is LTP. It is a lasting increase in amplitude of a population spike seen after a tetanizing or an above-threshold stimulation. Threshold stimulation results in a lasting increase in amplitude of a population spike identified as primed-burst potentiation. Primed-burst potentiation has been suggested to be a mechanism of induction of neural plasticity. Doses of DHEAS that induced increases in primed-burst potentiation (Diamond *et al.*, 1996) (24 and 48 mg/kg) also increased memory retention (Li *et al.*, 1995, 1997; Flood and Roberts, 1998). As reported in behavioral tests, DHEAS exhibited an inverted U-shaped dose–response curve. Intermediate doses (24 and 48 mg/kg, subcutaneously) are more effective than low (6 mg/kg, subcutaneously) or higher doses (96 mg/kg, subcutaneously) (Diamond *et al.*, 1996). Although Diamond *et al.* (1996) showed no effect of DHEAS on LTP, there is a reported enhancement of LTP with 10–30 mg/kg DHEAS. The excitatory post-synaptic potential (EPSP) slope and the population spike amplitude component of LTP were dose dependently increased by DHEAS (Yoo *et al.*, 1996). These effects on population spike amplitudes were correlated with an increase in firing rate of glutamatergic interneurons in the CA1 and the release of GABA-mediated inhibition in the dentate gyrus and CA1 (Steffensen, 1995; Meyer et al., 1999). The release of GABA inhibition led to entraining of neurons to a θ rhythm that is also correlated with learning and memory. DHEAS alters the electrical correlates of learning and memory by activating NMDA receptors and/or by inhibiting the GABAergic interneurons.

In summary, neurosteroids that are positive modulators of NMDA and σ_1-receptors and/or negative modulators at $GABA_A$ receptors enhance cognitive performance. This is in agreement with data that shows that NMDA and σ_1-agonists also enhance cognition, whereas antagonism of these sites leads to learning impairment. The memory-promoting effects of the sulfated neurosteroids PREGS and DHEAS in cognition are supported by behavioral and electrophysiological parameters of cognition. Although the exact role of neurosteroids in cognition is not fully understood, they are powerful endogenous modulators of cognition.

IV. Aggression

Aggressive behavior is important for species survival, but it must be tightly regulated to avoid adverse harm to the individual and the social group. Aggression is classified in animal models into two distinct types of behavior, predatory and affective aggression. Predatory aggression is interspecific, normally related to feeding, and is accompanied by minimal vocalization, stalking posture, and lethally directed attacks (e.g., at the back of the prey's neck). Affective aggression is intraspecific; it involves intense autonomic arousal, vocalizations, and threatening and defensive postures. Defensive aggression can be initiated by nocioceptive stimuli (direct painful stimulus) or exteroceptive stimuli (resident–intruder interactions).

By far, the receptor system most widely implicated in both predatory and affective aggression is serotonin (Olivier et al., 1990). Nearly all pharmacological manipulations that either increase or decrease 5-HT neurotransmission can inhibit offensive aggression (Miczek et al., 1995). These mixed results probably reflect differential effect on neural circuitry. For example, stimulation of the hypothalamic serotonergic system increases aggression, whereas ablation of amygdala serotonin decreases aggression (File et al., 1981). Furthermore, microdialysis data show decreased prefrontal cortex 5-HT levels in aggressive rats (van Erp and Miczek, 1997). The serotonin subtype most implicated in animal models of affective aggression is the $5\text{-}HT_{1B}$ receptor, autoreceptors that result in decreased 5-HT release. A class of substituted phenlypiperazine analogs that display remarkable antiaggressive activity in animal models have been termed serenics (Olivier *et al.,* 1990). Serenics specifically reduce offensive behavior without resulting in sedation, muscle relaxation, or motor stimulation. Lack of 5-HT_{1B} receptors in genetically engineered knockout mice results in high basal levels of aggression (Saudou *et al.,* 1994). Two serenics with the greatest specificity are TFMPP and RU 24969, phenlypiperazines with modest selectivity and high affinity for the 5-HT_{1B} receptor. The 5-HT_{1B}-receptor system has not been specifically studied in the context of $GABA_A$ mediated aggression or antiaggression; however, drug discrimination data show that 5-HT_{1B}-receptor activation is a component of the discriminative stimulus effects of the $GABA_A$-positive modulators ethanol and pentobarbital (Grant *et al.,* 1997; Bowen and Grant, 1999).

The effect of $GABA_A$-positive modulators on aggression is dose dependent, with lower doses increasing aggressive acts and higher doses decreasing aggression, probably as a result of sedation (Blanchard *et al.,* 1987, 1993; Miczek *et al.,* 1993, 1995). For example, low-dose effects of ethanol increase aggression (Weertz and Miczek, 1996; van Erp and Miczek, 1997). The heightened aggressive behavior occurs as prolonged "bursts" of aggressive acts and may be related to miscommunication between the aggressive animal and the opponent (Miczek *et al.,* 1997). However, increased aggression is not found in every animal tested, and a proportion of the population shows reduced aggression with the same doses of ethanol as animals showing aggression (van Erp and Miczek, 1997). Thus, there appears to be a large degree of individual variability in aggressive responses following treatment with $GABA_A$-positive modulators, an effect that is highly dependent on environmental circumstances.

Few studies have directly addressed the effects of neurosteroids on aggressive behavior. One of these investigation found the PROG derivative $3\alpha5\alpha$-TH PROG enhanced aggressive behavior toward intruders at low doses (10 mg/kg) and suppressed attack bites at higher doses (30 mg/kg) in mice (van Erp and Miczek, 1997). This bitonic effect was similar to the actions of benzodiazepines and ethanol in the mouse resident/intruder assay. In a different assay, the androgenic neurosteroid 3α-androstanediol also shows a bitonic effect on aggressive behavior exhibited during mating in female, ovariectomized rats (Frye *et al.,* 1996). Interestingly, DHEA inhibits aggressive behavior of castrated males in a resident/intruder assay even though this steroid has no efficacy as a $GABA_A$-positive modulator. Apparently, the antiaggressive effects is due to DHEA-induced decreases of PREGS concentrations and, presumably, the modulation of aggressive behavior through negative modulation of $GABA_A$ receptors (Robel *et al.,* 1995). Finally, offensive aggression in male mice is genetically correlated with the concentrations of the microsomal enzyme steroid sulfatase (STS) across 11 inbred strains of mice (Le Roy *et al.*, 1999). High STS activity is correlated with high aggression, and this correlation suggests a functional relationship, presumably through an increase in sulfation of neurosteroids, particularly DHEAS. Clearly, our understanding of the role that neurosteroids play in modulating aggressive behaviors is minimal. In addition to focusing on $GABA_A$ receptor systems, further studies should focus on the interaction between serotonin neurotransmission involved in aggressive behavior and neurosteroid activity.

V. Sleep, Feeding, and Reinforcement

Neurosteroids are involved in other regulatory behaviors, such as sleep, ingestion, and reinforcement. However, the number of studies that address these aspects of neurosteroid modulation of mammalian behavior are limited. The few that are available are reviewed here but are presented in abbreviated form. Because the database is not extensive, the conclusions that can be drawn about receptor mechanisms or external events that alter possible endogenous regulation by the neurosteroids must also be very limited.

Sleep is a behavior that is integral to maintaining the homeostasis of all vertebrates. The benzodiazepines and nonbenzodiazepines, such as zolpidem and zopiclone, are prescribed as hypnotics. Although these drugs effectively induce and increase sleep, there are some undesirable side effects, such as rebound wakefulness, decreased rapid eye movement (REM), impaired locomotor function, and impaired memory performance. In general, neurosteroids that are positive modulators of $GABA_A$ receptors are emerging as effective sleep aids without some of the undesirable side effects seen with currently available hypnotics. This effect is observed at doses approximately 10 times that for anxiolysis.

$3\alpha,5\beta$-TH PROG increased non-REM sleep without altering REM sleep in rats with no rebound wakefulness (Edgar *et al.*, 1997). $3\alpha,5\beta$-TH PROG showed less locomotor impairment at doses that induce hypnosis as compared to the benzodiazepine ligands. PREGS increased paradoxical sleep without any effects on wakefulness or slow-wave sleep in rats (Darnaudery *et al.*, 1999). In contrast to the memory-impairing effects of the benzodiazepine hypnotics, PREGS showed an enhanced memory performance at doses that increased paradoxical sleep. PREG improved the sleep efficiency index in human subjects (Steiger *et al.*, 1993). It also decreased intermittent wakefulness and slow-wave sleep without any change in REM sleep or blood cortisol and growth hormone (Steiger *et al.,* 1993). The overall sleep profile was more comparable to the nonbenzodiazepine hypnotics than the benzodiazepines. PROG decreases sleep latency; increases non-REM sleep; decreases pre-REM sleep; and, at the highest doses, decreases REM sleep (Steiger *et al.,* 1998; Lancel, 1999). $3\alpha,5\alpha$-TH PROG shortened sleep latency and increased pre-REM sleep. $3\alpha,5\alpha$-TH DOC also shortened sleep latency and increased non-REM sleep (Lancel, 1999). Overall, the neurosteroids appear to be effective sleep aids without changing sleep architecture and with less locomotor impairment. Although there remains many confounds in the study of the neurosteroids and sleep, these compounds may prove to be superior sleep aids as compared to those currently available.

Ingestion is a behavior common to all organisms, but it has evolved in mammals as a complex interaction of brain circuitry, endocrinological modulation, and gustatory neural and paracrine regulation. $3\alpha,5\alpha$-TH PROG, $3\alpha,5\alpha$-TH DOC, and alphaxalone increased feeding-related responses in *Hydra vulgaris,* one of the most primitive organisms to have a nervous system (Concas *et al.,* 1998). This suggests that neurosteroid regulation of feeding is a mechanism highly conserved through phylogeny. PROG, $3\alpha,5\alpha$-TH PROG, $3\alpha,5\beta$-TH PROG, alphaxalone, and the DBI receptor agonist 4'chlordiazepam increased food intake in rodents (Chen *et al.,* 1996; Reddy and Kulkarni, 1998, 1999). The effect of PROG, $3\alpha,5\alpha$-TH PROG, and 4'chlordiazepam on feeding was sensitive to picrotoxin but not flumazenil (Reddy and Kulkarni, 1998). Chen *et al.* (1996) reported that increased feeding induced by $3\alpha,5\beta$-TH PROG and alphaxalone, but not $3\alpha,5\alpha$-TH PROG, was sensitive to picrotoxin and all three were insensitive to flumazenil. Together, these data suggest that the effect of these neurosteroids on feeding is through nonbenzodiazepine-positive modulation of $GABA_A$ receptors. In contrast, Higgs and Cooper (1998) report no direct effect of $3\alpha,5\beta$-TH PROG on food intake or feeding architecture. However, the time spent eating a novel food was increased in response to $3\alpha,5\beta$-TH PROG, suggesting that the anxiolytic affect of the neurosteroid can affect some measures of feeding behavior (Higgs and Cooper, 1998). As opposed to neurosteroids that are positive modulators of $GABA_A$ receptors, PREGS and DHEAS produce an anorectic response (Chen *et al.,* 1996; Reddy and Kulkarni, 1998), suggesting that positive modulation of NMDA receptors repress feeding. Other factors, such as sex and menstrual cycle, may effect the ability of neurosteroids to regulate feeding behavior. Although $3\alpha,5\alpha$-TH PROG increased food intake in both male and female rats, the intake was more enhanced in the males (Reddy and Kulkarni, 1999). Furthermore, the female rats showed an effect of menstrual cycle phase and the ability of $3\alpha,5\alpha$-TH PROG to increase food intake (Reddy and Kulkarni, 1999), with $3\alpha,5\alpha$-TH PROG being most efficacious during the diestrous phase. The ability of DHEAS to inhibit feeding was independent of sex and menstrual cycle phase (Reddy and Kulkarni, 1999). Overall, these data support a role for neurosteroids in the regulation of feeding behavior, with neurosteroids that are positive modulators of $GABA_A$ receptors having a facilitating role and with the sulfated neurosteroids that are active at glutaminergic receptors being anorectic.

Finally, neurosteroids can themselves modulate processes of reinforcement. That is, some neurosteroids can function as reinforcers and maintain behaviors directed at acquisition and repeated self-administration. In this respect, particular neurosteroids share a potential abuse liability with barbiturates, benzodiazepines, and alcohol. Two procedures have been used to characterize the reinforcing effects of neurosteroids, the place preference procedure and self-administration procedures. The place preference paradigm measures the rewarding properties of a drug by pairing the drug effects with a particular environment. Place conditioning studies suggest that neurosteroids can alone act as incentive stimuli. Mice developed place preference for $3\alpha,5\alpha$-TH PROG in a dose-dependent fashion (Finn *et al.*, 1997a). Doses that produced place preference also increased spontaneous locomotor activity, suggesting that the low-dose-activating effects of the neurosteroid were producing the reinforcement. $3\alpha,5\beta$-TH PROG increased ethanol self-administration in rats (Janak *et al.*, 1998), suggesting that the neurosteroid may be able to enhance the reinforcing properties of other drugs of abuse. Alternatively, the neurosteroid may have blocked some of the rewarding properties of the ethanol. If this were the case, then a higher dose of ethanol would be required for the same effects to be experienced without the neurosteroid. $3\alpha,5\beta$-TH PROG was reinforcing in rhesus monkeys trained to press a lever for an injection of the barbiturate methohexital (0.01 mg/kg). $3\alpha,5\beta$-TH PROG (0.01–0.1 mg/kg, intraventricularly) maintained responding above that for saline; however, the reinforcing properties of $3\alpha,5\beta$-TH PROG was not robust in this assay (Rowlett *et al.*, 1999). It is important to recognize that the reinforcement profile described in this assay was on a barbiturate background. Thus, a different reinforcement profile for this neurosteroid may obtained in drug naïve animals. Overall, these data suggest that the neuroactive steroids that are positive modulators of $GABA_A$ receptors can act as reinforcers.

VI. Discriminative Stimulus Effects

Discriminative stimuli are events that specifically co-vary with the availability of reinforcement. Most commonly, discriminative stimuli are external environmental events, such as lights, tones, and smells that are perceived through sensory mechanisms and associated with the delivery of primary reinforcers, such as food. However, internally produced (interoceptive) stimulus effects of psychoactive drugs can also be conditioned to function as discriminative stimuli. In simple drug discrimination studies, the animal is trained, through differential reinforcement, to engage in a particular behavior in the presence of the internal effects of the drug and to engage in a different behavior in the absence of the internal effects of the drug. When given a novel drug or dose, the amount of responding that is appropriate to the drug training condition is then used as a measure of the similarity with the pharmacological actions of the training drug. Thus, the discrimination paradigm provides a measure of the association between interoceptive sensations and observable behaviors (Preston and Bigelow, 1991).

Drug discrimination procedures provide one of the most powerful avenues of research for characterizing the pharmacological aspects of the drug in relation to behavior. In particular, the reliable baseline of behavior produced by drug discrimination procedures allows a systematic approach to characterizing the influence of pharmacological variables (Colpaert, 1986). These data demonstrate that the discriminative stimulus effects of drugs vary along quantitative and qualitative dimensions and have characteristics indicative of receptor-mediated activity (Holtzman, 1990). The procedure has proved to be a reliable and valuable tool for screening substances for abuse potential, characterizing potential antagonists, identifying active metabolites, and establishing structure–activity relationships of psychoactive substances. The specificity of drug discrimination procedures is perhaps best illustrated by the ability of the procedure to separate the effects of ligands at the various sites on the $GABA_A$-receptor complex. Specifically, muscimol and 4,5,6,7-tetrahydroisoxazolo [4,5-c]pyrindin-3-ol (THIP) do not substitute for midazolam, diazepam, pentobarbital, or ethanol (Ator and Griffiths, 1989; Grech and Balster, 1993; Shelton and Balster, 1994). Likewise, pentobarbital and midazolam resulted in partial substitution for muscimol and THIP (Grech and Balster, 1997). These data imply that positive modulators of $GABA_A$ receptors produce stimulus effects that are fundamentally different from those produced by direct GABA agonists. In contrast to the GABA-site agonists, $GABA_A$-positive modulators, such as barbiturates and benzodiazepines, produce similar discriminative stimulus effects. However, the discriminative stimulus effects of benzodiazepines and barbiturates are not interchangeable. For example, benzodiazepines more likely to substitute for barbiturates compared to the ability of barbiturates to substitute for benzodiazepines (defined as an asymmetrical, or one-way, generalization (Ator and Griffiths, 1989, 1997). The asymmetrical generalizations between the different classes of positive $GABA_A$-receptor modulators are believed to be the result of different pharmacological mechanisms of action at $GABA_A$-receptor channels (Rowlett and Woolverton, 1996).

Molecular, biochemical, and electrophysiological data clearly show that benzodiazepines interact with a more discrete subset of $GABA_A$ receptors, based on subunit composition, compared with barbiturates (see Grobin et al., 1998). Thus, the asymmetrical generalization between barbiturates and benzodiazepines could be indicative of the discriminative stimulus effects of barbiturates encompassing receptor activity at all $GABA_A$ receptors that mediate the benzodiazepine stimulus effect as well as additional benzodiazepine-insensitive $GABA_A$ receptors. Within the benzodiazepine-sensitive receptors, discriminative stimulus effects reflect a further subdivision. Diazepam-sensitive receptors can be divided into BZ1 receptors, with high affinity for zolpidem, and BZ2 receptors with low affinity for zolpidem. Zolpidem results in only partial substitution in chlordiazepoxide discrimination (Depoortere et al., 1986), and benzodiazepines result in partial substitutions for zolpidem (Sanger and Zivkovic, 1986; Sanger, 1987). Differentiation among benzodiazepines, barbiturates, and BZ1-selective ligands can be further demonstrated with benzodiazepine vs barbiturate discriminations (Ator and Kautz, 2000). Thus, depending on the training drug and conditions, drug discrimination procedures are sensitive enough to differentiate BZ1 ligands from benzodiazepines that have activity at both BZ1 and BZ2 receptor subtypes.

Discrimination studies with neurosteroids have only recently been explored to any great extent, although PROG was one of the first drugs established as a discriminative stimulus (Stewart et al., 1967). In the only study reported of discrimination trained with a neurosteroid, $3\alpha,5\beta$-TH PROG, both pentobarbital and diazepam substituted, demonstrating cross generalization among these three compounds (Vanover, 1997). Taken together, these data suggest similarities between the receptor mechanisms of neurosteroids, barbiturates, and benzodiazepines; however, the overlap is not complete. The neurosteroid $3\alpha,5\alpha$-TH DOC produces midazolam-like (Deutsch and Mastropaolo, 1993) and diazepam-like (Ator et al., 1993) discriminative stimulus effects but does not produce lorazepam-like discriminative stimulus effects in rats (Ator et al., 1993). Both $3\alpha,5\alpha$-TH DOC and $3\alpha,5\alpha$-TH PROG produce pentobarbital-like discriminative stimulus effects in rats (Ator et al., 1993; Bowen et al., 1997; Bowen and Grant, 1999). Most extensively studied have been the effects of neurosteroids to produce ethanol-like discriminative stimulus effects. In general, neurosteroids that are positive modulators of $GABA_A$ ionophores substitute for ethanol in rats, including $3\alpha,5\alpha$-TH DOC, $3\alpha,5\alpha$-TH PROG, $3\alpha,5\beta$-TH PROG, and $3\beta,5\beta$-TH PROG (Bowen et al., 1999a). The finding that $3\beta,5\beta$-TH PROG can produce an ethanol-like discriminative stimulus effect is somewhat surprising given the low-efficacy nature of this neurosteroid at the $GABA_A$-receptor complex. In monkeys trained to discriminate alcohol, $3\alpha,5\alpha$-TH PROG and $3\alpha,5\beta$-TH PROG also produce ethanol-like effects; however, $3\beta,5\beta$-TH PROG fails to substitute for alcohol (K. A. Grant, unpublished data). The basis of this apparent species difference is not known and represents one of the few species differences in the characterization of ethanol's discriminative stimulus effects with $GABA_A$ ligands (Grant et al., 2000).

Macaque monkeys have also been used to show that the potency of $3\alpha,5\alpha$-TH PROG in producing discriminative stimulus effects varies with menstrual cycle phase. This finding has generally been attributed to endogenous levels of PROG and PROG derivatives associated with ovarian function. In nonstressed female rats, there is an estrus-cycle-dependent ovarian content and ovarian secretion of $3\alpha,5\alpha$-TH PROG (Holzbauer et al., 1975). Luteinizing hormone (LH) but not follicle-stimulating hormone produces a several-fold stimulation of PREG formation by the ovary (Ickikawa et al., 1972). Both ovarian-cultured theca/interstitial and luteal cell types produce $3\alpha,5\alpha$-TH PROG in response to LH stimulation (Dyer et al., 1994). $3\alpha,5\alpha$-TH PROG was found to be a major product of cultured theca/interstitial cells, whereas it was a relatively minor product of luteal cells. Moreover, theca/interstitial cell production of $3\alpha,5\alpha$-TH PROG was shown to be increased 200–300% by corticosterone and dexamethasone (10–100 nM) (Dyer et al., 1994). Thus, ovarian function, stressful events, or the interaction between the two could result in altered endogenous $3\alpha,5\alpha$-TH PROG levels.

Menstrual cycle phase can influence the sensitivity to the ethanol-like discriminative stimulus effects of both ethanol and $3\alpha,5\alpha$-TH PROG (Grant et al., 1996, 1997), but not midazolam (Green et al., 1999). These data are interpreted as an additive effect of endogenous $3\alpha,5\alpha$-TH PROG in the luteal phase of the menstrual cycle with exogenous administration of ethanol or $3\alpha,5\alpha$-TH PROG. The luteal phase of the menstrual cycle was defined as early-luteal (mean PROG 3.76 ng/ml) or mid-luteal (mean PROG 11.3 ng/nl), rather than premenstrual (late-luteal, PROG levels falling rapidly), based on menstrual cycle day (back-count Days 8–13) and average PROG levels (Grant et al., 1997; Green et al., 1999). That midazolam discriminative stimulus effects were not increased may be due to the relative contribution of $GABA_A$-receptor subtype to the discriminative stimulus effects of ethanol. In particular, the additive effects of endogenous $3\alpha,5\alpha$-TH PROG with exogenous $3\alpha,5\alpha$-TH PROG or ethanol may represent an effect through benzodiazepine-insensitive receptors. A recent study of $GABA_A$ subunit upregulation following PROG treatment supports this hypothesis (Smith et al., 1998). Specifically, α_4 subunits were increased following chronic (3-week) exposure to PROG in female rats. The increase in α_4 subunits was associated with increased susceptibility to seizures, increased pentobarbital sensitivity, and decreased benzodiazepine sensitivity, all of which were prevented by α_4 antisense oligonucleotide administration *in vivo*. In addition, blockade of $3\alpha,5\alpha$-TH PROG formation with indomethacin during PROG administration prevented benzodiazepine insensitivity, strongly suggesting that the α_4-subunit upregulation is in response to chronically elevated levels of $3\alpha,5\alpha$-TH PROG, not PROG per se (Smith *et al.*, 1998). An important factor in these studies was that measurements were taken 24 h following PROG withdrawal, presumably at the end of the rapidly falling PROG levels. Indeed, this approach has been suggested to be an animal model of premenstrual syndrome distress (Smith *et al.*, 1998).

The $GABA_A$-receptor mechanism can also be explored with antagonism studies. Briefly, the discriminative effects of pentobarbital and diazepam can be differentiated by antagonism with the benzodiazepine antagonist flumazenil and convulsants, such as bemegride (Ator, 1990). Antagonism of the discriminative stimulus effects of neurosteroids has not been reported. However, there are reports of using neurosteroids that are negative modulators at $GABA_A$ to block the discriminative stimulus effects of ethanol. The rationale for these tests is similar to using the partial inverse agonist Ro 15-4513. Although a binding site for Ro 15-4513 has been demonstrated on BZ-insensitive α_6 and α_4 subunits (Luddens *et al.*, 1990), Ro 15-4513 has activity at BZ-sensitive receptors and can block the discriminative stimulus effects of benzodiazepines (Rees and Balster, 1988; Hiltunen and Jarbe, 1989). In contrast, there are mixed reports about the efficacy of Ro 15-4513 to block the discriminative stimulus effects of pentobarbital and ethanol (Hiltunen and Jarbe, 1988, 1989; Rees and Balster, 1988; Emmett-Oglesby, 1990; Middaugh *et al.*, 1991; Gatto and Grant, 1997). To date, DHEA, DHEAS, and PREG have not antagonized the effects of ethanol, the only discrimination in which they have been tested as antagonists (Bowen *et al.*, 1999b; Bienkowski and Kostowski, 1997).

In summary, the drug discrimination procedure has been used to characterize the receptor mechanisms mediating the effects of $GABA_A$-positive modulators on a behavioral level of analysis. The data strongly support the hypothesis that the behavioral effects of particular neurosteroids are due to their positive modulation of $GABA_A$ activity. Future research should be focused on the discriminative stimulus effects of these neurosteroids themselves, rather than assessing them in barbiturate, benzodiazepine, or ethanol discriminations. In addition, physiological states, such as stress, reproductive function, and sleep deprivation, should be further explored as independent variables to determine the degree to which these receptor mechanisms are involved in mediating the stimulus effects of neurosteroids.

Acknowledgments

The preparation of this review was supported in part by P50AA11997 and T32AA07565.

References

Ator, N. A., and Griffiths, R. R. (1989). Asymmetrical cross-generalization in drug discrimination with lorazepam and pentobarbital training conditions. *Drug Dev. Res.* **16,** 355–364.
Ator, N. A. (1990). Drug discrimination and drug stimulus generalization with anxiolytics. *Drug Dev. Res.* **20,** 189–204.
Ator, N. A., Grant, K. A., Purdy, R. H., Paul, S. M., and Griffiths, R. R. (1993). Drug discrimination analysis of endogenous neuroactive steroids in rats. *Eur. J. Pharmacol.* **241,** 237–244.
Ator, N. A., and Griffiths, R. R. (1997). Selectivity in the generalization profile in baboons trained to discriminate lorazepam, benzodiazepines, barbiturates, and other sedative/anxiolytics. *J. PET.* **282,** 1442–1457.
Ator, N. A., and Kautz, M. A. (2000). Differentiating benzodiazepine- and barbiturate-like discriminative stimulus effects of lorazepem, diazepam, pentobarbital, imidazenil and zaleopon in two-versus three-lever procedures. *Behav. Pharmacol.* **11,** 1–14.
Barbaccia, M. L., Roscetti, G., Bolacchi, F., Concas, A., Mostallino, M. C., Purdy, R. H., and Biggio, G. (1996a). Stress-induced increase in brain neuroactive steroids: Antagonism by abecarnil. *Pharmacol. Biochem. Behav.* **54,** 205–210.
Barbaccia, M. L., Roscetti, G., Trabucchi, M., Mostallino, M. C., Concas, A., Purdy, R. H., and Biggio, G. (1996b). Time-dependent changes in rat brain neuroactive steroid concentrations and $GABA_A$ receptor function after acute stress. *Neuroendocrinology* **63,** 166–172.
Bernard, C. L. (1865). An Introduction to the Study of Experimental Medicine, (H. C. Green, trans.). Dover, New York, 1957.
Bienkowski, P., and Kostowski, W. (1997). Discriminative stimulus properties of ethanol in the rat: Effects of neurosteroids and picrotoxin. *Brain Res.* **753,** 348–352.
Bitran, D., Shiekh, M., and McLeod, M. (1995). Anxiolytic effect of progesterone is mediated by the neurosteroid allopregnanolone at brain $GABA_A$ receptors. *J. Neuroendocrinology* **7,** 171–177.
Blanchard, D. C., Veniegas, R., Elloran, T., and Blanchard, R. S. (1993). Alcohol and anxiety: Effects on offensive and defensive aggression. *J. Stud. Alc. (Suppl.)* **11,** 9–19.
Blanchard, R. S., Hori, K., Blanchard, D. C., and Hall, J. (1987). Ethanol effects on aggression of rats selected for different levels of aggression. *Pharmacol. Biochem. Behav.* **27,** 641–644.
Bowen, C. A., Gatto, G. J., and Grant, K. A. (1997). Assessment of the multiple discriminative stimulus effects of ethanol using an ethanol-pentobarbital-water discrimination in rats. *Behav. Pharmacol.* **8,** 339–352.
Bowen, C. A., and Grant, K. A. (1999). Increased specificity of ethanol's discriminative stimulus effects in an ethanol-pentobarbital-water discrimination in rats. *Drug Alcohol Depend.* **55,** 13–24.
Bowen, C. A., Purdy, R. H., and Grant, K. A. (1999a). Ethanol-like discriminative stimulus effects of endogenous neuroactive steroids: Effect of ethanol training dose and dosing procedure. *J. PET* **289,** 405–411.

Bowen, C. A., Purdy, R. H., and Grant, K. A. (1999b). An investigation of endogenous neuroactive steroid-induced modulation of ethanol's discriminative stimulus effects. *Behav. Pharmacol.* **10,** 297–311.

Brot, D. M., Akwa, Y., Purdy, R. H., Koob, G. F., and Britton, K. T. (1997). The anxiolytic-like effects of the neurosteroid allopregnanolone: Interactions with GABA$_A$ receptors. *Eur. J. Pharmacol.* **325,** 1–7.

Carboni, E., Wieland, S., Lan, N. C., and Gee, K. W. (1996). Anxiolytic properties of endogenously occurring pregnanediols in two rodent models of anxiety. *Psychopharmacology* **123,** 173–178.

Chen, S., Rodriguez, L., Davies, F. M., and Loew, G. H. (1996). The hyperphagic effect of 3α-hydroxylated pregnane steroids in male rats. *Pharmacol. Bichem. Behav.* **53,** 777–782.

Colpaert, F. C. (1986). Drug discrimination: Behavioral, pharmacological, and molecular mechanisms of discriminative drug effects. *In* "Behavioral Analysis of Drug Dependence" (S. R. Goldberg and I. P. Stolerman, Eds.), pp. 161–193. Academic Press, New York.

Concas, A., Pierobon, P., Mostallino, M. C., Porcu, P., Marino, G., Minri, R., and Biggio, G. (1998). Modulation of γ-aminobutyric acid (GABA) receptors and the feeding response by neurosteroids in *Hydra vulgaris. Neuroscience* **85,** 979–988.

Crawley, J. N., Glowa, J. R., Majewska, M. D., and Paul, S. M. (1986). Anxiolytic activity of an endogenous adrenal steroid. *Brain Res.* **398,** 382–385.

Darnaudery, M., Bouyer, J., Pallares, M., Le Moal, M., and Mayo, W. (1999). The promnesic neurosteroid pregnenolone sulfate increases paradoxical sleep in rats. *Brain Res.* **818,** 492–498.

Darnaudery, M., Koehl, M., Piazza, P., Le Moal, M., and Mayo, W. (2000). Pregnenolone sulfate increases hippocampal acetylcholine release and spatial recognition. *Brain Res.* **852,** 173–179.

Dazzi, L., Sanna, A., Cagetti, E., Concas, A., and Biggio, G. (1996). Inhibition by the neurosteroid allopregnanolone of basal and stress-induced acetylcholine release in the brain of freely moving rats. *Brain Res.* **710,** 275–280.

Deutsch, S. I., and Mastropaolo, J. (1993). Discriminative stimulus properties of midazolam are shared by a GABA-receptor positive steroid. *Pharmacol. Biochem. Behav.* **46,** 963–965.

Diamond, D. M., Branch, B. J., and Fleshner, M. (1996). The neurosteroid dehydroepiandrosterone sulfate (DHEAS) enhances hippocampal primed burst, but not long-term, potentiation. *Neuroscience Lett.* **202,** 204–208.

Depoortere, H., Zivkovic, B., Lloyd, K. G., Sanger, D. J., Perrault, G., Langer, S. Z., and Bartholini, G. (1986). Zolpidem, a novel nonbenzodiazepine hypnotic. I. Neuropharmacological and behavioral effects. *J. PET.* **237,** 649–658.

Dyer, C. A., Bonnet, D. J., and Curtiss, L. K. (1994). Ovarian theca-interstitial cell neurosteroid production is stimulated by LH and glucocorticoids. Abstract, Endocrine Society Annual Meeting, Anaheim, California, #1668.

Edgar, D. M., Seidel, W. F., Gee, K. W., Lan, N. C., Field, G., Xia, H., Hawkinson, J. E., Wieland, S., Carter, R. B., and Wood, P. L. (1997). CCD-3693: An orally bioavailable analog of the endogenous neuroactive steroid, pregnanolone, demonstrates potent sedative hypnotic actions in the rat. *J. PET* **282,** 420–429.

Emmett-Oglesby, M. W. (1990). Tolerance to the discriminative stimulus effects of ethanol. *Behav. Pharmacol.* **1,** 497–503.

Fernandez-Guasti, A., and Picazo, O. (1995). Flumazenil blocks the anxiolytic action of allopregnanolone. *Eur. J. Pharmacol.* **281,** 113–115.

File, S. E., James, T. A., and MacLeod, N. K. (1981). Depletion in amygdaloid 5-hydroxytryptamine concentration and changes in social and aggressive behaviour. *J. Neural Transmission* **50,** 1–12.

Finn, D. A., Phillips, T. J., Okorn, D. M., Chester, J. A., and Cunningham, C. L. (1997a). Rewarding effects of the neuroactive steroid 3α-hydroxy-5α-pregnan-20-one in mice. *Pharmacol. Biochem. Behav.* **56,** 261–264.

Finn, D. A., Roberts, A. J., Lotrich, F., and Gallaher, E. J. (1997b). Genetic differences in behavioral sensitivity to a neuroactive steroid. *J. PET* **280,** 820–828.

Flood, J. F., and Roberts, E. (1988). Dehydroepiandrosterone sulfate improves memory in aging mice. *Brain Res.* **448,** 178–181.

Flood, J. F., Smith, G. E., and Roberts, E. (1988). Dehydroepiandrosterone and its sulfate enhance memory retention in mice. *Brain Res.* **447,** 269–278.

Flood, J. F., Morley, J. E., and Roberts, E. (1992). Memory-enhancing effects in male mice of pregnenolone and steroids metabolically derived from it. *Proc. Natl. Acad. Sci. USA* **89,** 1567–1571.

Frye, C. A., Van Keuren, K. R., and Esrkine, M. S. (1996). Behavioral effects of 3a-androstanediol. I: Modulation of sexual receptivity and promotion of GABA-stimulated chloride flux. *Behav. Brain Res.* **79,** 109–118.

Gatto, G. J., and Grant, K. A. (1997). Attenuation of the discriminative stimulus effects of ethanol by the benzodiazepine partial inverse agonist Ro 15-4513. *Behav. Pharmacol.* **8,** 139–146.

Grant, K. A., Azarov, A., Bowen, C. A., Mirkis, S., and Purdy, R. H. (1996). Ethanol-like discriminative stimulus effects of the neurosteroid 3α-hydroxy-5α-pregnan-20-one in female *Macaca fascicularis* monkeys. *Psychopharm.* **124,** 340–346.

Grant, K. A., Azarov, A., Shively, C. A., and Purdy, R. H. (1997). Discriminative stimulus effects of ethanol and 3α-hydroxy-5α-pregnan-20-one in relation to menstrual cycle phase in cynomolgus monkeys (*Macaca fascicularis*). *Psychopharmacology* **130,** 59–68.

Grant, K. A., Waters, C. A., Green-Jordan, K., Azarov, A., and Szeliga, K. T. (2000). Characterization of the discriminative stimulus effects of GABA$_A$ receptor ligands in *Macaca fascicularis* monkeys under different ethanol training conditions. *Psychopharm.* **153,** 181–188.

Grech, D. M., and Balster, R. L. (1993). Pentobarbital-like discriminative stimulus effects of direct GABA agonists in rats. *Psychopharmacology* **110,** 295–301.

Grech, D. M., and Balster, R. L. (1997). The discriminative stimulus effects of muscimol in rats. *Psychopharmacology* **129,** 339–347.

Green, K. L., Azarov, A. V., Szeliga, K. T., Purdy, R. H., and Grant, K. A. (1999). The influence of menstrual cycle phase on sensitivity to the ethanol-like discriminative stimulus effects of GABA$_A$-positive modulator. *Pharmacol. Biochem. Behav.* **64,** 379–383.

Grobin, A. C., Roth, R. H., and Deutch, A. Y. (1992). Regulation of the prefrontal cortical dopamine system by the neuroactive steroid 3α,21-dihydroxy-5α-pregnane-20-one. *Brain Res.* **578,** 351–356.

Grobin, A. C., Matthews, D. B., Devaud, L. L., and Morrow, A. L. (1998). The role of GABA$_A$ receptors in the acute and chronic effects of ethanol. *Psychopharmacology* **139,** 2–19.

Higgs, S., and Cooper, S. J. (1998). Antiphobic effect of the neuroactive steroid 3α-hydroxy-5β-pregnan-20-one in male rats. *Pharmacol. Biochem. Behav.* **60,** 125–131.

Hiltunen, A. J., and Jarbe, T. U. C. (1988). Ro 15-4513 does not antagonize the discriminative stimulus- or rate-depressant effects of ethanol in rats. *Alcohol* **5,** 203–207.

Hiltunen, A. J., and Jarbe, T. U. C. (1989). Discriminative stimulus properties of ethanol: Effects of cumulative dosing and Ro 15-4513. *Behav. Pharmacol.* **1,** 133–140.

Holtzman, S. G. (1990). Discriminative stimulus effects of drugs: Relationship to potential for abuse. *In* "Modern Methods in Pharmacology, Testing and Evaluation of Drugs of Abuse". (M. W. Adler and A. Cowan, Eds.), Vol. 6, pp. 193–210. Wiley-Liss, New York.

Holzbauer, M. (1975). Physiological variations in the ovarian production of 5alpha-pregnane derivatives with sedative properties in the rat. *J. Steroid Biochem.* **6,** 1307–1310.

Holzbauer, M., Birmingham, M. K., DeNicola, A. F., and Oliver, J. T. (1985). *In vivo* secretion of 3α-hydroxy-5α-pregnan-20-one, a potent anaesthetic steroid, by the adrenal gland of the rat. *J. Steroid. Biochem.* **22,** 97–102.

Ickikawa, S., Morioka, H., and Sawada, T. (1972). Identification of neutral steroids in the ovarian venous plasma of LH stimulated rats. *Endocrinology* **88,** 372–383.

Isaacson, R. L., Varner, J. A., and de Wied, D. (1995). The effects of pregnenolone sulfate and ethylestrenol on retention of a passive avoidance task. *Brain Res.* **689,** 79–84.

Janak, P. H., Redfern, J. E. M., and Samson, H. H. (1998). The reinforcing effects of ethanol are altered by the endogenous neurosteroid, allopregnanolone. *Alcohol: Clin. Exp. Res.* **22,** 1106–1112.

Korneyev, A., Pan, B. S., Polo, A., Romeo, E., Guidotti, A., and Costa, E. (1993). Stimulation of brain pregnenolone synthesis by mitochondrial diazepam binding inhibitor receptor ligands *in vivo*. *J. Neurochem.* **61,** 1515–1524.

Lancel, M. (1999). Role of GABA$_A$ receptors in the regulation of sleep: Initial sleep responses to peripherally administered modulators and agonists. *Sleep* **22,** 33–42.

Le Foll, F., Louiset, E., Castel, H., Vaudry, H., and Cazin, L. (1997). Electrophysiological effects of various neuroactive steroids on the GABA$_A$ receptor in pituitary melanotrope cells. *Eur. J. Pharmacol.* **331,** 303–311.

Le Roy, I., Mortaud, S., Tordjman, S., Donsez-Darcel, E., Carlier, M., Degrelle, H., and Roubertoux, P. L. (1999). Genetic correlation between steroid sulfatase concentration and initiation of attack behavior in mice. *Behav. Gen.* **29,** 131–136.

Li, P., Rhodes, M. E., Burke, A. M., and Johnson, D. A. (1997). Memory enhancement mediated by the steroid sulfatase inhibitor (p-*O*-sulfamoyl)-*N*-tetradecanoyl tyramine. *Life Sci.* **60,** PL 45–51.

Li, P., Rhodes, M. E., Jagannathan, S., and Johnson, D. A. (1995). Reversal of scopolamine induced amnesia in rats by the steroid sulfatase inhibitor estrone-3-O-sulfamate. *Cognitive Brain Res.* **2,** 251–254.

Luddens, H., Pritchett, D. B., Kohler, M., Killisch, I., Keinanen, K., Monyer, H., Sprengel, R., and Seeburg, P. H. (1990). Cerebellar GABA$_A$ receptor selective for a behavioural alcohol antagonist. *Nature* **346,** 648–651.

Mathis, C., Paul, S. M., and Crawley, J. N. (1994). The neurosteroid pregnenolone sulfate blocks NMDA antagonist-induced deficits in a passive avoidance memory task. *Psychopharmacology* **116,** 201–206.

Mathis, C., Vogel, E., Cagniard, B., Criscuolo, F., and Ungerer, A. (1996). The neurosteroid pregnenolone sulfate blocks deficits induced by a competitive NMDA antagonist in active avoidance and lever-press learning tasks in mice. *Neuropharmacology* **35,** 1057–1064.

Maurice, T., Junien, J., and Privat, A. (1997). Dehydroepiandrosterone sulfate attenuates dizocilpine-induced learning impairment in mice via σ_1-receptors. *Behav. Brain Res.* **83,** 159–164.

Melchior, C. L., and Ritzmann, R. F. (1994). Pregnenolone and pregnenolone sulfate, alone and with ethanol, in mice on the plus-maze. *Pharmacol. Biochem. Behav.* **48,** 893–897.

Meyer, J. H., Lee, S., Wittenberg, G. F., Randall, R. D., and Groul, D. L. (1999). Neurosteroid regulation of inhibitory synaptic transmission in the rat hippocampus *in vitro*. *Neuroscience* **90,** 1177–1183.

Meziane, H., Mathis, C., Paul, S. M., and Ungerer, A. (1996). The neurosteroid pregnenolone sulfate reduces learning deficits induced by scopolamine and has promnestic effects in mice performing an appetitive learning task. *Psychopharmacology* **126,** 323–330.

Miczek, K. A., Weertz, E. M., and DeBold, J. F. (1993). Alcohol, benzodiazepine-GABA$_A$ receptor complex and aggression: Ethological analysis of individual differences in rodents and primates. *J. stud. Alcohol Suppl.* **11,** 170–178.

Miczek, K. A., Weerts, E. M., Vivian, J. A., and Barros, H. M. (1995). Aggression, anxiety and vocalizations in animals: $GABA_A$ and 5-HT anxiolytics. *Psychopharmacology* **121,** 38–56.

Miczek, K. A., DeBold, J. F., van Erp, A. M., and Tornatzky, W. (1997). Alcohol, $GABA_A$-benzodiazepine receptor complex, and aggression. *Recent Developments in Alcoholism* **13,** 139–171.

Middaugh, L. D., Bao, K., Becker, H. C., and Daniel, S. S. (1991). Effects of Ro 15-4513 on ethanol discrimination in C57BL/6 mice. *Pharmacol. Biochem. Behav.* **38,** 763–767.

Olivier, B., Mos, J., Tulp, M., Schipper, J., den Daas, S., and van Oortmerssen, G. (1990). Serotonergic involvement in aggressive behavior in animals. *In* "Violence and Suicidality" (H. M. van Pragg, R. Plutchik, and A. Apter, Eds.), pp. 79–137. Brunner/Mazel, New York.

Papadopoulos, V., Guarneri, P., Krueger, K. E., and Guidotti, A. (1992). Pregnenolone biosynthesis in C6-2B glioma cell mitochondria: Regulation by a mitochondrial diazepam binding inhibitor receptor. *Proc. Natl. Acad. Sci. USA* **89,** 5113–5117.

Patchev, V. K., Montkowski, A, Rouskova, D., Koranyi, L., and Holsboer, F. (1997). Neonatal treatment of rats with the neuroactive steroid tetrahydrodeoxycorticosterone (THDOC) abolishes the behavorial and neuroendocrine consequences of adverse early life events. *J. Clin. Invest.* **99,** 962–966.

Phan, V., Su, T., Privat, A., and Maurice, T. (1999). Modulation of steroidal levels by adrenalectomy/castration and inhibition of neurosteroid synthesis enzymes effect σ_1 receptor-mediated behavior in mice. *Eur. J. Neurosci.* **11,** 2385–2396.

Picazo, O., and Fernandez-Guasti, A. (1995). Anti-anxiety effects of progesterone and some of its metabolites: An evaluation using the burying behavior test. *Brain Res.* **680,** 135–141.

Pick, C. G., Peter, Y., Terkel, J., Gavish, M., and Weizman, R. (1996). Effect of the neuroactive steroid α-THDOC on staircase test behavior in mice. *Psychopharmacology* **128,** 61–66.

Pick, C. G., Peter, Y., Paz, L., Schreiber, S., Gavish, M., and Weizman, R. (1997). Effect of the pregnane-related GABA-active steroid alphaxalone on mice performance in the staircase test. *Brain Res.* **765,** 129–134.

Preston, K. L., and Bigelow, G. E. (1991). Subjective and discriminative effects of drugs. *Behav. Pharmacol.* **2,** 293–313.

Puia, G., Santi, M. R., Vicini, S., Pritchett, D. B., Purdy, R. H., Paul, S. M., Seeburg, P. H., and Costa, E. (1990). Neurosteroids act on recombinant human $GABA_A$ receptors. *Neuron* **4,** 759–765.

Puia, G., Ducic, I., Vicini, S., and Costa, E. (1993). Does neurosteroid modulatory efficacy depend on $GABA_A$ subunit composition?. *Receptors and Channels* **1,** 135–142.

Purdy, R. H., Morrow, A. L., Moore, P. H., and Paul, S. M. (1991). Stress-induced elevations of gamma-aminobutyric acid type A receptor- active steroids in the rat brain. *Proc. Natl. Acad. Sci. USA* **88,** 4553–4557.

Reddy, D. S., and Kulkarni, S. K. (1996). Role of GABA-A and mitochondrial diazepam binding inhibitor receptors in the anti-stress activity of neurosteroids in mice. *Psychopharmacology* **128,** 280–292.

Reddy, D. S., and Kulkarni, S. K. (1997a). Differential anxiolytic effects of neurosteroids in the mirrored chamber behavior test in mice. *Brain Res.* **752,** 61–71.

Reddy, D. S., and Kulkarni, S. K. (1997b). Reversal of benzodiazepine inverse agonist FG 7142-induced anxiety syndrome by neurosteroids in mice. *Meth. Find. Exp. Clin. Pharmacol.* **19,** 665–681.

Reddy, S. D., and Kulkarni, S. K. (1998). The role of GABA-A and mitochondrial diazepam-binding inhibitor receptors on the effects of neurosteroids on food intake in mice. *Psychopharmacology* **137,** 391–400.

Reddy, D. S., and Kulkarni, S. K. (1999). Sex and estrous cycle-dependent changes in neurosteroid and benzodiazepine effects on food consumption and plus-maze learning behaviors in rats. *Pharmacol. Bichem. Behav.* **62**, 53–60.

Rees, D. C., and Balster, R. L. (1988). Attenuation of the discriminative stimulus properties of ethanol and oxazepam, but not of pentobarbital, by Ro 15-4513 in mice. *J. PET* **244**, 592–598.

Robel, P., Young, J., Corpechot, C., Mayo, W., Perche, F., Haug, M., Simon, H., and Baulieu, E. E. (1995). Biosynthesis and assay of neurosteroids in rats and mice: Functional correlates. *J. Steroid Biochem. Molec. Biol.* **53**, 355–360.

Rowlett, J. K., and Woolverton, W. L. (1996). Assessment of benzodiazepine receptor heterogeneity *in vivo*: Apparent pA2 and pKB analyses from behavioral studies. *Psychopharmacology* **128**, 1–16.

Rowlett, J. K., Winger, G., Carter, R. B., Wood, P. L., Woods, J. H., and Woolverton, W. L. (1999). Reinforcing and discriminative stimulus effects of the neuroactive steroids pregnanolone and Co 8-7071 in rhesus monkeys. *Psychopharmacology* **145**, 205–212.

Sanger, D. J., and Zivkovic, B. (1986). The discriminative stimulus properties of zolpidem, a novel imidazopyridine hypnotic. *Psychopharmacology* **89**, 317–322.

Sanger, D. J. (1987). The effects of new hyponic drugs in rats trained to discriminate ethanol. *Behav. Pharmacol.* **8**, 287–292.

Saudou, F., Amara, D. A., Dierich, A., LeMeur, M., Ramboz, S., Segu, L., Buhot, M.-C., and Hen, R. (1994). Enhanced aggressive behavior in mice lacking 5-HT$_{1b}$ receptor. *Science* **265**, 1875–1878.

Selye, H. (1941). The anesthetic effect of steroid hormones. *Proc. Exp. Biol. Med.* **46**, 116–121.

Shelton, K. L., and Balster, R. L. (1994). Ethanol drug discrimination in rats: Substitution with GABA agonists and NMDA antagonists. *Behav. Pharmacol.* **5**, 441–450.

Simiand, J., Keane, P. E., and Morre, M. (1984). The staircase test in mice: A simple and efficient procedure for primary screening of anxiolytic agents. *Psychopharmacology* **84**, 48–53.

Smith, S. S., Gong, Q. H., Hsu, F.-C., Markowitz, R. S., French-Mullen, J. M. H., and Li, X. (1998). GABA$_A$ receptor a$_4$ subunit suppression prevents withdrawal properties of an endogenous steroid. *Nature* **392**, 926–930.

Steffensen, S. C. (1995). Dehydroepiandrosterone sulfate suppresses hippocampal recurrent inhibition and synchronizes neuronal activity to theta rhythm. *Hippocampus* **5**, 320–328.

Steiger, A., Trachsel, L., Guldner, J., Hemmeter, U., Rothe, B., Rupprecht, R., Vedder, H., and Holsboer, F. (1993). Neurosteroid pregnenolone induces sleep-EEG changes in man compatible with inverse agonistic GABA$_A$-receptor modulation. *Br. Res.* **615**, 267–274.

Steiger, A., Antonijevic, I. A., Bohlhalter, S., Frieboes, R. M., Friess, E., and Murck, H. (1998). Effects of hormones on sleep. *Horm. Res.* **49**, 125–130.

Stewart, H., Krebs, W. H., and Kaczender, E. (1967). State-dependent learning produced with steroids. *Nature* **216**, 1223–1224.

Vallée, M., Mayo, W., Darnaudery, M., Corpechot, C., Young, J., Koehl, M., Le Moal, M., Baulieu, E., Robel, P., and Simon, H. (1997). Neurosteroids: Deficient cognitive performance in aged rats depends on low pregnenolone sulfate levels in the hippocampus. *Proc. Natl. Acad. Sci. USA* **94**, 14865–14870.

van Erp, A. M. M., and Miczek, K. A. (1997). Increased aggression after ethanol self-administration in male resident rats. *Psychopharmacology* **131**, 287–295.

Vanover, K. E. (1997). Discriminative stimulus effects of the endogenous neuroactive steroid pregnanolone. *Eur. J. Pharmacol.* **327**, 97–101.

Vivian, J. A., Barros, H. M. T., Manitiu, A., and Miczek, K. A. (1997). Ultrasonic vocalizations in rat pups: Modulation at the γ-aminobutyric acid$_A$ receptor complex and the neurosteroid recognition site. *J. PET* **282,** 318–325.

Weertz, E. M., and Miczek, K. A. (1996). Primate vocalizations during social separation and aggression: Effects of alcohol and benzodiazepines. *Psychopharmacology* **127,** 255–264.

Wieland, S., Belluzzi, J., Hawkinson, J. E., Hogenkamp, D., Upasani, R., Srein, L., Wood, P. L., Gee, K. W., and Lan, N. C. (1997). Anxiolytic and anticonvulsant activity of a synthetic neuroactive steroid Co 3-0593. *Psychopharmacology* **134,** 46–54.

Yoo, A., Harris, J., and Dubrovsky, B. (1996). Dose–response study of dehydroepiandrosterone sulfate on dentate gyrus long term potentiation. *Exp. Neurol.* **137,** 151–156.

Zimmerberg, B., Brunelli, S. A., and Hofer, M. A. (1994). Reduction of rat pup ultrasonic vocalizations by the neuroactive steroid allopregnanolone. *Pharm. Biochem. Behav.* **47,** 735–738.

ETHANOL AND NEUROSTEROID INTERACTIONS IN THE BRAIN

A. Leslie Morrow,[1] Margaret J. VanDoren, Rebekah Fleming, and Shannon Penland

Bowles Center for Alcohol Studies and Departments of Psychiatry and Pharmacology
University of North Carolina at Chapel Hill, Chapel Hill
North Carolina 27599-7178

I. Introduction
II. Role of $3\alpha,5\alpha$-TH PROG in Ethanol Action
   A. Pregnane Neurosteroid Actions on $GABA_A$ Receptors
   B. Ethanol Actions on $GABA_A$ Receptors
   C. Ethanol Alters Brain Concentrations of GABAergic Neurosteroids
   D. $3\alpha,5\alpha$-TH PROG Contributes to Specific Effects of Ethanol
III. Role of Neurosteroids in Alcohol Reinforcement
   A. $3\alpha,5\alpha$-TH PROG and Related Steroids Alter Ethanol Self-Administration
IV. Role of Neurosteroids in Ethanol Tolerance
V. Role of Neurosteroids in Ethanol Dependence
   A. Ethanol-Dependent Rats Are Sensitized to GABAergic Neurosteroids
   B. Alcohol Withdrawal Seizure Sensitivity Associated with Altered GABAergic Neurosteroid Levels
   C. Relationship of Adaptations in Neurosteroid Levels and Neurosteroid Sensitivity of $GABA_A$ Receptors
   D. Relationship between Changes in Neurosteroid Levels and Altered $GABA_A$-Receptor Function
VI. Conclusions and Future Directions
   References

I. Introduction

Since the discovery of the endogenous neuroactive steroids 3α-hydroxy-5α-pregnan-20-one ($3\alpha,5\alpha$-TH PROG) and $3\alpha,21$-dihydroxy-5α-pregnan-20-one ($3\alpha,5\alpha$-TH DOC) in rat brain, there has been a tremendous interest in elucidating the physiological role of these molecules. Although they are synthesized in glial cells and found in brain tissue at pharmacologically relevant concentrations, it is unknown whether these pregnane steroids directly regulate neuronal excitability *in vivo*. Evidence suggests that endogenous

---

[1]Author to whom correspondence should be sent.

$3\alpha,5\alpha$-TH PROG normally influences γ-aminobutyric acid (GABA) neurotransmission, as blockade of $3\alpha,5\alpha$-TH PROG biosynthesis diminishes the electrophysiological and behavioral actions of GABA (Pinna et al., 2000). Studies of the modulation of neurosteroids under physiological conditions such as stress, pregnancy, or estrus cycling or following pharmacological challenges have also provided useful information on the role of neurosteroids in GABAergic transmission. Stress and the subsequent activation of the hypothalamic-pituitary-adrenal (HPA) axis increases brain levels of $3\alpha,5\alpha$-TH PROG and $3\alpha,5\alpha$-TH DOC (Purdy et al., 1991). In turn, $3\alpha,5\alpha$-TH PROG decreases corticotrophin-releasing factor (CRF) mRNA levels in the hypothalamus and CRF release into the portal system (Patchev et al., 1994, 1996). This response is proving to be a fundamental physiological adaptation to stress. Similar elevations in brain $3\alpha,5\alpha$-TH PROG levels are observed in response to several pharmacological agents, including fluoxetine (Uzunov et al., 1996; Uzunova et al., 1998), olanzapine (Marx et al., 2000), and alcohol (Morrow et al., 1998, 1999; VanDoren et al., 2000b).

Alcohol activates the HPA axis, increases concentrations of adrenal steroids in plasma and pregnane steroids in the brain, and enhances GABA-mediated inhibition in the brain. The interactions between alcohol and neurosteroids are interesting as a probe to understand both the physiological role of these steroids and the mechanisms of ethanol action. Because ethanol is so extensively used worldwide to promote relaxation and to reduce stress, this action may indicate that modulation of endogenous steroids has profound effects on central nervous system (CNS) function and behavior.

The mechanism of alcohol action in the CNS remains perplexing. Although there are many lines of indirect evidence that ethanol has anxiolytic and sedative effects and produces motor incoordination and cognitive impairment by enhancement of GABAergic inhibition in brain (Grobin et al., 1998), there is no convincing *direct* evidence for this hypothesis. Indeed, several studies have suggested that ethanol action on $GABA_A$ receptors may be indirect—involving intermediaries, such as protein kinases (Harris et al., 1995), GABA release (Freund and Palmer, 1997; Weiner et al., 1997b,c), or pregnane steroids (VanDoren et al., 2000b). The purpose of this chapter is to review evidence that endogenous pregnane steroids contribute to ethanol action, ethanol tolerance, and ethanol dependence. The evidence that pregnane steroids mediate the effects of ethanol in the brain underscores the potential physiological relevance of these compounds and provides important insights into the mechanisms of neurosteroid regulation of brain function.

II. Role of 3α,5α-TH PROG in Ethanol Action

A. PREGNANE NEUROSTEROID ACTIONS ON GABA$_A$ RECEPTORS

Neurosteroids are steroid hormones that are synthesized by enzymes in the brain. Unlike most hormones, which act by crossing the lipid bilayer and binding to intracellular steroid receptors, neuroactive steroids can bind and allosterically modulate ligand-gated ion channels on the cell surface. Two of these neurosteroids, 3α,5α-TH PROG and 3α,5α-TH DOC, act selectively at GABA$_A$ receptors to enhance Cl$^-$ channel function allosterically. Like benzodiazepines, 3α,5α-PROG and 3α,5α-TH DOC decrease the EC$_{50}$ for binding of compounds to the GABA and the benzodiazepine recognition sites. Electrophysiological studies have shown that these neurosteroid modulators increase the peak magnitude of agonist-stimulated Cl$^-$ flux and increase the rate of channel desensitization during prolonged applications of agonist (Simmonds *et al.*, 1984; Majewska *et al.*, 1986; Harrison *et al.*, 1987). Single-channel recordings have shown that the increase in Cl$^-$ flux is the result of increased frequency of GABA$_A$ Cl$^-$ channel opening (Puia *et al.*, 1990). Neurosteroids enhance the potency of muscimol in stimulating Cl$^-$ uptake in a concentration-dependent and stereospecific manner. 3α,5α-TH PROG and 3α,5α-TH DOC are 20 times more potent than benzodiazepines and 200 times more potent than barbiturates in augmenting GABA-receptor function (Morrow *et al.*, 1987). Structure–activity experiments demonstrate that the 3β-configuration, in either the 5β- or the 5α-reduced metabolites, is essential for pharmacological activity at GABA$_A$ receptors (Harrison *et al.*, 1987; Morrow *et al.*, 1990b). Moreover, 3α,5α-TH PROG, 3α,5α-TH DOC, and all other 3α,5α-reduced pregnane steroids *lack* significant activity at *N*-methyl-D-asparate (NMDA) or glycine receptors (Irwin *et al.*, 1994; Park-Chung *et al.*, 1999).

B. ETHANOL ACTIONS ON GABA$_A$ RECEPTORS

The behavioral effects of acute ethanol administration are remarkably similar to the effects of 3α,5α-reduced neurosteroids, benzodiazepines, and barbiturates—all modulators of GABA$_A$ receptors. Ethanol is anxiolytic, sedative/hypnotic, anticonvulsant, and motor incoordinating (Majchrowicz, 1975; Frye *et al.*, 1981). Ethanol also causes impairments in cognitive processing (Matthews *et al.*, 1995; Givens, 1997) and, at high concentrations, ethanol acts as an anesthetic and respiratory depressant (Sellers and Kalant,

1976). Although a precise mechanism of action for ethanol cannot be inferred from behaviors elicited by systemic ethanol administration, research using $GABA_A$-receptor modulators has identified this primary inhibitory neurotransmitter system as a key site for the behavioral effects of ethanol. A number of the behavioral effects of ethanol are enhanced by $GABA_A$-receptor agonists and attenuated by antagonists or inverse agonists (Lister and Linnoila, 1991). For example, picrotoxin, a potent $GABA_A$-receptor antagonist, and Ro 15-4513, a benzodiazepine receptor inverse agonist, reduce signs of ethanol intoxication in mice and rats (Suzdak et al., 1986; Lister, 1987; Hoffman et al., 1987). These two drugs also block the anxiolytic effects of ethanol (Glowa et al., 1988; Becker and Anton, 1990; Becker and Hale, 1991), whereas $GABA_A$-receptor agonists potentiate ethanol-induced anxiolysis (Goldstein, 1973). Therefore, many of the behavioral effects of ethanol mimic those of $GABA_A$-receptor agonists, suggesting the direct involvement of $GABA_A$ receptors in the action of ethanol.

Direct examination of ethanol interactions with $GABA_A$ receptors has suggested that ethanol increases $GABA_A$-receptor-mediated Cl^- influx in a variety of preparations. Physiologically relevant concentrations of ethanol (20–60 mM) potentiate muscimol-stimulated Cl^- uptake in cerebral cortical synaptoneurosomes of rats (Suzdak et al., 1986; Morrow et al., 1988), mouse cerebellar microsacs (Allan and Harris, 1987), and cultured spinal cord neurons (Ticku and Burch, 1980). Finally, ethanol potentiation of $GABA_A$-receptor-mediated Cl^- influx is blocked by Ro 15-4513 and $GABA_A$-receptor antagonists (Suzdak et al., 1986, 1988; Mehta and Ticku, 1988).

In contrast, electrophysiological investigations of ethanol effects on GABA responses *in vivo* raise questions about direct ethanol interactions with $GABA_A$ receptors. Acute ethanol administration potentiates GABA-mediated inhibition *in vivo* but only in specific brain regions or cell types (Celentano et al., 1988; Givens and Breese, 1990b; Proctor et al., 1992; Reynolds et al., 1992; Soldo et al., 1994; Frye et al., 1994; Aguayo et al., 1994; Sapp and Yeh, 1998). Systemic ethanol administration potentiates GABA-mediated inhibition in the cortex, medial septum, and inferior colliculus but not in the lateral septum or the hippocampus (Nestores, 1980; Mancillas et al., 1986; Siggins et al., 1987; Givens and Breese, 1990a; Simson et al., 1991; Criswell et al., 1993). Ethanol application by electro-osmosis potentiates GABA-mediated inhibition in several brain regions, but ethanol must by applied *before* GABA for the effect to be observed (Simson et al., 1991; Criswell et al., 1993). Investigations have shown that ethanol can potentiate $GABA_A$-receptor responses in the hippocampus under certain experimental conditions. Inclusion of intracellular adenosine triphosphate (ATP) (Weiner et al., 1994), cold shock treatment of hippocampal slices (Weiner et al., 1997a), or selective blockade of $GABA_B$ receptors by CPG-35348 (Wan *et al.*, 1996) have revealed ethanol potentiation of GABA-mediated inhibition in the hippocampus. Ethanol also potentiates GABA-mediated inhibition in hippocampal CA1 pyramidal cells following proximal, but not distal, stimulation (Weiner *et al.*, 1997c). In cerebellar *Purkinje* cells, β-adrenergic receptor stimulation sensitizes $GABA_A$ receptors to potentiation by ethanol (Lin *et al.*, 1991). However, none of these studies has conclusively demonstrated direct effects of ethanol on $GABA_A$ receptors under physiological conditions. Furthermore, direct interaction of ethanol with neuronal $GABA_A$ receptors using patch–clamp recording techniques has rarely been observed at physiological ethanol concentrations (Frye *et al.*, 1994; Crews *et al.*, 1996; Marszalec *et al.*, 1998). However, modulation of $GABA_A$-receptor function by ethanol has been reported in dorsal root ganglion cells (Nakahiro *et al.*, 1991). In outside-out patches from dorsal root ganglion cells, ethanol increased mean open time, frequency of opening, and burst duration while decreasing mean closed time of single $GABA_A$ receptors when it was co-applied with 1 μM GABA (Tatebayashi *et al.*, 1998). Other work has suggested that potentiation of GABAergic inhibitory post-synaptic potentials (IPSPs) by ethanol is indirect, resulting from inhibition of NMDA-mediated excitatory post-synaptic potentials (EPSPs) and subsequent release of Ca^{2+}-dependent modulation of the $GABA_A$ receptors (Marszalec *et al.*, 1998).

Interpretation of these discrepancies is complicated by the $GABA_A$ receptor's sensitivity to the effects of posttranslational modification. As mentioned β-adrenergic receptor stimulation and subsequent elevation of cyclic adenosine monophosphate (cAMP) and protein kinase A (PKA) activity are required to detect ethanol potentiation of GABA in a majority of cells. Furthermore, elevation of protein kinase C (PKC) in hippocampal slices increased the sensitivity of CA1 neurons to ethanol (Weiner *et al.*, 1997c). Because of the limitations of these model systems, (i.e., use of a deeply anesthetized animal or acute slice preparations), it is not clear whether low- or high-protein kinase activity is the normal state of the neuron. In other words, do these manipulations of protein PKA or PKC unmask a nonphysiological sensitivity to ethanol, or do they recreate the natural phosphorylation state of the $GABA_A$ receptor? These studies support the hypothesis that ethanol action on $GABA_A$ receptors is indirect and may involve distinct intermediates that are required for ethanol enhancement of GABA action in the CNS.

Ethanol potentiation of GABA responses has been clearly demonstrated in some recombinant $GABA_A$ receptors, but the relevance of this work to neuronal $GABA_A$ receptors is controversial (Harris *et al.*, 1998). Variability in ethanol responsive vs unresponsive cells might be a result of differences in the composition of $GABA_A$ isoreceptors or the presence of specific posttranslational modifiers. Studies in oocytes and stably transfected mouse

Ltk-cells showed that the $GABA_A$ receptor $\gamma 2L$ subunit was necessary for ethanol sensitivity (Wafford et al., 1991; Ducic et al., 1995), but other studies have not demonstrated a $\gamma 2L$ subunit requirement (Sigel et al., 1993; Kurata et al., 1993; Mihic et al., 1994; Sapp and Yeh, 1998). In addition, many recombinant $GABA_A$ receptors are insensitive to ethanol in any expression system. Studies using a series of chimeric GABA/glycine receptors have been used to demonstrate a binding pocket between TM2 and TM3 that controls ethanol responses as well as the dependence of these responses on alcohol chain length (Mihic et al., 1997; Wick et al., 1998). This model suggests that ethanol may interact directly with $GABA_A$ receptors at high millimolar concentrations. However, further work is still necessary to demonstrate direct ethanol interactions with neuronal $GABA_A$ receptors at intoxicating concentrations.

C. Ethanol Alters Brain Concentrations of GABAergic Neurosteroids

We have discovered that systemic ethanol administration (2 g/kg, intraperitoneally) elevates cerebral cortical $3\alpha,5\alpha$-TH PROG levels in male rats to 10–15 ng/g (Morrow et al., 1998), concentrations sufficient to potentiate $GABA_A$ receptors. Similar effects of ethanol on $3\alpha,5\alpha$-TH PROG and $3\alpha,5\alpha$-TH DOC were observed in Sardinian alcohol-preferring rats (Barbaccia et al., 1999). Therefore, it is conceivable that the effects of ethanol discussed previously could be mediated or facilitated by $3\alpha,5\alpha$-TH PROG produced by systemic ethanol administration or other experimental manipulations that increase pregnane steroid concentrations in the brain.

The effects of ethanol on cerebral cortical $3\alpha,5\alpha$-TH PROG levels in male rats are dose and time dependent (Fig. 1). The maximal effect of ethanol was observed at 2.5 g/kg ethanol. At higher ethanol concentrations (3.5–4.0 g/kg), the effect on cerebral cortical $3\alpha,5\alpha$-TH PROG levels diminished. This suggests that the effect of ethanol on brain $3\alpha,5\alpha$-TH PROG levels is biphasic, similar to the biphasic effects of ethanol on $GABA_A$-receptor-mediated Cl^- uptake in cerebral cortical synaptoneurosomes (Suzdak et al., 1986). The time course of the effect of acute ethanol administration (2 g/kg) suggests that $3\alpha,5\alpha$-TH PROG contributes to effects of systemic ethanol administration between 20 and 120 min following administration and reaches a plateau between 40 and 80 min after ethanol injection (Fig. 1B). The maximal effect was a $341 \pm 14\%$ increase in $3\alpha,5\alpha$-TH PROG levels above control. $3\alpha,5\alpha$-TH PROG levels were also slightly elevated in saline-injected control animals at 20 min but returned to control levels by 40 min. Hence, the contribution of the stress of injection to the effect of ethanol is minimal.

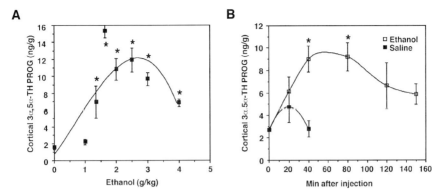

FIG. 1. Cortical 3α,5α-TH PROG levels are elevated following acute, systemic ethanol administration. (**A**) Cerebral cortical 3α,5α-TH PROG levels exhibit a biphasic response to increasing ethanol concentrations. Ethanol increased 3α,5α-TH PROG levels to pharmacologically active concentrations (*$P < 0.0001$ ANOVA, $P < 0.05$, Dunnet's post hoc) measured 60 min after ethanol administration. Data represent mean ± SEM of duplicate determinations in 6–10 rats/dose from two independent experiments. (**B**) Cerebral cortical 3α,5α-TH PROG levels peak between 40 and 80 min after an injection of ethanol (2 g/kg, intraperitoneally). Data shown are the mean ± SEM of a representative of two experiments ($n = 6$/time point; *$P < 0.05$ Dunnet's post hoc). (Adapted from M. J. VanDoren, D. B. Matthews, G. C. Janis, A. C. Grobin, L. L. Devaud, and A. L. Morrow. Neuroactive steroid 3α-hydroxy-5α-pregnan-20-one modulates electrophysiological and behavioral actions of ethanol. *Journal of Neuroscience* 2000; 20: 1982–1989.)

Plasma 3α,5α-TH PROG, progesterone, and corticosterone were also elevated after ethanol administration, but the levels of circulating steroids were not correlated with cortical levels of 3α,5α-TH PROG (Fig. 2). Therefore, the effects of ethanol administration on 3α,5α-TH PROG concentrations in the cerebral cortex may be mediated by mechanisms that are distinct from elevation of circulating 3α,5α-TH PROG or precursor.

Ethanol-induced levels of progesterone and corticosterone suggest a nonspecific activation of steroid biosynthesis and/or adrenal release. If pregnane steroid biosynthesis were activated nonspecifically by ethanol, then an increase in plasma or brain levels of dehydroepiandrosterone (DHEA) or its sulfated congener, DHEAS, would be predicted. However, systemic ethanol administration (2 g/kg, intraperitoneally) did not alter DHEA or DHEAS levels in several brain regions (Fig. 3). The lack of effect of ethanol on DHEA and DHEAS levels is consistent with the idea that androstane steroids are differentially regulated than pregnane steroids. Furthermore, these data emphasize the specificity of the effects of ethanol on 3α,5α-TH PROG and 3α,5α-TH DOC levels.

Because brain concentrations of 3α,5α-TH PROG above 5 ng/g would be expected to potentiate GABA-receptor-mediated Cl$^-$ flux (Morrow *et al.*,

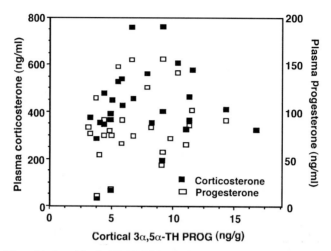

Fig. 2. Ethanol-induced 3α,5α-TH PROG levels do not correlate with plasma levels of corticosterone and progesterone. Plasma levels of corticosterone and progesterone were measured at several time points (20, 40, 80, 120 min) and compared with cortical levels of 3α,5α-TH PROG measured from the brains of the same animals. Circulating levels of corticosterone and progesterone did not correlate ($r = 0.179$, $p = 0.371$; $r = 0$, $P = 0.964$, respectively), with brain levels of 3α,5α-TH PROG.

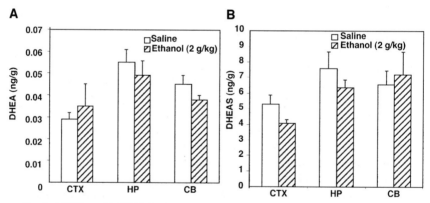

Fig. 3. DHEA and DHEAS levels in rat brain are not altered after systemic ethanol administration (2 g/kg, intraperitoneally, 20% ethanol in saline v/v). Male Sprague–Dawley rats were habituated to handling and sham injections for 5 days before the experiment. Rats were injected with saline or ethanol and decapitated 60 min later. (**A**) DHEA and (**B**) DHEAS levels were measured in the cerebral cortex (CTX), the hippocampus (HP), and the cerebellum (CB) by radioimmunoassay, each taken from the same animal. No significant changes in DHEA or DHEAS levels were found after ethanol administration ($n = 4$–5/group).

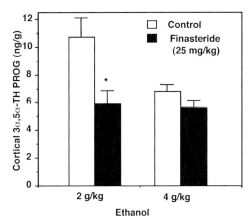

FIG. 4. Finasteride pre-administration blocks ethanol-induced $3\alpha,5\alpha$-TH PROG levels at low, but not high, doses. Groups of 6–10 well-handled and sham-injected male rats were given finasteride or vehicle 4 and 1.5 h before ethanol (2 or 4 g/kg, intraperitoneally, 20% ethanol in saline v/v) or saline injections. Brains were collected 60 min after ethanol injection, and $3\alpha,5\alpha$-TH PROG levels measured by radioimmunoassay. Finasteride pretreatment significantly reduced ethanol induction of $3\alpha,5\alpha$-TH PROG at the 2 g/kg dose ($P < 0.05$) but not at the higher 4 g/kg dose.

1987, 1990b), we hypothesized that ethanol induction of $3\alpha,5\alpha$-TH PROG in the brain could contribute to the pharmacological effects of ethanol *in vivo*. Ethanol could increase $3\alpha,5\alpha$-TH PROG levels via several mechanisms, including modulation of biosynthesis, catabolism, or release of an endogenous steroid pool. To investigate the mechanism by which ethanol increases $3\alpha,5\alpha$-TH PROG levels, specific steroid biosynthetic inhibitors were administered preceding ethanol administration to block subsequent $3\alpha,5\alpha$-TH PROG synthesis. Pre-administration of the 5α-reductase inhibitor finasteride (Normington and Russell, 1992) partially reversed the increase in cortical $3\alpha,5\alpha$-TH PROG levels induced by moderate ethanol (1.5–2.0 g/kg) doses (Fig. 4). This result indicates that ethanol modulation of *de novo* steroid biosynthesis may play a role in the mechanism by which moderate doses of ethanol elevate cerebral cortical $3\alpha,5\alpha$-TH PROG levels. However, finasteride pretreatment had no effect on elevations in $3\alpha,5\alpha$-TH PROG levels following hypnotic doses of ethanol (4.0 g/kg). Hence, it is likely that ethanol-induced elevation of brain $3\alpha,5\alpha$-TH PROG levels involves mechanisms other than *de novo* biosynthesis.

The 3α-hydroxysteroid oxidoreductase inhibitor indomethacin has been reported to block the biosynthesis of $3\alpha,5\alpha$-TH PROG (Smith *et al.*, 1998a). However, this biosynthetic enzyme inhibitor increased, rather than decreased, ethanol-induced $3\alpha,5\alpha$-TH PROG levels (Morrow *et al.*, 1998). The 3α-hydroxysteroid enzyme is responsible for the bidirectional conversion between 5α-dihydroprogesterone and 3α,5α-TH PROG (Costa *et al.*, 1994). Indomethacin pretreatment could increase 3α,5α-TH PROG concentrations by blocking oxidation of 3α,5α-TH PROG rather than blocking reduction of 5α-dihydroprogesterone to 3α,5α-TH PROG.

D. 3α,5α-TH PROG Contributes to Specific Effects of Ethanol

Having established a correlation between acute ethanol administration and 3α,5α-TH PROG levels, we investigated the hypothesis that 3α,5α-TH PROG contributes to the behavioral effects of ethanol. Sedative/hypnotic effects produced by high doses of ethanol are potentially mediated by $GABA_A$ receptors by either direct or indirect actions. These effects can be enhanced by GABA agonists and blocked by GABA-receptor antagonists (Martz *et al.*, 1983). A good correlation between cerebral cortical 3α,5α-TH PROG levels and ethanol sleep time was observed (VanDoren *et al.*, 2000b). However, finasteride had no effect on ethanol-induced cerebral cortical 3α,5α-TH PROG levels at hypnotic doses (Fig. 4) and, correspondingly, no effect on ethanol sleep-time (data not shown). Indeed, the correlation between cerebral cortical concentrations of 3α,5α-TH PROG and the duration of the loss of righting reflex was also observed in rats that were pretreated with finasteride (unpublished data).

Because finasteride pretreatment blocked ethanol-induced 3α,5α-TH PROG levels in the cerebral cortex using lower ethanol doses (2 g/kg, intraperitoneally), the ability of finasteride pretreatment to alter behavioral effects of ethanol was examined at this dose. Finasteride pretreatment had no effect on bicuculline-induced seizure thresholds in vehicle-injected control rats but completely prevented the anticonvulsant effect of ethanol at 40 min (VanDoren *et al.*, 2000b). Moreover, finasteride did not alter seizure thresholds 10 min after ethanol administration, a time when ethanol lacks an anticonvulsant effect. These data support the supposition that ethanol induction of 3α,5α-TH PROG levels in the brain mediates or modulates the anticonvulsant effects of ethanol via $GABA_A$ receptors.

Several behavioral effects of ethanol can be observed within minutes of systemic ethanol administration. For example, ethanol produces muscle relaxation and motor incoordination that can be investigated by measurement of the Majchrowicz Intoxication Scale (Majchrowicz, 1975) or the aerial righting reflex (Van Rijn *et al.*, 1990). To determine whether 3α,5α-TH PROG contributes to these actions of ethanol, the effect of finasteride pretreatment (50 mg/kg, 1.5 and 4 h prior to ethanol) was investigated in both of these tests. Finasteride did not block the effect of ethanol in these measures but enhanced ethanol effects at some time points. Because finasteride reduces some effects of the ethanol and increases others it appears that $3\alpha,5\alpha$-TH PROG may contribute to specific behavioral effects of ethanol. However, the mechanism for finasteride enhancement of specific actions of ethanol is unknown. Because $3\alpha,5\alpha$-TH PROG biosynthesis is differentially regulated throughout brain (Mellon and Deschepper, 1993), the effects of ethanol on $3\alpha,5\alpha$-TH PROG levels may differ in various brain regions. Alternatively, finasteride may lack effects on specific 5α-reductase isozymes (Russell and Wilson, 1994) that are responsible for the formation of $3\alpha,5\alpha$-TH PROG. Finally, finasteride administration may increase the formation of another $GABA_A$-receptor active neurosteroid, 3α-hydroxypregn-4-ene-20-one, which could be formed by alternate pathways when the major 5α-reductase pathway is inhibited. Hence, it is not yet clear whether $3\alpha,5\alpha$-TH PROG contributes to the motor-incoordinating effects of ethanol. Further studies are needed to resolve these issues.

The ability of $3\alpha,5\alpha$-TH PROG to mediate ethanol-induced effects on neurons in the brain was examined directly by monitoring the spontaneous firing rates of medial septum/diagonal band (MS/DB) of Broca neurons (VanDoren *et al.*, 2000b). Ethanol significantly reduces spontaneous firing rates 40–50% without altering the amplitude of action potentials. The effect was observed 15 min following ethanol injection and sustained for up to 55 min. This time course corresponds to the time course of ethanol-induced elevation in $3\alpha,5\alpha$-TH PROG levels in cerebral cortex. Finasteride (25–50 mg/kg) pretreatment had no effect on the spontaneous firing rate of MS/DB neurons but blocked the effect of ethanol on the spontaneous firing rate. Finasteride (50 mg/kg) reversed the effect of ethanol throughout the 60-min recording session, while the 25 mg/kg dose altered the effect of ethanol immediately following ethanol administration 15–20 min ($P < 0.05$), but not at later time points (35–60 min). To determine whether finasteride prevented inhibition of MS/DB neurons in a nonspecific manner, Matthews *et al.* (1995) demonstrated that the effect of exogenous $3\alpha,5\alpha$-TH PROG was not altered by finasteride pretreatment (data not shown), suggesting that finasteride pretreatment does not alter the ability of MS/DB neurons to respond to GABA-mediated inhibition. Therefore, it is likely that finasteride pretreatment prevents the formation of $3\alpha,5\alpha$-TH PROG that normally mediates or modulates the effect of ethanol on MS/DB neurons.

III. Role of Neurosteroids in Alcohol Reinforcement

The GABAergic system is important in ethanol self-administration and reinforcement of ethanol drinking. The reward pathway in the brain is thought to be a key site of action of several drugs of abuse. This pathway, composed of the nucleus accumbens, amygdala, and ventral tegmental area, is thought to modulate reinforcement, motivation, and self-administration of these drugs (Koob, 1996). $GABA_A$-receptor agonists increase both acquisition of ethanol-drinking behavior (Smith *et al.*, 1992; Petry, 1997) and volume of ethanol consumed (Pohorecky and Brick, 1988; Boyle *et al.*, 1993). Likewise, GABAergic inverse agonists decrease drinking behavior in nondependent rats (McBride *et al.*, 1988; Balakleevsky *et al.*, 1990) and suppress the reinforcing value of ethanol (Samson *et al.*, 1987; Rassnick *et al.*, 1993).

A. $3\alpha,5\alpha$-TH PROG AND RELATED STEROIDS ALTER ETHANOL SELF-ADMINISTRATION

The GABAergic neurosteroid $3\alpha,5\alpha$-TH PROG has effects similar to other GABAergic agonists on ethanol intake. $3\alpha,5\alpha$-TH PROG increased ethanol intake in nondependent rats (Janak *et al.*, 1998) and dose-dependently increased self-administration in nondependent, ethanol-preferring (P) rats (VanDoren *et al.*, 2000a). However, the effect of $3\alpha,5\alpha$-TH PROG depends on the drinking history of the animal. In ethanol-dependent P rats with high levels of ethanol consumption, $3\alpha,5\alpha$-TH PROG decreased ethanol self-administration. Therefore, $3\alpha,5\alpha$-TH PROG appears to increase drinking behavior in low-drinking rats and decrease ethanol self-administration in high-drinking, dependent rats. This suggests that endogenous neurosteroids could influence drinking behavior and genetic differences in alcohol preference. Further studies are needed to address this possibility.

Samson *et al.* (1998) have developed a method to distinguish the act of seeking ethanol from the subsequent drinking behavior in rats pretrained to lever-press to drink. Using this paradigm, they found that pre-administration of 3 mg/kg, but not 1 or 10 mg/kg, of $3\alpha,5\alpha$-TH PROG increased the number of times a rat pressed the lever to receive ethanol (Janak *et al.*, 1998). However, the higher dose of $3\alpha,5\alpha$-TH PROG (10 mg/kg) reduced rates of lever pressing for ethanol. Thus, in this paradigm, pre-administration of $3\alpha,5\alpha$-TH PROG modulated ethanol-seeking behavior in opposing directions, dependent on dose (Janak *et al.*, 1998).

Innate preference for ethanol may be related to the effects of ethanol on endogenous levels of GABAergic steroids. The Sardinian alcohol-preferring (sP) and non-preferring (sNP) rats are strains of rat selectively bred according to preference for alcohol consumption (Barbaccia *et al.*, 1999). When administered a low dose (1 g/kg) of ethanol, these strains show dramatically different ethanol-induced elevations in both $3\alpha,5\alpha$-TH PROG and $3\alpha,5\alpha$-TH DOC. Ethanol-induced levels of $3\alpha,5\alpha$-TH PROG are four times higher in the cortex and double in the hippocampus of sP rats over sNP rats (Barbaccia *et al.*, 1999). Levels of ethanol-induced 3α,5α-TH DOC are also higher in sP rats compared to sNP rats (Barbaccia *et al.*, 1999). Given the previous evidence that highlights the importance of neurosteroids in ethanol reward, this is further evidence that endogenous neurosteroids may be related to alcohol preference.

Ethanol self-administration can also be altered by stress. Both food restriction and social isolation stress have been reported to increase ethanol-seeking behavior, while food restriction increases the amount of ethanol self-administered (Piazza and Le Moal, 1998). Interestingly, 3α,5α-TH PROG has been reported to cause a hyperphagic effect at low doses (0.25 mg/kg), while the GABAergic antagonist DHEAS (5 mg/kg) induced an anoretic effect (Reddy and Kulkarni, 1999). Like 3α,5α-TH PROG effects on ethanol self-administration, this effect was sensitive enough to vary with the estrous cycle in female rats (Reddy and Kulkarni, 1999). In summary, GABAergic steroids play important roles in ethanol reward, self-administration, and preference in ethanol consumption. These steroids may contribute to regulation of behavioral as well as physiological responses to ethanol.

IV. Role of Neurosteroids in Ethanol Tolerance

Prolonged or repeated ethanol intake leads to the development of tolerance. Tolerance is demonstrated by a reduced behavioral response to ethanol at a particular cellular concentration, permitting alcoholic patients to remain sober with blood ethanol concentrations that produce severe damage to many organ systems. Tolerance results in a reduction in the anxiolytic, motor-incoordinating, and sedative/hypnotic effects of ethanol and a requirement for higher ethanol doses to elicit pharmacological effects (Boisse and Okamoto, 1980; Le *et al.*, 1986; Brown *et al.*, 1988). Alcohol withdrawal symptoms include agitation, anxiety, and minor tremors for less severe withdrawal and may result in generalized seizures with severe withdrawal (Sellers and Kalant, 1976). Prolonged ethanol exposure in animal models produces cellular adaptations similar to tolerance and dependence (i.e., a reduced response to ethanol and a requirement for ethanol to maintain normal function).

Although acute systemic ethanol administration elevates 3α,5α-TH PROG levels in the brain, chronic ethanol consumption does not alter steady-state 3α,5α-TH PROG levels (Janis *et al.*, 1998). Indeed, ethanol-dependent male rats show decreased levels of 3α,5α-TH PROG in the cerebral cortex as compared to pair-fed controls (Janis *et al.*, 1998). Using

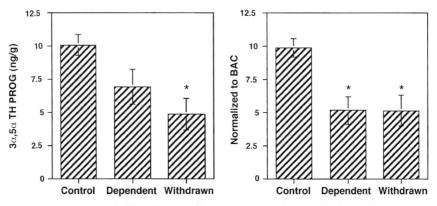

FIG. 5. Ethanol-induced 3α,5α-TH PROG levels are reduced in ethanol-dependent and withdrawn animals. Male Sprague Dawley rats were administered a liquid diet containing ethanol or equicaloric dextrose for 2 weeks using a pair-fed design. Dependent rats were tested on Day 15, while the ethanol diet was freely available. Withdrawn rats were tested 6–8 h following removal of ethanol diet from their cages. Rats were handled and sham injected during ethanol administration and, on the test day each control, dependent, or withdrawn animal received a challenge dose of ethanol (2 g/kg intraperitoneally, 20% ethanol in saline). Animals were sacrificed 60 min after ethanol challenge and cerebral cortical 3α,5α-TH PROG levels were measured along with plasma blood alcohol concentration (BAC). Ethanol-induced 3α,5α-TH PROG levels were significantly decreased in ethanol withdrawn animals ($P < 0.002$) compared to pair-fed controls. When adjusted for BAC, ethanol-induced 3α,5α-TH PROG levels were reduced in both ethanol-dependent and withdrawn animals ($P < 0.002$).

the same paradigm, withdrawn animals showed a return of cortical 3α,5α-TH PROG to baseline levels (Janis et al., 1998). However, when ethanol-dependent and withdrawn animals were given a 2-g/kg ethanol challenge, both ethanol-dependent and withdrawn animals showed reduced levels of ethanol-induced 3α,5α-TH PROG in the cerebral cortex (Fig. 5). In fact, when adjusted for blood alcohol content, the levels of ethanol-induced 3α,5α-TH PROG were significantly lower in both dependent and withdrawn rats. Hence, tolerance to ethanol may involve a loss of ethanol induction of 3α,5α-TH PROG and a corresponding reduction in various pharmacological effects of ethanol.

If 3α,5α-TH PROG contributes to the effects of ethanol, one would predict that these effects would be absent in ethanol-tolerant animals. Whether ethanol inhibition of spontaneous neural activity in the medial septum is lost in ethanol-dependent rats is unknown. However, tolerance to the anticonvulsant (Kokka et al., 1993) and hypnotic (Silveri and Spear, 1999) effects of ethanol are well established. Moreover, there are other lines of evidence that are consistent with this idea. Tolerance to ethanol can be environment dependent (Melchior and Tabakoff, 1985). If alcohol administration is repeatedly paired with a specific environment, then tolerance can be demonstrated in that environment but is absent if the animal is moved to a novel environment. Stress-induced elevation of $3\alpha,5\alpha$-TH PROG levels may contribute to this response, as novelty itself is a potent stressor (Morrow et al., 1998). In this model, stress activation of the HPA axis may elevate $3\alpha,5\alpha$-TH PROG levels and reverse tolerance by enhancing the effect of ethanol. Furthermore, tolerance can be maintained by vasopressin activation of the HPA axis (Hoffman, 1993). We propose that prolonged ethanol exposure reduces the ability of ethanol (and vasopressin) to induce $3\alpha,5\alpha$-TH PROG concentrations in brain, and the decrease in $3\alpha,5\alpha$-TH PROG concentrations could contribute to the reduced ethanol response (tolerance). This mechanism could further explain the important role of the HPA axis in tolerance development and the ability of vasopressin to maintain ethanol tolerance in the absence of ethanol.

In studies of the genetics of alcoholism, researchers found that sons of alcoholics are more likely to have a higher initial tolerance to ethanol than matched controls with no family history of alcoholism (Schuckit and Smith, 1996; Schuckit et al., 1996). Furthermore, the most robust predictor of the later development of alcoholism among family-history-positive individuals was changes in circulating adrenocorticotropic hormone and cortisol levels in response to ethanol (Schuckit, 1994). Cortisol and its rodent-equivalent corticosterone are the primary signaling molecules of HPA-axis activation and the systemic stress response. The reduced cortisol response in sons of alcoholics who later develop alcoholism may indicate that HPA axis activation plays an important role in alcohol tolerance and consumption.

Alterations in $GABA_A$-receptor function and expression may also contribute to the development of ethanol tolerance (for review see Grobin et al., 1998). Chronic ethanol exposure alters many of the properties of $GABA_A$ receptors in the brain. Ethanol tolerance and dependence are associated with a decrease in the sensitivity of $GABA_A$-receptor-mediated responses. Chronic ethanol administration differentially alters the expression of various $GABA_A$-receptor subunits, suggesting that alterations in the subunit expression of $GABA_A$ receptors may account for alterations in $GABA_A$-receptor function (Morrow et al., 1990a, 1992, Buck et al., 1991; Montpied et al., 1991; Mhatre and Ticku, 1992; Mhatre et al., 1993). Other evidence suggests that chronic ethanol administration may result in the functional uncoupling of GABA and benzodiazepine recognition sites (Devries et al., 1987; Weiner et al., 1997b) and that these effects may not be due to alterations in $GABA_A$-receptor gene expression (Klein et al., 1995). Although the exact mechanisms that account for alterations in $GABA_A$-receptor function following chronic ethanol administration are still under avid investigation, it is clear that adaptations in $GABA_A$ receptors play an important role in ethanol tolerance. These effects of ethanol on $GABA_A$-receptor function and gene expression could be mediated, at least in part, by $3\alpha,5\alpha$-TH PROG. There is also a growing body of evidence that neurosteroids may directly regulate $GABA_A$-receptor function and expression. Therefore, the interplay of the neuromodulating and genomic effects of $3\alpha,5\alpha$-TH PROG may play a key role in ethanol tolerance.

V. Role of Neurosteroids in Ethanol Dependence

A. ETHANOL-DEPENDENT RATS ARE SENSITIZED TO GABAERGIC NEUROSTEROIDS

Changes in $GABA_A$-receptor function are thought to underlie some of the consequences of ethanol dependence. In particular, the increase in anxiety and the decrease in seizure threshold seen in dependent animals and humans during ethanol withdrawal may be mediated by changes in GABAergic neurotransmission that are brought about by chronic exposure to a $GABA_A$-receptor-potentiating drug (Grobin et al., 1998). Because $3\alpha,5\alpha$-TH PROG protects against seizures induced by $GABA_A$-receptor antagonists (Belelli et al., 1989), it was hypothesized that $3\alpha,5\alpha$-TH PROG could block the increase in seizure susceptibility seen in rodents after withdrawal from chronic ethanol exposure. Administration of $3\alpha,5\alpha$-TH PROG blocks the decrease in bicuculline-induced seizure threshold seen in ethanol-withdrawn male and female rats after 14 days of exposure to ethanol in a liquid diet. The antagonism of bicuculline-induced seizures occurred at doses of $3\alpha,5\alpha$-TH PROG that do not increase seizure threshold in pair-fed control rats (Devaud et al., 1995a). Indeed, there is a dramatic sensitization to the anticonvulsant effects of $3\alpha,5\alpha$-TH PROG and $3\alpha,5\alpha$-TH DOC in ethanol-dependent rats (Devaud et al., 1996, 1998).

Behavioral sensitization to neurosteroids is accompanied by an increase in sensitivity of $GABA_A$ receptors to $3\alpha,5\alpha$-TH PROG and $3\alpha,5\alpha$-TH DOC in cerebral cortical synaptoneurosomes using a Cl^- flux assay (Devaud et al., 1996) and is not due to different endogenous levels of $3\alpha,5\alpha$-TH PROG in withdrawn vs control rats (Janis et al., 1998). Furthermore, neurosteroid sensitization during withdrawal is accompanied by an increase in $GABA_A$ receptor γ_1-, β_2- and β_3-receptor subunit mRNA and peptides (Devaud et al., 1996). These subunits have been shown to maintain or increase sensitivity of $GABA_A$ receptors to neurosteroid in recombinant expression systems (Puia et al., 1990, 1991).

Sensitization to neurosteroids in ethanol-dependent animals may have important clinical implications. Attempts to use the other $GABA_A$-receptor allosteric modulators, such as benzodiazepines, to alleviate the symptoms of ethanol withdrawal in humans are complicated by the fact that alcoholics often show cross tolerance to these substances. Because neurosteroids do not exhibit cross tolerance, but sensitization, they may be good candidates for treatment of the symptoms of ethanol withdrawal.

B. Alcohol Withdrawal Seizure Sensitivity Associated with Altered GABAergic Neurosteroid Levels

Increased sensitivity to the anticonvulsant properties of $3\alpha,5\alpha$-TH PROG in dependent animals has also been shown in mice. Furthermore, sensitivity to $3\alpha,5\alpha$-TH PROG has been linked to genetic resistance to seizures before ethanol exposure and during ethanol withdrawal. Two inbred mouse strains, C57BL/6J (B6) and DBA/2J (D2), differ substantially in their overall seizure susceptibility and in the severity of seizures during ethanol withdrawal. Ethanol naïve B6 mice are generally resistant to convulsant-induced seizures as compared to D2 mice (Kosobud and Crabbe, 1990). In addition, while handling-induced seizure sensitivity is increased during ethanol withdrawal in both B6 and D2 mice, the D2 mice exhibit a much greater increase in seizure susceptibility than do the B6 mice. In ethanol naïve animals, $3\alpha,5\alpha$-TH PROG dose dependently increased the dose of pentylenetetrazol (PTZ), a Cl^- channel blocker, required to induce seizure in both the B6 and the D2 strains (Finn *et al.*, 1997). However, there were differences in sensitivity to $3\alpha,5\alpha$-TH PROG between the strains, and these differences depended on the type of seizure assay used. In addition, Finn *et al.* (1997) showed that B6 and D2 mice have different sensitivities to $3\alpha,5\alpha$-TH PROG's anxiolytic, locomotor-stimulant, muscle relaxant, and ataxic properties. B6 mice were more sensitive than D2 mice to $3\alpha,5\alpha$-TH PROG's anxiolytic and locomotor-stimulant properties as measured by performance on an elevated plus-maze. In contrast D2 mice were more sensitive than B6 mice to the muscle-relaxant and ataxic properties of $3\alpha,5\alpha$-TH PROG. In all cases, plasma concentrations of $3\alpha,5\alpha$-TH PROG were similar in B6 and D2 mice, indicating that the differences in behavioral sensitivity are not due to differences in steroid metabolism (Finn *et al.*, 1997). Together, the behavioral and seizure threshold data are consistent with an increased sensitivity of $GABA_A$ receptors to neurosteroids in limbic structures in B6 vs D2 mice. If increased sensitivity to exogenous $3\alpha,5\alpha$-TH PROG indicates increased sensitivity to endogenous neurosteroid, then it is possible that the greater steroid sensitivity in B6 vs D2 mice is responsible for their relative seizure resistance.

When B6 and D2 mice are withdrawn from ethanol vapor administration, differences in seizure susceptibility between the two strains also occur. D2 mice experience much greater ethanol withdrawal severity, as measured by an increase in handling-induced convulsions, than B6 mice (Crabbe, 1998). Interestingly, under these conditions, the B6 vs D2 difference in sensitivity to exogenous neurosteroid is enhanced. When the animals are withdrawn from 72-h ethanol vapor inhalation, an increase in sensitivity to $3\alpha,5\alpha$-TH PROG's anticonvulsant effects, relative to air-exposed controls, was observed in B6 mice. D2 mice did not exhibit any sensitization to neurosteroid. This finding was independent of the convulsion endpoint used (Finn et al., 2000). Despite the increase in $3\alpha,5\alpha$-TH PROG's antagonism of PTZ-induced myoclonic twitch and face and forelimb clonus in B6 mice, no accompanying increase in sensitivity to $3\alpha,5\alpha$-TH PROG's anxiolytic or locomotor-activating properties was observed. This result indicates that increases in sensitivity to the anticonvulsant effects of $3\alpha,5\alpha$-TH PROG do not generalize to other behavioral effects of neurosteroids, at least in B6 mice. In summary, these data show that a genetic susceptibility to seizures during ethanol withdrawal is correlated with an inability to become sensitized to the anticonvulsant effect of $3\alpha,5\alpha$-TH PROG. Therefore, the greater sensitivity to endogenous neurosteroids may protect against seizures both in the drug-naïve state and during ethanol withdrawal.

Neurosteroid protection against withdrawal seizure severity does not generalize to all mouse strains. Withdrawal seizure prone (WSP) mice have been selectively bred for susceptibility to handling-induced convulsions during withdrawal from ethanol vapor inhalation, whereas withdrawal seizure resistant (WSR) mice exhibit fewer handling-induced convulsions during ethanol withdrawal (Crabbe et al., 1985). Ethanol naïve male WSP mice have a higher sensitivity to $3\alpha,5\alpha$-TH PROG than WSR mice. $3\alpha,5\alpha$-TH PROG dose dependently increased the dose of PTZ required to induce seizure in both the WSP and the WSR animals. However, the increases in seizure threshold for the myoclonic twitch and face and forelimb clonus convulsion endpoints were greater in WSP than in WSR mice (Finn et al., 1995). Further increases in neurosteroid sensitivity were seen in WSP mice during withdrawal from chronic ethanol exposure. $3\alpha,5\alpha$-TH PROG was more effective at blocking handling-induced convulsions in ethanol-withdrawn WSP mice relative to control, ethanol-naive, WSP mice (Finn et al., 1995).

In summary, for the WSP and the WSR animals, increased sensitivity to exogenous $3\alpha,5\alpha$-TH PROG did not correlate with a decreased seizure susceptibility. In this case, the seizure-prone animals were more sensitive to neurosteroid. In addition, unlike the D2 mice, WSP animals showed sensitization to $3\alpha,5\alpha$-TH PROG but still experienced severe withdrawal signs. Taken together, the B6/D2 and WSP/WSR data suggest that correlation of neurosteroid sensitivity with ethanol withdrawal seizure severity is dependent on the genetic background of the animals, implying that multiple mechanisms can contribute to seizure susceptibility. In some cases, sensitivity of $GABA_A$ receptors to endogenous neurosteroids may be a critical determinant of seizure severity (B6/D2 mice). In other cases, the contribution of neurosteroid sensitivity may be less important (WSP/WSR mice).

C. Relationship of Adaptations in Neurosteroid Levels and Neurosteroid Sensitivity of $GABA_A$ Receptors

The relationship between neurosteroid levels and the sensitivity of $GABA_A$ receptors to neurosteroids remains controversial. Although one study found no change in endogenous $3\alpha,5\alpha$-TH PROG levels in the cortex or plasma of ethanol-withdrawn rats (Janis *et al.*, 1998), other studies have reported decreases in neurosteroid levels in ethanol-withdrawing mice and humans. Plasma $3\alpha,5\alpha$-TH PROG levels were decreased in ethanol-withdrawn B6 and D2 mice, relative to controls (Finn *et al.*, 2000). Interestingly, the B6 mice showed a decrease in $3\alpha,5\alpha$-TH PROG levels of 15%, whereas there was a 50% decrease in D2 mice. The greater reduction in endogenous neurosteroid in withdrawn D2 mice could contribute to the greater withdrawal severity experienced by those animals. In alcoholics, decreased levels of both $3\alpha,5\alpha$-TH PROG and $3\alpha,5\alpha$-TH DOC have been reported during the early phase of ethanol withdrawal (Romeo *et al.*, 1996). This early phase of ethanol withdrawal is characterized by increases in anxiety and depression that can be measured by psychological tests. Later in withdrawal, when the anxiety and depression disappear, neurosteroid levels also return to normal. As in the mice, this study suggests that reduced levels of endogenous neurosteroids contribute to the negative consequences of ethanol withdrawal.

D. Relationship between Changes in Neurosteroid Levels and Altered $GABA_A$-Receptor Function

Withdrawal from progesterone, and associated decreases in $3\alpha,5\alpha$-TH PROG, result in behavioral changes that mimic the signs of ethanol withdrawal. Specifically, progesterone-withdrawn female rats exhibit increased anxiety (Gallo and Smith, 1993), increased sensitivity to convulsant drugs that act on $GABA_A$ receptors (Smith *et al.*, 1998a), and decreased sensitivity to the sedative effects of benzodiazepines (Moran and Smith, 1998). Whole-cell patch–clamp electrophysiology in dissociated neurons from CA1 of the hippocampus showed that the behavioral changes were accompanied by changes in the function of $GABA_A$ receptors (Smith *et al.*, 1998b). These changes are associated with increases in α_4-subunit peptide and mRNA during progesterone withdrawal (Smith *et al.*, 1998a,b). Furthermore, they show that blocking this increase using antisense α_4 infusion into the hippocampus during progesterone withdrawal eliminates the increase in seizure sensitivity (Smith *et al.*, 1998a), behavioral tolerance to benzodiazepines (Moran *et al.*, 1998), and the change in electrophysiological properties of $GABA_A$ receptors (Smith *et al.*, 1998a). Therefore, in this model, it seems that withdrawal from exposure to progesterone and $3\alpha,5\alpha$-TH PROG causes a change in $GABA_A$-receptor subunit composition that results in the changes in $GABA_A$ receptor function that are responsible for the behavioral signs of withdrawal. Therefore, it seems plausible that other situations that result in decreases in levels of endogenous $3\alpha,5\alpha$-TH PROG may result in similar changes in $GABA_A$-receptor function and similar behavioral signs.

Ethanol- and progesterone-withdrawing animals exhibit similar withdrawal signs, such as increased anxiety, increased seizure susceptibility, and decreased sensitivity to benzodiazepines. As discussed decreases in neurosteroid levels have been reported in ethanol-withdrawing mice (Finn *et al.*, 2000) and humans (Romeo *et al.*, 1996). Furthermore, decreases in cortical $3\alpha,5\alpha$-TH PROG have been observed in ethanol-dependent male rats relative to ethanol-naïve controls (Janis *et al.*, 1998). This decrease occurs at the same time as increases in α_4-subunit mRNA (Devaud *et al.*, 1995b) and peptide levels (Devaud *et al.*, 1997) in the cerebral cortices of ethanol-dependent rats. These adaptations are associated with alterations in $GABA_A$-receptor function, including reduced sensitivity to GABA and benzodiazepines and increased sensitivity to inverse agonists (see for review Grobin *et al.*, 1998). Therefore, it is possible that changes in endogenous neurosteroid levels that occur during chronic ethanol exposure and/or withdrawal contribute to the changes in $GABA_A$-receptor function that underlie ethanol dependence. However, ethanol withdrawal is associated with other changes in $GABA_A$-receptor expression and function that were not reported following progesterone withdrawal, including changes in β- and γ-subunit expression and sensitization to $3\alpha,5\alpha$-TH PROG (Devaud *et al.*, 1996, 1997). Therefore, further studies are needed to determine the role of neurosteroids in ethanol-induced adaptations of $GABA_A$ receptors.

VI. Conclusions and Future Directions

Alterations in neurosteroid levels in brain may be an important factor that controls the expression of $GABA_A$ receptors and the sensitivity of $GABA_A$ receptors to ethanol and other modulators. $3\alpha,5\alpha$-TH PROG can be oxidized rapidly to 5α-dihydroprogesterone, which has potent effects on nuclear progesterone-receptor-mediated gene expression (Rupprecht et al., 1994). Furthermore, ethanol directly elevates circulating progesterone levels as well as $3\alpha,5\alpha$-TH PROG (VanDoren et al., 2000b). *Therefore, the significance of this work could reach beyond the effects of acute ethanol administration and help to elucidate the mechanisms involved in the regulation of $GABA_A$ receptors by chronic ethanol administration.* The equilibrium between the function and expression of $GABA_A$ receptors and the concentrations of their endogenous modulators, the neuroactive steroids, may underlie the pharmacological effects of ethanol as well as adaptations to the effects of chronic ethanol exposure. Finally, this mechanism is likely to be relevant to human alcoholism, as ethanol effects on circulating steroid levels in young men (and other measures of high initial tolerance) are strong predictors of the eventual development of alcoholism (Schuckit, 1994).

Several key questions remain unanswered. If the research in this review convinces us of the synaptic availability of neurosteroids, then there is a compelling rationale to uncover the mechanisms that regulate the release of neurosteroids into the synapse. Are these compounds released from neurons, astrocytes, or oligodendrocytes? What is the trigger for release in basal conditions and in response to ethanol, HPA-axis activation, or other types of stimulation? What are the consequences of the dysregulation of neurosteroid biosynthesis and synaptic activity, and do these processes underlie pathological conditions, such as mental disorders and alcoholism? Clearly, these questions are of great interest to our field and will lead to significant advances in our understanding of the role of neurosteroids in brain function.

References

Aguayo, L. G., Pancetti, F. C., Klein, R. L., and Harris, R. A. (1994). Differential effects of GABAergic ligands in mouse and rat hippocampal neurons. *Brain Res.* **647,** 97–105.

Allan, A. M., and Harris, R. A. (1987). Acute and chronic ethanol treatments alter GABA receptor-operated chloride channels. *Pharmacol. Biochem. Behav.* **27,** 665–670.

Balakleevsky, A., Colombo, G., Fadda, F., and Gessa, G. L. (1990). Ro 19-4603, a benzodiazepine inverse agonist, attenuates voluntary ethanol consumption in rats selectively bred for high ethanol preference. *Alc. Alcohol.* **25,** 449.

Barbaccia, M. L., Affricano, D., Trabucchi, M., Purdy, R. H., Colombo, G., Agabio, R., and Gessa, G. L. (1999). Ethanol markedly increases "GABAergic" neurosteroids in alcohol-preferring rats. *Eur. J. Pharmacol.* **384,** R1–R2.

Becker, H. C., and Anton, R. F. (1990). Valproate potentiates and picrotoxin antagonizes the anxiolytic action of ethanol in a nonshock conflict task. *Neuropharmacology* **29,** 837–843.

Becker, H. C., and Hale, R. L. (1991). RO15-4513 antagonizes the anxiolytic effects of ethanol in a nonshock conflict task at doses devoid of anxiogenic activity. *Pharmacol. Biochem. Behav.* **39,** 803–807.

Belelli, D., Bolger, M. B., and Gee, K. W. (1989). Anticonvulsant profile of the progesterone metabolite 5α-pregnan-3α-ol-20-one. *Eur. J. Pharmacol.* **166,** 325–329.

Boisse, N. N., and Okamoto, M. (1980). Ethanol as a sedative-hypnotic: Comparison with barbiturate and non-barbiturate sedative-hypnotics. *In* "Alcohol Tolerance and Dependence" (H. Rigter and J. C. Crabbe, Eds.), pp. 265–292, Elsevier, Amsterdam.

Boyle, A. E., Segal, R., Smith, B. R., and Amit, Z. (1993). Bidirectional effects of GABAergic agonists and antagonists on maintenance of voluntary ethanol intake in rats. *Pharmacol. Biochem. Behav.* **46,** 179–182.

Brown, M. E., Anton, R. F., Malcolm, R., and Ballenger, J. C. (1988). Alcohol detoxification and withdrawal seizures: Clinical support for a kindling hypothesis. *Biol. Psych.* **23,** 507–514.

Buck, K. J., Hahner, L., Sikela, J., and Harris, R. A. (1991). Chronic ethanol treatment alters brain levels of gamma-aminobutyric acid$_A$ receptor subunit mRNAs: Relationship to genetic differences in ethanol withdrawal seizure severity. *J. Neurochem.* **57,** 1452–1455.

Celentano, J. J., Gibbs, T. T., and Farb, D. H. (1988). Ethanol potentiates GABA- and glycine-induced chloride currents in chick spinal cord neurons. *Brain Res.* **455,** 377–380.

Costa, E., Cheney, D. L., Grayson, D. R., Korneyev, A., Longone, P., Pani, L., Romeo, E., Zivkovich, E., and Guidotti, A. (1994). Pharmacology of neurosteroid biosynthesis: Role of the mitochondrial DBI receptor (MDR) complex. *Ann. NY Acad. Sci.* **746,** 223–242.

Crabbe, J. C., Kosobud, A., Young, E. R., Tam, B. R., and McSwigan, J. D. (1985). Bidirectional selection for susceptibility to ethanol withdrawal seizures in Mus musculus. *Behav. Genet.* **15,** 521–536.

Crabbe, J. C. (1998). Provisional mapping of quantitative trait loci for chronic ethanol withdrawal severity in BXD recombinant inbred mice. *J. Pharmacol. Exp. Ther.* **286,** 263–271.

Crews, F., Morrow, A. L., Criswell, H., and Breese, G. (1996). Effects of ethanol on ion channels. *In* "International Review of Neurobiology" (R. J. Bradley and R. A. Harris, Eds.), Vol. 39, pp. 283–367, Academic Press, New York.

Criswell, H. E., Simson, P. E., Duncan, G. E., McCown, T. J., Herbert, J. S., Morrow, A. L., and Breese, G. R. (1993). Molecular basis for regionally specific action of ethanol on gamma-aminobutyric acid$_A$ receptors: Generalization to other ligand-gated ion channels. *J. Pharmacol. Exp. Ther.* **267,** 522–537.

Devaud, L. L., Fritschy, J.-M., and Morrow, A. L. (1998). Influence of gender on chronic ethanol-induced alternations of GABA$_A$ receptors in rats. *Brain Res.* **796,** 222–230.

Devaud, L. L., Fritschy, J.-M., Sieghart, W., and Morrow, A. L. (1997). Bidirectional alterations of GABA$_A$ receptor subunit peptide levels in rat cortex during chronic ethanol consumption and withdrawal. *J. Neurochem.* **69,** 126–130.

Devaud, L. L., Purdy, R. H., Finn, D. A., and Morrow, A. L. (1996). Sensitization of γ-aminobutyric acid$_A$ receptors to neuroactive steroids in rats during ethanol withdrawal. *J. Pharmacol. Exp. Ther.* **278,** 510–517.

Devaud, L. L., Purdy, R. H., and Morrow, A. L. (1995a). The neurosteroid, 3α-hydroxy-5α-pregnan-20-one, protects against bicuculline-induced seizures during ethanol withdrawal in rats. *Alcohol. Clin. Exp. Res.* **19,** 350–355.

Devaud, L. L., Smith, F. D., Grayson, D. R., and Morrow, A. L. (1995b). Chronic ethanol consumption differentially alters the expression of γ-aminobutyric acid$_A$ receptor subunit mRNAs in rat cerebral cortex: Competitive, quantitative reverse transcriptase–polymerase chain reaction analysis. *Mol. Pharmacol.* **48,** 861–868.

Devries, D. J., Johnston, G. A. R., Ward, L. C., Wilce, P. A., and Shanley, B. C. (1987). Effects of chronic ethanol inhalation on the enhancement of benzodiazepine binding to mouse brain membranes by GABA. *Neurochem. Int.* **10,** 231–235.

Ducic, I., Caruncho, H. J., Zhu, W. J., Vicini, S., and Costa, E. (1995). Gamma-aminobutyric acid gating of Cl$^-$ channels in recombinant GABA$_A$ receptors. *J. Pharmacol. Exp. Ther.* **272,** 438–445.

Finn, D. A., Gallaher, E. J., and Crabbe, J. C. (2000). Differential change in neuroactive steroid sensitivity during ethanol withdrawal. *J. Pharmacol. Exp. Ther.* **292,** 394–405.

Finn, D. A., Roberts, A. J., and Crabbe, J. C. (1995). Neuroactive steroid sensitivity in withdrawal seizure-prone and -resistant mice. *Alcohol. Clin. Exp. Res.* **19,** 410–415.

Finn, D. A., Roberts, A. J., Lotrich, F., and Gallaher, E. J. (1997). Genetic differences in behavioral sensitivity to a neuroactive steroid. *J. Pharmacol. Exp. Ther.* **280,** 820–828.

Freund, R. K., and Palmer, M. R. (1997). Beta adrenergic sensitization of gamma-aminobutyric acid receptors to ethanol involves a cyclic AMP/protein kinase A second-messenger mechanism. *J. Pharmacol. Exp. Ther.* **280,** 1192–1200.

Frye, G. D., Chapin, R. E., Vogel, R. A., Mailman, R. B., Kilts, C. D., Mueller, R. A., and Breese, G. R. (1981). Effects of acute and chronic 1,3-butanediol treatment on central nervous system function: A comparison with ethanol. *J. Pharmacol. Exp. Ther.* **216,** 306–314.

Frye, G. D., Fincher, A. S., Grover, C. A., and Griffith, W. H. (1994). Interaction of ethanol and allosteric modulators with GABA$_A$-activated currents in adult medial septum/diagonal band neurons. *Brain Res.* **635,** 283–292.

Gallo, M. A., and Smith, S. S. (1993). Progesterone withdrawal decreases latency to and increases duration of electrified prod burial: A possible rat model of PMS anxiety. *Pharmacol. Biochem. Behav.* **46,** 897–904.

Givens, B. (1997). Effect of ethanol on sustained attention in rats. *Psychopharmacology* **129,** 135–140.

Givens, B. S., and Breese, G. R. (1990a). Electrophysiological evidence for an involvement of the medial septal area in the acute sedative effects of ethanol. *J. Pharmacol. Exp. Ther.* **253,** 95–103.

Givens, B. S., and Breese, G. R. (1990b). Site-specific enhancement of γ-aminobutyric acid-mediated inhibition of neural activity by ethanol in the rat medial septum. *J. Pharmacol. Exp. Ther.* **254,** 528–538.

Glowa, J. R., Crawley, J., Suzdak, P. D., and Paul, S. M. (1988). Ethanol and the GABA receptor complex: Studies with the partial inverse benzodiazepine receptor agonist Ro 15-4513. *Pharmacol. Biochem. Behav.* **31,** 767–772.

Goldstein, D. B. (1973). Alcohol withdrawal reactions in mice: Effects of drugs that modify neurotransmission. *J. Pharmacol. Exp. Ther.* **186,** 1–9.

Grobin, A. C., Matthews, D. B., Devaud, L. L., and Morrow, A. L. (1998). The role of GABA$_A$ receptors in the acute and chronic effects of ethanol. *Psychopharmacology* **139,** 2–19.

Harris, R. A., McQuilkin, S. J., Paylor, R., Abeliovich, A., Tonegawa, S., and Wehner, J. M. (1995). Mutant mice lacking the gamma isoform of protein kinase C show decreased behavioral actions of ethanol and altered function of gamma-aminobutyrate type A receptors. *Proc. Natl. Acad. Sci. USA* **92,** 3658–3662.

Harris, R. A., Mihic, S. J., and Valenzuela, C. F. (1998). Alcohol and benzodiazepines: Recent mechanistic studies. *Drug Alcohol Depend.* **51,** 155–164.

Harrison, N. L., Majewska, M. D., Harrington, J. W., and Barker, J. L. (1987). Structure–activity relationships for steroid interaction with the gamma-aminobutyric acid-A receptor complex. *J. Pharmacol. Exp. Ther.* **241,** 346–353.

Hoffman, P. L. (1993). The influence of neurohypophysial hormones on central nervous system processes of adaptation: Functional tolerance to ethanol. *Ann. NY Acad. Sci.* **689,** 300–308.

Hoffman, P. L., Tabakoff, B., Szabo, G., Suzdak, P. D., and Paul, S. M. (1987). Effect of an imidazobenzodiazepine, Ro 15-4513, on the incoordination and hypothermia produced by ethanol and pentobarbital. *Life Sci.* **41,** 611–619.

Irwin, R. P., Lin, S.-Z., Rogawski, M. A., Purdy, R. H., and Paul, S. M. (1994). Steroid potentiation and inhibition of N-methyl-D-aspartate receptor-mediated intracellular Ca^{++} responses: Structure activity studies. *J. Pharmacol. Exp. Ther.* **271,** 677–682.

Janak, P. H., Redfern, J. E. M., and Samson, H. H. (1998). The reinforcing effects of ethanol are altered by the endogenous neurosteroid, allopregnanolone. *Alcohol. Clin. Exp. Res.* **22,** 1106–1112.

Janis, G. C., Devaud, L. L., Mitsuyama, H., and Morrow, A. L. (1998). Effects of chronic ethanol consumption and withdrawal on the neuroactive steroid 3α-hydroxy-5α-pregnan-20-one in male and female rats. *Alcohol. Clin. Exp. Res.* **22,** 2055–2061.

Klein, R. L., Mascia, M. P., Whiting, P. J., and Harris, R. A. (1995). $GABA_A$ receptor function and binding in stably transfected cells: Chronic ethanol treatment. *Alcohol. Clin. Exp. Res.* **19,** 1338–1344.

Kokka, N., Sapp, D. W., Taylor, A. M., and Olsen, R. W. (1993). The kindling model of alcohol dependence: Similar persistent reduction in seizure threshold to pentylenetetrazol in animals receiving chronic ethanol or chronic pentylenetetrazol. *Alcohol. Clin. Exp. Res.* **17,** 525–531.

Koob, G. F. (1996). Drug addiction: The yin and yang of hedonic homeostasis. *Neuron* **16,** 893–896.

Kosobud, A. E., and Crabbe, J. C. (1990). Genetic correlations among inbred strains sensitivities to convulsions induced by 9 convulsant drugs. *Brain Res.* **526,** 8–16.

Kurata, Y., Marszalec, W., Hamilton, B. J., Carter, D. B., and Narahashi, T. (1993). Alcohol modulation of cloned $GABA_A$ receptor-channel complex expressed in human kidney cell lines. *Brain Res.* **631,** 143–146.

Le, A. D., Khanna, J. M., Kalant, H., and Grossi, F. (1986). Tolerance to and cross-tolerance among ethanol, pentobarbital and chlordiazepoxide. *Pharmacol. Biochem. Behav.* **24,** 93–98.

Lin, A. M.-Y., Freund, R. K., and Palmer, M. R. (1991). Ethanol potentiation of GABA-induced electrophysiological responses in cerebellum: Requirement for catecholamine modulation. *Neurosci. Lett.* **122,** 154–158.

Lister, R. G. (1987). The benzodiazepine receptor inverse agonists FG 7142 and Ro 15-4513 both reverse some of the behavioral effects of ethanol in a holeboard test. *Life Sci.* **41,** 1481–1489.

Lister, R. G., and Linnoila, M. (1991). Alcohol, the chloride ionophore and endogenous ligands for benzodiazepine receptors. *Neuropharmacology* **30,** 1435–1440.

Majchrowicz, E. (1975). Induction of physical dependence upon ethanol and the associated behavioral changes. *Psychopharmacologia* **43,** 245–254.

Majewska, M. D., Harrison, N. L., Schwartz, R. D., Barker, J. L., and Paul, S. M. (1986). Steroid hormone metabolites are barbiturate-like modulators of the GABA receptor. *Science* **232,** 1004–1007.

Mancillas, J. R., Siggins, J. R., and Bloom, F. E. (1986). Systemic ethanol: Enhancement of responses to acetylcholine and somatostatin in hippocampus. *Science* **231,** 161–163.

Marszalec, W., Aistrup, A. L., and Narahashi, T. (1998). Ethanol modulation of excitatory and inhibitory synaptic interactions in cultured cortical neurons. *Alcohol. Clin. Exp. Res.* **22,** 1516–1524.

Martz, A., Dietrich, R. A., and Harris, R. A. (1983). Behavioral evidence for the involvement of γ-aminobutyric acid in the actions of ethanol. *Eur. J. Pharmacol.* **89,** 53–62.

Marx, C. E., Duncan, G. E., Gilmore, J. H., Lieberman, J. A., and Morrow, A. L. (2000). Olanzapine increases allopregnanolone in the rat cerebral cortex. *Biol. Psychiatry* **47,** 1000–1004.

Matthews, D. B., Simson, P. E., and Best, P. J. (1995). Acute ethanol impairs spatial memory but not stimulus/response memory in the rat. *Alcohol. Clin. Exp. Res.* **19,** 902–909.

McBride, W. J., Murphy, J. M., Lumeng, L., and Li, T. K. (1988). Effects of Ro 15-4513, fluoxetine and desipramine on intake of ethanol water and food in alcohol preferring and non-preferring lines of rats. *Pharmacol. Biochem. Behav.* **30,** 1045–1050.

Mehta, A. K., and Ticku, M. K. (1988). Ethanol potentiation of GABAergic transmission in cultured spinal cord neurons involves γ-aminobutyric acid-gated chloride channels. *J. Pharmacol. Exp. Ther.* **246,** 558–564.

Melchior, C. L., and Tabakoff, B. (1985). Features of environment-dependent tolerance to ethanol. *Psychopharmacology* **87,** 94–100.

Mellon, S. H., and Deschepper, C. F. (1993). Neurosteroid biosynthesis: Genes for adrenal steroidogenic enzymes are expressed in the brain. *Brain Res.* **629,** 283–292.

Mhatre, M. C., Pena, G., Sieghart, W., and Ticku, M. K. (1993). Antibodies specific for GABA$_A$ receptor alpha subunits reveal that chronic alcohol treatment down-regulates alpha-subunit expression in rat brain regions. *J. Neurochem.* **61,** 1620–1625.

Mhatre, M. C., and Ticku, M. K. (1992). Chronic ethanol administration alters γ-aminobutyric acid$_A$ receptor gene expression. *Brain Res. Mol. Brain Res.* **42,** 415–422.

Mihic, S. J., Whiting, P. J., and Harris, R. A. (1994). Anaesthetic concentrations of alcohols potentiate GABA$_A$ receptor-mediated currents: Lack of subunit specificity. *Eur. J. Pharmacol. Mol. Pharmacol. Sect.* **268,** 209–214.

Mihic, S. J., Ye, Q., Wick, M. J., Koltchine, V. V., Krasowski, M. A., Finn, S. E., Mascia, M. P., Valenzuela, C. F., Hanson, K. K *et al.* (1997). Sites of alcohol and volatile anaesthetic action on GABA$_A$ and glycine receptors. *Nature* **389,** 385–389.

Montpied, P., Morrow, A. L., Karanian, J. W., Ginns, E. I., Martin, B. M., and Paul, S. M. (1991). Prolonged ethanol inhalation decreases gamma-aminobutyric acid$_A$ receptor α-subunit mRNAs in the rat cerebral cortex. *Mol. Pharmacol.* **39,** 157–163.

Moran, M. H., Goldberg, N., and Smith, S. S. (1998). Progesterone withdrawal: II Insensitivity to the sedative effects of a benzodiazepine. *Brain Res.* **807,** 91–100.

Moran, M. H., and Smith, S. S. (1998). Progesterone withdrawal: I Pro-convulsant effects. *Brain Res.* **807,** 84–90.

Morrow, A. L., Herbert, J. S., and Montpied, P. (1992). Differential effects of chronic ethanol administration on GABA$_A$ receptor α_1 and α_6 subunit mRNA levels in rat cerebellum. *Mol. Cell. Neurosci.* **3,** 251–258.

Morrow, A. L., Janis, G. C., VanDoren, M. J., Matthews, D. B., Samson, H. H., Janak, P. H., and Grant, K. A. (1999). Neurosteroids mediate pharmacological effects of ethanol: A new mechanism of ethanol action?. *Alcohol. Clin. Exp. Res.* **23,** 1933–1940.

Morrow, A. L., Montpied, P., Lingford-Hughes, A., and Paul, S. M. (1990a). Chronic ethanol and pentobarbital administration in the rat: Effects on GABA$_A$ receptor function and expression in brain. *Alcohol* **7,** 237–244.

Morrow, A. L., Pace, J. R., Purdy, R. H., and Paul, S. M. (1990b). Characterization of steroid interactions with gamma-aminobutyric acid receptor-gated chloride ion channels: Evidence for multiple steroid recognition sites. *Mol. Pharmacol.* **37**, 263–270.

Morrow, A. L., Suzdak, P. D., and Paul, S. M. (1987). Steroid hormone metabolites potentiate GABA receptor-mediated chloride ion flux with nanomolar potency. *Eur. J. Pharmacol.* **142**, 483–485.

Morrow, A. L., Suzdak, P. D., and Paul, S. M. (1988). Benzodiazepine, barbiturate, ethanol and hypnotic steroid hormone modulation of GABA-mediated chloride ion transport in rat brain synaptoneurosomes. *In* "Chloride Channels and Their Modulation by Neurotransmitters and Drugs" (G. Biggio and E. Costa, Eds.), pp. 247–261. Raven Press, New york.

Morrow, A. L., VanDoren, M. J., and Devaud, L. L. (1998). Effects of progesterone or neuroactive steroid? *Nature* **395**, 652–653.

Nakahiro, M., Arakawa, O., and Narahashi, T. (1991). Modulation of gamma-aminobutyic acid receptor-channel complex by alcohols. *J. Pharmacol. Exp. Ther.* **259**, 235–240.

Nestores, J. N. (1980). Ethanol specifically potentiates GABA-mediated neurotransmission in the feline cerebral cortex. *Science* **209**, 708–710.

Normington, K., and Russell, D. W. (1992). Tissue distribution and kinetic characteristics of rat steroid 5α-reductase isozymes. *J. Biol. Chem.* **267**, 19548–19554.

Park-Chung, M., Malayev, A., Purdy, R. H., Gibbs, T. T., and Farb, D. H. (1999). Sulfated and unsulfated steroids modulate gamma-aminobutyric acid$_A$ receptor function through distinct sites. *Brain Res.* **830**, 72–87.

Patchev, V. K., Hassan, A. H. S., Holsboer, F., and Almeida, O. F. X. (1996). The neurosteroid tetrahydroprogesterone attenuates the endocrine response to stress and exerts glucocorticoid-like effects on vasopressin gene transcription in the rat hypothalamus. *Neuropsychopharmacology* **15**, 533–540.

Patchev, V. K., Shoaib, M., Holsboer, F., and Almeida, O. F. X. (1994). The neurosteroid tetrahydroprogesterone counteracts corticotropin-releasing hormone-induced anxiety and alters the release and gene expression of corticotropin-releasing hormone in the rat hypothalamus. *Neuroscience* **62**, 265–271.

Petry, N. M. (1997). Benzodiazepine-GABA modulation of concurrent ethanol and sucrose reinforcement in the rat. *Exp. Clin. Psychopharm.* **5**, 183–194.

Piazza, P. V., and Le Moal, M. (1998). The role of stress in drug self-administration. *Trends Pharmacol. Sci.* **19**, 67–74.

Pinna, G., Uzunova, V., Matsumoto, K., Puia, G., Mienville, J. M., Costa, E., and Guidotti, A. (2000). Brain allopregnanolone regulates the potency of the GABA$_A$ receptor agonist muscimol. *Neuropharmacology* **39**, 440–448.

Pohorecky, L. A., and Brick, J. (1988). Pharmacology of ethanol. *Pharmacol Ther.* **36**, 335–427.

Proctor, W. R., Soldo, B. L., Allan, A. M., and Dunwiddie, T. V. (1992). Ethanol enhances synaptically evoked GABA$_A$ receptor-mediated responses in cerebral cortical neurons in rat brain slices. *Brain Res.* **595**, 220–227.

Puia, G., Santi, M., Vincini, S., Pritchett, D. B., Purdy, R. H., Paul, S. M., Seeburg, P. H., and Costa, E. (1990). Neurosteroids act on recombinant human GABA$_A$ receptors. *Neuron* **4**, 759–765.

Puia, G., Vincini, S., Seeburg, P. H., and Costa, E. (1991). Influence of recombinant GABA$_A$ receptor subunit composition on the action of allosteric modulators of GABA-gated Cl$^-$ currents. *Mol. Pharmacol.* **39**, 691–696.

Purdy, R. H., Morrow, A. L., Moore, P. H., Jr., and Paul, S. M. (1991). Stress-induced elevations of gamma-aminobutyric acid type A receptor-active steroids in the rat brain. *Proc. Natl. Acad. Sci. USA* **88**, 4553–4557.

Rassnick, S., D'Amico, E., Riley, E., and Koob, G. F. (1993). GABA antagonist and benzodiazepine partial inverse agonist reduce motivated responding for ethanol. *Alcohol. Clin. Exp. Res.* **17,** 124–130.

Reddy, D. S., and Kulkarni, S. K. (1999). Sex and estrous cycle-dependent changes in neurosteroid and benzodiazepine effects on food consumption and plus-maze learning behaviors in rats. *Pharmacol. Biochem. Behav.* **62,** 53–60.

Reynolds, J. N., Prasad, A., and MacDonald, J. F. (1992). Ethanol modulation of GABA receptor-activated Cl currents in neurons of the chick, rat and mouse central nervous system. *Eur. J. Pharmacol.* **224,** 173–181.

Romeo, E., Brancati, A., De Lorenzo, A., Fucci, P., Furnari, C., Pompili, E., Sasso, G. F., Spalletta, G., Troisi, A., and Pasini, A. (1996). Marked decrease of plasma neuroactive steroids during alcohol withdrawal. *J. Pharmacol. Exp. Ther.* **247,** 309–322.

Rupprecht, R., Reul, J., Trapp, T., van Steensel, B., Wetzel, C., Damm, K., Zieglgansberger, W., and Holsboer, F. (1994). Progesterone receptor-mediated effects of neuroactive steroids. *Neuron* **11,** 523–530.

Russell, D. W., and Wilson, J. D. (1994). Steroid 5α-reductase: Two genes/two enzymes. *Ann. Rev. Biochem.* **63,** 26–61.

Samson, H. H., Slawecki, C. J., Sharpe, A. L., and Chappell, A. (1998). Appetitive and consummatory behaviors in the control of ethanol consumption: A measure of ethanol seeking behavior. *Alcohol. Clin. Exp. Res.* **22,** 1783–1787.

Samson, H. H., Tolliver, G. A., Pfeffer, A. U., Sadeghi, K. G., and Mills, F. G. (1987). Oral ethanol reinforcement in the rat: Effect of the partial inverse benzodiazepine agonist Ro15-4513. *Pharmacol. Biochem. Behav.* **27,** 517–519.

Sapp, D. W., and Yeh, H. H. (1998). Ethanol-GABA$_A$ receptor interactions: A comparison between cell lines and cerebellar *Purkinje* cells. *J. Pharmacol. Exp. Ther.* **284,** 768–776.

Schuckit, M. A. (1994). Low level of response to alcohol as a predictor of future alcoholism. *Amer. J. Psychiatry* **151,** 184–189.

Schuckit, M. A., and Smith, T. L. (1996). An 8-year follow-up of 450 sons of alcoholic and control subjects. *Arch. Gen. Psychiatry* **53,** 202–210.

Schuckit, M. A., Tsuang, J. W., Anthenelli, R. M., Tipp, J. E., and Nurnberger, J. I., Jr. (1996). Alcohol challenges in young men from alcoholic pedigrees and control families: A report from the COGA project. *J. Stud. Alcohol* **57,** 368–377.

Sellers, E. M., and Kalant, H. (1976). Alcohol intoxication and withdrawal. *New Engl. J. Med.* **294,** 757–762.

Sigel, E., Baur, R., and Malherbe, P. (1993). Recombinant GABA$_A$ receptor function and ethanol. *FEBS Lett.* **324,** 140–142.

Siggins, G. R., Pittman, Q. J., and French, E. D. (1987). Effects of ethanol on CA1 and CA3 pyramidal cells in the hippocampal slice preparation: An intracellular study. *Brain Res.* **414,** 22–34.

Silveri, M. M., and Spear, L. P. (1999). Ontogeny of rapid tolerance to the hypnotic effects of ethanol. *Alcohol. Clin. Exp. Res.* **23,** 1180–1184.

Simmonds, M. A., Turner, J. P., and Harrison, N. L. (1984). Interactions of steroids with the GABA$_A$ receptor complex. *Neuropharmacology* **23,** 877–878.

Simson, P. E., Criswell, H. E., and Breese, G. R. (1991). Ethanol potentiates γ-aminobutyric acid-mediated inhibition in the inferior colliculus: Evidence for local ethanol/γ-aminobutyric acid interactions. *J. Pharmacol. Exp. Ther.* **259,** 1288–1293.

Smith, B. R., Robidoux, J., and Amit, Z. (1992). GABAergic involvement in the acquisition of voluntary ethanol intake in laboratory rats. *Alcohol* **27,** 227–231.

Smith, S. S., Gong, Q. H., Hsu, F.-C., Markowitz, R. S., French-Mullen, J. M. H., and Li, X. (1998a). GABA$_A$ receptor α_4 subunit suppression prevents withdrawal properties of an endogenous steroid. *Nature* **392,** 926–930.

Smith, S. S., Gong, Q. H., Li, X., Moran, M. H., Bitran, D., Frye, C. A., and Hsu, F. (1998b). Withdrawal from 3α-OH-5α-pregnan-20-one using a pseudopregnancy model alters the kinetics of hippocampal GABA$_A$-gated current and increases the GABA$_A$ receptor α_4 subunit in association with increased anxiety. *J. Neurosci.* **18,** 5275–5284.

Soldo, B. L., Proctor, W. R., and Dunwiddie, T. V. (1994). Ethanol differentially modulates GABA$_A$ receptor-mediated chloride currents in hippocampal, cortical, and septal neurons in rat brain slices. *Synapse* **18,** 94–103.

Suzdak, P. D., Glowa, J. R., Crawley, J. N., Schwartz, R. D., Skolnick, P., and Paul, S. M. (1986). A selective imidazobenzodiazepine antagonist of ethanol in the rat. *Science* **234,** 1243–1247.

Suzdak, P. D., Schwartz, R. D., Skolnick, P., and Paul, S. M. (1986). Ethanol stimulates γ-aminobutyric acid receptor-mediated chloride transport in rat brain synaptoneurosomes. *Proc. Natl. Acad. Sci. USA.* **83,** 4071–4075.

Suzdak, P. D., Paul, S. M., and Crawley, J. N. (1988). Effects of Ro 15-4513 and other benzodiazepine inverse agonists on alcohol induced intoxication in the rat. *J. Pharmacol. Exp. Ther.* **245,** 880–886.

Tatebayashi, H., Motomura, H., and Narahashi, T. (1998). Alcohol modulation of single GABA$_A$ receptor-channel kinetics. *Neuroreport* **9,** 1769–1775.

Ticku, M. K., and Burch, T. (1980). Alterations in γ-aminobutyric acid receptor sensitivity following acute and chronic ethanol treatments. *J. Neurochem.* **34,** 417–423.

Uzunov, D. P., Cooper, T. B., Costa, E., and Guidotti, A. (1996). Fluoxetine-elicited changes in brain neurosteroid content measured by negative ion mass fragmentography. *Proc. Natl. Acad. Sci. USA.* **93,** 12599–12604.

Uzunova, V., Sheline, Y., Davis, J. M., Rasmusson, A., Uzunov, D. P., Costa, E., and Guidotti, A. (1998). Increase in the cerebrospinal fluid content of neurosteroids in patients with unipolar major depression who are receiving fluoxetine or fluvoxamine. *Proc. Natl. Acad. Sci. USA* **95,** 3239–3244.

Van Rijn, C. M., Willems-van Bree, E., Van der Velden, T. J. A. M., and Rodrigues de Miranda, J. F. (1990). Binding of the cage convulsant, [^{3}H]TBOB, to sites linked to the GABA$_A$ receptor complex. *Eur. J. Pharmacol.* **179,** 419–425.

VanDoren, M. J., Johnson, C. J., and Morrow, A. L. (2000a). Alcohol preference in ethanol preferring P rats is associated with altered HPA axis and neurosteroid responses to alcohol. *Alcohol. Clin. Exp. Res.* **24,** 49A.

VanDoren, M. J., Matthews, D. B., Janis, G. C., Grobin, A. C., Devaud, L. L., and Morrow, A. L. (2000b). Neuroactive steroid 3α-hydroxy-5α-pregnan-20-one modulates electrophysiological and behavioral actions of ethanol. *J. Neurosci.* **20,** 1982–1989.

Wafford, K. A., Burnett, D. M., Leidenheimer, N. J., Burt, D. R., Wang, J. B., Kofuji, P., Dunwiddie, T. V., Harris, R. A., and Sikela, J. M. (1991). Ethanol sensitivity of the GABA$_A$ receptor expressed in *Xenopus* oocytes requires 8 amino acids contained in the gamma2L subunit. *Neuron* **7,** 27–33.

Wan, F.-J., Berton, F., Madamba, S. G., Francesconi, W., and Siggins, G. R. (1996). Low ethanol concentrations enhance GABAergic inhibitory postsynaptic potentials in hippocampal pyramidal neurons only after block of GABA$_B$ receptors. *Proc. Natl. Acad. Sci. USA.* **93,** 5049–5054.

Weiner, J. L., Zhang, L., and Carlen, P. L. (1994). Potentiation of GABA$_A$-mediated synaptic current by ethanol in hippocampal CA1 neurons: Possible role of protein kinase C. *J. Pharmacol. Exp. Ther.* **268,** 1388–1395.

Weiner, J. L., Gu, C., and Dunwiddie, T. V. (1997a). Differential ethanol sensitivity of subpopulations of $GABA_A$ synapses onto rat hippocampal CA1 pyramidal neurons. *J. Neurophysiol.* **77,** 1306–1312.

Weiner, J. L., Svoboda, K. R., Lupica, C. R., and Dunwiddie, T. V. (1997b). Pre- and post-synaptic actions of ethanol on $GABA_A$ receptor-mediated synaptic transmission in rat hippocampal CA1 pyramidal neurons. *Soc. Neurosci. Abst.* **23,** 959.

Weiner, J. L., Valenzuela, C. F., Watson, P. L., Frazier, C. J., and Dunwiddie, T. V. (1997). Elevation of basal protein kinase C activity increases ethanol sensitivity of $GABA_A$ receptors in rat hippocampal CA1 pyramidal neurons. *J. Neurochem.* **68,** 1949–1959.

Wick, M. J., Mihic, S. J., Ueno, S., Mascia, M. P., Trudell, J. R., Brozowski, S. J., Ye, Q., Harrison, N. L., and Harris, R. A. (1998). Mutations of gamma-aminobutyric acid and glycine receptors change alcohol cutoff: Evidence for an alcohol receptor? *Proc. Natl. Acad. Sci. USA.* **95,** 6504–6509.

PRECLINICAL DEVELOPMENT OF NEUROSTEROIDS AS NEUROPROTECTIVE AGENTS FOR THE TREATMENT OF NEURODEGENERATIVE DISEASES

Paul A. Lapchak [*,†,‡,1] and Dalia M. Araujo[†]

*Department of Neuroscience, University of California–San Diego,
La Jolla, California 92093-0624; †VASDHS, San Diego,
California 92161; and ‡ Veterans Medical Research
Foundation, San Diego, California 92161

I. Neurosteroids and the Brain
II. Synthesis of Central Nervous System Neurosteroids
III. Receptor Signaling Pathways
   A. GABA Receptors
   B. NMDA Receptors
IV. Neurosteroids and Central Nervous System Plasticity
V. Neurosteroids and Neuroprotection
   A. Ischemia and Stroke
   B. Alzheimer's Disease and Aging
VI. Conclusions
   References

Recent literature has emphasized the unique role that the neurosteroid subclass of steroids, which includes dehydroepiandrosterone (DHEA) and dehydroepiandrosterone sulfate (DHEAS), play in the developing and adult central nervous system (CNS). Both DHEA and DHEAS are found in abundance in the CNS (Majewska, 1995), and both can be synthesized and metabolized in the brain of many species (Baulieu, 1981, 1998; Corpéchot et al., 1981, 1983; Zwain and Yen, 1999). DHEA and DHEAS have been implicated as potential signaling molecules for neocortical organization during neuronal development, suggesting that they have trophic factor-like activity (neurotrophic or neurotropic) or can interact with various neurotransmitter systems to promote neuronal remodeling (Compagnone and Mellon, 1998; Mao and Barger, 1998). Consistent with a neurotrophic role for these steroids, studies have shown that DHEAS protects certain neuronal populations against neurotoxic insults inflicted by the excitatory amino acid glutamate (Kimonides et al., 1998; Mao and Barger, 1998). This finding suggests that DHEAS may be useful in treating neurodegenerative diseases in which excitotoxicity is believed to be the underlying cause or a major contributor to cell death. Moreover, because DHEA and DHEAS are multifunctional

---

[1]Author to whom correspondence should be sent.

and exhibit a variety of properties in the CNS, including memory consolidation, neuroprotection, and reduction of neurodegeneration (Majewska, 1992, 1995; Lapchak *et al.*, 2000), their potential therapeutic benefits may be extended to include the treatment of other neurodegenerative diseases not directly linked to excitotoxicity. © 2001 Academic Press.

I. Neurosteroids and the Brain

In 1981, Baulieu observed that many different neurosteroids, including pregnenolone, dehydroepiandrosterone (DHEA), and dehydroepiandrosterone sulfate (DHEAS), were more concentrated in the rat central nervous system (CNS) particularly in the brain, than in plasma. Furthermore, levels of these steroids in the CNS remained elevated even after adrenalectomy (Corpéchot *et al.*, 1981, 1983), suggesting that they were most likely intrinsic to the CNS and not of peripheral origin. These findings were among the first to propose the radical idea that certain steroids might be synthesized *de novo* in the CNS, thus leading to their renaming and classification as neurosteroids to differentiate steroids endogenous to the brain from the more classical "steroid" peripheral organs. The remarkable concept of neurosteroids has held up ever since, and great advances in the field of neurosteroid research have led to a greater understanding of CNS function and the potential usefulness of neurosteroids as therapeutic agents.

II. Synthesis of Central Nervous System Neurosteroids

In the intervening years, the intense interest in neurosteroids has spearheaded extensive work geared toward uncovering the synthetic pathways involved in the production of neurosteroids in the CNS. In general, the biosynthetic pathway was demonstrated to consist of steroidogenic enzymes similar to those found in the periphery, such as for adrenal steroid synthesis (Kroboth *et al.*, 1999; Mensah-Nyagan *et al.*, 1999). In fact, numerous enzymes are involved in neurosteroidogenesis and can be loosely classified into two subcategories: (1) the cytochrome P450 group of heme-binding oxidases that catalyze conversion of steroids; and (2) the non-P450 group, which includes various hydroxysteroid dehydrogenases, sulfotransferases, and sulfatases (Mensah-Nyagan *et al.*, 1999). In contrast to the periphery, the brain also contains steroid-metabolizing enzymes, including sulfotransferases and sulfohydrolases, which convert classic steroid hormones into a variety of neuroactive compounds, such as DHEA, DHEAS, allopregnanolone, and progesterone (Mensah-Nyagan *et al.*, 1999). In the CNS, the expression of steroidogenic enzymes is determined by a variety of factors, including developmental stage, brain region specificity, and even cellular specificity (Mensah-Nyagan *et al.*, 1999). Ultimately, the regulation of the neurosteroid expression pattern is a finely tuned process that ensures the proper biosynthesis of the appropriate neurosteroid for different cell types during different stages of the individual's life span.

III. Receptor Signaling Pathways

Certain neurosteroids, such as DHEA and DHEAS, have been shown to modify neuronal excitability very rapidly via mechanisms distinct from the activation of classical steroid nuclear receptors (Harrison and Simmonds, 1984; Debonnel *et al.*, 1996a; Debonnel *et al.*, 1996b; Debonnel and de Montigny, 1996). In particular, DHEA and DHEAS are potent allosteric modulators of γ-aminobutyric acid type A (GABA$_A$) (Barker *et al.*, 1986; Majewska *et al.*, 1986; Park-Chung *et al.*, 1997; Schmid *et al.*, 1998; Shen *et al.*, 1999; Sousa and Ticku, 1997), σ (Monnet *et al.*, 1995), N-methyl-D-aspartate (NMDA), and non-NMDA-receptor binding site function (Gibbs *et al.*, 1999). In view of their abundance and widespread localization in the brain, the findings demonstrating nongenomic action of neurosteroids are particularly significant and further underscore the potential importance of the neurosteroids' pharmacological effects in the CNS via interaction with diverse receptive substances. The following sections detail the interaction of neurosteroids with classical neurotransmitter receptors:

A. GABA RECEPTORS

For some time, it has been known that neurosteroids are allosteric modulators of GABA$_A$ receptors (Harrison and Simmonds, 1984; Majewska *et al.*, 1986; Purdy *et al.*, 1990; Kiessling and Gass, 1994; Lambert *et al.*, 1995; Park-Chung *et al.*, 1997, 1999; Sousa and Ticku, 1997; Shen *et al.*, 1999). In the mid-1980s, two elegant studies provided important evidence for the ability of neurosteroids to modulate neurotransmitter function (Harrison and Simmonds, 1984; Majewska *et al.*, 1986). A later study showing definitively that allopregnanolone is a potent GABA$_A$-receptor modulator (Majewska and Vaupel, 1991) was followed by reports that metabolites related to progesterone, such as dihydroprogesterone, pregnenolone, and pregnenolone sulfate, also bind to $GABA_A$ receptors at sites different from the benzodiazepine and barbiturate binding sites (Lambert et al., 1995; Sousa and Ticku, 1997). However, the exact mechanism(s) involved in the modulation of GABA receptors by neurosteroids remains a hotly debated subject, although the use of recombinant $GABA_A$-receptor subunits transfected into non-neuronal cells has allowed for a better understanding of neurosteroid-mediated GABA-receptor modulation. The $GABA_A$-receptor is composed of a pentameric structure that includes various combinations of multiple α, β and γ subunits (Luddens and Korpi, 1995; Luddens et al., 1995). In vitro recombination studies indicate that neurosteroid-mediated modulation of $GABA_A$ receptors requires the presence of both α and γ subunits (Puia et al., 1990, 1993). Even so, different neurosteroids may function either as positive or negative modulators of $GABA_A$ receptors and may display different binding characteristics (Puia et al., 1990; Lan et al., 1991a,b; Majewska and Vaupel, 1991; Shingai et al., 1991; Majewska, 1992, 1995; Sapp et al., 1992; Korpi and Luddens, 1993; Schmid et al., 1998), depending on the neuronal population studied.

Increasing evidence has demonstrated that $GABA_A$ receptors in different regions of the brain may be differentially affected by neurosteroids and that the resulting behavioral effects of neurosteroids may be related to the localization of $GABA_A$ receptors in the CNS. For example, many hippocampal $GABA_A$ receptors that are sensitive to benzodiazepines are located on noradrenergic neurons, whereas cerebellar receptors are on benzodiazepine-insensitive glutaminergic neurons (Schmid et al., 1996). Schmid et al. (1998) also have shown that neurosteroids differentially affect two subtypes of $GABA_A$ receptors that have distinct pharmacology, neuronal localizations, and functions. Even within the same brain region, differences that can be attributed to anatomical coordinates may be noted. Accordingly, in the pituitary gland, disparate responses to DHEAS in the posterior and anterior lobes correlate with the localization of $GABA_A$ receptors, with a high affinity to DHEAS apparent in the former compared with the latter subregion (Hansen et al., 1999). Thus, because the role of DHEAS in the pituitary gland is thought to be control of neuropeptide secretion, the different responses elicited in different anatomical locales by $GABA_A$ receptors may provide a mechanism for refined adjustment of neuropeptide secretion.

Evidence suggests that differential behavioral effects may occur via modulation of a variety of neurotransmitter systems. The majority of neurosteroids that actively modulate $GABA_A$ receptors in animal models are sedatives, anticonvulsants, or anxiolytics (reviewed in Melchior and Ritzmann, 1994; Reddy and Kulkarni, 1997a,b; Baulieu, 1998). There is also evidence that neurosteroids modulate feeding behavior via a $GABA_A$ mechanism (Reddy and Kulkarni, 1998).

B. NMDA RECEPTORS

The glutamate receptor family is subdivided into NMDA and α-amino-3-hydroxy-5-methyl-4-isoxazoleproprionate/kainate receptors (Doble, 1995; Nakanishi et al., 1998). NMDA receptors are composed of ε_1-and ε_2-receptor subunits, which are expressed in hippocampal CA1 (Sakimura et al., 1995) and CA3 neurons (Ito et al., 1997). The ε_1-NMDA receptor subunit has been implicated in learning processes because ε_1 knockout mice have deficits in hippocampal CA1 LTP and spatial learning (Sakimura et al., 1995). Neurosteroids like allopregnanolone, allopregnanolone sulfate, DHEA, and DHEAS are allosteric modulators of NMDA-receptor function (Wu et al., 1991; Bowlby, 1993; Park-Chung et al., 1994; Fahey et al., 1995; Compagnone and Mellon, 1998). DHEA and DHEAS are positive allosteric modulators of NMDA receptors, whereas allopregnanolone sulfate has been shown to be a negative allosteric modulator of the NMDA receptor (Park-Chung et al., 1994). However, as discussed in the section about GABA receptors, the biochemical basis for positive and negative allosteric NMDA-receptor regulation is not completely understood.

Another class of receptor linked to the activity of NMDA receptors, the σ receptor, has been characterized and postulated to be involved in the CNS activity of neurosteroids (Debonnel, 1993; Debonnel et al., 1996a,b; Debonnel and de Montigny, 1996). Evidence has postulated that the function of certain σ receptors also can be modulated by neurosteroids. Moreover, it has been proposed that certain neurosteroids acting as potentiators of NMDA receptors may be endogenous ligands for σ receptors. To date, three different σ-receptor subtypes, denoted as σ_1, σ_2, and σ_3 have been identified (Quirion et al., 1992; Booth et al., 1993; Park-Chung et al., 1994). DHEA and DHEAS have been shown to induce NMDA-mediated norepinephrine release from hippocampal slices via activation of σ_1 receptors (Monnet et al., 1995), a finding that was significant because the function and physiological role of σ receptors was unclear. A σ-receptor-mediated potentiation of NMDA responses in hippocampal CA3 neurons also has been demonstrated (Debonnel et al., 1996a,b; Debonnel and de Montigny, 1996).

The behavioral consequences of neurosteroid modulation of NMDA receptors have been investigated at multiple levels. Behavioral effects of DHEAS, such as memory improvement, that are associated with NMDA-receptor activity appear to be related to a pharmacological action at σ sites (Bergeron et al., 1996). Consistent with an interaction of DHEAS with σ_1-receptors, the neurosteroid was shown to attenuate dizocilpine-induced learning impairment in mice (Maurice et al., 1997). Moreover, DHEAS is capable of enhancing spatial memory in mice, an action thought to result from positive modulation of hippocampal NMDA receptors (Mathis et al., 1996).

These results strongly support the hypothesis that neurosteroids regulate learning and memory via modulation of hippocampal NMDA receptors. In addition, regulation of food intake by neurosteroid activation of NMDA receptors has been reported, although the complexities of this mechanism have yet to be deciphered (Reddy and Kulkarni, 1998).

IV. Neurosteroids and Central Nervous System Plasticity

DHEA and DHEAS also have been described as modulators of synaptic plasticity in the CNS (Bologa et al., 1987; Compagnone and Mellon, 1998) because both affect neurite outgrowth in various brain regions (Bologa et al., 1987; Compagnone and Mellon, 1998). However, it appears that the two neurosteroids have distinct roles in specific brain regions. For instance, in the cerebral cortex, DHEA promoted axonal growth and synaptic sprouting, whereas DHEAS increased dendritic complexity and branching (Compagnone and Mellon, 1998), indicating that DHEA and DHEAS differentially regulate processes involved in neuronal growth and plasticity. Because of their dissimilar activities, it is likely that the two neurosteroids modify distinct aspects of the growth process, which may depend on the relative ratio of DHEA:DHEAS concentrations in the brain region in question, which in turn may be a direct reflection of the temporal expression of sulfohydrolase and sulfotransferase activities. The "neurotropic"- like effects of neurosteroids in the brain are not exclusive to DHEA and DHEAS, as other neurosteroids, such as allopregnanolone, are potent inducers of axonal regression in the developing hippocampus (Brinton, 1994). In addition, neurosteroids appear to protect neurons subjected to adverse conditions. In embryonic cortical cells *in vitro*, DHEAS was neuroprotective under anoxic conditions (Marx et al., 2000), suggesting that DHEAS may be of significance in disorders of neurodevelopment that involve anoxia. Overall, the resultant effects, many of which are mediated by NMDA-receptor activation, of neurosteroid expression in the developing brain translate into refined axonal growth and stimulation of specific neuronal networks.

V. Neurosteroids and Neuroprotection

In humans, during development and later in life, DHEA and DHEAS are among the most abundant circulating neurosteroids (Kroboth et al., 1999), but the levels of the neurosteroids rapidly decline with age (Orentreich et al.,

1984, 1992; Carlstrom *et al.*, 1988; Birkenhager-Gillesse *et al.*, 1994; Sulcova *et al.*, 1997). As humans age, they become increasingly vulnerable to neurodegenerative processes, particularly diseases like Alzheimer's disease and ischemia/stroke, that are intimately associated with neuronal death (Agid, 1995; Graeber *et al.*, 1998). Therefore, based on preclinical evidence, it has been hypothesized that neurosteroids may have important clinical applications as neurotrophic and/or neuroprotective agents in the treatment of neurodegenerative disorders. The following sections detail some of the reported effects of neurosteroids in models of neurodegenerative diseases.

A. ISCHEMIA AND STROKE

Ischemia activates a cascade of events (del Zoppo, 1997; Hickenbottom and Grotta, 1998; Amar and Levy, 1999) that leads to the induction and expression of genes in a variety of cell types throughout the CNS (Yamagata *et al.*, 1993; Kiessling and Gass, 1994; Sharp and Sagar, 1994; Wang *et al.*, 1995; Feuerstein *et al.*, 1997). The excitatory amino acid glutamate, which is released immediately following an ischemic event (del Zoppo, 1997), is known to produce neuronal damage via long-term activation of NMDA receptors (Benveniste *et al.*, 1984; Choi *et al.*, 1988). A few studies have reported that neurosteroids are potent neuroprotective agents for hippocampal neurons treated with glutamate (Weaver *et al.*, 1997; Kimonides *et al.*, 1998; Cardounel *et al.*, 1999). One study in particular showed that DHEAS exerted a neuroprotective effect that was associated with activation of NFκB transcriptional activity (Mao and Barger, 1998). Thus, it is evident that DHEAS and other neurosteroids activate diverse molecular entities to exert their neuroprotective effects, such as counteracting excitotoxin-induced neuronal degeneration.

In our laboratory, we have studied the pharmacological effects of DHEAS in the reversible rabbit spinal cord ischemia model (RSCIM) (Lapchak *et al.*, 2000). In this model, ischemia is produced by occlusion of the aorta at the level of the renal arteries by pulling on a catheter around the aorta for a fixed interval. Varying the duration of occlusion can produce all grades of ischemic damage ranging from full recovery to permanent paraplegia (Zivin *et al.*, 1982; Lapchak *et al.*, 2000). Thus, short periods of occlusion (<20 min) produce no apparent neurological deficit, whereas long durations of occlusion (>30 min) result in paraplegia, with extensive lesions throughout the gray matter from the upper lumbar region to the end of the sacral segments of the spinal cord. The therapeutic potential of a drug is statistically analyzed and demonstrated using computer reconstruction of an ischemic "dose–response" curve that generates an ED_{50} value. This is defined as the duration of occlusion that produces permanent paraplegia

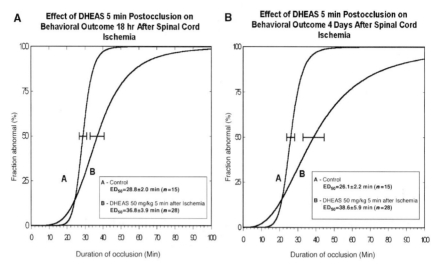

FIG. 1. Fraction abnormal (paraplegic) as a function of the duration of ischemia (aortic occlusion) in minutes. **(A)** Eighteen hours following ischemia: The curve labeled Control **(A)** shows the development of paraplegia in vehicle-treated rabbits. The curve labeled DHEAS **(B)** shows that DHEAS (50 mg/kg) increases the tolerance of aortic occlusion (ischemia) by approximately 8 min. **(B)** Four days following ischemia. The curve labeled Control **(A)** shows the development of paraplegia in vehicle-treated rabbits: The curve labeled DHEAS **(B)** shows that DHEAS (50 mg/kg) increases the tolerance of aortic occlusion (ischemia) by approximately 12 min. The results show that the effect of DHEAS to promote behavioral recovery (i.e., be neuroprotective) is durable and lasts 4 days following a single injection.

in 50% of the animals in a group (Zivin and Waud, 1992). When DHEAS (50 mg/kg) was administered 5 min after the initiation of occlusion and behavioral analysis was done 18 h later, DHEAS caused a significant shift to the right in the quantal analysis curve (Fig. 1A), suggesting that DHEAS was neuroprotective. When behavioral analysis was done 4 days following occlusion and DHEAS injection (Fig. 1B), there was still a significant shift in the quantal analysis curves, demonstrating a durable effect of DHEAS in this model. In contrast, when given 30 min after the initiation of occlusion, the neuroprotective effect of DHEAS was no longer evident, suggesting that DHEAS may be modulating the release of a rapid neurotransmitter immediately following the onset of ischemia.

Thus, there is convincing evidence that DHEAS modulates GABA- and NMDA-mediated neurotransmission. In view of this evidence and the demonstrated important role of GABAergic mechanisms in the spinal cord (Madden, 1994), it was important to assess whether the effects of DHEAS in the RSCIM were mediated by GABA receptors. For this, DHEAS was

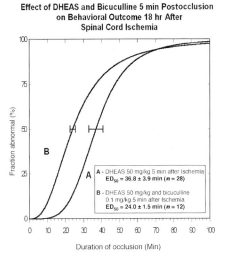

FIG. 2. Fraction abnormal (paraplegic) as a function of the duration of ischemia (aortic occlusion) 18 hours following ischemia (in minutes). The curve labeled DHEAS (A) shows the development of paraplegia in DHEAS-treated rabbits. The curve labeled DHEAS and Bicuculline (B) shows that bicuculline (0.1 mg/kg) reverses the beneficial effects of DHEAS. The P_{50} for the drug combination curve is similar to a control curve ED_{50}. The curve labeled DHEAS (A) shows the development of paraplegia in DHEAS-treated rabbits. The curve labeled DHEAS and Bicuculline (B) shows that, once there is blockade of DHEAS activity, the paraplegia is permanent.

administered concomitantly with the $GABA_A$ antagonist bicuculline to occluded rabbits, and behavior was analyzed 18 h thereafter (Fig. 2). Because a significant shift in the quantal analysis curve was not observed, as had been shown for DHEAS alone, it could be concluded that the neuroprotective effect of DHEAS was mediated by modulating GABA neurotransmisson. Overall, our evidence suggests that DHEAS may be useful to treat ischemia and stroke because it increases the tolerance to ischemia in the RSCIM.

Weaver *et al.* (1997) also showed that the neurosteroid pregnanolone hemisuccinate ($3\alpha5\beta HS$) was neuroprotective in a rat model of stroke in which the middle cerebral artery (MCA) was occluded for 2 h, resulting in global ischemia to the brain, specifically the cerebral cortex and the caudate-putamen (Fig. 3). When $3\alpha5\beta HS$ was given immediately after the initiation of occlusion, the infarct volume in the cortex and subcortical structures like the caudate-putamen were significantly decreased (Fig. 3 A and B). In contrast, delayed administration (30 min) produced significant neuroprotection only in the cortex (Fig. 3C). Because the *in vitro* studies indicated that $3\alpha5\beta HS$ was a potent inhibitor of NMDA receptors

A Administration immediately post ischemia

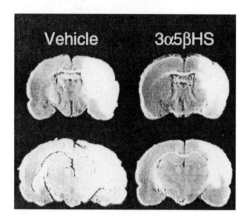

B Administration immediately post ischemia

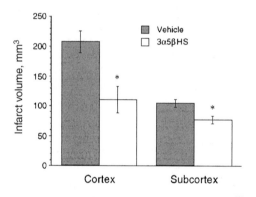

C Administration 30 min post ischemia

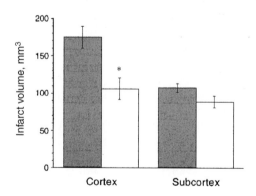

(Weaver *et al.*, 1997), the *in vivo* effect in the stroke model also was postulated to involve NMDA-receptor blockade. In view of the effectiveness of the neurosteroid $3\alpha5\beta$HS in the MCA occlusion model, the results suggest that neurosteroids that attenuate or block NMDA receptors may be useful to treat ischemia and stroke.

B. ALZHEIMER'S DISEASE AND AGING

Alzheimer's disease (AD) is a progressive neurodegenerative disease characterized by degeneration of the cholinergic system in the basal forebrain, including the septohippocampal pathway and the basalocortical pathways (Araujo *et al.*, 1988), but AD also involves alterations of other neurotransmitters throughout the brain (Olney *et al.*, 1997; Kalaria, 2000). Clinically and pathologically, AD presents itself as a disease with deterioration of both cognitive and mnemonic abilities, accompanied by widespread morphological and anatomical deficits (Bartus *et al.*, 1982; Whitehouse *et al.*, 1982; Terry and Katzman, 1983). Although less evidence on the effects of AD on neurosteroids is available, reduced DHEAS levels in patients with AD have been reported (Nasman *et al.*, 1991).

Preclinical testing of DHEAS in animal models of AD has resulted in some interesting findings. DHEAS enhancement of memory retention (Flood and Roberts, 1988; Flood *et al.*, 1988) and long-term memory (Roberts *et al.*, 1987) as well as improved performance in passive avoidance tests and plus-mazes (Reddy and Kulkarni, 1998) have been documented. Nitric-oxide-mediated mechanisms have been implicated in these effects (Reddy and Kulkarni, 1998). More detailed analyses of neurosteroid improvements in spatial/reference, working, and long-term memory in rodents revealed that DHEAS was most effective at decreasing distances to hidden platforms in the Morris water maze and increasing the percentage of correct choices in the Y-maze (Frye and Sturgis, 1995; Frye and Lacey, 1999). The memory-boosting effect has been proposed to be the result of

---

FIG. 3. Effect of pregnanolone hemisuccinate ($3\alpha5\beta$HS) on infarct volume measured in rodents following MCA occlusion. (**A**) Representative sections that were stained with 2,3,5-triphenyl tetrazolium. Note the large infarct (white areas) in the cortex and striatum of the vehicle-treated rats and less of an infarct in the cortex and the striatum of drug-treated rats. (**B**) Quantitative aspects of the neuroprotective effects of $3\alpha5\beta$HS when administered immediately following the occlusion. (**C**) The results from a delayed administration regimen. $3\alpha5\beta$HS was effective at reducing the cortical infarct when immediate or delayed administration was used (*$P < 0.005$ compared to vehicle control). Reproduced with permission of Weaver *et al.*, *Proc. Natl. Acad. Sci. USA* **94,** 10450–10454 (1997).

increased hippocampal bursting patterns and the induction of neuroplasticity by DHEAS (Diamond *et al.*, 1995, 1996, 1997).

The improvement in memory tasks subsequent to DHEAS treatment observed in laboratory animals under a variety of circumstances suggests that restoration of DHEAS serum levels may be a useful therapeutic tool for patients with memory impairment. In older (over 65 years of age) men and women, low plasma DHEAS levels have been reported to correlate with dyspnea, depression, poor life outlook, increased short-term mortality, and even death occurring within 2–4 years of the baseline measurement (Berr *et al.*, 1996). Because of the significant decline of DHEAS levels during aging (Orentreich *et al.*, 1984, 1992; Carlstrom *et al.*, 1988; Birkenhager-Gillesse *et al.*, 1994; Sousa and Ticku, 1997), numerous clinical tests of the effects of DHEA or DHEAS replacement in humans during aging have been attempted. In a notable trial in which DHEA was given to healthy individuals (60–79 years old), DHEAS increased libido and skin status in women (Baulieu *et al.*, 2000), suggesting that DHEAS normalized some of the effects of aging. In other studies, DHEA was found to improve responses to stress (Kudielka *et al.*, 1998) and to increase the sense of well-being and mood in aged women, but not men (Wolf *et al.*, 1997). There also was a trend for better performance in a cognitive test that relied on picture memory (Wolf *et al.*, 1997). In a study of patients with multi-infarct dementia, daily intravenous administration of DHEAS increased serum and cerebrospinal fluid levels of DHEAS, an effect that appeared to translate into significantly improved daily activities and emotional state (Azuma *et al.*, 1999). Although there was some normalization of electroencephalogram in some of the patients, direct measurements of memory were not conducted. However, in studies that used more sophisticated memory tests, such as the Mini-Mental State Examination, no significant association between DHEAS levels and cognitive impairment and decline were observed (Kalmijn *et al.*, 1998; Yaffe *et al.*, 1998). In-depth evaluation of the effects of DHEAS on CNS activity showed a correlation between increased serum levels and modulation of the P3 component of event-related potentials, a reflection of improved short-term memory consolidation, but overall memory or mood was not affected (Wolf *et al.*, 1998a; Wolf *et al.*, 1998b; Barnhart *et al.*, 1999; Flynn *et al.*, 1999; Wolf and Kirschbaum, 1999; Huppert *et al.*, 2000).

Considering the positive effects of DHEAS on cognition in animal studies, the rather modest results obtained in the human studies were disappointing. The latter served only to illustrate the continued controversy surrounding the "beneficial" aspects of DHEAS supplementation in humans. However, certain promising results should be taken into account in determining whether a thorough evaluation of DHEAS as a memory-enhancer in a well-controlled clinical setting is warranted.

VI. Conclusions

This chapter has reviewed the importance of neurosteroids and their regulatory mechanisms in both the induction of specific neuronal networks and the modulation of various pathways in the CNS. Neurosteroids appear to be of particular importance during aging and under circumstances during which neurons are exposed acutely to stressful stimuli. The decline in DHEAS levels in response to aging and AD further highlights the significance of this neurosteroid in the regulation of various neuronal processes that may involve GABA- and NMDA-receptor mechanisms. Work from our laboratory supports the hypothesis that DHEAS may have clinical therapeutic value for the treatment of ischemia and, possibly, stroke. In conclusion, neurosteroids eventually may be useful to treat neurodegenerative diseases and to counteract some of the effects of normal aging.

References

Agid, Y. (1995). Aging, disease and nerve cell death. *Bull. Acad. Natl. Med.* **179,** 1193–1203.
Amar, A. P., and Levy, M. L. (1999). Pathogenesis and pharmacological strategies for mitigating secondary damage in acute spinal cord injury. *Neurosurgery* **44,** 1027–1039.
Araujo, D. M., Lapchak, P. A., Robitaille, Y., Gauthier, S., and Quirion, R. (1988). Differential alteration of various cholinergic markers in cortical and subcortical regions of human brain in Alzheimer's disease. *J. Neurochem.* **50,** 1914–1923.
Azuma, T., Nagai, Y., Saito, T., Funauchi, M., Matsubara, T., and Sakoda, S. (1999). The effect of dehydroepiandrosterone sulfate administration to patients with multi-infarct dementia. *J. Neurol. Sci.* **162,** 69–73.
Barker, J. L., Harrison, N. L., Meyers, D. E., and Majewska, M. D. (1986). Steroid modulation of $GABA_A$ receptor-coupled Cl^- conductance. *Clin. Neuropharmacol.* **9,** 392–394.
Barnhart, K. T., Freeman, E., Grisso, J. A., Rader, D. J., Sammel, M., Kapoor, S., and Nestler, J. E. (1999). The effect of dehydroepiandrosterone supplementation to symptomatic perimenopausal women on serum endocrine profiles, lipid parameters, and health-related quality of life. *J. Clin. Endocrinol. Metab.* **84,** 3896–3902.
Bartus, R. T., Dean, R. L. D., Beer, B., and Lippa, A. S. (1982). The cholinergic hypothesis of geriatric memory dysfunction. *Science* **217,** 408–414.
Baulieu, E. E. (1981). Steroid hormones in the brain: Several mechanisms? *In* "Steroid Hormone Regulation of the Brain" (K. Fuxe, J. A. Gustafons, and L. Wetterbrg, Eds.), pp. 3–14. Pergamon Press, Oxford, United Kingdom.
Baulieu, E. E. (1998). Neurosteroids: a novel function of the brain. *Psychoneuroendocrinology* **23,** 963–987.
Baulieu, E. E., Thomas, G., Legrain, S., Lahlou, N., Roger, M., Debuire, B., Faucounau, V., Girard, L., Hervy, M. P., Latour, F., Leaud, M. C., Mokrane, A., Pitti-Ferrandi, H., Trivalle, C., de Lacharriere, O., Nouveau, S., Rakoto-Arison, B., Souberbielle, J. C., Raison, J., Le Bouc, Y., Raynaud, A., Girerd, X., and Forette, F. (2000). Dehydroepiandrosterone (DHEA), DHEA sulfate, and aging: Contribution of the DHEAge Study to a sociobiomedical issue. *Proc. Natl. Acad. Sci. USA* **97,** 4279–4284.

Benveniste, H., Drejer, J., Schousboe, A., and Diemer, N. H. (1984). Elevation of the extracellular concentrations of glutamate and aspartate in rat hippocampus during transient cerebral ischemia monitored by intracerebral microdialysis. *J. Neurochem.* **43,** 1369–1374.
Bergeron, R., de Montigny, C., and Debonnel, G. (1996). Potentiation of neuronal NMDA response induced by dehydroepiandrosterone and its suppression by progesterone: Effects mediated via sigma receptors. *J. Neurosci.* **16,** 1193–1202.
Berr, C., Lafont, S., Debuire, B., Dartigues, J. F., and Baulieu, E. E. (1996). Relationships of dehydroepiandrosterone sulfate in the elderly with functional, psychological, and mental status, and short-term mortality: A French community-based study. *Proc. Natl. Acad. Sci. USA* **93,** 13410–13415.
Birkenhager-Gillesse, E. G., Derksen, J., and Lagaay, A. M. (1994). Dehydroepiandrosterone sulphate (DHEAS) in the oldest old, aged 85 and over. *Ann. NY. Acad. Sci.* **719,** 543–552.
Bologa, L., Sharma, J., and Roberts, E. (1987). Dehydroepiandrosterone and its sulfated derivative reduce neuronal death and enhance astrocytic differentiation in brain cell cultures. *J. Neurosci. Res.* **17,** 225–234.
Booth, R. G., Wyrick, S. D., Baldessarini, R. J., Kula, N. S., Myers, A. M., and Mailman, R. B. (1993). New sigma-like receptor recognized by novel phenylaminotetralins: Ligand binding and functional studies. *Mol. Pharmacol.* **44,** 1232–1239.
Bowlby, M. R. (1993). Pregnenolone sulfate potentiation of N-methyl-D-aspartate receptor channels in hippocampal neurons. *Mol. Pharmacol.* **43,** 813–819.
Brinton, R. D. (1994). The neurosteroid 3-alpha-hydroxy-5-alpha-pregnan-20-one induces cytoarchitectural regression in cultured fetal hippocampal neurons. *J. Neurosci.* **14,** 2763–2774.
Cardounel, A., Regelson, W., and Kalimi, M. (1999). Dehydroepiandrosterone protects hippocampal neurons against neurotoxin- induced cell death: Mechanism of action. *Proc. Soc. Exp. Biol. Med.* **222,** 145–149.
Carlstrom, K., Brody, S., Lunnell, N. O., Lagrelius, A., Mollerstrom, G., Pousette, A., Rannevik, G., Stege, R., and von Schoultz, B. (1988). Dehydroepiandrosterone sulphate and dehydroepiandrosterone in serum: Differences related to age and sex. *Maturitas* **10,** 297–306.
Choi, D. W., Koh, J. Y., and Peters, S. (1988). Pharmacology of glutamate neurotoxicity in cortical cell culture: Attenuation by NMDA antagonists. *J. Neurosci.* **8,** 185–196.
Compagnone, N. A., and Mellon, S. H. (1998). Dehydroepiandrosterone: A potential signalling molecule for neocortical organization during development. *Proc. Natl. Acad. Sci. USA* **95,** 4678–4683.
Corpéchot, C., Robel, P., Axelson, M., Sjovall, J., and Baulieu, E. E. (1981). Characterization and measurement of dehydroepiandrosterone sulfate in rat brain. *Proc. Natl. Acad. Sci. USA* **78,** 4704–4707.
Corpéchot, C., Synguelakis, M., Talha, S., Axelson, M., Sjovall, J., Vihko, R., Baulieu, E. E., and Robel, P. (1983). Pregnenolone and its sulfate ester in the rat brain. *Brain Res.* **270,** 119–125.
Debonnel, G. (1993). Current hypotheses on sigma receptors and their physiological role: Possible implications in psychiatry. *J. Psychiatry. Neurosci.* **18,** 157–172.
Debonnel, G., Bergeron, R., and de Montigny, C. (1996a). Potentiation by dehydroepiandrosterone of the neuronal response to N-methyl-D-aspartate in the CA3 region of the rat dorsal hippocampus: An effect mediated via sigma receptors. *J. Endocrinol.* **150 Suppl,** S33–S42.
Debonnel, G., Bergeron, R., Monnet, F. P., and De Montigny, C. (1996b). Differential effects of sigma ligands on the N-methyl-D-aspartate response in the CA1 and CA3 regions of the dorsal hippocampus: Effect of mossy fiber lesioning. *Neuroscience* **71,** 977–987.

Debonnel, G., and de Montigny, C. (1996)). Modulation of NMDA and dopaminergic neurotransmissions by sigma ligands: Possible implications for the treatment of psychiatric disorders. *Life Sci.* **58,** 721–734.
del Zoppo, G. J. (1997). Microvascular responses to cerebral ischemia/inflammation. *Ann. NY. Acad. Sci.* **823,** 132–147.
Diamond, D. M., Branch, B. J., and Fleshner, M. (1996). The neurosteroid dehydroepiandrosterone sulfate (DHEAS) enhances hippocampal primed burst, but not long-term, potentiation. *Neurosci. Lett.* **202,** 204–208.
Diamond, D. M., Branch, B. J., Fleshner, M., and Rose, G. M. (1995). Effects of dehydroepiandrosterone sulfate and stress on hippocampal electrophysiological plasticity. *Ann. NY. Acad. Sci.* **774,** 304–307.
Diamond, D. M., Fleshner, M., and Rose, G. M. (1999). The enhancement of hippocampal primed burst potentiation by dehydroepiandrosterone sulfate (DHEAS) is blocked by psychological stress. *Stress* **3,** 107–121.
Doble, A. (1995). Excitatory amino acid receptors and neurodegeneration. *Therapie* **50,** 319–337.
Fahey, J. M., Lindquist, D. G., Pritchard, G. A., and Miller, L. G. (1995). Pregnenolone sulfate potentiation of NMDA-mediated increases in intracellular calcium in cultured chick cortical neurons. *Brain Res.* **669,** 183–188.
Feuerstein, G. Z., Wang, X., and Barone, F. C. (1997). Inflammatory gene expression in cerebral ischemia and trauma. Potential new therapeutic targets. *Ann. NY Acad. Sci.* **825,** 179–193.
Flood, J. F., and Roberts, E. (1988). Dehydroepiandrosterone sulfate improves memory in aging mice. *Brain Res.* **448,** 178–181.
Flood, J. F., Smith, G. E., and Roberts, E. (1988). Dehydroepiandrosterone and its sulfate enhance memory retention in mice. *Brain Res.* **447,** 269–278.
Flynn, M. A., Weaver-Osterholtz, D., Sharpe-Timms, K. L., Allen, S., and Krause, G. (1999). Dehydroepiandrosterone replacement in aging humans. *J. Clin. Endocrinol. Metab.* **84,** 1527–1533.
Frye, C. A., and Lacey, E. H. (1999). The neurosteroids DHEA and DHEAS may influence cognitive performance by altering affective state. *Physiol. Behav.* **66,** 85–92.
Frye, C. A., and Sturgis, J. D. (1995). Neurosteroids affect spatial/reference, working, and long-term memory of female rats. *Neurobiol. Learn. Mem.* **64,** 83–96.
Gibbs, T. T., Yaghoubi, N., Weaver, C. E., Park-Chung, M., Russek, S. J., and Farb, D. H. (1999). Modulation of ionotropic glutamate receptors by neuroactive steroids. *In* "Contemporary Endocrinology: Neurosteroids: A New Regulatory Function in the Nervous System"(E.-E. Baulieu, P. Robel, and M. Schumacher, Eds.), pp. 167–190. Humana Press, Totowa, NJ.
Graeber, M. B., Grasbon-Frodl, E., Eitzen, U. V., and Kosel, S. (1998). Neurodegeneration and aging: role of the second genome. *J. Neurosci. Res.* **52,** 1–6.
Hansen, S. L., Fjalland, B., and Jackson, M. B. (1999). Differential blockade of gamma-aminobutyric acid type A receptors by the neuroactive steroid dehydroepiandrosterone sulfate in posterior and intermediate pituitary. *Mol. Pharmacol.* **55,** 489–496.
Harrison, N. L., and Simmonds, M. A. (1984). Modulation of the GABA receptor complex by a steroid anaesthetic. *Brain Res.* **323,** 287–292.
Hickenbottom, S. L., and Grotta, J. (1998). Neuroprotective therapy. *Semin. Neurol.* **18,** 485–492.
Huppert, F. A., Van Niekerk, J. K., and Herbert, J. (2000). Dehydroepiandrosterone (DHEA) supplementation for cognition and well-being. *Cochrane Database Syst. Rev.* **2.**
Ito, I., Futai, K., Katagiri, H., Watanabe, M., Sakimura, K., Mishina, M., and Sugiyama, H. (1997). Synapse-selective impairment of NMDA receptor functions in mice lacking NMDA receptor epsilon 1 or epsilon 2 subunit. *J. Physiol. (London)* **500,** 401–408.

Kalaria, R. N. (2000). The role of cerebral ischemia in Alzheimer's disease. *Neurobiol. Aging* **21**, 321–330.

Kalmijn, S., Launer, L. J., Stolk, R. P., de Jong, F. H., Pols, H. A., Hofman, A., Breteler, M. M., and Lamberts, S. W. (1998). A prospective study on cortisol, dehydroepiandrosterone sulfate, and cognitive function in the elderly. *J. Clin. Endocrinol. Metab.* **83**, 3487–3492.

Kiessling, M., and Gass, P. (1994). Stimulus-transcription coupling in focal cerebral ischemia. *Brain. Pathol.* **4**, 77–83.

Kimonides, V. G., Khatibi, N. H., Svendsen, C. N., Sofroniew, M. V., and Herbert, J. (1998). Dehydroepiandrosterone (DHEA) and DHEA-sulfate (DHEAS) protect hippocampal neurons against excitatory amino acid-induced neurotoxicity. *Proc. Natl. Acad. Sci. USA* **95**, 1852–1857.

Korpi, E. R., and Luddens, H. (1993). Regional gamma-aminobutyric acid sensitivity of t- butylbicyclophosphoro[^{35}S]thionate binding depends on gamma-aminobutyric acid$_A$ receptor alpha subunit. *Mol. Pharmacol.* **44**, 87–92.

Kroboth, P. D., Salek, F. S., Pittenger, A. L., Fabian, T. J., and Frye, R. F. (1999). DHEA and DHEA-S: A review. *J. Clin. Pharmacol.* **39**, 327–348.

Kudielka, B. M., Hellhammer, J., Hellhammer, D. H., Wolf, O. T., Pirke, K. M., Varadi, E., Pilz, J., and Kirschbaum, C. (1998). Sex differences in endocrine and psychological responses to psychosocial stress in healthy elderly subjects and the impact of a 2-week dehydroepiandrosterone treatment. *J. Clin. Endocrinol. Metab.* **83**, 1756–1761.

Lambert, J. J., Belelli, D., Hill-Venning, C., and Peters, J. A. (1995). Neurosteroids and GABA$_A$ receptor function. *Trends Pharmacol. Sci.* **16**, 295–303.

Lan, N. C., Bolger, M. B., and Gee, K. W. (1991a). Identification and characterization of a pregnane steroid recognition site that is functionally coupled to an expressed GABA$_A$ receptor. *Neurochem. Res.* **16**, 347–356.

Lan, N. C., Gee, K. W., Bolger, M. B., and Chen, J. S. (1991b). Differential responses of expressed recombinant human gamma-aminobutyric acid$_A$ receptors to neurosteroids. *J. Neurochem.* **57**, 1818–1821.

Lapchak, P. A., Chapman, D. F., Nunez, S. Y., and Zivin, J. A. (2000). Dehydroepiandrosterone sulfate is neuroprotective in a reversible spinal cord ischemia model: Possible involvement of GABA(A) receptors. *Stroke* **31**, 1953–1956.

Luddens, H., and Korpi, E. R. (1995). GABA antagonists differentiate between recombinant GABAA/benzodiazepine receptor subtypes. *J. Neurosci.* **15**, 6957–6962.

Luddens, H., Korpi, E. R., and Seeburg, P. H. (1995). GABA$_A$/benzodiazepine receptor heterogeneity: Neurophysiological implications. *Neuropharmacology* **34**, 245–254.

Madden, K. P. (1994). Effect of gamma-aminobutyric acid modulation on neuronal ischemia in rabbits. *Stroke* **25**, 2271–2274.

Majewska, M. D. (1992). Neurosteroids: Endogenous bimodal modulators of the GABA$_A$ receptor. Mechanism of action and physiological significance. *Prog. Neurobiol.* **38**, 379–395.

Majewska, M. D. (1995). Neuronal actions of dehydroepiandrosterone. Possible roles in brain development, aging, memory, and affect. *Ann. NY. Acad. Sci.* **774**, 111–120.

Majewska, M. D., Harrison, N. L., Schwartz, R. D., Barker, J. L., and Paul, S. M. (1986). Steroid hormone metabolites are barbiturate-like modulators of the GABA receptor. *Science* **232**, 1004–1007.

Majewska, M. D., and Vaupel, D. B. (1991). Steroid control of uterine motility via gamma-aminobutyric acid$_A$ receptors in the rabbit: A novel mechanism? *J. Endocrinol.* **131**, 427–434.

Mao, X., and Barger, S. W. (1998). Neuroprotection by dehydroepiandrosterone-sulfate: Role of an NFκB- like factor. *Neuroreporter* **9**, 759–763.

Marx, C. E., Jarskog, L. F., Lauder, J. M., Gilmore, J. H., Lieberman, J. A., and Morrow, A. L. (2000). Neurosteroid modulation of embryonic neuronal survival *in vitro* following anoxia. *Brain Res.* **871,** 104–112.

Mathis, C., Vogel, E., Cagniard, B., Criscuolo, F., and Ungerer, A. (1996). The neurosteroid pregnenolone sulfate blocks deficits induced by a competitive NMDA antagonist in active avoidance and lever-press learning tasks in mice. *Neuropharmacology* **35,** 1057–1064.

Maurice, T., Junien, J. L., and Privat, A. (1997). Dehydroepiandrosterone sulfate attenuates dizocilpine-induced learning impairment in mice via sigma 1-receptors. *Behav. Brain Res.* **83,** 159–164.

Melchior, C. L., and Ritzmann, R. F. (1994). Dehydroepiandrosterone is an anxiolytic in mice on the plus maze. *Pharmacol. Biochem. Behav.* **47,** 437–441.

Mensah-Nyagan, A. G., Do-Rego, J. L., Beaujean, D., Luu-The, V., Pelletier, G., and Vaudry, H. (1999). Neurosteroids: Expression of steroidogenic enzymes and regulation of steroid biosynthesis in the central nervous system. *Pharmacol. Rev.* **51,** 63–81.

Monnet, F. P., Mahe, V., Robel, P., and Baulieu, E. E. (1995). Neurosteroids, via sigma receptors, modulate the [^{3}H]norepinephrine release evoked by N-methyl-D-aspartate in the rat hippocampus. *Proc. Natl. Acad. Sci. USA* **92,** 3774–3778.

Nakanishi, S., Nakajima, Y., Masu, M., Ueda, Y., Nakahara, K., Watanabe, D., Yamaguchi, S., Kawabata, S., and Okada, M. (1998). Glutamate receptors: Brain function and signal transduction. *Brain Res. Brain Res. Rev.* **26,** 230–235.

Nasman, B., Olsson, T., Backstrom, T., Eriksson, S., Grankvist, K., Viitanen, M., and Bucht, G. (1991). Serum dehydroepiandrosterone sulfate in Alzheimer's disease and in multi-infarct dementia. *Biol. Psychiatry* **30,** 684–690.

Olney, J. W., Wozniak, D. F., and Farber, N. B. (1997). Excitotoxic neurodegeneration in Alzheimer disease. New hypothesis and new therapeutic strategies. *Arch. Neurol.* **54,** 1234–1240.

Orentreich, N., Brind, J. L., Rizer, R. L., and Vogelman, J. H. (1984). Age changes and sex differences in serum dehydroepiandrosterone sulfate concentrations throughout adulthood. *J. Clin. Endocrinol. Metab.* **59,** 551–555.

Orentreich, N., Brind, J. L., Vogelman, J. H., Andres, R., and Baldwin, H. (1992). Long-term longitudinal measurements of plasma dehydroepiandrosterone sulfate in normal men. *J. Clin. Endocrinol. Metab.* **75,** 1002–1004.

Park-Chung, M., Malayev, A., Purdy, R. H., Gibbs, T. T., and Farb, D. H. (1999). Sulfated and unsulfated steroids modulate gamma-aminobutyric acid$_A$ receptor function through distinct sites. *Brain Res.* **830,** 72–87.

Park-Chung, M., Wu, F. S., and Farb, D. H. (1994). 3 Alpha-Hydroxy-5 beta-pregnan-20-one sulfate: A negative modulator of the NMDA-induced current in cultured neurons. *Mol. Pharmacol.* **46,** 146–150.

Park-Chung, M., Wu, F. S., Purdy, R. H., Malayev, A. A., Gibbs, T. T., and Farb, D. H. (1997). Distinct sites for inverse modulation of N-methyl-D-aspartate receptors by sulfated steroids. *Mol. Pharmacol.* **52,** 1113–1123.

Puia, G., Ducic, I., Vicini, S., and Costa, E. (1993). Does neurosteroid modulatory efficacy depend on GABA$_A$ receptor subunit composition? *Receptors Channels* **1,** 135–142.

Puia, G., Santi, M. R., Vicini, S., Pritchett, D. B., Purdy, R. H., Paul, S. M., Seeburg, P. H., and Costa, E. (1990). Neurosteroids act on recombinant human GABA$_A$ receptors. *Neuron* **4,** 759–765.

Purdy, R. H., Morrow, A. L., Blinn, J. R., and Paul, S. M. (1990). Synthesis, metabolism, and pharmacological activity of 3-alpha-hydroxy steroids which potentiate GABA-receptor-mediated chloride ion uptake in rat cerebral cortical synaptoneurosomes. *J. Med. Chem.* **33,** 1572–1581.

Quirion, R., Bowen, W. D., Itzhak, Y., Junien, J. L., Musacchio, J. M., Rothman, R. B., Su, T. P., Tam, S. W., and Taylor, D. P. (1992). A proposal for the classification of sigma binding sites. *Trends Pharmacol. Sci.* **13,** 85–86.

Reddy, D. S., and Kulkarni, S. K. (1997a). Differential anxiolytic effects of neurosteroids in the mirrored chamber behavior test in mice. *Brain Res.* **752,** 61–71.

Reddy, D. S., and Kulkarni, S. K. (1997b). Neurosteroid coadministration prevents development of tolerance and augments recovery from benzodiazepine withdrawal anxiety and hyperactivity in mice. *Methods Find. Exp. Clin. Pharmacol.* **19,** 395–405.

Reddy, D. S., and Kulkarni, S. K. (1998). The role of GABA-A and mitochondrial diazepam-binding inhibitor receptors on the effects of neurosteroids on food intake in mice. *Psychopharmacology (Berlin)* **137,** 391–400.

Roberts, E., Bologa, L., Flood, J. F., and Smith, G. E. (1987). Effects of dehydroepiandrosterone and its sulfate on brain tissue in culture and on memory in mice. *Brain Res.* **406,** 357–362.

Sakimura, K., Kutsuwada, T., Ito, I., Manabe, T., Takayama, C., Kushiya, E., Yagi, T., Aizawa, S., Inoue, Y., and Sugiyama, H., *et al.* (1995). Reduced hippocampal LTP and spatial learning in mice lacking NMDA receptor epsilon 1 subunit. *Nature* **373,** 151–155.

Sapp, D. W., Witte, U., Turner, D. M., Longoni, B., Kokka, N., and Olsen, R. W. (1992). Regional variation in steroid anesthetic modulation of [^{35}S]TBPS binding to gamma-aminobutyric acid$_A$ receptors in rat brain. *J. Pharmacol. Exp. Ther.* **262,** 801–808.

Schmid, G., Bonanno, G., and Raiteri, M. (1996). Functional evidence for two native GABA$_A$ receptor subtypes in adult rat hippocampus and cerebellum. *Neuroscience* **73,** 697–704.

Schmid, G., Sala, R., Bonanno, G., and Raiteri, M. (1998). Neurosteroids may differentially affect the function of two native GABA(A) receptor subtypes in the rat brain. *Naunyn. Schmiedebergs Arch. Pharmacol.* **357,** 401–407.

Sharp, F. R., and Sagar, S. M. (1994). Alterations in gene expression as an index of neuronal injury: Heat shock and the immediate early gene response. *Neurotoxicology* **15,** 51–59.

Shen, W., Mennerick, S., Zorumski, E. C., Covey, D. F., and Zorumski, C. F. (1999). Pregnenolone sulfate and dehydroepiandrosterone sulfate inhibit GABA-gated chloride currents in *Xenopus* oocytes expressing picrotoxin-insensitive GABA(A) receptors. *Neuropharmacology* **38,** 267–271.

Shingai, R., Sutherland, M. L., and Barnard, E. A. (1991). Effects of subunit types of the cloned GABA$_A$ receptor on the response to a neurosteroid. *Eur. J. Pharmacol.* **206,** 77–80.

Sousa, A., and Ticku, M. K. (1997). Interactions of the neurosteroid dehydroepiandrosterone sulfate with the GABA(A) receptor complex reveals that may act via the picrotoxin site. *J. Pharmacol. Exp. Ther.* **282,** 827–833.

Sulcova, J., Hill, M., Hampl, R., and Starka, L. (1997). Age and sex related differences in serum levels of unconjugated dehydroepiandrosterone and its sulphate in normal subjects. *J. Endocrinol.* **154,** 57–62.

Terry, R. D., and Katzman, R. (1983). Senile dementia of the Alzheimer type. *Ann. Neurol.* **14,** 497–506.

Wang, X., Yue, T. L., Young, P. R., Barone, F. C., and Feuerstein, G. Z. (1995). Expression of interleukin-6, c-fos, and zif268 mRNAs in rat ischemic cortex. *J. Cereb. Blood Flow Metab.* **15,** 166–171.

Weaver, C. E., Jr., Marek, P., Park-Chung, M., Tam, S. W., and Farb, D. H. (1997). Neuroprotective activity of a new class of steroidal inhibitors of the *N*-methyl-D-aspartate receptor. *Proc. Natl. Acad. Sci. USA* **94,** 10450–10454.

Whitehouse, P. J., Price, D. L., Struble, R. G., Clark, A. W., Coyle, J. T., and Delon, M. R. (1982). Alzheimer's disease and senile dementia: loss of neurons in the basal forebrain. *Science* **215,** 1237–1239.

Wolf, O. T., and Kirschbaum, C. (1999). Actions of dehydroepiandrosterone and its sulfate in the central nervous system: Effects on cognition and emotion in animals and humans. *Brain Res. Brain Res. Rev.* **30,** 264–288.

Wolf, O. T., Kudielka, B. M., Hellhammer, D. H., Hellhammer, J., and Kirschbaum, C. (1998a). Opposing effects of DHEA replacement in elderly subjects on declarative memory and attention after exposure to a laboratory stressor. *Psychoneuroendocrinology* **23,** 617–629.

Wolf, O. T., Naumann, E., Hellhammer, D. H., and Kirschbaum, C. (1998b). Effects of dehydroepiandrosterone replacement in elderly men on event-related potentials, memory, and well-being. *J. Geronol. A. Biol. Sci. Med. Sci.* **53,** M385–M390.

Wolf, O. T., Neumann, O., Hellhammer, D. H., Geiben, A. C., Strasburger, C. J., Dressendorfer, R. A., Pirke, K. M., and Kirschbaum, C. (1997). Effects of a two-week physiological dehydroepiandrosterone substitution on cognitive performance and well-being in healthy elderly women and men. *J. Clin. Endocrinol. Metab.* **82,** 2363–2367.

Wu, F. S., Gibbs, T. T., and Farb, D. H. (1991). Pregnenolone sulfate: A positive allosteric modulator at the N-methyl-D-aspartate receptor. *Mol. Pharmacol.* **40,** 333–336.

Yaffe, K., Ettinger, B., Pressman, A., Seeley, D., Whooley, M., Schaefer, C., and Cummings, S. (1998). Neuropsychiatric function and dehydroepiandrosterone sulfate in elderly women: A prospective study. *Biol. Psychiatry* **43,** 694–700.

Yamagata, K., Andreasson, K. I., Kaufmann, W. E., Barnes, C. A., and Worley, P. F. (1993). Expression of a mitogen-inducible cyclooxygenase in brain neurons: Regulation by synaptic activity and glucocorticoids. *Neuron* **11,** 371–386.

Zivin, J. A., DeGirolami, U., and Hurwitz, E. L. (1982). Spectrum of neurological deficits in experimental CNS ischemia. A quantitative study. *Arch. Neurol.* **39,** 408–412.

Zivin, J. A., and Waud, D. R. (1992). Quantal bioassay and stroke. *Stroke* **23,** 767–773.

Zwain, I. H., and Yen, S. S. (1999). Dehydroepiandrosterone: Biosynthesis and metabolism in the brain. *Endocrinology* **140,** 880–887.

CLINICAL IMPLICATIONS OF CIRCULATING NEUROSTEROIDS

Andrea R. Genazzani,[1] Patrizia Monteleone, Massimo Stomati,
Francesca Bernardi, Luigi Cobellis,* Elena Casarosa, Michele Luisi, S. Luisi,
and Felice Petraglia*

Department of Reproductive Medicine and Child Development, Division of Gynecology and
Obstetrics, University of Pisa
*Obstetrics and Gynecology, University of Siena, 53100 Siena, Italy

I. Introduction
   A. Assays for Allopregnanolone and DHEA in Humans
   B. Sources and Targets of Neurosteroids
II. Changes and Possible Role of Neurosteroids in Humans
   A. Puberty
   B. Fertile Life
   C. Menopause
   D. Aging and Related Disorders
   References

I. Introduction

Clinical studies on neurosteroids in humans mainly derive from the measurement of circulating forms, including pregnenolone, dehydroepiandrosterone (DHEA) and their sulfate esters, and the tetrahydroderivative compounds of progesterone, such as 3α-hydroxy-5α-pregnan-20-one (allopregnanolone), and tetrahydrodeoxycorticosterone ($3\alpha,5\alpha$-TH DOC) (Majewska, 1992; Corpéchot *et al.*, 1993; Thijssen and Nieuwenhuyse, 1999). This chapter focuses on the studies investigating allopregnanolone and DHEA concentrations in human tissues, serum, and other biological fluids as well as on the clinical effects of DHEA administration.

A. Assays for Allopregnanolone and DHEA in Humans

In the 1990s, some highly sensitive assays for allopregnanolone were developed. In 1990, Purdy *et al.* described a radioimmunoassay (RIA) using policlonal antibodies raised in rabbits against 3α-hydroxy-5α-pregnan-11α-yl

[1] Author to whom correspondence should be sent.

carboxymethyl ether coupled to bovine serum albumin (BSA). Allopregnanolone was purified from ether extracts of human plasma by high-performance liquid chromatography (HPLC). The sensitivity of the assay was 25 pg, with intra- and interassay coefficients of variation of 6.5% and 8.5%, respectively.

In 1995, Cheney et al. proposed coupling HPLC with gas chromatography–mass fragmentography because they considered the RIA method, even when combined with HPLC, unreliable for measuring neurosteroid content with sufficient specificity. They suggested identifying steroids (1) by their HPLC retention times, (2) by their gas chromatography retention times, and (3) by their unique fragmentation spectra following derivatization with heptafluorobutyric anhydride or methoxyamine hydrochloride.

Our recent assay for allopregnanolone levels in humans uses polyclonal antiserum raised by Purdy and Sluss in sheep against the aforementioned BSA conjugate. The sensitivity of the assay was 20 pg/tube, and the intra- and interassay coefficients of variation were 7.2% and 9.1%, respectively. The antibody obtained in sheep resulted in a more specific finding than that obtained in rabbits. In fact, this new method does not require HPLC but, rather, a purification by C18 Sep Pak cartridges to reduce cross-reactivities (Bernardi et al., 1998).

To date, the determination of DHEA in humans is based on the extraction of serum samples by ether purification through a C18 Sep-Pak cartridge and then assay by a specific RIA (Stomati et al., 1999).

B. Sources and Targets of Neurosteroids

The brain is the first known source of neurosteroids in humans. In a study by Bixo et al. (1997), postmortem concentrations of allopregnanolone were measured in 17 brain areas and in serum in five fertile and five postmenopausal women. Allopregnanolone was detected in the brain cortex, amygdala, hippocampus, caudate nucleus, putamen, and thalamus; the highest levels were observed in the substantia nigra and the basal hypothalamus. According to the authors, the regional differences in brain steroid levels imply different local mechanisms for steroid uptake and binding. Brain concentrations of allopregnanolone were significantly higher in fertile women in the luteal phase than in postmenopausal controls; these may depend on ovarian steroid production, indicating that the secretion pattern during the menstrual cycle is reflected in the brain. Δ^5-Androgens have also been measured in specific regions of cadavers brain. The central/plasmatic ratio of DHEA demonstrated a higher concentration of the steroid within the brain. In the same study, DHEA sulphatase and sulphotransferase activities were also detected (Majewska et al., 1990; Mellon, 1994). Allopregnanolone is also measurable in cerebrospinal fluid (CSF). In an elegant study, Uzunova et al. (1998) evaluated allopregnanolone content in four cisternal-lumbar fractions of CSF before and 8–10 weeks after treatment with fluoxetine or fluvoxamine in 15 patients with unipolar major depression. Most interestingly, the concentration of allopregnanolone was about 60% lower in patients with major unipolar depression, while, in the same patients, fluoxetine or fluvoxamine treatment normalized the CSF allopregnanolone content.

Other possible sources of allopregnanolone in women were investigated by evaluating the response of serum allopregnanolone to functional endocrine tests [gonadotropin-releasing-hormone (GnRH) test, corticotropin-releasing factor (CRF) test, and adrenocorticotropin hormone (ACTH) test] (Genazzani et al., 1998) (Fig. 1). Both the GnRH test, which activates ovarian function, and the CRF or ACTH tests, which stimulate adrenal

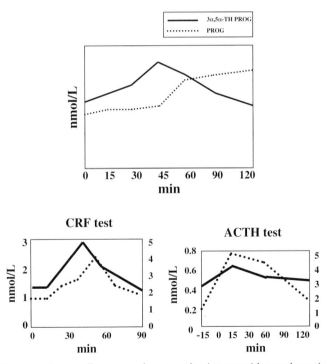

FIG. 1. Response of serum allopregnanolone to endocrine tests with gonadotropin-releasing hormone (GnRH), corticotropin-releasing factor (CRF), and adrenocorticotropin hormone (ACTH). The GnRH, CRF, and ACTH injections stimulate the allopregnanolone release, suggesting that both the ovaries and the adrenal glands are sources of this steroid.

cortex function, induce a significant rise in circulating allopregnanolone, suggesting that both the ovaries and the adrenal glands are sources of this steroid (Genazzani *et al.*, 1998). When the GnRH test is performed, serum allopregnanolone levels rise to points greater than those observed for progesterone, showing a response that is more prone toward allopregnanolone than toward progesterone release (Genazzani *et al.*, 1998). CRF prompts a delayed allopregnanolone response that is with respect to ACTH, probably related to the time required for CRF to elicit the pituitary hormone secretion. Finally, dexamethasone leads to a marked reduction in serum allopregnanolone, thus confirming that the adrenal cortex largely contributes to its circulating levels (Genazzani *et al.*, 1998).

There is little information concerning allopregnanolone levels in men. The adrenal glands seem to produce allopregnanolone. In fact, our unpublished data indicate that allopregnanolone levels increase following CRF or ACTH stimuli. Further origins of allopregnanolone cannot be excluded; in particular, testis, adipose tissue, and liver are under investigation as possible origins.

II. Changes and Possible Role of Neurosteroids in Humans

A. PUBERTY

In circulating progesterone of both boys and girls, DHEA and allopregnanolone levels increase around age 7 and continue to rise in correlation with Tanner's stage (Fadalti *et al.*, 1999). The age-related increase in circulating allopregnanolone during the pubertal period suggests a possible role for this hormone in the maturational process of both hypothalamic-pituitary-gonadal and hypothalamic-pituitary-adrenal (HPA) axes.

Increased levels of adrenal androgens seem to have a role in sexual development during the early stages of puberty, constituting the so-called adrenarche (Hopper and Yen, 1975). Just before gonadal maturation, the human adrenal gland undergoes morphological and functional changes that give rise to a prominent zona reticularis, characterized by the activation of DHEA sulfotransferases, the absence of β hydroxysteroid dehydrogenase, and enhancement of DHEA/dehydroepiandrosterone sulfate (DHEAS) production (Parker, 1999). In this period serum DHEA and DHEAS levels induce pubic and axillary hair growth. The increase in DHEA levels begins before that of testosterone and is not regulated by ACTH or cortisol. The mechanism for the enhanced androgen secretion during the early phase of puberty remains obscure, but it is clear that this event is also correlated with behavioral changes (Reiter et al., 1977; Grumbach et al., 1978).

It is well known that pubertal transition affects the psychological distress of adolescents. The onset of puberty itself is a novel event that calls for psychological adaptation, physical maturation, and fertility (Xiaojia et al., 1996). Neurosteroids may be involved in the complex neuroendocrine modifications occurring during this period.

B. FERTILE LIFE

1. *Menstrual Cycle and Related Disorders*

In normal fertile women, circulating allopregnanolone levels vary much like those of progesterone throughout the menstrual cycle; their circulating levels are much higher in the luteal phase than in the follicular phase (Purdy et al., 1990; Mellon, 1994; Bicikova et al., 1995; Genazzani et al., 1998). DHEA is not influenced by gonadotropin levels because only random fluctuations in plasma DHEA levels have been observed during the cycle (Vermeulen and Verdonck, 1976).

a. Premenstrual Syndrome. Premenstrual syndrome (PMS) is the cyclical recurrence of physical, psychological, or behavioral symptoms that appear after ovulation and resolve within a few days after the onset of menstruation (World Health Organization, 1996). Initial studies showed no variations of serum allopregnanolone levels in the luteal phase of PMS patients with respect to controls, although higher luteal phase allopregnanolone levels were associated with improved symptom ratings in PMS patients (Schmidt et al., 1994; Wang et al., 1996). Bicikova et al. (1998) found low circulating levels of both progesterone and allopregnanolone in PMS patients in the follicular phase only. Rapkin et al. (1997) reported that women with this syndrome had low circulating levels of allopregnanolone in the luteal phase. Our data (Monteleone et al., 2000) sustain the low levels of allopregnanolone in the luteal phase of women with PMS when compared to controls, while progesterone levels seem to be significantly lower in both phases of the menstrual cycle in PMS patients. The reason that some women may have allopregnanolone deficiency remains unexplained. However, our data are indicative of an impaired synthesis of this hormone on behalf of the corpus luteum or other steroidogenic organs (adrenal glands, liver, and adipose tissue). Women with PMS may have an inadequate luteal phase resulting from the underdevelopment of the large luteal cells of the granulosa, which are responsible for producing the greater part of progesterone and probably even allopregnanolone. Conversely, their estradiol levels are normal, which supports the concept that the small luteal cell apparatus is intact. Our patients had low basal levels of progesterone and normal to high estradiol levels; in cases like these, a large luteal cell deficiency may be diagnosed and may be an intrinsic characteristic of such individuals (Hinney et al., 1996). We also found that a GnRH injection during the luteal phase prompted a low allopregnanolone and progesterone response in women with PMS compared with controls (Monteleone et al., 2000). These data reinforce the concept that some sort of deficiency in corpus luteum function and in the synthesis of allopregnanolone and progesterone may underlie PMS. In fact, the low basal levels of allopregnanolone and progesterone found in our PMS patients as well as their impaired response to GnRH suggest an impaired GABA-mediated anxiolytic effect and thus a reduced sense of well-being during the luteal phase. In addition, reports support that the biological key to the initiation of PMS may also be found in the altered "genomic" effects related to this neurosteroid at the central level. Smith and colleagues (1998) demonstrated that, in rats, anxiety and seizure susceptibility are associated with sharp declines in circulating levels of allopregnanolone in the brain. By using progesterone withdrawal, they reproduced PMS and postpartum syndrome in rats and found that these effects were the result of reduced levels of allopregnanolone. This neuroactive steroid enhances transcription of the gene encoding the α_4 subunit of the $GABA_A$ receptor. These authors advanced the hypothesis that PMS symptoms may be partly due to alterations in the expression of $GABA_A$-receptor subunits as a result of progesterone withdrawal. Another hypothesis is that progesterone metabolism may be altered in the brain. This ovarian steroid may be transformed preferentially into an anxiogenic neurosteroid rather than into the anxiolytic allopregnanolone in women with PMS (Berga, 1998). However, allopregnanolone deficiency may not be the only explanation for the onset of distressing symptoms in the luteal phase. Sundstrom et al. (1998) showed that administration of increasing doses of intravenous pregnanolone, a neuroactive precursor of allopregnanolone, causes a decrease in saccadic eye movement only in the follicular phase of women with PMS and not in the luteal phase, although the response is positive in both phases in normal women. PMS may be associated with decreased $GABA_A$-receptor sensitivity to pregnanolone in brain areas controlling saccadic eye movements in PMS patients in the late luteal phase. Therefore, PMS patients may have lowered sensitivity of critical central nervous system (CNS) sites to allopregnanolone.

One study indicated a possible involvement of androgens in the pathophysiology of PMS. DHEA levels were found to be significantly higher in PMS subjects than in controls around the time of ovulation (Eriksson et al., 1992) and, thus, may be in part responsible for premenstrual irritability and dysphoria.

In conclusion, the onset of premenstrual anxiety symptoms may result from the interplay of neurosteroids with opposite GABA$_A$-receptor activity within the CNS.

b. Hypothalamic Amenorrhea. Hypothalamic amenorrhea (HA) is a functional disorder characterized by altered GnRH pulsatility (Meczekalski *et al.*, 2000) improper pulsatile gonadotropin secretion (follicle-stimulating hormone and luteinizing hormone), and, in turn, altered ovarian steroidogenesis (Liu, 1990). It is known that psychological and physical stress may lead to amenorrhea by influencing GnRH pulsatility (Xia *et al.*, 1992). Several neuroendocrine changes occur in HA, such as modifications of the opioid peptide, melatonin (Russell *et al.*, 1984; Laatikainen *et al.*, 1986; Chrousos and Gold, 1992), or corticotropin-releasing hormone (CRH) levels (Vale *et al.*, 1981; Mortola *et al.*, 1993). CRH a 41-amino acid peptide, inhibits GnRH secretion directly or indirectly by increasing β-endorphin secretion, thus suppressing the electrophysiological activity of the GnRH pulse generator (Farrud-Coune *et al.*, 1986; Gambacciani *et al.*, 1986; Saffran and Schally, 1955). Women with HA, despite high ACTH and cortisol levels and therefore, HPA axis hyperactivity, are characterized by low allopregnanolone basal levels, which are probably the result of an impairment of both adrenal and ovarian synthesis. The blunted response of ACTH, allopregnanolone, and cortisol to CRH indicates that women with HA have a reduced sensitivity and/or expression of the CRH receptor (Petraglia *et al.*, 1987).

Anorexia nervosa is one of the leading causes of HA. Anorexia nervosa is, in fact, accompanied by disorders in GnRH pulsatile secretion; (Reddy and Kulkarni, 1998). *In vitro* and *in vivo* animal studies have demonstrated that allopregnanolone reduces GnRH secretion. Furthermore, when administered to rats, it determines hyperphagia (Chen *et al.*, 1996; Calogero *et al.*, 1998). Our data show that serum allopregnanolone levels are significantly higher in patients with anorexia nervosa than in normal fertile women (Vincens *et al.*, 1994). Allopregnanolone may be involved in appetite control; the elevation of allopregnanolone levels in women with anorexia nervosa may represent a specific stimulus for hunger and therefore food intake. However, high circulating allopregnanolone may be the effect of an intense stress response, which is typical of anorexia nervosa.

2. *Pregnancy*

Maternal serum allopregnanolone and progesterone levels rise significantly throughout gestation in healthy pregnant women (Luisi *et al.*, 2000). No marked changes have been found at delivery, except for a significant decrease of maternal and cord serum allopregnanolone levels during emergency cesarean section. Furthermore, serum allopregnanolone levels are significantly higher in patients with chronic hypertension, with or without preeclampsia, than in healthy pregnant women. The changes in these two hormones in maternal and umbilical cord serum are correlated with gestation, but not with parturition, neither in healthy women nor in hypertensive patients.

Serum allopregnanolone levels increase during pregnancy parallel with serum progesterone levels, suggesting a common origin. The increased production rate of progesterone during pregnancy provides the intrauterine tissues with more substrate for the formation of 5α-dihydroprogesterone and allopregnanolone (Dombroski et al., 1997). However, some discrepancy is observed at delivery and in hypertensive patients, indicating that physiological or pathological events reveal other source(s) for allopregnanolone. The lack of correlation between maternal and cord serum progesterone and allopregnanolone levels at delivery, independent from the mode of parturition, suggests a different source(s) for allopregnanolone in both compartments. Indeed, an adrenal cortex or pituitary origin has been shown in healthy nonpregnant women (Bernardi et al., 1998).

In cases of emergency cesarean section, both maternal and umbilical serum allopregnanolone and progesterone concentrations behaved differently from the other modes of parturition, and these changes may be explained by fetal distress. The significant decrease of maternal serum and umbilical cord serum allopregnanolone levels observed during emergency cesarean section cannot be explained exclusively by the surgery because it was absent during elective cesarean section. The role of allopregnanolone in pregnancy remains uncertain. Allopregnanolone may play a role in uterine kinetics (Majewska et al., 1989; Majewska and Vaupel, 1991; Putnam et al., 1991), although an *in vitro* study showed that progesterone metabolites are not crucial for uterine contractions (Lofgren et al., 1988). Maternal anxiety may be related to serum allopregnanolone levels, given the potent anxiolytic effect of this neuroactive steroid (Goascogne et al., 1991; Bitrain et al., 1995).

The evidence of increased levels of serum allopregnanolone in hypertensive pregnant women is of great interest in view of the possible role of progesterone in vascular tension during pregnancy. Evidence suggests that the control of systemic blood pressure during pregnancy may be modulated by steroid hormones (Lofgren et al., 1988; Gangula et al., 1997; Lofgren and Bäckstrom 1997; Masilamani and Heesch, 1997). Indeed, progesterone modulates the antihypertensive effect of calcitonin-gene-related peptide postpartum in the rat (Gangula et al., 1997). Allopregnanolone may contribute to the adaptation of baroflex control of sympathetic outflow and heart rate in pregnant rats (Gangula et al., 1997). It is likely that, in pregnant women with chronic hypertension, the elevation of allopregnanolone levels may be compensatory to this condition and represent an attempt on behalf of the maternal or fetal organism to lower maternal blood pressure levels. Moreover, direct vascular effects of estradiol and progesterone on isolated omental artery from normotensive and preeclamptic pregnant women supported the idea of a direct *in vitro* vasodilator activity of progesterone, partially dependent on the endothelium (Belfort *et al.*, 1996). However, plasma progesterone and its metabolite levels did not show any significant difference in women at risk of developing pregnancy-induced hypertension (Dawood, 1976; Fuchs and Fuchs, 1984). In contrast, a significantly high amniotic fluid progesterone concentration has been reported in pregnant women who later developed preeclampsia (Jarczok *et al.*, 1987).

In conclusion, there is a significant increase in maternal serum allopregnanolone levels during pregnancy, with high levels in chronic hypertension. The lack of correlation between progesterone and allopregnanolone over parturition and in hypertensive patients may be the result of an extraplacental source and/or of a different role of allopregnanolone in the behavioral and cardiovascular adaptive response during pregnancy and delivery.

The fetal adrenal gland expresses large amounts of DHEA sulfotransferase and minimal amounts, at least until the very end of gestation, of 3-β-hydroxysteroid dehydrogenase. This pattern of enzyme expression favors substantial secretion of DHEA and DHEAS with minimal production of cortisol; the DHEA/DHEAS serves as the major precursor for placental estrogen formation in human pregnancy (Parker, 1999). In a longitudinal study, the changes in plasma concentrations of DHEAS were followed in 10 normal pregnant women throughout gestation. The mean plasma concentration of DHEAS decreased in early gestation to minimum levels at Week 38. At delivery, there was a twofold increase in plasma DHEAS. The peak levels of DHEAS on admission to the delivery room reflect increased maternal and fetal stress with the onset of labor (Buckwalter *et al.*, 1999). Distressed infants have lower serum levels of DHEA and higher levels of corisol than control infants (Peter *et al.*, 1994). It also seems that higher levels of DHEA are associated with a better mood during pregnancy (Parker *et al.*, 1993).

C. MENOPAUSE

Studies in postmenopausal women suggest that the reduction in 17,20-desmolase activity, the enzyme that governs the biosynthesis of the Δ^5-adrenal pathway, may provoke modifications in DHEA(S) synthesis (Liu *et al.*, 1990).

Few studies have been conducted in humans to investigate the physiological brain effect of DHEA(S). It is unknown whether changes in plasma

DHEA(S) levels can directly affect CNS functions. In experimental animals, DHEA treatment induced a memory-enhancing effect (Thijssen and Nieuwenhuyse, 1999). *In vitro* studies suggest a neurotrophic effect on neurons and glial cells (Thijssen and Nieuwenhuyse, 1999).

In postmenopausal women, the daily administration of DHEA (50 mg/day) for 3 months induced an improvement in psychological and physical well-being (Morales *et al.*, 1994). It has been shown that the oral administration of DHEA (50 mg/daily) for 3 months determines an increase in well-being and mood in women but not in men (Wolf *et al.*, 1997).

To clarify the effects on the neuroendocrine functions, our group has investigated the effects of DHEA and DHEAS supplementation on the opiatergic tonus in postmenopausal women. In particular, the attention was focused on β-endorphin (β-EP), an important and biologically active endogenous opioid peptide that has behavioral, analgesic, thermoregulatory, and neuroendocrine properties.

Clinical studies demonstrated that variations in circulating β-EP levels may be considered one of the markers of neuroendocrine function (Wolf *et al.*, 1997; Wolkowitz *et al.*, 1997). In postmenopausal women, sex steroid hormone withdrawal modifies the neuroendocrine equilibrium by changing neuroactive transmitters, and a decrease in plasma β-EP levels has been demonstrated (Genazzani *et al.*, 1990). This reduction in circulating β-EP was suggested to have a role in the mechanisms of hot flushes and sweats and in the pathogenesis of mood, behavior, and nociceptive modifications (Petraglia *et al.*, 1993). Experimental and clinical studies have shown that β-EP synthesis and release is modulated by noradrenaline, dopamine, serotonin, acetylcholine, GABA, and CRF (Petraglia *et al.*, 1993). In fertile subjects, bolus injection with naloxone, an opioid receptor antagonist, and clonidine, an α_2-presynaptic receptor agonist, increases β-EP levels. In postmenopause, a response of β-EP to naloxone and to clonidine is lacking, and these findings suggest postmenopausal impairment of adrenergic and opiatergic receptors in modulating β-EP release. Hormone replacement therapy (HRT) restores basal plasma β-EP levels to those present in fertile women as well as the β-EP response to clonidine and naloxone (Petraglia *et al.*, 1993). In a preliminary trial, postmenopausal women ($n = 6$;) (age range: 52–56 years), received oral DHEA (100 mg/day) for 7 days. Women underwent a clonidine test (0.150 mg intraventricularly), before and after 7 days of treatment. Following DHEA treatment, a significant increase of plasma DHEA, DHEAS, A, T, E_1 and E_2 levels was found. On the contrary, basal plasma β-EP levels were not significantly modified after short-term DHEA administration. After the treatment, a significant increase of plasma β-EP levels ($P < 0.01$ at 15 and 30 min) was observed in response to adrenergic activation with clonidine.

Further data have been obtained in postmenopausal women with basal plasma DHEA levels <5 nmol/L (Stomati et al., 1999). All of the women received DHEAS (50 mg orally/day). Subjects were observed monthly during the 3 months of therapy. Blood was drawn for the determination of basal plasma DHEA, DHEAS, A, 17-OHP, T, E1 and E2 sex-hormone-binding globulin (SHBG), cortisol, and β-EP levels. Before and after 3 months of therapy, β-EP levels were evaluated in response to three neuroendocrine tests: clonidine (0.150 mg), naloxone (4 mg intraventricularly), and fluoxetine (30 mg by month). DHEA, DHEAS, androstenedione, testosterone, estradiol, and estrone levels increased significantly and progressively during the 3 months of treatment ($P < 0.05$). SHBG, cortisol, and 17-OHP levels did not show significant variations. From the first month of therapy, a significant increase in basal plasma β-EP levels was observed ($P < 0.05$). Although no response to the three tests was observed before treatment, a significant increase in plasma β-EP levels was shown after 3 months in response to clonidine and naloxone tests.

According to the literature, both studies confirm the significant increase of basal plasma levels of androgens and estrogens after DHEA(S) supplementation, indicating that the Δ^5-androgen and its sulphate conjugated ester may be converted into active steroids (Yamaji and Ibayashi, 1969; Mortola and Yen, 1990; Berr et al., 1996; Thijssen and Nieuwenhuyse, 1999). The increase of basal plasma β-EP levels after the first month of DHEAS therapy support an estrogen-like effect of the molecule on CNS. However, no similar data were observed, which might be consequence of the short-term DHEA administration. The treatment with DHEA(S) induces a restoration of the β-EP response to clonidine, naloxone, or fluoxetine tests, respectively. These findings suggest that DHEA(S) restores the neuroendocrine control of α_2-adrenergic, opioidergic, and serotonine receptors on the anterioro-pituitary β-EP secretion. The modulation of the neuroendocrine pathways after DHEAS supplementation may be mediated by a specific estrogenic action of DHEAS metabolites or, alternatively, by a similar receptorial specificity of DHEAS and estro-progestin compounds on opiatergic and adrenergic neurons. In conclusion, the findings suggest that both of the Δ^5-androgens, and or their metabolites, may be considered one of neuroendocrine correlates of the DHEA(S)-induced psychological and physical improvements. However, previous studies and present data on DHEA(S) need further investigation to identify the possible use of the molecules in the treatment of adrenopause and climateric syndrome.

In postmenopausal women, levels of allopregnanolone are similar to follicular phase levels present in fertile women. The age-related decrease in progesterone levels is not paralleled by a similar decrease in allopregnanolone levels, suggesting that ovarian progesterone is not the major determinant of circulating allopregnanolone in this phase of a woman's life (Genazzani *et al.*, 1998).

D. AGING AND RELATED DISORDERS

1. *Aging*

Serum allopregnanolone levels show an age-related decrease in men, with the lowest levels after 60 years of age (Fig. 2). Preliminary data show that, in postmenopausal women, allopregnanolone levels are similar to those observed in fertile women during the follicular phase and are significantly lower than those of women during the luteal phase because of ovarian inactivity (Genazzani *et al.*, 1998) (Fig. 2). However, in women, allopregnanolone concentrations do not vary with increasing age (Genazzani *et al.*, 1998).

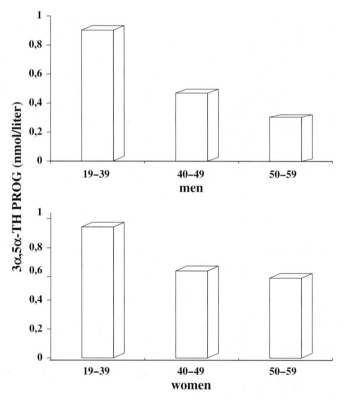

FIG. 2. Age-related changes in allopregnanolone levels in men and in women.

Clinical experiments, performed on a large number of patients, are ongoing to confirm this evidence and to clarify the effects of HRT in early and late postmenopause and in the elderly.

It is well known that DHEA levels decrease with age both in men and in women (Liu *et al.*, 1990; Berr *et al.*, 1996; Parker, 1999). In humans, DHEA and DHEAS are the major adrenal cortex products, and the circulating levels of both Δ^5-steroids show selective changes during the course of a life time, reaching their highest values around 25 years of age. After the third decade of life, a progressive decline of the synthesis of the Δ^5-androgens is described in both sexes; after 70 years of age, DHEA(S) levels are maintained at 20% or less of the maximum plasma concentrations, while cortisol levels remain unchanged (Yamaji *et al.*, 1969; Liu *et al.*, 1990). The decrease of adrenal DHEA(S) secretion is not related to cortisol changes. The decline in Δ^5-androgens and the parallel increase in the cortisol/DHEA(S) ratio has been proven to be in part responsible of the physiological and/or pathological age-associated changes. Epidemiological studies have shown a relationship between the progressive decrease of circulating DHEA(S) levels and the increase of cardiovascular morbidity in men (Barrett-Connor *et al.*, 1986), the risk of breast cancer in women (Bulbrook *et al.*, 1975; Helzlsouer *et al.*, 1992) and the impairment of the immune competence in both sexes (Thoman and Weigle, 1989). Clinical studies have investigated the effects of DHEA(S) administration in men and women of different ages. The first trials reported metabolic effects, which are different and not comparable both in young and old men and women treated with high doses of DHEA (1600 mg/day) (Mortola and Yen, 1990). In other studies, the use of a lower dose of DHEA (50–100 mg/day) led to a restoration in circulating DHEA and DHEAS levels to physiological values (Morales *et al.*, 1994), associated with a twofold increase in circulating levels of androgens (androstenedione, testosterone, and dihydrotestosterone) and estrogens (estradiol and estrone) and no changes in SHBG levels in women (Stomati *et al.*, 1999). A single oral dose (500 mg) of DHEA 60 min before sleeping induced a significant increase in rapid eye movement sleep, which appears to be implicated in memory processes (Wolf *et al.*, 1997; Thijssen and Nieuwen huyse, 1999).

2. *Dementia*

Patients with Alzheimer's dementia (AD) and vascular dementia (VD) have lower serum allopregnanolone levels than age-matched healthy controls, confirming that serum DHEA levels are low in dementia (Nasman *et al.*, 1991; Miller *et al.*, 1998). It is known that DHEA and DHEAS improve long-term memory and diminish amnesia in mice and enhance neuronal and glial survival and differentiation in cultures of embryonic mouse brain cells (Roberts, 1990). The administration of DHEA improves physical and psychological well-being and cognitive performances in aged subjects (Liu *et al.*, 1990). The mechanism(s) of DHEA-mediated neuroprotection are unclear: DHEA probably antagonizes some deleterious effects of cortisol, which causes progressive hippocampal damage and cognitive impairment in dementia (Sapolsky *et al.*, 1986; Svec and Lopez, 1989; Leblhuber *et al.*, 1991; Wolkowitz *et al.*, 1992, 1993). A low DHEA: cortisol ratio may be deleterious to hippocampal function (Leblhuber *et al.*, 1991). The hippocampus has a leading role in the HPA axis regulation, but it is also an important brain site in the memorization processing and in the connections between intellectual capacity and psychoaffective conditions; moreover, it is vulnerable to the hypoxic-anoxic stimuli.

Experimental data in rats and in primates showed that the increase in the glucocorticoid concentrations determines loss of specific receptors and neural damage in the hippocampus. Patients who underwent prolonged and high-dose corticosteroid treatments showed memory and logical capacity deficits. It can be hypothesized that functional alterations of hippocampus glucocorticoid receptors for the negative feedback lead to a hyperactive HPA axis function, which in turn may cause further neural damage in hippocampus; this could explain the correlation between cortisol levels and degree of brain vascular aging.

The finding that cortisol levels in AD and VD are higher than in control patients has been reported (Gottfries *et al.*, 1994; Hatzinger *et al.*, 1995; Nasman *et al.*, 1996; Hartmann *et al.*, 1997; Swanwick *et al.*, 1998), and a correlation between the severity of dementia and the increase of cortisol levels has been described (Balldin *et al.*, 1994; Nasman *et al.*, 1996; Weiner *et al.*, 1997; Miller *et al.*, 1998). In addition, data have demonstrated that allopregnanolone circulating levels are reduced in patients with AD or VD. The hypothesis that allopregnanolone might be involved in modulating cognitive function is supported by the evidence that conditions characterized by modifications in behavior, mood, and cognitive performances, such as menstrual cycle, pregnancy, and aging, are associated with changes in allopregnanolone levels (Majewska, 1992; Schmidt *et al.*, 1994; Genazzani *et al.*, 1998). It is possible to hypothesize that, in dementia, the low allopregnanolone levels are a consequence of the low CRF levels, which may also explain the low DHEA levels observed. In contrast, both the AUC of DHEA and the allopregnanolone response to CRF have been described to be reduced in dementia, indicating that these patients have an impairment in the total neuroendocrine balance capacity involving DHEA and allopregnanolone secretion. The different pattern observed between allopregnanolone and DHEA could reflect different metabolic means involved in their secretion and could indicate a greater or more precocious involvement of DHEA secretion in dementia with respect to allopregnanolone. This report has also indicated that cortisol secretion is ultra-sensitive in response to CRF stimulation. In fact, cortisol response to the CRF test was higher both as AUC and as % max increase in AD and VD subjects than in controls. Cortisol response to the CRF test in AD has been described as similar (Martignoni *et al.*, 1990a,b) and lower than in controls; however, data are consistent with our findings of a higher cortisol response to the CRF test (Nasman *et al.*, 1996) and the ACTH test (O'Brien *et al.*, 1996) in AD patients. In contrast, Dodt *et al.* (1991) described a similar response of DHEA to the CRF test in elderly and young subjects, regardless of their mental state; however, the simultaneous administration of vasopressin in association with CRF (half dose compared with this study) makes the results incomparable. The reason for the difference between cortisol and DHEA basal and stimulated levels in dementia remains unclear. In AD patients, an impairment of the DHEA:cortisol ratio appears more clearly in response to the CRF test than at basal levels, indicating that, in dementia, contrarily to controls, CRF determines a stimulation of cortisol secretion and with a radical DHEA release. Because the adrenal androgen-stimulating factor and the alterations in adrenal androgen synthesis have been proposed for explaining the enhanced cortisol and reduced DHEA secretion in the elderly (Dodt *et al.*, 1991), such factors might also play a role in dementia. In addition, the kind of dementia seems to influence neither allopregnanolone, DHEA, and cortisol levels nor their response to CRF (Fig. 3).

In conclusion, this evidence supports the concept that dementia per se, independent of nature, may be the result of an altered stress response; the modifications in the circulating levels of allopregnanolone, a stress-related steroid hormone, might enter in this view. Therefore, it will be interesting to study the role of allopregnanolone in stress and cognition, possibly via the interplay between brain CRF and CRF-related peptides (Fig. 3).

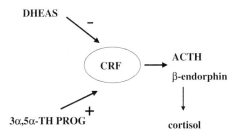

FIG. 3. Control mechanism of allopregnanolone and DHEA(S) on the hypothalamic-pituitary-adrenal axis.

References

Balldin, J., Blennow, K., Brane, G., Gottfries, C. G., Karlsson, I., Regland, B., and Wallin, A. (1994). Relationship between mental impairment and HPA axis activity in dementia disorders. *Dementia* **5,** 252–256.

Barrett-Connor, E., Khaw, K., and Yen, S. S. C. (1986). A prospective study of DS mortality and cardiovascular disease. *N. Engl. J. Med.* **315,** 1519–1524.

Belfort, M. A., Saade, G. R., Suresh, M., and Vedernikov, Y. P. (1996). Effects of estradiol-17 beta and progesterone on isolated human omental artery from premenopausal nonpregnant women and from normotensive and preeclamptic pregnant women. *Am. J. Obstet. Gynecol.* **174,** 246–253.

Berga, S. L. (1998). Understanding Premenstrual syndrome. *Lancet* **351,** 465–466.

Bernardi, F., Salvestroni, C., Casarosa, E., Nappi, R. E., Lanzone, A., Luisi, S., Purdy, R. H., Petraglia, F., and Genazzani, A. R. (1998). Aging is associated with changes in allopregnanolone concentrations in brain, endocrine glands and serum in male rats. *Eur. J. Endocrinol.* **138,** 316–321.

Berr, C., Lafont, S., Debuire, B., Dartigues, J.-F., and Baulieu, E. E. (1996). Relationship of dehydroepiandrosterone sulphate in the elderly functional, and mental status, and short-term mortality: A French community-based study. *Proc. Natl. Acad. Sci. USA* **93,** 13410–13415.

Bicikova, M., Dibbelt, L., Hill, M., Hampl, R., *et al.* (1998). Allopregnanolone in women with premenstrual syndrome. *Horm. Metab. Res.* **30,** 227–230.

Bicikova, M., Lapcik, O., and Hampl, R., *et al.* (1995). A novel redioimmunoassay of allopregnanolone. *Steroids* **60,** 210–213.

Bitran, D., Shiekh, M., and MacLeod, M. (1995). Anxiolytic effect of progesterone is mediated by the neurosteroid allopregnanolone at brain $GABA_A$ receptors. *J. Neuroendocrinol.* **7,** 171–177.

Bixo, M., Andersson, A., Winblad, B., Purdy, R. H., and Backstrom, T. (1997). Progesterone, 5alpha-pregnane-3,20-dione and 3alpha-hydroxy-5alpha-pregnane-20-one in specific regions of the human female brain in different endocrine states. *Brain Res.* **764,** 173–178.

Buckwalter, J. G, Stanczyk, F. Z., McCleary, C. A., Bluestein, B. W., Buckwalter, D. K., Rankin, K. P., Chang, L., and Goodwin, T. M. (1999). Pregnancy, the postpartum, and steroid hormones: Effects on cognition and mood. *Psychoneuroendocrinology* **24,** 69–84.

Bulbrook, R. D., Hayward, J. L., and Spicer, C. C. (1975). Relation between urinary androgen and corticoid secretion and subsequent breast cancer. *Lancet* **2,** 395–398.

Calogero, A. E., Palumbo, M. A., Bosboom, A. M., Burrello, N., Ferrara, E., Palumbo, G., Petraglia, F., and D'Agata, R. (1998). The neuroactive steroid allopregnanolone suppresses hypothalamic gonadotropin-releasing hormone release through a mechanism mediated by the gamma-aminobutyric $acid_A$ receptor. *J. Endocrinol.* **158,** 121–125.

Chen, S. W., Rodriguez, L., Davies, M. F., and Loew, G. H. (1996). The hyperphagic effect of 3alpha-hydroxylated pregnane steroids in male rats. *Pharmacol. Biochem. Behav.* **53,** 777–782.

Cheney, D. L., Uzunov, D., Costa, E., and Guidotti, A. (1995). Gas chromatographic–mass fragmentographic quantitation of 3alpha-hydroxy-5 alpha-pregnan-20-one (allopregnanolone) and its precursors in blood and brain of adrenalectomized and castrated rats. *J. Neurosci.* **15,** 4641–4650.

Chrousos, G. P., and Gold, P. W. (1992). The concepts of stress and stress system disorders. *JAMA* **267,** 1244–1252.

Corpéchot, C., Young, J., Calvel, M., Wehrey, C., Veltz, J. N., Touyer, G., Mouren, M., Prasad, V. V. K., Banner, C., Sjovall, J., Baulieu, E. E., and Robel, P. (1993). Neurosteroids: 3α-Hydroxy-5α-Pregnan-20-one and its precursors in the brain, plasma, and steroidogenic glands of male and female rats. *Endocrinology* **133,** 1003–1009.

Dawood, M. Y. (1976). Circulating maternal serum progesterone in high-risk pregnancies. *Am. J. Obstet. Gynecol.* **125,** 832–840.

Dodt, C., Dittmann, J., Hruby, J., Spath-Schwalbe, E., Born, J., Schuttler, R., and Fehm, H. L. (1991). Different regulation of adrenocorticotropin and cortisol secretion in young, mentally healthy elderly and patients with senile dementia of Alzheimer's type. *J. Clin. Endocrinol. Metab.* **72,** 272–276.

Dombroski, R. A., Casey, M. L., and MacDonald, P. C. (1997). 5-Alpha-dihydroprogesterone formation in human placenta from 5alpha-pregnan-3beta/alpha-ol-20-ones and 5-pregnan-3beta-yl-20-one sulfate. *J. Steroid. Biochem. Mol. Biol.* **63,** 155–163.

Eriksson, E., Sundblad, C., Lisjo, P., Modigh, K., and Andersch, B. (1992). Serum levels of androgens are higher in women with premenstrual irritability and dysphoria than in controls. *Psychoneuroendocrinology* **17,** 195–204.

Fadalti, M., Petraglia, F., Luisi, S., Bernardi, F., Casarosa, E., Ferrari, E., Luisi, M., Saggese, G., Genazzani, A. R., and Bernasconi, S. (1999). Changes of serum allopregnanolone levels in the first 2 years of life and during pubertal development. *Pediatr. Res.* **46,** 323–327.

Favrod-Coune, C. A., Gaillard, R. C., Langerin, H., Jaquier, M. C., and Dolci, W. (1986). Anatomical localization of corticotropin-releasing activity in the humans brain. *Life. Sci.* **39,** 2475–2481.

Fuchs, A. R., and Fuchs, P. (1984). Endocrinology of human parturition: A review. *Br. J. Obstet. Gynecol.* **91,** 948–967.

Gambacciani, M., Yen, S. S. C., and Rasmussen, D. D. (1986). GnRH release from the mediobasal hypothalamus: *In vitro* inhibition by corticotropin releasing factor. *Neuroendocrinology* **43,** 533–536.

Gangula, P. R., Wimalawansa, S. J., and Yallampalli, C. (1997). Progesterone up-regulates vasodilator effects of calcitonin gene-related peptide in N(G)-nitro-L-arginine methyl ester-induced hypertension. *Am. J. Obstet. Gynecol.* **176,** 894–900.

Genazzani, A. R., Petraglia, F., Bernardi, F., Casarosa, E., Salvestroni, C., Tonetti, A., Nappi, R. E., Luisi, S., Palumbo, M., Purdy, R. H., and Luisi, M. (1998). Circulating levels of allopregnanolone in humans: Gender, age and endocrine influences. *J. Clin. Endocrinol. Metab.* **83,** 2099–2210.

Genazzani, A. R., Petraglia, F., and Mercuri, N., *et al.* (1990). Effect of steroid hormones and antihormones on hypothalamic beta-endorphin concentrations in intact and castrated female rats. *J. Endocrinol. Invest.* **13,** 91–96.

Goascogne, C., Jo, D. H., Corpechot, C., Simon, P., Baulieu, E. E., and Robel, P. (1991). Neurosteroids: Biosynthesis, metabolism and function of pregnenolone and dehydroepiandrosterone in the brain. *J. Steroid Biochem. Molec. Biol.* **40,** 71–81.

Gottfries, C. G., Balldin, J., Blennow, K., Brane, G., Karlsson, I., Regland, B., and Wallin, A. (1994). Hypothalamic dysfunction in dementia. *J. Neural. Transm. Suppl.* **43,** 203–209.

Grumbach, M. M., Richards, G. E., Conte, F. A., and Kaplan, S. L. (1978). Clinical disorders of adrenal function and puberty: An assessment of the role of the adrenal cortex in normal and abnormal puberty in man and evidence for an ACTH-like pituitary adrenal androgen stimulating hormone. *In* "The Endocrine Function of the Human Adrenal Cortex" (M. Serio, Ed.), pp. 583–612. Academic Press, New York.

Hartmann, A., Veldhius, J. D., Deuschle, M., Standhardt, H., and Heuser, I. (1997). Twenty-four hour cortisol release profiles in patients with Alzheimer's and Parkinson's disease compared to normal controls: Ultradian secretory pulsatility and diurnal variation. *Neurobiol. Aging.* **18,** 285–289.

Hatzinger, M., Z'Braun, A., Hemmeter, U., Seifritz, E., Baumann, F., Holsboer-Trachsler, E., and Heuser, I. J. (1995). Hypothalamic-pituitary-adrenal system function in patients with Alzheimer's Disease. *Neurobiol. Aging.* **16,** 205–209.

Helzlsouer, K. J., Gordon, G. B., Alberg, A., Bush, T. L., and Comstock, G. W. (1992). Relationship of prediagnostic serum levels of DHEA and DS to the risk of developing premenopausal breast cancer. *Cancer. Res.* **52,** 1–4.

Hinney, B., Henze, C., and Kuhn, W., *et al.* (1996). The corpus luteum deficiency: A multifactorial disease. *J. Clin. Endocrinol. Metab.* **81,** 365–370.

Hopper, B. R., and Yen, S. C. (1975). Circulating concentrations of dehydroepiandrosterone and dehydroepiandrosterone sulfate during puberty. *J. Clin. Endocrinol. Metab.* **40,** 458–461.

Jarczok, T., Zwirner, M., and Schindler, A. E. (1987). Progesterone and 5 alpha-pregnane-3,20-dione in human amniotic fluid. *Exp. Clin. Endocrinol.* **89,** 39–47.

Laatikainen, T., Virtanen, T., and Apter, D. (1986). Plasma immunoreactive beta-endorphin in exercise-associated amenorrhea. *Am. J. Obstet. Gynecol.* **154,** 94–97.

Leblhuber, F., Windhager, E., Neubauer, C., Weber, J., Reisecker, F., and Dienstl, E. (1991). Antiglucocorticoid effects of DHEAS in Alzheimer's disease (letter). *Am. J. Psychiatr.* **30,** 684–696.

Liu, C. H., Laughlin, G. A., Fischer, U. G., and Yen, S. S. C. (1990). Marked attenuation of ultradian and circadian rhythms of dehydroepiandrosterone in postmenopausal women: Evidence for a reduced 17,20-desmolase enzymatic activity. *J. Clin. Endocrinol. Metab.* **71,** 900–906.

Liu, J. H. (1990). Hypothalamic amenorrhea: Clinical perspectives, pathophysiology and management. *Am. J. Obstet. Gynecol.* **163,** 1732–1736.

Lofgren, M., and Bäckstrom, T. (1997). High progesterone is related to effective human labor. Study of serum progesterone and 5alpha-pregnane-3,20-dione in normal and abnormal deliveries. *Acta Obstet. Gynecol. Scand.* **76,** 423–430.

Lofgren, M., Bäckstrom, T., and Joelsson, I. (1988). Decrease in serum concentration of 5alpha-pregnane-3,20-dione prior to spontaneous labor. *Acta Obstet. Gynecol. Scand.* **67,** 467–470.

Luisi, S., Petraglia, F., Benedetto, C., Nappi, R. E., Bernardi, F., Fadalti, M., Reis, F. M., Luisi, M., and Genazzani, A. R. (2000). Serum allopregnanolone levels in pregnant women: Changes during pregnancy, at delivery and in hypertensive patients. *J. Clin. Endocrinol. Metab.* **83,** 2733–2749.

Majewska, M. D., Falkay, G., and Baulieu, E. E. (1989). Modulation of uterine GABA$_A$ receptors during gestation and by tetrahydroprogesterone. *Eur. J. Pharmacol.* **174,** 43–47.

Majewska, M. D., Demirgoren, S., Spivak, C. E., and London, E. D. (1990). The neurosteroid DHEA is an allosteric antagonist of the GABA A receptor. *Brain. Res.* **526,** 143–146.

Majewska, M. D., and Vaupel, D. B. (1991). Steroid control of uterine motility via gamma-aminobutyric acidA receptors in the rabbit: A novel mechanism? *J. Endocrinol.* **13,** 427–434.

Majewska, M. D. (1992). Neurosteroids: Endogenous bimodal modulators of the GABA-A receptor. Mechanism of action and physiological significance. *Prog. Neurobiol.* **38,** 379–395.

Martignoni, E., Petraglia, F., Costa, A., Bono, G., Genazzani, A. R., and Nappi, G. (1990a). Dementia of the Alzheimer type and hypothalamus-pituitary adrenocortical axis: Changes in cerebrospinal fluid corticotropin releasing factor and plasma cortisol levels. *Acta Neurol. Scand.* **81,** 452–456.

Martignoni, E., Petraglia, F., Costa, A., Monzani, A., Genazzani, A. R., and Nappi, G. (1990b). Cerebrospinal fluid corticotropin releasing factor levels and stimulation test in dementia of the Alzheimer type. *J. Clin. Lab. Anal.* **4,** 5–8.

Masilamani, S., and Heesch, C. M. (1997). Effects of pregnancy and progesterone metabolities on arterial baroreflex in conscious rats. *Am. J. Physiol.* **272,** R924–R934.

Meczekalski, B., Tonetti, A., Monteleone, P., Bernardi, F., Luisi, S., Stomati, M., Luisi, M., Petraglia, F., and Genazzani, A. R. (2000). Hypothalamic amenorrhea with normal body weight: ACTH, allopregnanolone and cortisol responses to corticotropin-releasing hormone test. *Eur. J. Endocrinol.* **142,** 280–285.

Mellon, S. H. (1994). Neurosteroids: Biochemistry, modes of action, and clinical relevance. *J. Clin. Endocrinol. Metab.* **78,** 1003–1100.

Miller, T. P., Taylor, J., Rogerson, S., Mauricio, M., Kennedy, Q., Schatzberg, A., Tinklenberg, J., and Yesavage, J. (1998). Cognitive and noncognitive symptoms in dementia patients: Relationship to cortisol and dehydroepiandrosterone. *Int. Psychogeriatr.* **10,** 85–96.

Monteleone, P., Giardina, L., Morittu, A., Luisi, S., Tonetti, A., Stomati, D., Luchi, C., Santuz, M., Gasparini, M., and Genazzani, A. R. (1999, March 4–20). High circulating levels of allopregnanolone in women with anorexia nervosa. The 4th Congress of the European society for Gynecologic and Obstetric Investigation, Madonna di Campiglio (Italy).

Monteleone, P., Luisi, S., Tonetti, A., Bernardi, F., Genazzani, A. D., Luisi, M., Petraglia, F., and Genazzani, A. R. (2000). Allopregnanolone concentrations and premenstrual syndrome. *Eur. J. Endocrinol.* **142,** 269–273.

Morales, A. J., Nolan, J. J., Nelson, J. C., and Yen, S. S. C. (1994). Effects of replacement dose of dehydroepiandrosterone in men and women of advancing age. *J. Clin. Endocrinol. Metab.* **78,** 1360–1370.

Mortola, J. F., Laughlin, G. A., and Yen, S. S. C. (1993). Melatonin rhythms in women with anorexia nervosa and bulimia nervosa. *J. Clin. Endocrinol. Metab.* **77,** 1540–1544.

Mortola, J. F., and Yen, S. S. C. (1990). The effects of oral dehydroepiandrosterone on endocrine–metabolic parameters in postmenopausal women. *J. Clin. Endocrinol. Metab.* **71,** 696–704.

Nasman, B., Olsson, T., Backstrom, T., Eriksson, S., Grankvist, K., Viitanen, M., and Bucht, G. (1991). Serum DHEA sulfate in Alzheimer's disease and in multi-infarct dementia. *Biol. Psychiatry.* **30,** 684–690.

Nasman, B., Olsson, T., Fagerlund, M., Eriksson, S., Viitanen, M., and Carlstrom, K. (1996). Blunted adrenocorticotropin and increased adrenal steroid response to human corticotropin-releasing hormone in Alzheimer's disease. *Biol. Psychiatry.* **39,** 311–318.

O'Brien, J. T., Ames, D., Schweitzer, I., Mastwyk, M., and Colman, P. (1996). Enhanced Adrenal sensitivity to adrenocorticotrophic hormone (ACTH) is evidence of HPA axis hyperactivity in Alzheimer's disease. *Psychol. Med.* **26,** 7–14.

Parker Jr., C. R., Favor, J. K., Carden, L. G., and Brown, C. H. (1993). Effects of intrapartum stress on fetal adrenal function. *Am. J. Obstet. Gynecol.* **169,** 1407–1411.

Parker Jr., C. R. (1999). Dehydroepiandrosterone and dehydroepiandrosterone sulfate production in the human adrenal during development and aging. *Steroids* **64,** 640–647.

Peter, M., Dorr, H. G., and Sippell, W. G. (1994). Changes in the concentrations of dehydroepiandrosterone sulfate and estriol in maternal plasma during pregnancy: A longitudinal study in healthy women throughout gestation and at term. *Horm. Res.* **42,** 278–281.

Petraglia, F., Sutton, S., Vale, W., and Plotsky, P. (1987). Corticotropin releasing factor decreases plasma luteinizing hormone levels in female rats by inhibiting GnRH release into hypophyseal-portal circulation. *Endocrinology* **120,** 1083–1088.

Petraglia, F., Comitini, G., and Genazzani, A. R., *et al.* (1993). β-Endorphin in human reproduction. *In* "Opiods II" (A. Herz, Ed.), pp. 763–780. Springer-Verlag, Berlin.

Purdy, R. H., Moore, P. H. Jr., Rao, N., Hagino, N., Yamaguchi, Schmidt, P., Rubinow, D. R., Morrow, A. L., and Paul, S. M. (1990). Radioimmunoassay of 3α-hydroxy-5α-pregnan-20-one in rat and in human plasma. *Steroids* **55,** 290–295.

Putnam, C. D., Brann, D. W., Kolbeck, R. C., and Mahesh, V. B. (1991). Inhibition of uterine contractility by PROG and PROG metabolites: Mediation by progesterone and gamma amino butyric acid$_A$ recpetor systems. *Biol. Reprod.* **45**, 266–272.

Rapkin, A. J., Morgan, M., and Goldman, L., *et al.* (1997). Progesterone metabolite allopregnanolone in women with premenstrual syndrome. *Obstet. Gynecol.* **90**, 709–714.

Reddy, D. S., and Kulkarni, S. K. (1998). The role of GABA-A and mitochondrial diazepam-binding inhibitor receptors on the effects of neurosteroids on food intake in mice. *Psychopharmacology* **137**, 391–400.

Reiter, E. O., Fuldauer, V. G., and Root, A. W. (1977). Secretion of the adrenal androgen, DHEA sulfate, during normal infancy, childhood, and adolescence, in sick infants, and in children with endocrinologic abnormalities. *J. Pediatr.* **90**, 766–770.

Roberts, E. (1990). Dehydroepiandrosterone and its sulphate (DHEAS) as neural facilitators: Effects on brain tissue in culture and on memory in young and old mice: A cyclic GMP hypothesis of action of DHEA and DHEAS in nervous system and other tissues. *In* "The biologic role of dehydroepiandrosterone" (M. Kalimi and W. Regelson, Eds.), pp. 13–42. Walter de Gruyter, Berlin.

Russell, J. B., Mitchell, D. E., and Collins, D. C. (1984). The role of beta-endorphin and catechol estrogens on hypothalamic-pituitary axis in female athletes. *Fertil. Steril.* **42**, 690–695.

Saffran, M., and Schally, A. V. (1955). The release of corticotropin by anterior pituitary in culture. *Can. J. Biochem. Physiol.* **33**, 408–412.

Sapolsky, R. M., Krey, L. C., and McEwen, B. S. (1986). The neuroendocrinology of stress and aging: The glucocorticoid cascade hypothesis. *Endocr. Rev.* **7**, 284–301.

Schmidt, P. J., Purdy, R. H., Moore, P. H., *et al.* (1994). Circulating levels of anxiolytic steroids in the luteal phase in women with premenstrual syndrome and in control subjects. *J. Clin. Endocrinol. Metab.* **79**, 1256–1260.

Smith, S. S., Gong, Q. H., Hsu, F.-C, *et al.* (1998). GABA$_A$ receptor $\alpha 4$ subunit suppression prevents withdrawal properties of an endogenous steroid. *Nature.* **392**, 926–930.

Stomati, M., Rubino, S., Spinetti, A., Parrini, D., Luisi, S., Casarosa, E., Petraglia, F., and Genazzani, A. R. (1999). Endocrine, neuroendocrine and behavioural effects of oral dehydroepiandrosterone sulphate supplementation in postmenopausal women. *Gynecol. Endocrinol.* **13**, 15–25.

Sundstrom, I., Andersson, A., Nyberg, S., *et al.* (1998). Patients with Premenstrual syndrome have a different sensitivity to a neuroactive steroid during the menstrual cycle compared to control subjects. *Neuroendocrinology* **7**, 126–138.

Swanwick, G. R., Kirby, M., Bruce, I., Buggy, F., Coen, R. F., Coakley, D., and Lawlor, B. A. (1998). Hypothalamic-pituitary-adrenal axis dysfunction in Alzheimer's disease: Lack of association between longitudinal and cross-sectional findings. *Am. J. Psychiatry.* **155**, 286–289.

Svec, F., and Lopez, A. (1989). Antiglucocorticoid actions of DHEA and low concentrations in Alzheimer's disease (letter). *Lancet* **2**, 1335–1336.

Thijssen, J. H. H., and Nieuwenhuyse, H. (1999). "DHEA: A Comprehensive Review." Parthenon Publishing Group, New York.

Thoman, M. L., and Weigle, W. O. (1989). The cellular and subcellular bases of immunosenescence. *Adv. Immunol.* **46**, 221–261.

Uzunova, V., Sheline, Y., Davis, J. M., Rasmusson, A., Uzunov, D. P., Costa, E., and Guidotti, A. (1998). Increase in the cerebrospinal fluid content of neurosteroids in patients with unipolar major depression who are receiving fluoxetine or fluvoxamine. *Proc. Natl. Acad. Sci. USA* **95**, 3239–3244.

Vale, W., Spies, J., Rivier, C., and Rivier, J. (1981). Characterization of a 41 residue ovine hypothalamic peptide that stimulates secretion of corticotropin and beta-endorphin. *Science* **213**, 1394–1397.

Vermeulen, A., and Verdonck, L. (1976). Plasma androgen levels during the menstrual cycle. *Am. J. Obstet. Gynecol.* **125,** 491–494.

Vincens, M., Li, S. Y., and Pelletier, G. (1994). Inhibitory effect of 5beta-pregnan-3alpha-o1-20-one on gonadotropin-releasing hormone gene expression in the male rat. *Eur. J. Pharmacol.* **260,** 157–162.

Wang, M., Seippel, L., Purdy, R. H., and Backstrom, T. (1996). Relationship between symptom severity and steroid variation in women with premenstrual syndrome: Study on serum PREG sulfate, 5α-pregnan-3, 20-dione and 3α-hydroxy-5α-pregnan-20-one. *J. Clin. Endocrinol. Metab.* **81,** 1076–1082.

Weiner, M. F., Vobach, S., Olsson, K., Svetlik, D., and Risser, R. C. (1997). Cortisol secretion and Alzheimer's disease progression. *Biol. Psychiatry.* **42,** 1030–1038.

Wolf, O. T., Neumenn, O., Helhammer, D. H., Geiben, A. C., Strasburger, C. J., *et al.* (1997). Effects of a two-week physiological dehydroepiandrosterone substitution on cognitive performance and well-being in healthy elderly women and men. *J. Clin. Endocrinol. Metab.* **82,** 2363–2367.

Wolkowitz, O. M., Reus, V. I., Manfredi, F., and Roberts, E. (1992). Antiglucocorticoid effects in Alzheimer's disease. *Am. J. Psychiatry.* **149,** 1126.

Wolkowitz, O. M., Reus, V. I., and Roberts, E. (1993). Role of DHEA and DHEAS in Alzheimer's disease. *Am. J. Psychiatry.* **150,** 9.

Wolkowitz, O. M., Reus, V. I., Roberts, E., Manfredi, F., Chan, T., Raum, W. J., Ormiston, S., *et al.* (1997). Dehydroepiandrosterone (DHEA) treatment of depression. *Biol. Psychiatry.* **41,** 311–318.

World Health Organization. (1996). "Mental, Behavioural and Developmental Disorders. Tenth revision of the International Classification of Diseases (ICD-10)," Geneva.

Xia, L., Van Vught, D., Alston, E. J., Luckhaus, J., and Ferin, M. (1992). A surge of GnRH accompanies the estradiol induced gonadotropin surge in the Rhesus monkey. *Endocrinology* **131,** 2812–2820.

Xiaojia, G., Conger, R. D., and Elder, G. H. (1996). Coming of age too early: Pubertal influences on girls' vulnerability to psychological distress. *Child Develop.* **67,** 3386–3400.

Yamaji, T., and Ibayashi, H. (1969). Serum deydroepiandrosterone sulphate in normal and pathological conditions. *J. Clin. Endocrinol. Metab.* **29,** 273–278.

NEUROACTIVE STEROIDS AND CENTRAL NERVOUS SYSTEM DISORDERS

Mingde Wang, Torbjörn Bäckström,[1] Inger Sundström, Göran Wahlström, Tommy Olsson, Di Zhu, Inga-Maj Johansson, Inger Björn, and Marie Bixo

Department of Obstetrics and Gynecology, Department of Medicine
Department of Pharmacology, University of Umeå, S-901 87 Umeå, Sweden

I. Neuroactive Steroids and the Central Nervous System
II. Concentrations of Neuroactive Steroids in the Brain
III. Sensory-Motor and Cognitive Function
IV. Estrogen and Alzheimer's Disease
V. Neuroactive Steroids and Menstrual-Cycle-Linked Mood Changes
  A. Endocrine Profile of PMS
  B. Ovarian Steroids and Symptom Severity in PMS
  C. Neurosteroids and PMS
VI. Side Effects of Oral Contraceptives
VII. Neuroactive Steroids and Menopause
  A. Menopause
  B. Postmenopausal Symptoms
VIII. Side Effects of Hormone Replacement Therapy
IX. Neuroactive Steroids and Epilepsy
  A. Brain Excitability
  B. Catamenial Epilepsy
  C. Hormones and Epilepsy in Men
  D. Neuroactive Steroids in Epilepsy Treatment
  References

Steroid hormones are vital for the cell life and affect a number of neuroendocrine and behavioral functions. In contrast to their endocrine actions, certain steroids have been shown to rapidly alter brain excitability and to produce behavioral effects within seconds to minutes. In this article we direct attention to this issue of neuroactive steroids by outlining several aspects of current interest in the field of steroid research. Recent advances in the neurobiology of neuroactive are described along with the impact of advances on drug design for central nervous system (CNS) disorders provoked by neuroactive steriods. The theme was selected in association with the clinical aspects and therapeutical potentials of the neuroactive steroids in CNS disorders. A wide range of topics relating to the neuroactive steroids are outlined, including steroid concentrations in the brain, premenstrual syndrome, estrogen and Alzheimer's disease, side effects

[1]Author to whom correspondence should be sent.

of oral contraceptives, mental disorder in menopause, hormone replacement therapy, Catamenial epilepsy, and neuractive steriods in epilepsy treatment. © 2001 Academic Press.

I. Neuroactive Steroids and the Central Nervous System

The brain is a target organ of steroid hormones (Selye, 1942; Holzbauer, 1971). Blood-borne signals from the peripheral endocrine organs are input signals modulating brain function. Progesterone is quickly taken up by and metabolized in the cerebral cortex (Appelgren, 1967). The brain concentrations of estradiol and progesterone closely follow peripheral hormone variation. In general, estradiol exerts excitatory actions, and progesterone exerts inhibitory effects on the central nervous system (CNS). Ovarian steroids modulate several CNS functions, such as memory and learning (Sherwin, 1997), balance (Hammar et al., 1996), movement disorders (Kompoliti, 1999), and pain perception (Amandusson et al., 1996). Furthermore, various CNS disorders are affected by steroids [e.g., stress (Dubrovsky, 2000), epilepsy (Bäckström and Rosciszewska, 1997), Alzheimer's disease (Ohkura et al., 1995), migraine (Fettes, 1997)]. Ovarian steroids regulate neuroendocrine, reproductive, and behavioral functions via a number of cellular mechanisms. In the classical genomic mechanism of steroids, estrogen and progesterone induce relatively long-term actions on neurons by activating specific intracellular receptors that modulate transcription and protein synthesis. Both estradiol and progesterone receptors are found in the CNS (McEwen and Wooley, 1994).

As a result of the investigations of Baulieu and colleagues (Baulieu and Robel, 1990), it is now clear that the brain can synthesize steroids *de novo*. Baulieu introduced the term "neurosteroid" in 1981 to designate a steroid intermediate, dehydroepiandrosterone sulfate, which was found in the brain at a concentration independent of peripheral resource. The term was later proposed to refer to pregnenolone, pregnenolone sulfate, progesterone, and allopregnanolone synthesized in the brain. A more general definition includes all steroids that are synthesized *de novo* in the brain. Using this definition, estrogen can be classified as a neurosteroid because it can be synthesized from testosterone in various regions of the brain. The term "neuroactive steroids" refers to steroid hormones that are active on neuronal tissue. They may therefore be synthesized endogenously in the brain or by peripheral endocrine organs, but they act on neuronal tissues.

The profound effects of estrogen on different tissues may involve at least two estrogen receptor (ER) subtypes, ER_α and the more recently discovered ER_β. Where and how the two ER subtypes differentially or cooperatively mediate estrogen actions, however, are still unknown. ER_α mRNA is widely distributed in both female and male reproductive organs and the CNS, while ER_β mRNA is more widely distributed in the female than in the male brain. ER_α predominates in the uterus and the mammary gland, whereas ER_β has significant roles in the CNS, cardiovascular and immune systems, urogenital tract, bone, kidney, and lung (Couse et al., 1997; Pau and Spies, 1998; Register et al., 1998; Gustafsson, 2000).

In addition, estrogen can also exert very rapid effects in the brain that cannot be attributed to genomic mechanisms. These nongenomic actions influence a variety of neuronal properties, including electrical excitability, synaptic functioning, and morphological features, and they are involved in many of the physiological functions and clinical effects of estrogen in the brain. Estrogen may utilize direct membrane mechanisms, such as activation of ligand-gated ion channels and G-protein-coupled second messenger systems and regulation of neurotransmitter transporters (Wong et al., 1996; Kelly and Wagner, 1999). The existence of an ER on the plasma membrane has been supported by data emerging from numerous laboratories since the 1980s. A cell membrane protein that could bind and rapidly respond to 17β-estradiol was isolated recently. The nongenomic actions of 17β-estradiol can thus be mediated through the plasma membrane ER (Levin, 1999).

Estrogen modulates synaptic plasticity in the hippocampus (Shughrue and Merchenthaler, 2000), where application of 17β-estradiol rapidly enhances the amplitude of kainate-induced currents of CA1 neurons (Wong and Moss, 1994; Moss and Gu, 1999). The potentiation resulted from a cyclic adenosine monophosphate-dependent phosphorylation process rather than a direct allosteric modulation of α-amino-3-hydroxy-5-methyl-4-isoxazole-propionate (AMPA)/kainate receptors (Moss and Gu, 1999). The binding sites responsible for the potentiation are genetically and pharmacologically distinct from both ER_α and ER_β (Wong et al., 1996; Moss and Gu, 1999). The effects of 17β-estradiol on both N-methyl-D-aspartate (NMDA)-receptor-mediated excitatory postsynaptic potentials (EPSPs) through intracellular recordings and long-term potentiation (LTP) through extracellular recordings were investigated. The effects of 17β-estradiol on NMDA-receptor-mediated activity were excitatory; concentrations >10 nM induced seizure activity, and lower concentrations (1 nM) markedly increased the amplitude of NMDA-mediated EPSPs (Foy et al., 1999). In extracellular experiments, slices perfused with 17β-estradiol (100–700 pM) exhibited a pronounced, persisting, and significant enhancement of LTP (Wang et al.,

1997; Foy *et al.*, 1999). These data demonstrate that estrogen enhances NMDA-receptor-mediated currents and promotes an enhancement of LTP magnitude.

In the adult female rat, the densities of dendritic spines and synapses on hippocampal CA1 pyramidal cells are dependent on estradiol; moreover, spine and synapse density fluctuate naturally as ovarian steroid levels vary across the estrous cycle (McEwen *et al.*, 1995, 1997). According to Brinton *et al.* (1997), 17β-estradiol can enhance the growth and viability of select populations of neocortical neurons, and the growth-promoting effects of 17β-estradiol can be blocked by an antagonist to the NMDA-glutamate receptor and not by an antagonist to the estrogen nuclear receptor. Murphy and Segal (1996) studied the effects of estrogen on dendritic spines in cultured hippocampal neurons. Exposure to 17β-estradiol caused up to a twofold increase in dendritic spine density in these neurons. The effect of estradiol was stereospecific and blocked by the steroid antagonist tamoxifen. The estradiol-induced rise in spine density was also blocked by the NMDA antagonist, but not by the AMPA/kainate antagonist. These results, in parallel with estradiol-induced increases in NMDA-receptor binding (Wolley and McEwen, 1994; Woolley *et al.*, 1997), indicated that estradiol treatment increases sensitivity of hippocampal neurons to NMDA-receptor-mediated synaptic input; furthermore, sensitivity to NMDA-receptor-mediated synaptic input is well correlated with dendritic spine density (Woolley and McEwen, 1994; Woolley *et al.*, 1997). Thus, it is concluded that estradiol exerts its effect on hippocampal dendritic spine density via a mechanism requiring activation of NMDA-receptors specifically.

The γ-aminobutyric acid (GABA) transmitter system is the major inhibitory system in the mammalian CNS, with GABA being abundant in approximately 30% of the brain synapses. Certain progesterone metabolites modulate the $GABA_A$-receptor as agonists (Gee, 1988). Among the progesterone metabolites, 3α-hydroxy-5α-pregnan-20-one (allopregnanolone) and its 5β-stereoisomer, 3α-hydroxy-5β-pregnan-20-one (pregnanolone), are the most potent modulators on the $GABA_A$-receptor (Korkmaz and Wahlström, 1997). It has been suggested that allopregnanolone modulates $GABA_A$-receptor function through an independent and specific binding site at the $GABA_A$-receptor (Puia *et al.*, 1990; Lan *et al.*, 1991). When injected intravenously, allopregnanolone and pregnanolone have anesthetic, antiepileptic, and anxiolytic effects on both humans and animals (Bäckström *et al.*, 2000). In addition, it has been suggested that allopregnanolone plays a role as an endogenous stress-protective compound (Morrow *et al.*, 1995; Barbaccia *et al.*, 1998).

II. Concentrations of Neuroactive Steroids in the Brain

The effects of neurosteroids on the CNS are concentration dependent. Conjugated or nonconjugated pregnenolone and dehydroepiandrosterone and 5α- and 5β-reduced isomers of progesterone and desoxycorticosterone are shown to exist in large quantities in the brain of rats, mice, pigs, guinea pigs, monkeys, and humans. The progesterone concentrations in the brain are in the range of 10–100 nM. The estradiol concentrations in the brain are much lower than that of progesterone (see Table I). An increased level of estradiol and progesterone is observed in the human and rat brain at the time of high ovarian production of the hormone (Table I), (Bixo et al., 1986, 1995, 1997; Bixo and Bäckström, 1990; Purdy et al., 1990; Corpéchot et al., 1993; Wang et al., 1996). In fertile women, the serum level of allopregnanolone increases from 2 nM in the follicular phase to ~4 nM in the luteal phase (Wang et al., 1996). The corresponding increase of pregnanolone is from 1.6 to 2.4 nM (Sundström and Bäckström, 1998). During the third trimester of pregnancy, allopregnanolone and pregnanolone concentrations in the serum are ~30 ng/ml or 100 nM (Paul and Purdy, 1992). Allopregnanolone concentrations in fertile women are highly correlated with progesterone in the serum (Wang et al., 1996), and it is conceivable that progesterone and progesterone metabolites produced by the corpus

TABLE I
AVERAGE SERUM AND BRAIN CONCENTRATIONS OF NEUROACTIVE STEROIDS AT FOLLICULAR AND LUTEAL PHASES OF THE MENSTRUAL CYCLE IN FERTILE WOMEN AND IN POSTMENOPAUSAL WOMEN

Steroid	Plasma (nmol/l)		Brain (nmol/kg)	
	Follicular	Luteal	Postmenopause	Luteal
Progesterone	2.0^a	35^a	65^b	137^b
Allopregnanolone	1.8^a	3.6^a–7.8^e	47^b	66^b
Pregnanolone	1.6^d	2.3^d	—	—
Pregnenolone sulfate	11^a	15^a	—	100^e
Estradiol	0.32^a	0.31^a	0.29^f	0.70^f

[a] Wang et al. (1996).
[b] Bixo et al. (1997).
[c] Purdy et al. (1990).
[d] Sundström et al. (1998).
[e] Lanthier and Patwardhan (1986).
[f] Bixo et al. (1995).

luteum are the major sources of progesterone metabolites in the brain (Bäckström et al., 1986; Bixo et al., 1997).

In Rhesus monkeys, a substantial amount of the ovarian progesterone production (60%) is taken up by the brain during the luteal phase (Billiar et al., 1975). The average brain concentration of allopregnanolone in the male rat is around 3 nM. This is significantly higher than its plasma concentration, which is barely detectable (Cheney et al., 1995). It has been inferred that the reduction of progesterone also takes place in human brain, but this has been difficult to show. Both 5α-reductase and 3α-hydroxysteroid oxidoreductase exist in the brain of animals (Karavolas et al., 1984; Melcangi et al., 1990), and allopregnanolone/pregnanolone can be reduced from progesterone in cultured neurons and glia (Barnea et al., 1990), thus it is likely that a similar process can also occur in humans. Postmortem studies in fertile and postmenopausal women have indicated that allopregnanolone is accumulated in the brain. The highest levels of allopregnanolone were found in the substantia nigra and hypothalamus, with concentrations ranging from 35 to 40 ng/g (Bixo et al., 1997).

The estradiol and progesterone concentrations in the brain of the rat are not evenly distributed, but rather an accumulation in specific regions is observed. For example, the progesterone concentration in the cerebral cortex is about 300 times higher in the luteal phase compared with the follicular phase, while in the hypothalamus the increase is eight times and in plasma only two times (Bixo et al., 1986). There is a significant correlation between the plasma and the cortex concentration of progesterone but not between the hypothalamus and plasma (Bixo et al., 1986). The regional difference of steroid concentrations were also noted in the human brain when Bixo et al. (1995, 1997) measured the postmortem concentrations of progesterone, pregnanedione allopregnanolone, pregnanolone, and 17β-estradiol in six fertile and five postmenopausal women. Generally, the highest levels of estradiol were found in the hypothalamus, preoptic area, and substantia nigra. The highest levels of progesterone were noted in the amygdala, cerebellum, hypothalamus, and nucleus accumbens. For pregnanedione and pregnanolone the highest levels were detected in the substantia nigra, hypothalamus, and amygdala.

Measurement of pregnenolone and pregnenolone sulfate also revealed a nonuniform distribution in the monkey brain (Robel et al., 1987). Their concentrations tended to be higher in the hypothalamus (27 nM), hippocampus (33 nM), and rhinencephalon (40 nM) than those in the cerebellum and the cerebral cortex (<14 nM). The brain pool of pregnenolone sulfate is larger than that of pregnenolone (Robel and Baulieu, 1995). The enzyme activity of 5α-reductase also varies between brain regions as shown by measurement of the pregnanedione: progesterone ratio in the rat brain

(Bixo and Bäckström, 1990). The highest ratio was detected in the medulla oblongata, hippocampus, and striatum. The synthesis and metabolism of neurosteroids are thus region dependent in the brain.

Purdy et al. (1991) and Barbaccia et al. (1998) have studied the brain concentrations of allopregnanolone after exposure of male rats to swim stress. A rapid (<5 min) and robust (4- to 20-fold) increase of this steroid was observed in both brain and plasma. The brain concentrations after acute stress were approximately 10–30 nM, which were within the range of this steroid shown to modulate $GABA_A$-receptor function *in vitro*. During anesthesia induced by intravenous infusion of progesterone in the male rats, the average brain concentration of allopregnanolone at anesthesia was found to be in the micromolar range. A 5000- to 6000-fold increase in progesterone concentrations was detected in the striatum and hypothalamus compared with those of female rats in the follicular phase (Bixo et al., 1986; Bixo and Bäckström, 1990). Allopregnanolone concentrations in the cerebral cortex and hypothalamus were about 4000 and 6000 times higher at anesthesia induced by intravenous injection of this steroid than the basal levels in male rats (Table II), (Wang et al., 1995). The average serum and brain concentrations of neuroactive steroids during different endocrine states in fertile and postmenopausal women are described in the Table I. The average serum and brain concentrations of allopregnanolone + pregnanolone in physiological state, pregnancy, sedation, and anesthesia in humans are described in Table II.

TABLE II

AVERAGE SERUM AND BRAIN CONCENTRATIONS OF PREGNANOLONE + ALLOPREGNANOLONE AT PHYSIOLOGICAL STATE, PREGNANCY, SEDATION, AND ANESTHESIA IN HUMAN AND RAT

Treatment	Plasma (nmol/l)	Brain (nmol/kg)
Physiological, luteal phase	6–12[a,b,c]	60–130[d]
Pregnancy, late	100[e]	—
Saccade influence	80–110[c]	—
Sedation	160–200[b]	—
Anesthesia	530–1700[f]	31000–60000[g]

[a] Wang et al. (1996).
[b] Purdy et al. (1990).
[c] Sundström et al. (1998).
[d] Bixo et al. (1997).
[e] Paul and Purdy (1992).
[f] Carl et al. (1990).
[g] Wang et al. (1995).

III. Sensory-Motor and Cognitive Function

Behavioral studies have suggested that the estrogen-dominated follicular phase is associated with increased alertness and activity (Asso and Braier, 1982, (e.g., increase in rapidly alternating tasks, such as walking, finger tapping frequency, typing, and word fluency) (Broverman et al., 1968). In general, these tasks involve movement and control of the limbs and digits. Increased levels of estradiol are also associated with increased two-point discrimination, fine touch, perception, hearing, and smell as well as visual signal detection (Zimmerman and Parlee, 1973). An improvement in postural balance has been shown in postmenopausal women using estrogen replacement therapy (ERT) compared with women without ERT (Hammar et al., 1996). There is an increase in fracture incidence shortly after menopause in women without ERT that is not related to the degree of osteoporosis. It has been suggested that this increase is the result of a change in balance and limb coordination secondary to an estrogen decrease at the beginning of menopause (Naessén et al., 1990).

Sex steroid hormones play an important role in the developmental organization of the brain and in activation across the life span. Study of the cognitive effects of these steroids has focused on estrogen. High levels of estrogen have been associated with enhanced verbal fluency, articulation (Hampson and Kimura, 1988; Hampson, 1990), and memory (Sherwin, 1988; Phillips and Sherwin, 1992a,b), but the reports are inconsistent (Barrett-Connor and Kritz-Silverstein, 1993). Visuospatial abilities may be decreased when levels of estrogen are high (Hampson, 1990), although inconsistencies are again reported (Schmidt et al., 1996; Resnick et al., 1997).

To provide exploratory analyses of associations between levels of several sex hormones and cognitive performance in elderly women, Drake et al. (2000) compared the cognitive performance across or between conditions associated with different hormone levels, such as phases of the menstrual cycle, surgical menopause, and ERT, and suggested that conditions with higher levels of estrogen are associated with better verbal memory and possibly worse visuospatial ability. High estradiol levels were associated with better delayed verbal memory and retrieval efficiency, whereas low levels were associated with better immediate and delayed visual memory. Levels of testosterone were related positively to verbal fluency. Levels of progesterone and androstenedione were unrelated to cognitive performance. In some small, open studies, beneficial effects of estrogen on memory function and performance have been shown in Alzheimer's disease (AD) patients of mild dementia type; however, in severely affected patients, no effect of estrogen was noted (Fillit et al., 1986; Honjo et al., 1989). The hippocampus is a brain region closely related to memory function. The number of synapses in the hippocampal CA1 region decreased significantly only 6 days after oophorectomy in female rats. The effect reached a maximum at 40 days after the operation. At 24 h after onset of ERT, the number of synapses was found to be restored (Wooley and McEwen, 1992, 1993). The cerebral circulation was also shown to increase 10 min after an intravenous estradiol injection (Goldman *et al.*, 1976; Paganini-Hill, 1996).

Some studies are investigating the clinical effect of ERT on memory. Three studies have shown an improvement in visual and verbal memory function with estrogen treatment (Fedor-Freybergh, 1977; Sherwin, 1988; Phillips and Sherwin 1992a,b), while other studies have not been able to show any effects of estrogen at the dosages used in postmenopausal therapy (Rauramo *et al.*, 1975; Vanhulle and Demol, 1976; Barrett-Connor and Kritz-Silverstein, 1993). Two of the studies that have shown estrogen effect were conducted in oophorectomized patients who were surgically postmenopausal (Sherwin, 1988; Phillips and Sherwin, 1992a). In one of the studies not showing an effect (Vanhulle and Demol, 1976), a weak estradiol preparation was used. One interesting study from Sherwin (1994) in Canada showed a deterioration of memory function in patients receiving gonadotrophin-releasing hormone (GnRH)-agonists for treatment of endometriosis. These patients were given ERT or placebo in a randomized control study as "add back." The estrogen add back restored the memory function, while the patients on placebo continued to show deteriorated memory function (Sherwin, 1994). The presence of ovaries might be of importance for estrogen's effect on the memory.

IV. Estrogen and Alzheimer's Disease

Approximately 6–8% of all persons ages >65 years have AD, the most prevalent cause of dementia. The prevalence of the disease is increasing. It affects twice as many women as men. In postmenopausal women, both the aging process and hypoestrogenism resulting from the loss of ovarian function seem to be related to the progressive impairment of cognitive functions and to a higher risk of developing AD. Laboratory evidence suggests that estrogen may increase cholinergic transmission, have neurotrophic effects (Xu *et al.*, 1997), have antiamyloidogenic properties (McEwen *et al.*, 1997), affect β-amyloid metabolism (Jaffe *et al.*, 1994), reduce oxidative stress or cardiovascular risk, improve cerebral blood flow, stimulates neuron or gliacyte (see reviews by Skoog and Gustafson, 1999), and interact with genetic factors involving cognitive decline (Yaffe *et al.*, 2000). This has made estrogen an attractive candidate for both prevention and treatment of AD (see reviews by van Duijn, 1999; Palacios *et al.*, 2000).

Several studies have suggested that estrogen use in elderly women improves cognition (Jacobs *et al.*, 1998; Asthana *et al.*, 1999), prevents development of dementia (Paganini-Hill and Henderson, 1994; Tang *et al.*, 1996; Kawas *et al.*, 1997) and lessens the severity of dementia (Ohkura *et al.*, 1995). In a meta-analysis of 10 studies, including eight case-control and two prospective observational studies, Yaffe *et al.* (1998) determined that the risk of AD among postmenopausal women taking ERT was reduced by 29% compared with women never given estrogen. Two additional case-control studies reported that women using estrogen have a 58–72% reduced risk of AD (Baldereschi *et al.*, 1998; Waring *et al.*, 1999). However, according to these observational studies, women on ERT may be better educated, have higher socioeconomic status, and have better access to medical care than those who are not treated with ERT.

Other studies have not found an association between ERT and improved cognition or dementia prevention (Broe *et al.*, 1990; Barrett-Connor and Kritz-Silverstein, 1993). Three large, double-blind, placebo-controlled trials of estrogen in women with AD found no role for long-term treatment of conjugated equine estrogen (CEE) in preventing or delaying the onset of AD in postmenopausal women (Henderson *et al.*, 2000; Mulnard *et al.*, 2000; Wang *et al.*, 2000). In an editorial review, Marder and Sano (2000) analyzed why these three studies were negative. Laboratory studies suggest that adequate brain levels of estradiol are necessary to improve cognition. Long-term, steady-state exposure to estrogen can downregulate estrogen receptors. These laboratory findings are supported by the benefit seen with estrogen delivery via patch (Asthana *et al.*, 1999), which may yield higher bioavailability of estrogen than oral preparations that undergo first-pass metabolism. The effect was seen after only brief exposure (8 weeks). However, observational studies using oral preparations (Yaffe *et al.*, 1998) show greater risk reduction for AD with longer continuous exposure, arguing against receptor downregulation.

In summary, the results of randomized clinical trials do not support a role for estrogen in the treatment of AD. The decision to take ERT remains an individual calculation of potential risks and benefits. There are many other benefits of ERT, and we may yet learn that estrogen has a salutory influence on brain function.

V. Neuroactive Steroids and Menstrual-Cycle-Linked Mood Changes

Premenstrual syndrome (PMS) or premenstrual dysphoric disorder (PMDD) is a psychoneuroendocrine disorder. Fluctuations in gonadal hormones, psychiatric vulnerability traits, and psychosocial factors act in concert to provoke symptoms in affected women during the luteal phase of the menstrual cycle. There is great confusion, and sometimes controversy, among women and clinicians about what PMS is and how to name it. This confusion is largely the result of a failure to appreciate that, although most women experience mild mood and somatic symptoms premenstrually, a small but significant number (2–6%) are severely disabled by the symptoms (Sveindottir and Bäckström, 2000). The terms PMS and PMDD are used interchangeably by some researchers. Others claim that there is a distinction between PMS and PMDD, with the term PMS reserved for milder somatic symptoms, such as breast tenderness, bloating, headache, and minor mood changes.

Menstrual-cycle-linked changes that are negative symptoms are commonly encountered in the western world and have a substantial effect on several aspects of the daily lives of women (Sveindottir and Bäckström, 2000). The disorder is defined by the cyclical recurrence of a cluster of negative mood and physical symptoms. Symptoms develop in the luteal phase of the menstrual cycle and remit within a few days after the onset of menstrual bleeding. The American Psychiatric Association's Diagnostic and Statistical Manual of Mental Disorders (fourth edition; DSM-IV, 1994) defines diagnostic criteria for PMDD. To fulfill the criteria for PMS (or PMDD), a patient needs to present with at least five of the listed symptoms during the premenstrual week. At least one of these symptoms must be a mood symptom, such as depressed mood, anxiety, affective lability, or irritability. Other symptoms described by the DSM-IV criteria are decreased interest in usual activities, difficulty in concentrating, marked lack of energy, hypersomnia, or insomnia. Physical symptoms are also included among the listed symptoms in the diagnostic criteria. Common physical symptoms encountered among PMS patients are breast tenderness, swelling, joint pain, or headaches. In addition to the presence of a number of typical symptoms, the DSM-IV criteria also state that symptoms must interfere with usual activities (school, work performance, or interpersonal relationships) to provide some measure on the severity of symptoms. Furthermore, patients must be devoid of symptoms in the follicular phase to ensure that the premenstrual complaint is not merely an exacerbation of an underlying mood disorder. Finally, unlike other DSM-IV criteria, a diagnosis of PMS must be confirmed by prospective ratings for at least 2 months. As none of the symptoms of PMS are unique to the syndrome, patients need to keep a daily diary of symptoms to establish the temporal relationship between onset of symptoms and the premenstrual period. Furthermore, at least 2 months of symptom ratings are needed to confirm the consistency of premenstrual symptoms over time.

A. ENDOCRINE PROFILE OF PMS

Despite numerous efforts to identify endocrine disturbances in PMS patients, there are few consistent endocrine findings. The most important finding by far is the necessity of ovulation and consequent corpus luteum formation for symptom development in the luteal phase. During spontaneous anovulatory cycles, the cyclicity in symptoms disappears (Hammarbäck et al., 1991). Anovulation can be induced by GnRH agonists, which in several studies have been proven to be beneficial for premenstrual complaints (Hammarbäck and Bäckström, 1988; Brown et al., 1994; Mortola et al., 1991; Mezrow et al., 1994). Other ways of manipulating the menstrual cycle to achieve anovulation have also been reported. Estradiol implants in doses high enough to cause anovulation are effective in abolishing the cyclical mood changes. Danazol is also able to inhibit ovulation and consequently has an effect on the cyclicity of symptoms (Magos et al., 1986; Hahn et al., 1995). Pregnancy, characterized by high, nonfluctuating levels of estradiol and progesterone, represents another anovulatory state. During pregnancy, a large number of PMS patients reported symptom relief (Hammarbäck and Bäckström, 1989). Oophorectomy, the first treatment described for PMS, has a marked effect on PMS symptoms but is, for obvious reasons, rarely recommended today. Hysterectomy, leaving the ovaries intact, does not relieve PMS. Thus, the presence of symptoms in the late luteal phase can not be solely the result of anticipation of an approaching menstrual period but rather on some factor produced by the corpus luteum (Bäckström et al., 1985).

The symptom development in PMS coincides with the luteal phase of the menstrual cycle, in which levels of estradiol and progesterone are high. The intensity gradually increases during the luteal phase, and the maximum symptom severity occurs 3–5 days after the progesterone peak in the luteal phase. The remission of symptoms does not occur until a few days after the onset of menstrual bleeding, when gonadal hormone levels have declined (Bäckström et al., 1983a,b). Although the temporal relationship with the progesterone peak in the luteal phase is obvious, PMS patients do not have higher or lower peripheral levels of progesterone than healthy women. There is a general opinion on the absence of peripheral markers of hypothalamus-pituitary-gonadal axis dysfunction in PMS. Hence, plasma levels of progesterone and estradiol do not differ between PMS patients and control subjects in the luteal phase, nor do luteinizing hormone (LH), follicle-stimulating hormone (FSH), dehydroepiandrostendione sulfate, or dihydrotestosterone (Bäckström et al., 1983b; Rubinow et al., 1988).

B. Ovarian Steroids and Symptom Severity in PMS

Although no differences in progesterone and estradiol plasma levels have been demonstrated between PMS patients and control subjects, these steroids appear to have an impact on symptom severity within patients. Particular attention has been given to progesterone as a symptom-provoking factor in PMS. Evidence to substantiate this assumption is found from sequential hormone replacement therapy (HRT) in postmenopausal women. The estrogen/gestogen sequential replacement therapy resembles the hormonal variations during an ovulatory menstrual cycle, and the estrogen-only treatment is similar to an anovulatory cycle. Women receiving gestogen in the last part of the treatment period respond with a significant cyclicity in their mood and physical signs, whereas those receiving only estrogen did not show any deterioration of mood at the end of the treatment cycle (Hammarbäck et al., 1985). Add-back HRT during GnRH-agonist treatment in PMS (to relieve the hypoestrogenic symptoms) induced progestogenic side effects similar to the underlying condition, although the efficacy of the GnRH treatment on mood symptoms was still evident (Mortola et al., 1991). However, not only progesterone is symptom provoking. Increased estradiol and progesterone plasma levels during the luteal phase were found to be related to more severe symptoms when compared to cycles in the same individuals with lower luteal phase estradiol and progesterone levels. The symptom severity was particularly related to the luteal estradiol levels (Hammarbäck et al., 1989). In addition, a subgroup of patients with higher luteal phase estradiol concentrations showed more severe symptoms compared with patients with lower luteal phase estradiol (Seippel and Bäckström, 1997). Moreover, estradiol treatment during the luteal phase induced more negative symptoms than placebo in a group of PMS patients (Dhar and Murphy, 1990). In postmenopausal women receiving HRT, women having higher plasma estradiol concentrations during the progestogen period experienced more symptoms than women with lower concentrations (Klaiber et al., 1997). However, estradiol alone cannot provoke symptoms. During the follicular phase of the menstrual cycle, when estrogen predominates and also reaches it highest levels, these women feel best. Obviously, estradiol and progesterone acting together seem to induce another response in the CNS than when they act separately.

C. Neurosteroids and PMS

The plasma levels of allopregnanolone and pregnanolone appear not to differ between PMS patients and control subjects in either phase of the menstrual cycle (Schmidt et al., 1994; Wang et al., 1996; Sundström and Bäckström 1998). One study was, however, able to detect significantly lower plasma levels of Allopregnanolone in the late luteal phase of PMS patients as compared to the control group (Rapkin et al., 1997). Whereas healthy women increased their pregnanolone levels in the luteal phase compared with the follicular phase, pregnanolone concentrations remained unaltered throughout the menstrual cycle in PMS patients (Sundström and Bäckström, 1998). Again, concentrations of neurosteroids seem to have an impact on symptom severity. When symptom scores were compared among PMS patients, a worsening of symptoms was experienced during cycles with high levels of estradiol, pregnenolone, and pregnenolone sulfate (Wang et al., 1996). In contrast, cycles with increased levels of allopregnanolone were associated with improved symptom ratings. The symptom peak occurred with a delay of 3–4 days from the plasma progesterone, pregnenolone, and allopregnanolone peaks. The peak in pregnenolone sulfate appeared the same day or 1 day ahead of the symptom peak (Wang et al., 1996). Given the anxiolytic actions demonstrated in laboratory animals, one would expect that allopregnanolone and pregnanolone would have beneficial effects on PMS. These compounds have not been used in clinical trials, but oral micronized progesterone treatment, resulting in supraphysiological levels of allopregnanolone and pregnanolone, has been tried. The use of oral micronized progesterone was based on the fact that vaginal administrations of progesterone, as well as synthetic progestins, do not result in the formation of $GABA_A$-receptor active metabolites (De Lignieres et al., 1995). The progesterone treatment has had diverging results and has not always alleviated the overall premenstrual distress. In one study, oral micronized progesterone was shown to have a beneficial effect over placebo, whereas other studies have been unable to detect any symptom improvement despite achieving high concentrations of allopregnanolone (Dennerstein et al., 1985; Vanselow et al., 1996).

Although plasma levels of neurosteroids do not differ between PMS patients and controls, they still might play a role in the symptom provocation of PMS. In PMS patients, the sedative response to pregnanolone was reduced compared with controls (Sundström et al., 1998). In addition, patients with severe symptoms were less sensitive to pregnanolone compared with patients with more moderate symptoms (Sundström et al., 1998). Possible mechanisms by which neurosteroids may be involved in the pathophysiology of PMS is the subject of speculation. Our results indicate that PMS patients are less prone to experience anxiolysis from increasing plasma levels of neurosteroids, as a result of a reduced sensitivity to these substances in the CNS. Hence, PMS patients would have less benefit from endogenous tranquilizers in coping with stressful events of everyday life. It is, however, difficult to judge whether the decreased sensitivity to pregnanolone provocation is of physiological significance to symptom relief/provocation at endogenous levels during the luteal phase.

VI. Side Effects of Oral Contraceptives

Negative mood symptoms are well-known side effects of oral contraceptives (OC) in women of fertile ages. Approximately 30% of women state mental side effects as reason for interrupting OC use (Milsom et al., 1991). Women on OC who experience adverse mood symptoms appear to be the same women as those who have PMS (Cullberg, 1972). Clinical trials on OC as possible therapeutics for PMS have yielded mostly negative findings. It seems that women with mild PMS might benefit from OC, whereas women with more severe PMS develop negative mood symptoms (Bäckström, 1996). Cullberg (1972) noted in his OC study that only the women who had suffered from PMS reacted badly on the pill. This suggests that women with PMS are more sensitive to hormonal provocation than women without PMS. There are also signs of a different sensitivity in brain functions to ovarian hormones in the women with PMS compared with the controls (Bäckström et al., 1985). CNS transmitter system sensitivity also seems to differ between PMS patients and controls. The maximal velocity of the saccade—the rapid eye movement when visual fields are changed—is dependent on the activity in the $GABA_A$-receptor. The sensitivity of the $GABA_A$-receptor is indirectly measured by giving an intravenous dosage of benzodiazepine or pregnanolone. In such studies, PMS patients react less than controls, indicating less sensitive $GABA_A$-receptor activity (Sundström et al., 1997a, 1998). This finding is important because it suggests that it would be possible to predict who would react with negative mood changes on HRT and OC (i.e., the women who in their fertile life had PMS). It may be of importance that the differences were eliminated when PMS patients were treated with serotonin reuptake inhibitors (SSRI) (Sundström et al., 1998). There is also a relationship between affective disorders and sensitivity to sex hormones (Bäckström and Hammarbäck, 1991), and SSRI treatment has had significant positive effects on sex-steroid-induced mood changes (Sundblad, 1992; Korzekwa and Steiner, 1997).

All progestogens seem to be able to induce negative mood changes (Bäckström, 1996). There are, however, few comparative studies being reported. The dose relationship is complex, as women seem to show more symptoms when taking OC with lower progestogen content (Cullberg, 1972). Women taking triphasic pills seem to develop more symptoms compared to those using monophasic pills, which contain a higher dose of progestogen (Bancroft *et al.*, 1987). Sherwin (1991) has investigated the effects of various doses of estrogen and progestogen on psychological functioning and sexual behavior of 48 healthy, naturally menopausal women. The groups that received CEE and medroxyprogesterone acetate had more negative moods and more psychological symptomatology during treatment compared with those who were receiving CEE and placebo. Women receiving 0.625 mg CCE for Days 1–25 and 5 mg medroxyprogesterone acetate from Days 15–25 had more negative mood changes and more psychological sympotomatology compared with those receiving 1.25 mg CEE and placebo instead of medroxyprogesterone acetate.

These findings demonstrate that the effects of progestogen on the CNS are reflected by an increase in psychological symptomatology. A few studies suggested a similarity of the symptoms experienced on OC and those experienced in PMS. Women on OC continue to show cyclical mood changes but have a different pattern during the treatment cycle compared to the natural cycle (Walker and Bancroft, 1990; Bäckström *et al.*, 1992). The negative responses of PMS patients to OC were much stronger than those who did not suffer from PMS (Cullberg, 1972). This adverse mood effect of OC could also relate to the type of progestogen used. A study on OC treatment of PMS revealed that desogestrel-containing pills provoked less change in mood parameters than levogestrel-containing pills (Bäckström *et al.*, 1992). A study by Smith *et al.* (1994) using Moos Menstrual Distress Questionnaire as a tool of mood analysis indicated that progestogens vary in the type of symptoms they cause. Norethisterone acetate (NETA) is more likely to cause symptoms from the Moos pain symptom cluster than either medroxyprogesterone or dydrogesterone, but NETA is less likely to cause negative affect symptom cluster symptoms. The relative levels of estrogen and progestogen may influence the severity of progestogenic symptoms (Bäckström, 1996).

VII. Neuroactive Steroids and Menopause

A. Menopause

The aging of the ovary starts before the age of 40, as noted by the increase of early-follicular and mid-cycle levels of FSH (Sherman *et al.*, 1976). In the meantime, the ovary becomes less responsive to gonadotrophin stimulation (Sherman and Korenman, 1975). An increase in LH occurs later, and it is not as pronounced as the increase in FSH. The gradual decrease of ovarian estrogen and inhibin production are a possible reason for the elevated FSH level in the early-follicular phase (Monroe *et al.*, 1972; MacNaughton *et al.*, 1992) to the periodic pulsatile pattern of LH and FSH secretion remains in postmenopausal women. The pulses of gonadotropin occur at approximately 60- to 90-min intervals (Yen *et al.*, 1972). As the woman approaches menopause, the menstrual cycles become irregular, and the number of ovulatory cycles decrease (Metcalf, 1983). The number of follicles in the ovary is low, or the ovaries are empty of premodial follicles after menopause (Richardson *et al.*, 1987).

Ovarian estrogen production after menopause is usually small, and ovarian removal is not accompanied by any large decrease in circulating estrogen (Judd *et al.*, 1974). However, the ovary continues to be endocrinologically active in terms of production of androgens. Testosterone and androstenedione are produced by the ovarian stromal and hilus cells (Judd *et al.*, 1974). Androstenedione is also produced by the adrenal gland. The androgens can be converted to estrogen outside either the ovary or the adrenal gland. This extraglandular aromatization is an important concept for the production of estrogen in postmenopausal women (Patkai *et al.*, 1974). Fat, liver, kidney, and brain are sites where extraglandular aromatization can occur. The extraglandular aromatization is influenced by age and weight. Heavy women have higher conversion rates and higher circulating estrogen concentrations than slender women. There is a close relationship between body mass index and estrogen concentration in blood (Boman *et al.*, 1990). This indicates that a woman's weight is of great importance for the expression of postmenopausal symptoms.

The median age for menopause is 51 years in western countries. Smokers reach menopause about 2 years earlier than nonsmokers. In about 10% of women, menopause arrives abruptly, while, in the rest, menopause is preceded by a period of up to 4 years of irregular menstrual bleeding (McKinlay *et al.*, 1992). Up to 70% of postmenopausal women experience symptoms, such as hot flushes and perspiration. These symptoms occur mainly during the first 4–5 years but decrease in severity during the last years. About 5 years after menopause, only 20% still have hot flushes (McKinlay *et al.*, 1992).

B. POSTMENOPAUSAL SYMPTOMS

1. *Hot Flush*

The hot flush is conventionally considered to be the characteristic clinical manifestation of the climacteric. Approximately 35% of women will seek medical help for their postmenopausal symptoms. Hot flushes are more common in women who smoke. There is great variability among women from different ethnic groups. The hot flush is described as a sensation of pressure in the head, much like a headache. This effect progresses in intensity until the actual flush occurs. The actual flush begins in the head and neck areas and passes, often in waves, over the entire body; it is described as a feeling of heat or burning. This period is followed by an outbreak of sweating involving the entire body but particularly marked in the head, neck, chest, and back. The entire episode can vary from 0.5 min to 30 min but is on average about 10 min. Shortly after the onset of a flush sensation, there is an increase in hand and forearm blood flow as well as an increase in pulse rate. This increase in blood flow and pulse rate persists for minutes after the sensation of the flush disappears. At the same time, the finger temperature starts to increase and reaches a maximum about 5 min after the end of the sensation (Molnar, 1975; Ginsburg et al., 1981).

The exact mechanism behind the occurrence of the hot flush is still not fully understood. The phenomenon is clearly related to a withdrawal of estrogen, and the symptoms disappear after ERT (Jaszman, 1976). After the onset of ERT, the number of hot flushes decreases gradually over a period of up to 20 days. During the next 2 months, the number of hot flushes will stabilize to a level related to the doses of ERT as well as the effect on decreasing FSH concentration (Holst et al., 1989).The estrogen add back also stimulates endometrial proliferation, which will cause endometrial hyperplasy with increased risk of endometrial carcinoma (Ziel and Finkle, 1975). Therefore, sequential estrogen/progestogen replacement therapy or combined estrogen progestogen treatments are preferred. Sequential treatment induces regular endometrial shedding or gives an atrophic endometrium in the combined therapy, as progesterone inhibits the proliferation. The risk of endometrial cancer is therefore abolished (Persson et al., 1989).

2. *Mood Changes*

Mood changes during the menopause are frequently discussed in the lay press and among patients. Several longitudinal studies have *not* been able to confirm an increase of clinical depression or severe mood change in relation to the menopausal transition (Hällström and Samuelsson, 1985; Kaufert et al., 1992; Avis et al., 1994). Hällström and Samuelsson (1985) investigated 899 women on two occasions within 6 years. They could not detect any increased risk for mental illness at menopause. Milder mood changes, such as decreased well-being, appear postmenopausally (Hunter, 1992; Kaufert et al., 1992). Cross-sectional studies on a general population sample in Norway have shown a decrease in subjective well-being in at least 10% of the women; however, no changes in major depression were noticed (Holte, 1992). This has been confirmed in longitudinal studies in North America and Europe (Hunter, 1992; Kaufert et al., 1992). Other cross-sectional studies have not been able to show any difference in well-being among postmenopausal women compared to premenopausal women (McKinlay et al., 1987, Ballinger, 1990; Dennerstein et al., 1993; Collins and Landgren, 1994).

In women with natural menopause, estrogen add back as a treatment for clinical depression has been tested in several controled, randomized clinical trials. Three studies revealed that estrogen at doses used for HRT had no effect on clinical depression (Campbell, 1976; Thompson and Oswald, 1977; Coope, 1981). In an open study (Klaiber *et al.*, 1979), estrogen was shown to be effective in severely depressed psychiatric patients at a higher dose than HRT. This study has not been replicated in a controlled group. A recent double-blind, placebo-controlled study revealed, however, that estrogen in a higher dosage than HRT had significant beneficial effect in women with severe postpartum depression (Gregoire *et al.*, 1996).

Results are different in depressed patients after oophorectomy. In four controlled studies, estrogen add back improved depression symptoms significantly measured by using the Beck's Depression Scale or the Hamilton's Rating Scale for depression (Campbell, 1976; Dennerstein *et al.*, 1979; Montgomery *et al.*, 1987; Ditkoff, 1991). It was noticed in these studies that a higher estrogen dose had a significantly better effect than a lower dose. In another study (Sherwin and Gelfand, 1985), a dose-related effect of estrogen on depression treatment was documented.

VIII. Side Effects of Hormone Replacement Therapy

It is common for women receiving HRT to experience adverse symptoms while taking cyclical progestogen (Fig. 1). Some women on HRT experience negative mood changes (depression and irritability) related to the progestogen component of sequential HRT (Björn and Bäckström, 1999). In these women, the symptoms begin shortly after progestogen has been added to the treatment and continue to rise in severity until the progestogen treatment has been ended. They usually abate within 2 or 3 days after the end of the progestogen treatment (Fig. 1) (Hammarbäck *et al.*, 1985; Björn *et al.*, 2000). There seems to be a dose-dependent increase in negative mood with increasing dosage of the progestogen (Magos *et al.*, 1986). Negative mood changes are a well-known clinical side effect of HRT and are one of major problems concerning compliance with the HRT (Bäckström *et al.*, 1996). The exact mechanism behind the induction of negative mood is still not understood. It is, however, clear that it is related to the addition of progestogen to the postmenopausal replacement therapy.

When women with an intact uterus are treated with estrogen, progestogen is added to protect the endometrium from hyperplasia and malignancies. The length of time that progestogen is given is of importance. It has been suggested that 12 days of sequential treatment is adequate to protect the endometrium. The meta-analysis of depression and mood changes

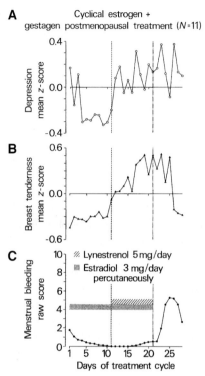

FIG. 1. Mean depression (A), breast tender z-scores (B), and menstrual bleeding scores (C) each day of treatment cycle in 11 postmenopausal women during sequential estrogen/progestogen replacement therapy. (Reprinted with permission from S. Hammarbäck, T. Bäckström, J. Holst, B von Schultz, and S. Lyrenäs. Cyclical mood changes as in the premenstrual tension syndrome during sequential estrogen–progestagen postmenopausal replacement treatment. *Acta Obstet. Gynecol. Scand.* 1985;64:393–397.)

in women receiving HRT by Zweifel and O'Brien (1997) revealed that estrogen was associated with symptom improvement, whereas progesterone given in combination with estrogen counteracted the beneficial effect of estrogen. The reasons for poor compliance with HRT and, in particular, drug-related reasons, have not yet been fully elucidated. In a cohort study of peri- or postmenopausal women, Björn and Bäckström (1999) investigated why some women never start or discontinue HRT, even when great effort has been made to inform and fulfill the demands of the patient. Of 356 women receiving the questionnaire, 92% replied. Two percent never started the therapy, but 75% continued the therapy for more than 3 years. Reasons for discontinuing HRT were negative side effects (35%), desire to find out if climacteric symptoms had ended (26%), fear of cancer and thrombosis (25%), weariness of bleeding (19%), and a wish to deal with the problems "naturally" (15%). The authors concluded that compliance with HRT can be high if adequate information is given and follow-up appointments made. The main reason for poor compliance was negative side effects, most likely progestogen related. There seem to be a relationship between negative mood while on progestogens and reports of earlier PMS. Women with PMS during their fertile life reacted with more negative mood symptoms while taking progestogen than women without earlier PMS (Björn *et al.*, 2000). The challenge will be to minimize the negative side effects of HRT.

The negative effect of progestogen by itself has, however, been debated. In one study, in surgically postmenopausal women, no adverse side effects were noted with sequential oral medroxyprogesterone acetate (MPA) given in conjunction with transdermal estrogen, regardless of a history of PMS (Kirkham *et al.*, 1991). This study was, however, conducted only during the first treatment cycle, and the positive effect of the HRT treatment on postmenopausal symptoms increases continuously during the first treatment cycle (Holst *et al.* 1989) and might blunt a negative effect of MPA in the end of a sequential treatment; the cyclical pattern would be the shield. In another study, Prior *et al.* (1994) found no adverse effects of MPA treatment without estrogen in postmenopausal women. In this study, no estrogen was added, and it seemed to be important. It has been shown that PMS can be treated with MPA, but the beneficial effect of the therapy was a consequence of total inhibition of hormone production rather than a result of only ovulation suppression (West, 1990). When the menstrual cycle is totally disrupted, both estradiol and progesterone production from the ovary are inhibited but, when only ovulation is inhibited, the estradiol production is present. This finding suggests that MPA together with estrogen has a different, negative-mood-provoking effect compared with MPA in the absence of estrogen. Norethisterone showed no beneficial effects compared to placebo, but some women withdrew because of adverse effects (West, 1990). In a study in which patients with PMS were given estrogen or placebo during the luteal phase, the symptoms were aggravated during the estrogen treatment compared with placebo (Dhar and Murphy, 1990).

IX. Neuroactive Steroids and Epilepsy

A. Brain Excitability

Reproductive hormones influence neuronal excitability by specific mechanisms that alter cell metabolism and neural transmission (Herzog,

1987; Smith, 1989). Estradiol lowers many seizure thresholds, while progesterone and some of its metabolites exert antiseizure effects (Herzog, 1991). Seizure frequency in women may increase in relation to the menstrual cycle (catamenial epilepsy) as a result of the withdrawal of progesterone or the elevation of the serum estradiol:progesterone ratio (Laidlaw, 1956; Bäckström, 1976). The rapid premenstrual withdrawal of the antiseizure effect of progesterone is considered to be a factor in the premenstrual exacerbation of seizures (Laidlaw, 1956). Abnormally low progesterone secretion during the luteal phase is considered to be a factor in seizure exacerbation that occupies the entire second half of the menstrual cycle (Laidlaw, 1956; Bäckström, 1976; Mattson *et al.*, 1981; Herzog, 1991).

Estradiol has been shown to decrease the seizure threshold in a dose-dependent fashion. At 5 min after an intravenous injection of 40 mg Premarine, an estrogen product, in women with partial epilepsy, an increased frequency of epileptic discharges from an epileptic focus was noted in 11 of 14 women. A grand mal seizure was provoked within 15 min after the injection in 4 of 14 patients (Logothetis *et al.*, 1959). Although such a high dose of estrogen is not used in HRT, the result demonstrates an excitatory effect of estrogen on brain activity. It is also possible to induce an epileptic focus by applying estradiol directly on the cerebral cortex (Marcus *et al.*, 1968). Estrogen application has been used as a model for epilepsy. The focus gives a "petit mal like" activity, which resembles a 3-per-second spike-and-wave activity (Marcus *et al.*, 1968).

On the contrary, progesterone has been shown to increase the threshold of electroshock seizure in animals and protect against metrazol-induced seizures (Spiegel and Wycis, 1945). In oophorectomized cats, progesterone can significantly decrease the interictal spikes frequency from epileptic foci (Landgren *et al.*, 1978). Bäckström *et al.* (1984) have studied women with partial epilepsy and well-defined epileptic foci. A continuous intravenous injection of progesterone at 4.0–12.0 mg/h for 2 h significantly reduced the spike frequency monitored by continuous electroencephalogram (EEG). Four of seven patients had a significant reduction in interictal spike frequency shortly after intravenous progesterone infusion (Fig. 2). Reduction of epileptic seizure frequency in women after intramuscular injection of MPA was also reported (Mattson and Cramer, 1985; Herzog, 1995). The effect of progesterone is probably mediated by its A-ring reduced metabolites allopregnanolone and pregnanolone. When micronized progesterone is given orally, allopregnanolone is one of the major metabolites in the blood. Experiments in oophorectomized cats with a penicillin-induced epileptic foci revealed that allopregnanolone was more potent in inhibiting epileptic activity than clonazepam (Landgren *et al.*, 1987).

A relationship between corticosteroids and seizures was found in both experimental and clinical studies. Dubrovsky (2000) has reviewed the effect of

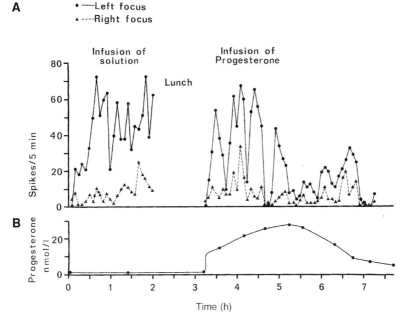

FIG. 2. The effect of progesterone infusion in a patient with partial epilepsy. A continuous electroencephalogram recording was made to count the epileptic discharges. (**A**) The frequency of epileptic discharges from two separate foci in the left and right frontotemporal lobes. (**B**) The plasma progesterone concentration expressed in nmol/L during infusion of progesterone solution. The infusion rate was 8.0 mg/hr of progesterone after a initial dose of 0.5 mg progesterone. (Reprinted with permission from T. Bäckström, B. Zetterlund, S. Blom, and M. Romano. Effects of continuous progesterone infusion on the epileptic discharge frequency in women with partial epilepsy. *Acta Neurol. Scand.* 1984;69:240–248.)

stress and neuroactive steroids produced by the adrenal gland. The adrenal gland produces both pro- and anticonvulsant steroids. Aird and Gordon (1951) showed as early as 1950 that desoxycorticosterone has an anticonvulsant effect in patients with epilepsy. $3\alpha,5\alpha$-Tetrahydrodeoxycorticosterone, an A-ring-reduced metabolite of deoxycorticosterone, has been shown as a $GABA_A$-receptor agonist similar to allopregnanolone (Kraulis *et al.*, 1975; Paul and Purdy, 1992). In contrast, cortisol increases brain excitabillity (Verandakis and Woodbury, 1963).

B. CATAMENIAL EPILEPSY

One half of women with epilepsy may have some form of reproductive dysfunction: amenorrhea, oligomenorrhea, or abnormally prolonged or shortened menstrual cycle intervals (Herzog *et al.*, 1986). These menstrual disorders are characterized by inadequate luteal phase cycles (i.e., abnormally low progesterone secretion during the second half of the menstrual cycle) (Herzog, 1991). Inadequate luteal phase cycles have abnormally elevated serum estradiol:progesterone ratios, which are associated with greater seizure frequency (Bäckström, 1976; Mattson et al., 1981). Natural progesterone is considered to be the treatment of choice for inadequate luteal phase cycles (Jones, 1976).

Several clinical studies show a change in seizure frequency with different phases in the menstrual cycle in a subset of women with epilepsy (Dickerson, 1941; Laidlaw, 1956; Bäckström, 1976; Rosciszewska, 1980; Mattson and Cramer, 1985; Herkes et al., 1993; Herzog et al., 1997). There are, however, discrepancies in the literature regarding seizure frequency exacerbation related to menstruation in the population of women with epilepsy. Almquist (1955) found almost no evidence of such a relationship, while others described menstrual influence on seizures in 10%–72% of women with epilepsy (Dickerson, 1941; Liadlaw, 1956). In a prospective study, 69 women were observed for 4 years, and an increase in seizure incidence 2 days before or during the first day of menstruation was found in 63% of patients (Rosciszewska, 1980). In a few women, seizures occur only during the days immediately preceding menstruation or during menstrual flow. Only 4 of 226 female epileptic patients were found whose seizures occurred only in relation to menstruation (Rosciszewska, 1975). The term catamenial epilepsy should be limited to such cases.

In a prospective study of women with partial epilepsy, a relationship between seizure and hormonal variations during the menstrual cycle is noted (Bäckström, 1976). The total number of generalized seizures showed two periods of increased frequency. The first occurred shortly after the rapid decrease of progesterone during the menstruation. The second peak of seizures occurred during elevating preovulatory estrogen. During high-progesterone periods, the number of generalized seizures was very low. Patients without ovulation all showed increasing number of seizures related to estrogen peaks (Bäckström, 1976; Bäckström and Jorpes, 1979). Estrogen seems to have an activating effect, seen both in ovulatory and especially in anovulatory periods. These findings are in accordance with those reported by Laidlaw (1956) and Mattson and Cramer (1985), who found a similar variation in the seizure frequency during the menstrual cycle and findings similar to those of Logothetis and colleagues (1959), who provoked seizures with estrogen injections in women with epilepsy.

The results indicate that there exist antiepileptic factors during the luteal phase that inhibit the spread of epileptic activity, as the decrease in seizure frequency was noted especially for generalized seizures. These factors could be the CNS-active progesterone metabolites, such as allopregnanolone. The increase in seizure frequency during menstruation could be explained by a rebound effect after a rapid decrease in antiepileptic factors premenstrually, similar to that seen when the intake of antiepileptic drugs is ended abruptly. There were no changes in antiepileptic drug concentrations during the menstrual cycle, so a change in the pharmacokinetics of the antiepileptic drugs is not an explanation for the menstrual-cycle-linked changes in this group (Bäckström and Jorpes, 1979). In another study, a decrease in phenytoin levels was found 2 days before menstruation, parallel to a rise in the number of seizures (Rosciszewska *et al.*, 1986). A clinical trial with medroxyprogesterone also has shown beneficial effects on seizure frequency (Mattson *et al.*, 1984). MPA is, however, weaker in its CNS action than progesterone, and further studies, perhaps using some of the aforementioned, more potent progesterone metabolites, will show if this is a new way to treat epilepsy.

Herzog (1986, 1995) has studied the natural progesterone suppository treatment of catamenially exacerbated complex partial seizures in two open trials. He assessed the effects of adjunctive progesterone therapy on seizure frequency in women with catamenial exacerbation of complex partial (CPS) and secondary generalized motor (SGMS) seizures of temporal lobe origin. progesterone was well tolerated; 72% of patients experienced a decline in seizure frequency during a 3-month treatment period compared with the 3 months before therapy. Average daily CPS frequency declined by 54%, while SGMS declined by 58%.

Different types of epilepsy seem to show different cyclical pattern of seizures. In the aforementioned cases, the patients had partial epilepsy and secondary generalized seizures. In petit mal epilepsy, patients seem to have a different pattern of seizure distribution during the menstrual cycle. In a few cases of patients with petit mal epilepsy, repeated 24-h EEG recordings and hormone analyses were done over one menstrual cycle. These patients with petit mal epilepsy showed the highest seizure frequency during the luteal phase, with a peak during the last 5 premenstrual days and with a rapid decrease in the seizure frequency shortly after the onset of menstrual bleeding (Bäckström *et al.*, 1983a). The pattern is very similar to the pattern of mood changes during the menstrual cycle seen in patients with PMS (Bäckström *et al.*, 1983b). Neuroactive steroids can exacerbate GABA-induced absence seizures in rats (Banerjee and Snead, 1998), and progesterone can exacerbate typical absence seizures in women with petit mal epilepsy (Grunewald *et al.*, 1992). Menstrual-cycle-linked seizure variation may be associated with PMS. In women with catamenial seizures, PMS coexisted in 84%, while in women with no catamenial exacerbation only 22% had PMS (Rosciszewska, 1980). Primary generalized tonic-clonic seizures are also more frequent in the premenstrual period than in the rest of the cycle (Rosciszewska, 1980).

Bäckström, T., Andersson, A., Baird, D. T., and Selstam, G. (1986). The human corpus luteum secretes 5α-pregnane-3,20-dione. *Acta Endocrinologica* **111,** 116–121.
Bäckström, T., and Hammarbäck, S. (1991). Premenstrual syndrome—Psychiatric or gynecological disorder? *Ann. Med.* **23,** 625–633.
Bäckström, T . (1996). Side effects of contraceptives and gonadal hormones. *Bailliere's Clinical Psychiatry* **2,** 715–726.
Bäckström, T., Lindhe, B. Å., Cavalli-Björkman, B., Nordenström, S., and Hansson, Y. (1992). Effects of oral contraceptives on mood: A randomized comparison of three phasic and monophasic preparations. *Contraception* **46,** 253–268.
Bäckström, T., Bixo, M., Seippel, L., Sundström, I., and Wang, M. D. (1996). Progestins and behavior. *In* "The Brain: Source and Target for Sex Steroid Hormones"(A. Genazzani, Ed.), pp. 277–291. Parthenon Publishing, London.
Bäckström, T., and Rosciszewska, D. (1997). Effects of hormones on seizure expression. *In* "Epilepsy: A Comprehensive Textbook" (J. Engel, Jr. and T. A. Pedley, Eds.), pp. 2003–2012. Lippincott-Raven Publishing, Philadelphia.
Bäckström, T., Appelblad, P., Bixo, M., Haage, D., Johansson, S., Landgren, S., Seippel, L., Sundström, I., Wang, M. D., and Wahlström, G. (2000). Female sex steroids, the brain and behaviour. *In* "Mood Disorders in Women" (M. Steiner K. Yonkers, and E. Eriksson, eds.), pp. 189–205. Martin Dunitz Publishers, London.
Badalian, L. O., Temin, P. A., and Muhin, K. I. U. (1991). Obtained effect of testenat treatment of epilepsy in men. *Zh. Nevropatol. Psikhiatr.* **91,** 44–47.
Baldereschi, M., Di Carlo, A., Lepore, V., Bracco, L., Maggi, S., Grigoletto, F., Scarlato, G., and Amaducci, L. (1998). Estrogen-replacement therapy and Alzheimer's disease in the Italian Longitudinal Study on Aging. *Neurology* **50,** 996–1002.
Ballinger, C. B. (1990). Psychiatric aspects of the menopause. *Br. J. Psychiatry* **156,** 773–787.
Bancroft, J., Sanders, D., Warner, P., and Loudon, N. (1987). The effects of oral contraceptives on mood and sexuality: Comparison of triphasic and combined preparations. *J. Psychosom. Obstet. Gynaecol.* **7,** 1–8.
Banerjee, P. K., and Snead, O. C. (1998). Neuroactive steroids exacerbate gamma-hydroxybutyric acid-induced absence seizures in rats. *Eur. J. Pharmacol.* **359,** 41–48.
Barbaccia, M. L., Concas, A., Serra, M., and Biggio, G. (1998). Stress and neurosteroids in adult and aged rats. *Exp. Gerontol.* **33,** 697–712.
Barnea, A., Hajibeigi, A., Trant, J. M., and Mason, I. (1990). Expression of steroid metabolizing enzymes by aggregating fetal brain cells in culture: A model for developmental regulation of the progesterone 5α-reductase pathway. *Endocrinology* **127,** 500–502.
Barrett-Connor, E., and Kritz-Silverstein, D. (1993). Estrogen replacement therapy and cognitive function in older women. *JAMA* **269,** 2637–2641.
Baulieu, E. E., and Robel, P. (1990). Neurosteroids: a new brain-function. *J. Steroid Biochem. Mol. Biol.* **37,** 395–403.
Belelli, D., Bolger, M. B., and Gee, K. W. (1989). Anticonvulsant profile of the progesterone metabolite 5α-pregnan-3α-ol-20-one. *Eur. J. Pharmacol.* **166,** 325–329.
Billiar, R. B., Little, B., Kline, I., Reier, P., Takaoka, Y., and White, R. J. (1975). The metabolic clearance rate, head and brain extractions, and brain distribution and metabolism of progesterone in the anesthetized, female monkey (*Macaca mulatta*). *Brain Res.* **94,** 99–113.
Bixo, M., Bäckström, T., Winblad, B., Selstam, G., and Andersson, A. (1986). Comparison between pre- and postovulatory distributions of oestradiol and progesterone in the brain of the PMSG-treated rat. *Acta Physiol. Scand.* **128,** 241–246.
Bixo, M., and Bäckström, T. (1990). Regional distribution of progesterone and 5α-pregnane-3,20-dione in rat brain during progesterone-induced "anesthesia." *Psychoneuroendocrinology* **15,** 159–162.

30% and 50% and some patients became seizure free. Three patients in Mattson's (1984) study did not show any changes in seizure frequency, and they all had menstrual bleeding suggestive of fluctuating estrogen, while the women with amenorrhea showed a significant change in seizure frequency. MPA inhibits ovulation and, in higher doses, it can also inhibit the estradiol production. MPA thus inhibits fluctuations in plasma hormone concentrations and probably the excitability of the CNS during the menstrual cycle. For this reason, it may be easier for the patient to become seizure free with ordinary antiepileptic therapy. In patients with petit mal epilepsy, there may be a different origin for the effect in the CNS than in partial epilepsy and this discussion may not be applicable for these patients. A woman with antiepileptic therapy needs 150 mg Depo-Provera at 1- or 2-month intervals to develop amenorrhea. A common side effect is weight increase, mainly caused by increased appetite and irregular menstrual bleeding. If the women also need a contraceptive, Depo-Provera is a good choice.

Positive effects of natural progesterone applied in suppositories was observed by Herzog (1986) in six of eight women with CPS and an inadequate luteal phase. Average monthly seizure frequency decreased in 68%. In an open trial, Herzog (1995) assessed the effects of adjunctive progesterone therapy on seizure frequency in 25 women with catamenial exacerbation of CPS and SGMS. Eighteen women (72%) experienced a decline in seizure frequency during a 3-month treatment period compared with the 3 months before therapy. In another study, a decrease was noted in 23 of 34 epileptic females treated with synthetic progesterone. The per-month decrease occurred both in tonic-clonic seizure frequency (average 62%) and in CPS (average 70%). The remaining eight cases were not improved and, in three women, the seizure frequency increased (Feder, 1984). The effect of intravenous progesterone infusion on epileptic discharge frequency was studied in seven women with partial epilepsy. In four, a significant decrease in spike frequency was shown (Bäckström *et al.*, 1984). The effect was related to the concentration of plasma progesterone binding capacity (Bäckström *et al.*, 1984). No therapeutic results of progesterone were reported by Dana-Heari and Richens (1983) in nine patients with catamenial seizures. None of the aforementioned studies were randomized and placebo controlled. It is therefore difficult to evaluate the clinical value of such treatments, and controlled studies are needed.

Progesterone metabolites like allopregnanolone may be more suitable to use as antiepileptic therapy in the future (Belelli *et al.*, 1989). Althesin is a combination of two synthetic steroids, alphaxalone and alphadelone, that are structurally close to allopregnanolone. Althesin has been successfully used to treat drug-resistant status epilepticus (Munari *et al.*, 1979). Eight of 11 patients with status epilepticus resistant to barbiturates and benzodiazepines were successfully treated with Althesin (Munari *et al.*, 1979).

Althesin also decreased intracranial cranial pressure (Turner, 1973) and reduced oxygen consumption in the gray matter of 40% (Pickerodt et al., 1972). The neuroactive progesterone metabolites and some syntethic derivatives might be useful as antiepileptic therapeutics in the future (Gasior et al., 1997).

References

Aird, R., and Gordon, G. (1951). Anticonvulsive properties to desoxycorticosterone. *JAMA* **145**, 715–779.

Almquist, R. (1955). The rhythm of epileptic attacks and its relationship to the menstrual cycle. *Acta Neuro. Psychiatr. Scand.* **30** (Suppl 105), 7–116.

Amandusson, Å., Hermansson, O., and Blomqvist, A. (1996). Colocalization of oestrogen receptor immunoreactivity and proenkephalin mRNA expression to neurons in the superficial laminae of the spinal and medullary dorsal horn of rats. *Eur. J. Neurosci.* **8**, 2440–2445.

American Psychiatric Association. (1994). Diagnostic and Statistical Manual of Mental Disorders, 4th edition, pp 714–718. U.S. Department of Health and Human Services, Washington, DC.

Appelgren, L. E. (1967). Sites of steroid hormone formation. Autoradiographic studies using labelled precursors. *Acta Physiol. Scand. Suppl.* **301**, 1–108.

Asso, D., and Braier, J. R. (1982). Changes with the menstrual cycle in psychophysiological and self-report measures of activation. *Biol. Psychol.* **15**, 95–107.

Asthana, S., Craft, S., Baker, L. D., Raskind, M. A., Birnbaum, R. S., Lofgreen, C. P., Veith, R. C., and Plymate, S. R. (1999). Cognitive and neuroendocrine response to transdermal estrogen in postmenopausal women with Alzheimer's disease: Results of a placebo-controlled, double-blind, pilot study. *Psychoneuroendocrinology* **24**, 657–677.

Avis, N. E., Brambilla, D., McKinlay, S. M., and Vass, K. (1994). A longitudinal analysis of the association between menopause and depression. Results from the Massachusetts Women's Health Study. *Ann. Epidemiol.* **4**, 214–220.

Bäckström, T. (1976). Epileptic seizures in women in relation to variations of plasma estrogen and progesterone during the menstrual cycle. *Acta Neurol. Scand.* **54**, 321–347.

Bäckström, T., and Jorpes, P. (1979). Serum phenytoin, phenobarbital, carbamazepine, albumin, and plasma estradiol and progesterone concentrations during the menstrual cycle in women with epilepsy. *Acta Neurol. Scand.* **59**, 63–71.

Bäckström, T., Baird, D. T., Bancroft, J., Bixo, M., Hammarbäck, S., Sanders, D., Smith, S., and Zetterlund, B. (1983a). Endocrinological aspects of cyclical mood changes during the menstrual cycle or the premenstrual syndrome. *J. Psychosom. Obst. Gyn.* **2**, 8–20.

Bäckström, T., Sanders, D., Leask, R., Davidson, D., Warner, P., and Bancroft, J. (1983b). Mood, sexuality, hormones, and the menstrual cycle. II. Hormone levels and their relationship to the premenstrual syndrome. *Psychosom. Med.* **45**, 503–507.

Bäckström, T., Zetterlund, B., Blom, S., and Romano, M. (1984). Effects of continuous progesterone infusion on the epileptic discharge frequency in women with partial epilepsy. *Acta Neurol. Scand.* **69**, 240–248.

Bäckström, T., Smith, S., Lothian, H., and Baird, D. T. (1985). Prolonged follicular phase and depressed gonadotropins following hysterectomy and corpus lute-ectomy in women with premenstrual tension syndrome. *Clin. Endocrinol.* **22**, 723–732.

Bäckström, T., Andersson, A., Baird, D. T., and Selstam, G. (1986). The human corpus luteum secretes 5α-pregnane-3,20-dione. *Acta Endocrinologica* **111,** 116–121.

Bäckström, T., and Hammarbäck, S. (1991). Premenstrual syndrome—Psychiatric or gynecological disorder? *Ann. Med.* **23,** 625–633.

Bäckström, T . (1996). Side effects of contraceptives and gonadal hormones. *Bailliere's Clinical Psychiatry* **2,** 715–726.

Bäckström, T., Lindhe, B. Å., Cavalli-Björkman, B., Nordenström, S., and Hansson, Y. (1992). Effects of oral contraceptives on mood: A randomized comparison of three phasic and monophasic preparations. *Contraception* **46,** 253–268.

Bäckström, T., Bixo, M., Seippel, L., Sundström, I., and Wang, M. D. (1996). Progestins and behavior. *In* "The Brain: Source and Target for Sex Steroid Hormones"(A. Genazzani, Ed.), pp. 277–291. Parthenon Publishing, London.

Bäckström, T., and Rosciszewska, D. (1997). Effects of hormones on seizure expression. *In* "Epilepsy: A Comprehensive Textbook" (J. Engel, Jr. and T. A. Pedley, Eds.), pp. 2003–2012. Lippincott-Raven Publishing, Philadelphia.

Bäckström, T., Appelblad, P., Bixo, M., Haage, D., Johansson, S., Landgren, S., Seippel, L., Sundström, I., Wang, M. D., and Wahlström, G. (2000). Female sex steroids, the brain and behaviour. *In* "Mood Disorders in Women" (M. Steiner K. Yonkers, and E. Eriksson, eds.), pp. 189–205. Martin Dunitz Publishers, London.

Badalian, L. O., Temin, P. A., and Muhin, K. I. U. (1991). Obtained effect of testenat treatment of epilepsy in men. *Zh. Nevropatol. Psikhiatr.* **91,** 44–47.

Baldereschi, M., Di Carlo, A., Lepore, V., Bracco, L., Maggi, S., Grigoletto, F., Scarlato, G., and Amaducci, L. (1998). Estrogen-replacement therapy and Alzheimer's disease in the Italian Longitudinal Study on Aging. *Neurology* **50,** 996–1002.

Ballinger, C. B. (1990). Psychiatric aspects of the menopause. *Br. J. Psychiatry* **156,** 773–787.

Bancroft, J., Sanders, D., Warner, P., and Loudon, N. (1987). The effects of oral contraceptives on mood and sexuality: Comparison of triphasic and combined preparations. *J. Psychosom. Obstet. Gynaecol.* **7,** 1–8.

Banerjee, P. K., and Snead, O. C. (1998). Neuroactive steroids exacerbate gamma-hydroxybutyric acid-induced absence seizures in rats. *Eur. J. Pharmacol.* **359,** 41–48.

Barbaccia, M. L., Concas, A., Serra, M., and Biggio, G. (1998). Stress and neurosteroids in adult and aged rats. *Exp. Gerontol.* **33,** 697–712.

Barnea, A., Hajibeigi, A., Trant, J. M., and Mason, I. (1990). Expression of steroid metabolizing enzymes by aggregating fetal brain cells in culture: A model for developmental regulation of the progesterone 5α-reductase pathway. *Endocrinology* **127,** 500–502.

Barrett-Connor, E., and Kritz-Silverstein, D. (1993). Estrogen replacement therapy and cognitive function in older women. *JAMA* **269,** 2637–2641.

Baulieu, E. E., and Robel, P. (1990). Neurosteroids: a new brain-function. *J. Steroid Biochem. Mol. Biol.* **37,** 395–403.

Belelli, D., Bolger, M. B., and Gee, K. W. (1989). Anticonvulsant profile of the progesterone metabolite 5α-pregnan-3α-ol-20-one. *Eur. J. Pharmacol.* **166,** 325–329.

Billiar, R. B., Little, B., Kline, I., Reier, P., Takaoka, Y., and White, R. J. (1975). The metabolic clearance rate, head and brain extractions, and brain distribution and metabolism of progesterone in the anesthetized, female monkey (*Macaca mulatta*). *Brain Res.* **94,** 99–113.

Bixo, M., Bäckström, T., Winblad, B., Selstam, G., and Andersson, A. (1986). Comparison between pre- and postovulatory distributions of oestradiol and progesterone in the brain of the PMSG-treated rat. *Acta Physiol. Scand.* **128,** 241–246.

Bixo, M., and Bäckström, T. (1990). Regional distribution of progesterone and 5α-pregnane-3,20-dione in rat brain during progesterone-induced "anesthesia." *Psychoneuroendocrinology* **15,** 159–162.

Bixo, M., Bäckström, T., Winblad, B., and Andersson, A. (1995). Estradiol and testosterone in specific regions of the human female brain in different endocrine states. *J. Steroid Biochem. Mol. Biol.* **55,** 297–303.

Bixo, M., Andersson, A., Winblad, B., Purdy, R. H., and Bäckström, T. (1997). Progesterone, 5α-pregnane-3,20-dione and 3α-hydroxy-5α-pregnane-20-one in specific regions of the human female brain in different endocrine states. *Brain Res.* **764,** 173–178.

Björn, I., and Bäckström, T. (1999). Compliance to estrogen and progestins in treatment of peri and postmenopausal women. The significance of negative side effects. *Maturitas* **32,** 77–86.

Björn, I., Bixo, M., Nöjd-Strandberg, K., Nyberg, S., and Bäckström, T. (2000). Negative mood changes during hormonal replacement therapy—A comparison between two different progestins. *Am. J. Obstet. Gyn* **183,** 1419–1426.

Boman, K., Bäckström, T., Gerdes, U., and Stendahl, U. (1990). Oestrogens and clinical characteristics in endometrial carcinoma. *Anticancer Res.* **10,** 247–251.

Brinton, R. D., Tran, J., Proffitt, P., and Montoya, M. (1997). 17 Beta-estradiol enhances the outgrowth and survival of neocortical neurons in culture. *Neurochem. Res.* **22,** 1339–1351.

Broe, G. A., Henderson, A. S., Creasey, H., McCusker, E., Korten, A. E., Jorm, A. F., Longley, W., and Anthony, J. C. (1990). A case-control study of Alzheimer's disease in Australia. *Neurology* **40,** 1698–1707.

Broverman, D. M, Klaiber, E. L., Kobayashi, Y., and Vogel, W. (1968). Roles of activation and inhibition in sex differences in cognitive abilities. *Psychol. Rev.* **75,** 23–50.

Brown, C. S., Ling, F. W., Andersen, R. N., Farmer, R. G., and Areheart, K. L. (1994). Efficacy of depot leuprolide in premenstrual syndrome: Effect of symptom severity and type in a controlled trial. *Obstet. Gynecol.* **84,** 779–786.

Campbell, S. (1976). Double blind psychometric studies on the effects of natural oestrogens on postmenopausal women. *In* "Managment of the Menopause and the Postmenopausal Years" (S. Campbell, Ed.), pp. 149–158.

Cheney, D. L., Uzunov, D., Costa, E., and Guidotti, A. (1995). Gas chromatographic–mass fragmentographic quantitation of 3α-hydroxy-5α-pregnan-20-one (allopregnanolone) and its precursors in blood and brain of adrenalectomized and castrated rats. *J. Neurosci.* **15,** 4641–4650.

Collins, A., and Landgren, B. M. (1994). Reproductive health, use of estrogen and experience of symptoms in perimenopausal women: A population-based study. *Maturitas* **20,** 101–111.

Coope, J. (1981). Is oestrogen effective in the treatment of menopausal depression? *J. R. Coll. Gen. Pract.* **31,** 134–140.

Corpéchot, C., Young, J., Calvel, M., Wehrey, C., Veltz, J. N., Touyer, G., Mouren, M., Prasad, V. V., Banner, C., Sjovall, J., Baulieu, E. E., and Robel, P. (1993). Neurosteroids: 3α-Hydroxy-5α-pregnan-20-one and its precursors in the brain, plasma, and steroidogenic glands of male and female rats. *Endocrinology* **133,** 1003–1009.

Couse, J. F., Lindzey, J., Grandien, K., Gustafsson, J. A., and Korach, K S. (1997). Tissue distribution and quantitative analysis of estrogen receptor-alpha (ERalpha) and estrogen receptor-beta (ERbeta) messenger ribonucleic acid in the wild-type and ERalpha-knockout mouse. *Endocrinology* **138,** 4613–4621.

Cullberg, J. (1972). Mood changes and menstrual symptoms with different gestagen/estrogen combinations. A double blind comparison with placebo. *Acta Psychiat. Scand. Suppl.* **236,** 1–84.

Dana-Haeri, J., and Richens, A. (1983). Effect of norethisterone on seizures associated with menstruation. *Epilepsia* **24,** 377–381.

De Lignieres, B., Dennerstein, L, and Bäckström, T. (1995). Influence of route of administration on progesterone metabolism. *Maturitas* **21,** 251–257.

Dennerstein, L., Burrows, G. D., Hyman, G. J., and Sharpe, K. (1979). Hormone therapy and affect. *Maturitas* **1,** 247–259.
Dennerstein, L., Spencer-Gardner, C., Gotts, G., Brown, J. B., Smith, M. A., and Burrows, G. D. (1985). Progesterone and the premenstrual syndrome: A double blind crossover trial. *Br. Med. J.* **290,** 1617–1621.
Dennerstein, L., Smith, A. M., Morse, C., Burger, H., Green, A., Hopper, J., and Ryan, M. (1993). Menopausal symptoms in Australian women. *Med. J. Aust.* **159,** 232–236.
Dhar, V., and Murphy, B. E. (1990). Double-blind randomized crossover trial of luteal phase estrogens (Premarin) in the premenstrual syndrome (PMS). *Psychoneuroendocrinology* **15,** 489–493.
Dickerson, W. (1941). The effect of menstruation on seizure incidence. *J. Nerv. Ment. Dis.* **94,** 160–169.
Ditkoff, E. C., Crary, W. G., Cristo, M., and Lobo, R. A. (1991). Estrogen improves psychological function in asymptomatic postmenopausal women. *Obstet. Gynecol.* **78,** 991–995.
Drake, E. B., Henderson, V. W., Stanczyk, F. Z., McCleary, C. A., Brown, W. A., Smith, C. A., Rizzo, A. A., Murdock, G. A., and Buckwalter, J. G. (2000). Associations between circulating sex steroid hormones and cognition in normal elderly women. *Neurology* **54,** 599–603.
Dubrovsky, B. (2000). The specificity of stress responses to different nocuous stimuli: Neurosteroids and depression. *Brain Res. Bull.* **51,** 443–455.
Feder, H. H. (1984). Hormones and sexual behavior. *Ann. Rev. Psychol.* **35,** 165–200.
Fedor-Freybergh, P. (1977). The influence of oestrogens on the well-being and mental performance in climacteric and postmenopausal women. *Acta Obstet. Gynecol. Scand.* **64,** 1–91.
Fenwick, P., Wheeler, M., Brouwn, S., and Toone, B. (1989). The effect of subcutaneous testosterone implants in patients with deficiency and epilepsy. *Abstract Book: 18th International Congress of New Delhi*.
Fettes, I. (1997). Menstrual migraine. Methods of prevention and control. *Postgrad. Med.* **101,** 67–70, 73–7.
Fillit, H., Weinreb, H., Cholst, I., Luine, V., McEwen, B., Amador, R., and Zabriskie, J. (1986). Observations in a preliminary open trial of estradiol therapy for senile dementia-Alzheimer's type. *Psychoneuroendocrinology* **11,** 337–345.
Foy, M. R., Xu, J., Xie, X., Brinton, R. D., Thompson, R. F., and Berger, T. W. (1999). 17-beta-estradiol enhances NMDA receptor-mediated EPSPs and long-term potentiation. *J. Neurophysiol.* **81,** 925–929.
Gasior, M., *et al.* (1997). Anticonvulsant and behavioral effects of neuroactive steroids alone and in conjunction with diazepam. *J. Pharmacol. Exp. Ther.* **282,** 543–553.
Gee, K.W. (1988). Steroid modulation of the GABA/benzodiazepine receptor-linked chloride ionophore. *Mol. Neurobiol.* **2,** 291–317.
Ginsburg, J., Swinhoe, J., and O'Reilly, B. (1981). Cardiovascular responses during the menopausal hot flush. *Br. J. Obstet. Gynaecol.* **88,** 925–930.
Goldman, H., Skelley, E. B., Sandman, C. A., Kastin, A.J., and Murphy, S. (1976). Hormones and regional brain blood flow. *Pharmacol. Biochem. Behav* **5** (Suppl 1), 165–169.
Gregoire, A. J, Kumar, R., Everitt, B., Henderson, A. F., and Studd, J. W. (1996). Transdermal oestrogen for treatment of severe postnatal depression. *Lancet* **347,** 930–933.
Grudzinska, B., and Rosciszewska, D. (1980). Dynamics of EEG changes during the menstrual cycle in epileptic women. Abstract Book of XI Meeting of Polish Neurological Association in Bydgoszoz.
Grunewald, R. A., Aliberti, V., and Panayiotopoulos, C. P. (1992). Exacerbation of typical absence seizures by progesterone. *Seizure* **1,** 137–138.
Gustafsson, J. A. (2000). An update on estrogen receptors. *Semin Perinatol.* **24,** 66–69.

Hahn, P. M., Van Vugt, D. A., and Reid, R. L. (1995). A randomized, placebo-controlled, crossover trial of danazol for the treatment of premenstrual syndrome. *Psychoneuroendocrinology* **20,** 193–209.

Hall, S. (1995). Treatment of menstrual epilepsy with a progesterone-only oral contraceptive. *Epilepsia* **18,** 235–236.

Hällström, T., and Samuelsson, S. (1985). Mental health in the climacteric. The longitudinal study of women in Gothenburg. *Acta Obstet. Gynecol. Scand. Suppl.* **130,** 13–18.

Hammar, M., Lindgren, R., Berg, G. E., Moller, C. G., and Niklasson, M. K. (1996). Effects of hormonal replacement therapy on the postural balance among postmenopausal women. *Obstet. Gynecol.* **88,** 955–960.

Hammarbäck, S., Bäckström, T., Holst, J., von Schoultz, B., and Lyrenäs, S. (1985). Cyclical mood changes as in the premenstrual tension syndrome during sequential estrogen-progestagen postmenopausal replacement treatment. *Acta Obstet. Gynecol. Scand.* **64,** 393–397.

Hammarbäck, S., and Bäckström, T. (1988). Induced anovulation as treatment of premenstrual tension syndrome—A double-blind crossover study with GnRH-agonist versus placebo. *Acta Obstet. Gynaecol. Scand.* **67,** 159–166.

Hammarbäck, S., and Bäckström, T. (1989). A demographic study in subgroups of women seeking help for premenstrual syndrome. *Acta Obstet. Gynecol. Scand.* **68,** 247–253.

Hammarbäck, S., Damber, J.-E., and Bäckström, T. (1989). Relationship between symptom severity and hormone changes in women with premenstrual syndrome. *J. Clin. Endocrinol. Metab.* **68,** 125–130.

Hammarbäck, S., Ekholm, U. B., and Bäckström, T. (1991). Spontaneous anovulation causing disappearance of cyclical symptoms in women with the premenstrual syndrome. *Acta Endocrinol. (Copenhagen)*, **125,** 132–137.

Hampson, E., and Kimura, D. (1988). Reciprocal effects of hormonal fluctuations on human motor and perceptual-spatial skills. *Behav. Neurosci.* **102,** 456–459.

Hampson, E. (1990). Estrogen-related variations in human spatial and articulatory motor skills. *Psychoneuroendocrinology* **15,** 97–111.

Henderson, V. W., Paganini-Hill, A., Miller, B. L., Elble, R. J., Reyes, P. F., Shoupe, D., McCleary, C. A., Klein, R. A., Hake, A. M., and Farlow, M. R. (2000). Estrogen for Alzheimer's disease in women: Randomized, double-blind, placebo-controlled trial. *Neurology* **54,** 295–301.

Herkes, G., Eadie, M., Sharbrough, F., and Moyer, T. (1993). Patterns of seizure occurrence in catamenial epilepsy. *Epilepsy Res.* **15,** 47–52.

Herzog, A. G. (1986). Intermittent progesterone therapy and frequency of complex partial seizures in women with menstrual disorders. *Neurology* **36,** 1607–1610.

Herzog, A. G. (1987). Progesterone in seizure therapy [reply to letter]. *Neurology* **37,** 1433.

Herzog, A. G., Seibel, M. M., Schomer, D. L., Vaitukaitus, J. L., and Geschwind, N. (1986). Reproductive endocrine disorders in women with partial seizures of temporal lobe origin. *Arch. Neurol.* **43,** 341–346.

Herzog, A. G. (1991). Reproductive endocrine considerations and hormonal therapy in men with epilepsy. *Epilepsia Suppl.* **32,** 834–837.

Herzog, A. G. (1995). Progesterone therapy in women with complex partial and secondary generalized seizures. *Neurology* **45,** 1660–1662.

Herzog, A. G., Klein, P., and Ransil, B. J. (1997). Three patterns of catamenial epilepsy. *Epilepsia* **38,** 1082–1088.

Holst, J., Bäckström, T., Hammarbäck, S., and von Schoultz, B. (1989). Progesterone addition during oestrogen replacement therapy effects on vasomotor symptoms and mood. *Maturitas* **11,** 13–20.

Holte, A. (1992). Influences of natural menopause on health complaints: A prospective study of healthy Norwegian women. *Maturitas* **14,** 127–141.

Holzbauer, M. (1971). In vivo production of steroids with central depressant action by the ovary of the rat. *Br. J. Pharmacol.* **43,** 560–569.

Honjo, H., Ogino, Y., Naitoh, K., Urabe, M., Kitawaki, J., Yasuda, J., Yamamoto, T., Ishihara, S., Okada, H., and Yonezawa, T. (1989). In vivo effects by estrone sulfate on the central nervous system–Senile dementia (Alzheimer's type). *J. Steroid Biochem.* **34,** 521–525.

Hunter, M. S. (1992). The S.E. England longitudinal study of the climacteric and postmenopause. *Maturitas* **14,** 117–126.

Isojärvi, J., Pakarinen, A., and Myllyla, U. (1988). Effects of carbamazepin therapy on serum sex hormone levels in male patients with epilepsy. *Epilepsia* **29,** 781–786.

Jacobs, D. M., Tang, M. X., Stern, Y., Sano, M., Marder, K., Bell, K. L., Schofield, P., Doonelef, G., Gurland, B., and Mayeux, R. (1998). Cognitive function in nondemented older women who took estrogen after menopause. *Neurology* **50,** 368–373.

Jaffe, A. B., Toran-Allerand, C. D., Greengard, P., and Gandy, S. E. (1994). Estrogen regulates metabolism of Alzheimer amyloid beta precursor protein. *J. Biol. Chem.* **269,** 13065–13068.

Jaszman, L. (1976). In "The Management of the Menopause and Postmenopausal Years" (S. Campbell, Ed.), pp. 11–23. MTP Press, Lancaster.

Jones, G. S. (1976). The luteal phase defect. *Fertil. Steril.* **27,** 351–356.

Judd, H. L., Judd, G. E., Lucas, W. E., and Yen, S. S. C. (1974). Endocrine function of the postmenopausal ovary: Concentrations of androgens and estrogens in ovarian and peripheral vein blood. *J. Clin. Endocrinol. Metab.* **39,** 1020–1024.

Karavolas, H. J., Bertics, P. J., Hodges, D., and Rudie, N. (1984). Progesterone processing by neuroendocrine structures. In "Metabolism of Hormonal Steroids in the Neuroendocrine Structures" (F. Celotti, ed.), pp. 149–170. Raven Press, New York.

Kaufert, P. A., Gilbert, P., and Tate, R. (1992). The Manitoba Project: A re-examination of the relationship between menopause and depression. *Maturitas* **14,** 143–156.

Kawas, C., Resnick, S., Morrison, A., Brookmeyer, R., Corrada, M., Zonderman, A., Bacal, C., Lingle, D. D., and Metter, E. (1997). A prospective study of estrogen replacement therapy and the risk of developing Alzheimer's disease: The Baltimore Longitudinal Study of Aging. *Neurology* **48,** 1517–1521.

Kelly, M. J., and Wagner, E. J. (1999). Estrogen modulation of G-protein-coupled receptors. *Trends Endocrinol. Metab.* **10,** 369–374.

Kirkham, C., Hahn, P. M., Van Vugt, D. A., Carmichael, J. A., and Reid, R. L. (1991). A randomized, double-blind, placebo-controlled, cross-over trial to assess the side effects of medroxyprogesterone acetate in hormone replacement therapy. *Obstet. Gynecol.* **78,** 93–97.

Klaiber, E. L., Broverman, D. M., Vogel, W., and Kobayashi, Y. (1979). Estrogen replacement therapy for severe persistent depression in women. *Arch. Gen. Psychiatry* **36,** 550–554.

Klaiber, E. L, Broverman, D. M., Vogel, W., Peterson, L. G., and Snyder, M. B. (1997). Relationships of serum estradiol levels, menopausal duration and mood during hormonal replacement therapy. *Psychoneuroendocrinology* **22,** 549–558.

Kompoliti, K. (1999). Estrogen and movement disorders. *Clin. Neuropharmacol.* **22,** 318–326.

Korkmaz, S., and Wahlström, G. (1997). The EEG burst suppression threshold test for the determination of CNS sensitivity to intravenous anaesthetics in rats. *Brain Research (Protocols)* **1,** 378–384.

Korzekwa, M. I., and Steiner, M. (1997). Premenstrual syndromes. *Clin. Obstet. Gynecol.* **40,** 564–576.

Kraulis, I., Foldes, G., Traikov, H., Dubrovsky, B., and Birmingham, M. K. (1975). Distribution, metabolism and biological activity of deoxycorticosterone in the central nervous system. *Brain Res.* **88,** 1–14.

Laidlaw, J. (1956). Catamenial epilepsy. *Lancet* **271,** 1235–1237.

Lan, N. C., Gee, K.W., Bolger, M. B., and Chen, J. S. (1991). Differential responses of expressed recombinant human γ-aminobutyric acid A receptors to neurosteroids. *J. Neurochem.* **57,** 1818–1821.

Landgren, S., Bäckström, T., and Kalistratov, G. (1978). The effect of progesterone on the spontaneous interictal spike evoked by topical application of penicillin to the cat's cerebral cortex. *J. Neurol. Sci.* **36,** 119–133.

Landgren, S., Aasly, J., Bäckström, T., Dubrowsky, B., and Danielsson, E. (1987). The effect of progesterone and its metabolites on the interictal epileptiform discharge in the cat's cortex. *Acta Physiol. Scand.* **131,** 33–42.

Levin, E. R. (1999). Cellular functions of the plasma membrane estrogen receptor. *Trends Endocrinol. Metab.* **10,** 374–377.

Livingstone, S. (1966). Drug Therapy. Charles C Thomas, Springfield, IL.

Löfgren, M., Bäckström, T., and Joelsson, I. (1988). Decrease in serum concentration of 5α-pregnane-3,20-dione prior to spontaneous labour. *Acta Obstet. Gynecol. Scand.* **67,** 467–470.

Logothetis, J., Harner, R., Morrell, F., and Torres, F. (1959). The role of estrogens in catamenial exacerbation of epilepsy. *Neurology* **9,** 352–360.

MacNaughton, J., Bangah, M., McCloud, P., Hee, J., and Burger, H. G. (1992). Age related changes in follicle stimulating hormone, luteinizing hormone, estradiol and imunoreactive inhibin in women of reproductive age. *Clin. Endocrinol.* **36,** 339–345.

Magos, A. L., Brewster, E., Sing, R., O'Dowd, T. M., and Studd, J. W. (1986). The effect of norethisterone in postmenopausal women on oestrogen therapy: A model for the premenstrual syndrome. *Br. J. Obstet. Gynaecol.* **93,** 1290–1296.

Magos, A. L., Brincat, M., and Studd, J. W. (1986). Treatment of the premenstrual syndrome by subcutaneous estradiol implants and cyclical oral norethisterone: Placebo controlled study. *Br. Med. J.* **292,** 1629–1633.

Marcus, E. M., Watson, C. W., and Simon, S. A. (1968). An experimental model of some varieties of petit mal epilepsy. Electrical-behavioral correlations of acute bilateral epileptogenic foci in cerebral cortex. *Epilepsia* **9,** 233–248.

Marder, K., and Sano, M. (2000). Estrogen to treat Alzheimer's disease: too little, too late? So what's a woman to do? *Neurology* **54,** 2036–2037.

Mattson, R., Cramer, J., Caldwell, B., and Siconolfi, B. (1984). Treatment of seizures with medroxyprogesterone acetate: Preliminary report. *Neurology* **34,** 1255–1258.

Mattson, R., and Cramer, J. (1985). Epilepsy, sex hormones and antiepileptic drugs. *Epilepsia* **26** (Suppl. 1), S40–S51.

Mattson, R. H., Kamer, J. A., Caldwell, B. V., and Cramer, J. A. (1981). Seizure frequency and the menstrual cycle: a clinical study [Abstract]. *Epilepsia* **22,** 242.

McEwen, B. S., Gould, E., Orchinik, M., Weiland, N. G., and Woolley, C. S. (1995). Oestrogens and the structural and functional plasticity of neurons: Implications for memory, ageing and neurodegenerative processes. *Ciba Foundation Symposium* **191,** 52–66.

McEwen, B. S., Alves, S. E., Bulloch, K., and Weiland, N. G. (1997). Ovarian steroids and the brain: implications for cognition and aging. *Neurology* **48** (Suppl 7), S8–S15.

McEwen, B., and Wooley, C. S. (1994). Estradiol and progesterone regulate neuronal structure and synaptic connectivity in adult as well as developing brain. *Exp. Gerontol.* **29,** 431–436.

McKinlay, J. B., McKinlay, S. M., and Brambilla, D. (1987). The relative contributions of endocrine changes and social circumstances to depression in mid-aged women. *J. Health Soc. Behav.* **28,** 345–363.

McKinlay, S. M., Brambilla, D. J., and Posner, J. G. (1992). The normal menopause transition. *Maturitas* **14,** 103–115.

Melcangi, R. C, Celotti, F., Ballabio, M., Castano, P., Massarelli, R., Poletti, A., and Martini, L. (1990). 5α-Reductase activity in isolated and cultured neuronal and glial cells of the rat. *Brain Res.* **516,** 229–236.

Metcalf, M. G . (1983). Incidence of ovulation from the menarche to the menopause. *NZ Med. J.* **96,** 645–648.

Mezrow, G., Lobo, R., Shoupe, D., Leung, B., Spicer, D., and Pike, M. (1994). Depot leuprolide acetate with estrogen and progestin add-back for long-term treatment of premenstrual syndrome. *Fertil. Steril.* **62,** 932–937.

Milsom, I., Sundell, G., and Andersch, B. (1991). A longitudinal study of contraception and pregnancy outcome in a reprecentative sample of young Swedeish women. *Contraception* **43,** 111–119.

Molnar, G. W. (1975). Body temperature during menopausal hot flashes. *J. Appl. Physiol.* **1,** 499–503.

Monroe, S. E., Jaffe, R. B., and Midgley, A. R. (1972). Regulation of human gonadotropins. XIII. Changes in serum gonadotropins in menstruating women in response to oophorectomy. *J. Clin. Endocrinol. Metab.* **34,** 420–422.

Montgomery, J. C., Appelby, L., Brincat, M., Versi, E., Tapp, A., Ferwick, P. B., and Studd, J. W. (1987). Effect of oestrogen and testosterone implants on psychological disorders in the climacteric. *Lancet* **7,** 297–299.

Morrow, A. L., Devaud, L. L., Purdy, R. H., and Paul, S. M. (1995). Neuroactive steroid modulators of the stress response. *Ann. NY Acad. Sci.* **771,** 257–272.

Mortola, J. F., Girton, L., and Fischer, U. (1991). Successful treatment of severe premenstrual syndrome by combined use of gonadotropin-releasing hormone agonist and estrogen/progestin. *J. Clin. Endocrinol. Metab.* **71,** 252A–252F.

Moss, R. L., and Gu, Q. (1999). Estrogen: Mechanisms for a rapid action in CA1 hippocampal neurons. *Steroids* **64,** 14–21.

Motta, E., and Rosciszewska, D. (1980). The effect of gonadal hormones in the treatment of epilepsy. *Proceedings of XIV Meeting of Polish neurological association*. Bydgoszcz: Abstract book; 121.

Mulnard, R., Cotman, C. W., Kawas, C., van Dyck, C. H., Sano, M., Doody, R., Koss, E., Pfeiffer, E., Jin, S., Garnst, A., Grundman, M., Thomas, R., and Thal, L. J. (2000). Estrogen replacement therapy for treatment of mild to moderate Alzheimer's disease: A 1-year randomized controlled trial. *JAMA* **283,** 1007–1015.

Munari, C., Casaroli, D., Matteuzzi, G., and Pacifico, L. (1979). The use of Althesin in drug-resistant status epilepticus. *Epilepsia* **20,** 475–484.

Murphy, D. D., and Segal, M. (1996). Regulation of dendritic spine density in cultured rat hippocampal neurons by steroid hormones. *J. Neurosci.* **16,** 4059–4068.

Naessén, T., Persson, J., Adami, H., Bergström, R., and Bergkvist, L. (1990). Hormone replacement therapy and the risk for first hip fracture. *Ann. Intern. Med.* **113,** 95–103.

Ohkura, T., Isse, K., Akazawa, K., Hamamoto, M., Yaoi, Y., and Hagino, N. (1994). Low-dose estrogen replacement therapy for Alzheimer-disease in women. *Menopause* **1,** 125–130.

Ohkura, T., Isse, K., Akazawa, K., Hamamoto, M., Yaoi, Y., and Hagino, N. (1995). Long-term estrogen replacement therapy in female-patients with dementia of the alzheimer-type-7 case-reports. *Dementia* **6,** 99–107.

Paganini-Hill, A., and Henderson, V. W. (1994). Estrogen deficiency and risk of Alzheimer's disease in women. *Am. J. Epidemiol.* **140,** 256–261.

Paganini-Hill, A. (1996). Oestrogen replacement therapy and Alzheimer's disease. *Br. J. Obstet. Gynaecol.* **103** (Suppl 13), 80–86.

Palacios, S., Cifuentes, I., Menendez, C., and von Helde, S. (2000). The central nervous system and HRT. *Int. J. Fertil. Womens Med.* **45,** 13–21.

Patkai, P., Johannson, G., and Post, B. (1974). Mood, alertness and sympathetic-adrenal medullary activity during the menstrual cycle. *Psychosom. Med.* **36,** 503–512.

Pau, K. Y., and Spies, H. G. (1998). Putative estrogen receptor beta and alpha mRNA expression in male and female Rhesus macaques. *Mol. Cell. Endocrinol.* **25,** 59–68.

Paul, S. M., and Purdy, R. H. (1992). Neuroactive steroids. *FASEB J.* **6,** 2311–2322.
Persson, I., Adami, H. O., Bergkvist, L., Lindgren, A., Pettersson, B., Hoover, R., and Schairer, C. (1989). Risk of endometrial cancer after treatment with oestrogens alone or in conjunction with progestogens: Results of a prospective study. *Br. Med. J.* **298,** 147–151.
Phillips, S. M., and Sherwin, B. B. (1992a). Effects of estrogen on memory function in surgically menopausal women. *Psychoneuroendocrinology* **17,** 485–495.
Phillips, S. M., and Sherwin, B.B. (1992b). Variations in memory functions and sex steroid hormones across the menstrual cycle. *Psychoneuroendocrinology* **17,** 497–506.
Pickerodt, V., McDowal, D. G, Coroneos, N. J., and Keaney, N. P. (1972). Effect of Althesin on cerebral perfusion, cerebral metabolism and intracranial pressure in anestesized baboon. *Br. J. Anaesth.* **44,** 751–757.
Prior, J. C., Alojado, N., McKay, D. W., and Vigna, Y. M. (1994). No adverse effects of medroxyprogesterone treatment without estrogen in pastmenopausal women: Double blind, placebo-controlled, crossover trial. *Obstet. Gynecol.* **83,** 24–28.
Puia, G., Santi, M., Vicini, S., Pritchett, D. B., Purdy, R. H., Paul, S. M., Seeburg, P. H., and Costa, E. (1990). Neurosteroids act on recombinant human $GABA_A$ receptors. *Neuron* **4,** 759–765.
Purdy, R. H., Moore, P. H., Jr., Rao, P. N., Hagino, N., Yamaguchi, T., Schmidt, P., Rubinow, D. R., Morrow, A. L., and Paul, S. M. (1990). Radioimmunoassay of 3α-hydroxy-5α-pregnan-20-one in rat and human plasma. *Steroids* **55,** 290–296.
Purdy, R. H., Morrow, A. L., Moore, P. H., Jr., and Paul, S. M. (1991). Stress-induced elevations of gamma-aminobutyric acid type A receptor-active steroids in the rat brain. *Proc. Nat. Acad. Sci. USA* **88,** 4553–4557.
Rapkin, A. J., Morgan, M., Goldman, L., Brann, D.W., Simone, D., and Mahesh, V. B. (1997). Progesterone metabolite allopregnanolone in women with premenstrual syndrome. *Obstet. Gynecol.* **90,** 709–714.
Rauramo, L., Lagerspetz, K., Engholm, P., and Punnonen, R. (1975). The effect of castration and peroral estrogen therapy on some psychological functions. *Front. Horm. Res.* **3,** 94–104.
Register, T. C., Shively, C. A., and Lewis, C. E. (1998). Expression of estrogen receptor alpha and beta transcripts in female monkey hippocampus and hypothalamus. *Brain Res.* **30,** 320–322.
Resnick, S. M., Metter, E. J., and Zonderman, A. B. (1997). Estrogen replacement therapy and longitudinal decline in visual memory. A possible protective effect? *Neurology* **49,** 1491–1497.
Richardson, S. J., Senikas, V., and Nelson, J. F. (1987). Follicular depletion during the menopausal transition: evidence for accelerated loss and ultimate exhaustion. *J. Clin. Endocrinol. Metab.* **65,** 1231–1237.
Robel, P., Bourreau, E., Corpéchot, C., Dang, D. C., Halberg, F., Clarke, C., Haug, M., Schlegel, M. L., Synguelakis, M., Vourch, C., and Baulieu, E. E. (1987). Neuro-steroids: 3α-Hydroxy-Δ_5-derivatives in rat and monkey brain. *J. Steroid Biochem.* **27,** 649–655.
Robel, P., and Baulieu, E. E. (1995). Neurosteroids: Biosynthesis and function. *Critical Reviews in Neurobiology* **9,** 383–394.
Rosciszewska, D. (1975). Clinical course of epilepsy in women dependent on puberty, maturity, and climacterium. *Neurol. Neurochir. Pol.* **9,** 217–222.
Rosciszewska, D. (1980). Analysis of seizures dispersion during menstrual cycle in women with epilepsy. *In* "Epilepsy. A Clinical and Experimental Research" (J. Majkowaki, Ed.), pp. 280–284. Basel: Karger.
Rosciszewska, D., Buntner, B., Guz, I., and Zawisza, L. (1986). Ovarian hormones, anticonvulsant drugs, and seizures during the menstrual cycle in women with epilepsy. *J. Neurol. Neurosurg. Psychiatry* **149,** 47–51.

Rubinow, D. R., Hoban, M. C., Grover, G. N., Galloway, D. S., Roy-Burne, P., Andersen, R., and Merriam, G. R. (1988). Changes in plasma hormones across the menstrual cycle in patients with menstrually related mood disorder and in control subjects. *Am. J. Obstet. Gynecol.* **158**, 5–11.

Schmidt, P. J., Purdy, R. H., Moore, P. H., Paul, S. M., and Rubinow, D. R. (1994). Circulating levels of anxiolytic steroids in the luteal phase in women with premenstrual syndrome and in control subjects. *J. Clin. Endocrinol. Metab.* **79**, 1256–1260.

Schmidt, R., Fazekas, F., Reinhart, B., Kapeller, P., Fazekas, G., Offenbacher, H., Eber, B., Schumacher, M., and Freidl, W. (1996). Estrogen replacement therapy in older women: Androstenedione neuropsychological and brain MRI study. *J. Am. Geriatr. Soc.* **44**, 1307–1313.

Seippel, L., and Bäckström, T. (1997). Luteal phase estradiol relates to symptom severity between patients with premenstrual syndrome. *J. Clin. Endocrinol. Metab.* **83**, 1988–1992.

Selye, H. (1942). Correlations between the chemical structure and the pharmacological actions of the steroids. *Endocrinology* **30**, 437–453.

Sherman, B. W., and Korenman, S. G. (1975). Hormonal characteristics of the human menstrual cycle throughout reproductive life. *J. Clin. Invest.* **55**, 699–706.

Sherman, B. W., West, J. H., and Korenman, S. G. (1976). The menopausal transition: Analysis of LH, FSH, estradiol and progesterone concentrations during menstrual cycles of older women. *J. Clin. Endocrinol. Metab.* **42**, 629–636.

Sherwin, B. B., and Gelfand, M. M. (1985). Sex steroids and affect in the surgical menopause: A double-blind cross-over study. *Psychoneuroendocrinology* **10**, 325–335.

Sherwin, B. B. (1988). Estrogen and/or androgen replacement therapy and cognitive functioning in surgically menopausal women. *Psychoneuroendocrinology* **13**, 345–357.

Sherwin, B. B. (1991). The impact of different doses of estrogen and progestin on mood and sexual behavior in postmenopausal women. *J. Clin. Endocrinol. Metab.* **72**, 336–343.

Sherwin, B. B. (1994). Estrogenic effects on memory in women. *Ann. NY Acad. Sci.* **743**, 213–230.

Sherwin, B. B. (1997). Estrogen effects on cognition in menopausal women. *Neurology* **48**, 21–26.

Shughrue, P. J., and Merchenthaler, I. (2000). Estrogen is more than just a "sex hormone": Novel sites for estrogen action in the hippocampus and cerebral cortex. *Front. Neuroendocrinol.* **21**, 95–101.

Skoog, I., and Gustafson, D. (1999). HRT and dementia. *J. Epidemiol. Biostat.* **4**, 227–251.

Smith, R. N., Holland, E. F., and Studd, J. W. (1994). The symptomatology of progestogen intolerance. *Maturitas* **18**, 87–91.

Smith, S. S. (1989). Estradiol administration increases neuronal responses to excitatory amino acid as a long-term effect. *Brain Res.* **503**, 354–357.

Södergård, R., Bäckström, T., Shanbagh, V., and Carstensen, H. (1982). Calculation of free and protein bound fractions of testosterone and estradiol-17β to human plasma proteins at body temperature. *J. Steroid Biochem.* **16**, 801–816.

Spiegel, E., and Wycis, H. (1945). Anticonvulsant effects of steroids. *J. Lab. Clin. Med.* **30**, 947–953.

Sundblad, C., Modigh, K., Andersch, B., *et al.* (1992). Clomipramine effectively reduces premenstrual irritability and dysphoria: A placebo-controlled trial. *Acta Psychiatr. Scand.* **85**, 39–47.

Sundström, I., Nybergm, S., and Bäckström, T. (1997a). Patients with premenstrual syndrome have reduced sensitivity to midazolam compared to control subjects. *Neuropsychopharmacology* **17**, 370–381.

Sundström, I, Aschbrook, D., and Bäckström, T. (1997b). Reduced benzodiazepine sensitivity in patients with premenstrual syndrome, a pilot study. *Psychoneuroendocrinology* **22**, 25–38.

Sundström, I., and Bäckström, T. (1998). Patients with premenstrual syndrome have decreased saccadic eye velocity compared to control subjects. *Biol. Psychiatry* **44,** 755–764.

Sundström, I., Andersson, A., Nyberg, S., Ashbrook, D., Purdy, R. H., and Bäckström, T. (1998). Patients with premenstrual syndrome have a different sensitivity to a neuroactive steroid during the menstrual cycle compared to control subjects. *Neuroendocrinology* **67,** 126–138.

Sveindottir, H., and Bäckström, T. (2000). Prevalence of menstrual cycle symptom cyclicity and premenstrual dysphoric disorder in a random sample of women using and not using oral contraceptives. *Acta Obstet. Gyn. Scand.* **79,** 405–413.

Tang, M. X., Jacobs, D., Stern, Y., Marde, K., Schofield, P., Gurland, B., Andrews, H., and Mayeux, R. (1996). Effect of oestrogen during menopause on risk and age at onset of Alzheimer disease. *Lancet* **348,** 429–432.

Thompson, J., and Oswald, I. (1977). Effect of oestrogen on the sleep, mood and anxiety of menopausal women. *Br. Med. J.* **2,** 1317–1319.

Timiras, P. S., and Hill, H. F. (1980). Hormones and epilepsy. *In* "Antiepileptic Drugs, Mechanism of Action: Advances in Neurology" (G. H. Glaser, J. K. Penry, and D. M. Woodbury, Eds.), Vol. 27, pp. 655–666. Raven Press, New York.

Toone, B. (1986). Hyposexuality among male epileptic patients: Clinical and hormonal correlates. *In* "Aspects of Epilepsy and Psychiatry" (M. R. Trimble and T. G. Bolwig, Eds.), pp. 61–74. John Wiley and Son Ltd, London.

Turner, J. M., Coroneos, N. Y., Gibson, R. M., Powell, D., Ness, M. A., and McDowall, D. G. (1973). The effect of Althesin on intracranial pressure in man. *Br. J. Anaesth.* **45,** 186–172.

van Duijn, C.M. (1999). Hormone replacement therapy and Alzheimer's disease. *Maturitas* **31,** 201–205.

Vanhulle, R., and Demol, R. (1976). A double-blind study into the influence of estriol on a number of psychological tests in post-menopausal women. *In* "Consensus on Menopause Research" (P. A. van Keep, Ed.), pp. 94–99. Lancaster, England: MTP.

Vanselow, W., Dennerstein, L., Greenwood, K.M., and De Lignieres, B. (1996). Effect of progesterone and its 5α- and 5β-metabolites on symptoms of premenstrual syndrome according to route of administration. *J. Psychosom. Obstet. Gynaecol.* **17,** 29–38.

Verandakis, A., and Woodbury, D. (1963). Effect of cortisol on electroshock seizure threshold in developing rats. *J. Pharmacol. Exp. Ther.* **139,** 110–113.

Walker, A., and Bancroft, J. (1990). Relationship between premenstrual symptoms and oral contraceptive use: A controlled study. *Psychosom. Med.* **52,** 86–96.

Wang, M. D, Landgren, S., and Bäckström, T. (1997). The effects of allopregnanolone, pregnenolone sulphate and pregnenolone on the CA1 population spike of the rat hippocampus after 17β-oestradiol priming. *Acta Physiol. Scand.* **159,** 343–344.

Wang, M. D., Seippel, L., Purdy, R. H., and Bäckström, T. (1996). Relationship between symptom severity and steroid variation in women with premenstrual syndrome: Study on serum pregnenolone, pregnenolone sulfate, 5α-pregnan-3, 20-dione and 3α-hydroxy-5α-pregnan-20-one. *J. Clin. Endocrinol. Metab.* **81,** 1076–1082.

Wang, M. D., Wahlström, G., Gee, K. W., and Bäckström, T. (1995). Potency of lipid and protein formulation of 5α-pregnanolone at induction of anaesthesia and the corresponding regional brain distribution. *Br. J. Anaesth.* **74,** 553–557.

Wang, P. N., Liao, S. Q., Liu, R. S., Liu, C. Y., Chao, H. T., Lu, S. R., Yu, H. Y., Wang, S. J., and Liu, H. C. (2000). Effects of estrogen on cognition, mood, and cerebral blood flow in AD: A controlled study. *Neurology* **54,** 2061–2066.

Waring, S. C., Rocca, W. A., Petersen, R. C., O'Brien, P. C., Tangalos, E. G., and Kokmen, E. (1999). Postmenopausal estrogen replacement therapy and risk of AD: A population-based study. *Neurology* **52,** 965–970.

Werboff, L. H., and Havlena, J. (1969). Audiogenic seizures in adult male rats treated with various hormones. *Gen. Comp. Endocrinol.* **3,** 389–397.

West, C. P. (1990). Inhibition of ovulation with oral progestins—effectiveness in premenstrual syndrome. *Eur J. Obstet. Gynecol. Reprod. Biol.* **34,** 119–128.

Wong, M., and Moss, R. L. (1994). Patch–clamp analysis of direct steroidal modulation of glutamate receptor-channels. *J. Neuroendocrinol.* **6,** 347–355.

Wong, M., Thompson, T.L., and Moss, R.L. (1996). Nongenomic actions of estrogen in the brain: Physiological significance and cellular mechanisms. *Critical Reviews in Neurobiology* **10,** 189–203.

Wooley, C. S., and McEwen, B. S. (1992). Estradiol mediates fluctuation in hippocampal synapse density during the estrous cycle in the adult rat. *J. Neurosci.* **12,** 2549–2554.

Woolley, C. S., and McEwen, B. S. (1993). Roles of estradiol and progesterone in regulation of hippocampal dendritic spine density during the estrous cycle in the rat. *J. Comp. Neurol.* **336,** 293–306.

Woolley, C. S., and McEwen, B. S. (1994). Estradiol regulates hippocampal dendritic spine density via an N-methyl-D-aspartate receptor-dependent mechanism. *J. Neurosci.* **14,** 7680–7687.

Woolley, C. S., Weiland, N. G., McEwen, B. S., and Schwartzkroin, P. A. (1997). Estradiol increases the sensitivity of hippocampal CA1 pyramidal cells to NMDA receptor-mediated synaptic input: Correlation with dendritic spine density. *J. Neurosci.* **17,** 1848–1859.

Xu, H., Sweeney, D., Wang, R., Thinakaran, G., Lo, A. C., Sisodia, S. S., Greengard, P., and Gandy, S. (1997). Generation of Alzheimer beta-amyloid protein in the trans-Golgi network in the apparent absence of vesicle formation. *Proc. Natl. Acad. Sci. USA* **94,** 3748–3752.

Yaffe, K., Sawaya, G., Lieberburg, I., and Grady, D. (1998). Estrogen therapy in postmenopausal women: effects on cognitive function and dementia. *JAMA* **279,** 688–695.

Yaffe, K., Haan, M., Byers, A., Tangen, C., and Kuller, L. (2000). Estrogen use, APOE, and cognitive decline: Evidence of gene–environment interaction. *Neurology* **54,** 1949–1954.

Yen, S. S., Tsai, C. C., Naftolin, F., Vanderberg, G., and Ajabar, L. (1972). Pulsatile patterns of gonadotropin release in subjects with and without ovarian function. *J. Clin. Endocrinol Metab.* **34,** 671–675.

Ziel, H. K., and Finkle, W. D. (1975). Increased risk of endometrial carcinoma among users of conjugated estrogens. *N. Engl. J. Med.* **293,** 1167–1170.

Zimmerman, A. W., Holden, K. R., Reiter, E. O., and Dekaban, A. S. (1973). Medroxyprogesterone acetate in the treatment of seizures associated with menstruation. *J. Pediatr.* **83,** 959–963.

Zimmerman, E., and Parlee, M. B. (1973). Behavioral changes associated with the menstrual cycle: An experimental investigation. *J. Appl. Soc. Psychol.* **3,** 335–344.

Zweifel, J. E., and O'Brien, W. H. (1997). A meta-analysis of the effect of hormone replacement therapy upon depressed mood. *Psychoneuroendocrinology* **22,** 189–212.

NEUROACTIVE STEROIDS IN NEUROPSYCHOPHARMACOLOGY

Rainer Rupprecht*,[1] and Florian Holsboer[†]

*Department of Psychiatry, Ludwig-Maximilians-University of Munich
80336 Munich, Germany; and †Max-Planck-Institute of Psychiatry
80804 Munich, Germany

I. Introduction
II. Sources and Biosynthesis of Neuroactive Steroids
III. Steroid Modulation of $GABA_A$ Receptors
IV. Steroid Modulation of Other Neurotransmitter Receptors
V. A Putative Specific Steroid-Binding Site on Ligand-Gated Ion Channels
VI. Genomic Effects of Neuroactive Steroids
VII. Neuropsychopharmacological Properties of Neuroactive Steroids
VIII. Modulation of Endogenous Neuroactive Steroids as a Pharmacological Principle
IX. Outlook
    References

Steroids influence neuronal function through binding to intracellular receptors, which may act as transcription factors in the regulation of gene expression. In addition, certain so-called neuroactive steroids are potent modulators of an array of ligand-gated ion channels and of distinct G-protein-coupled receptors via nongenomic mechanisms. Neuroactive steroids may modulate an array of neurotransmitter receptors and regulate gene expression. This intracellular cross-talk between genomic and nongenomic steroid effects provides the basis for their neuropsychopharmacological potential with regard to both clinical effects and side effects. These compounds may influence sleep and memory. Moreover, they may play a role in the response to stress and the treatment of neuropsychiatric disorders, such as epilepsy, depression, and anxiety disorders. Neuroactive steroids affect a broad spectrum of behavioral functions through their unique molecular properties and may constitute an unexploited class of drugs. However, particular attention must be drawn to putative side effects that are inherent to their molecular diversity. Moreover, it must be determined whether synthetic steroid compounds really offer an advantage over already known drugs and whether the modulation of endogenous neuroactive steroids might constitute a useful alternative strategy for pharmacological intervention. © 2001 Academic Press.

---

[1]Author to whom correspondence should be sent.

I. Introduction

Steroid hormone action involves binding of the steroids to their respective intracellular receptors (Evans, 1988; Truss and Beato, 1993). These receptors subsequently change their conformation by dissociation from chaperone molecules (e.g., the heat shock proteins) and translocate to the nucleus, where they bind as homo- or heterodimers to the respective response elements that are located in the regulatory regions of target promoters. Alternatively, these ligand-activated receptors may influence transactivation through protein–protein interaction with other transcription factors. Thus, steroid hormone receptors act as transcription factors in the regulation of gene expression (Evans, 1988; Truss and Beato, 1993; Rupprecht, 1997). Since the 1990s, considerable evidence has emerged that certain steroids may alter neuronal excitability via the cell surface through interaction with certain neurotransmitter receptors (Majewska et al., 1986; Paul and Purdy, 1992; Lambert et al., 1995; Rupprecht, 1997). The term neuroactive steroids has been coined for steroids with these particular properties (Paul and Purdy, 1992). Although the action of steroids at the genome requires a time period from minutes to hours that is limited by the rate of protein biosynthesis (McEwen, 1991), the modulatory effects of neuroactive steroids are fast-occurring events requiring only milliseconds to seconds (McEwen, 1991). Thus, genomic and nongenomic steroid effects within the central nervous system provide the molecular basis for a broad spectrum of steroid action on neuronal function and plasticity. The molecular mechanisms of neuroactive steroid action and the neuropsychopharmacological potential based on their molecular profile have been reviewed (Rupprecht and Holsboer, 1999).

A major issue for pharmacological steroid research is the issue of specificity. It must be determined whether steroids are really specific for distinct intracellular receptors or distinct neurotransmitter receptors. A further question is whether certain steroids act via either the genomic or the nongenomic pathway. It has long been believed that steroid hormones act exclusively through the classical genomic pathway, whereas certain neuroactive steroids that do not bind to known steroid receptors [e.g., 3α-reduced metabolites of progesterone and deoxycorticosterone, such as $3\alpha, 5\alpha$-tetrahydroprogesterone ($3\alpha,5\alpha$-TH PROG; 3α-hydroxy-5α-pregnan-20-one; allopregnanolone) and $3\alpha,5\alpha$-tetrahydrodeoxycorticosterone ($3\alpha,5\alpha$-TH DOC; $3\alpha,21$-dihydroxy-5α-pregnan-20-one; allotetrahydrodeoxycorticosterone)], pregnenolone sulfate (PREGS), or dehydroepiandrosterone sulfate (DHEAS) are allosteric modulators of specific neurotransmitter receptors, such as γ-aminobutyric acid type A (GABA$_A$) receptors (Evans, 1988; Paul and Purdy, 1992). This conceptual view has changed with the identification of binding sites for classical steroid hormones (e.g., progesterone) (Ramirez and Zheng, 1996), estradiol (Pappas *et al.*, 1995; Ramirez and Zheng, 1996), testosterone (Ramirez and Zheng, 1996), glucocorticoids (Orchinik *et al.*, 1991), and aldosterone (Wehling, 1997) at membranes of cells or tissues and with the identification of a large number of signal transduction pathways involved in steroid hormone action (Wehling, 1997). In addition, a variety of steroid hormones have been identified that interact with different neurotransmitter receptors and thus also need to be defined as neuroactive steroids. Moreover, neuroactive steroids that do not bind to steroid receptors themselves may nevertheless influence gene expression via steroid receptors through the generation of metabolites that are ligands of steroid receptors (Rupprecht *et al.*, 1993; Rupprecht, 1997). However, effects of steroids at the genomic level may also influence neuronal excitability by interfering with the expression patterns of neurotransmitter receptor subunits (Weiland and Orchinik, 1995) and voltage-gated ion channels (Zakon, 1998). Moreover, the modulation of ligand-gated ion channels or G-protein-coupled receptors by steroids may alter the activity of intracellular kinases, which consequently affects the expression patterns of downstream genes [e.g., via the cyclic adenosine monophosphate (cAMP) protein kinase A–cAMP-reponsive element binding protein pathway] (Wehling, 1997; Zakon, 1998). A further scenario for nongenomic steroid effects that are independent of neurotransmitter receptors is provided by the neuroprotective effects of estrogens in that these steroids may themselves serve as scavengers for free radicals (Behl *et al.*, 1997). Meanwhile, it is realized that steroid effects in the brain are highly complex and cannot be attributed to a single mode of action.

II. Sources and Biosynthesis of Neuroactive Steroids

Owing to the lipophilic nature of steroids produced in various endocrine organs, they can easily cross the blood–brain barrier. However, a variety of neuroactive steroids may be synthesized in the brain itself without the aid of peripheral sources (Baulieu, 1998). These steroids that are formed within the brain from cholesterol are often defined as neurosteroids (Baulieu, 1998). Direct evidence for steroid synthesis within the brain has come from experiments demonstrating the formation of steroids in rat glial cells following treatment of these cultures with radioactive precursors and derivatives of cholesterol (Rupprecht, 1997; Baulieu, 1998). The biosynthesis of steroids in the brain and the expression of the enzymes involved is reviewed in detail elsewhere (Mellon, 1994; Baulieu, 1998). Progesterone may be formed from pregnenolone by the 3β-hydroxysteroid dehydrogenase/Δ^5–Δ^4-isomerase.

The 5α-reductase catalyzes the reduction of progesterone and deoxycorticosterone into the 5α-pregnane steroids 5α-dihydroprogesterone (5α-DH PROG) and 5α-dihydrodeoxycorticosterone (5α-DH DOC), respectively, and the 5β-reductase reduces progesterone to 5β-dihydroprogesterone (5β-DH PROG). These reactions are irreversible in mammalian cells (Celotti *et al.*, 1992). The pregnane steroids may be further reduced to the neuroactive steroids 3α,5α-TH PROG, 3α,5β-tetrahydroprogesterone (3α,5β-TH PROG; 3α-hydroxy-5β-pregnan-20-one; pregnanolone) and 3α,5α-TH DOC by the 3α-hydroxysteroid oxidoreductase (Rupprecht, 1997). This reaction may work both in the reductive and in the oxidative direction depending on the co-factors present in the environment (Rupprecht *et al.*, 1993). Pregnenolone is also a precursor for dehydroepiandrosterone (DHEA). These two steroids also exist as conjugated sulfate esters (e.g., PREGS and DHEAS) and fatty acid esters at concentrations frequently exceeding those of the free steroids (Baulieu, 1998). Both progesterone and DHEA are converted to androstenedione, which is a precursor of testosterone. Estradiol is formed by the aromatase either from testosterone or from androstenedione via estrone (Fig. 1). It must be considered that the formation of steroids in the brain

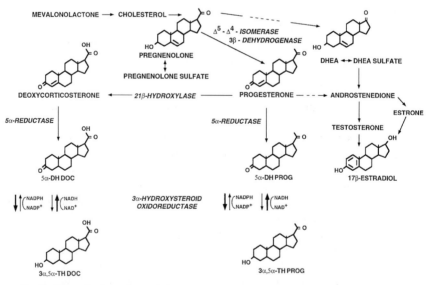

FIG. 1. Biosynthesis and metabolism of neuroactive steroids. (Reprinted from *Trends in Neurosciences*, Vol. 22: Rupprecht, R., and Holsboer, F. Neuroactive steroids: Mechanisms of action and neuropsychopharmacological perspectives, pp. 410–416. Copyright © 1999, with permission from Elsevier Science.)

(e.g., of estradiol via the aromatase pathway) may be rapidly regulated by steroids and neurotransmitters (Balthazart and Ball, 1998), which in turn affects steroid concentrations and consequently neurotransmitter receptor function.

III. Steroid Modulation of GABA$_A$ Receptors

In 1986, it was shown for the first time that the neuroactive steroids $3\alpha,5\alpha$-TH PROG and $3\alpha,5\alpha$-TH DOC may modulate neuronal excitability via their interaction with GABA$_A$ receptors (Majewska et al., 1986). GABA$_A$ receptors consist of various subunits that form ligand-gated ion channels with considerable homology to glycine, nicotinic acetylcholine, and serotonin type 3 (5-HT$_3$) receptors (Paul and Purdy, 1992; Lambert et al., 1995; Wetzel et al., 1998). The steroids $3\alpha,5\alpha$-TH PROG and $3\alpha,5\alpha$-TH DOC were able not only to displace t-butylbicyclophosphorothionate from the choride channel but also to enhance the GABA-evoked chloride current (Majewska et al., 1986) (Fig. 2). In addition, these steroids may enhance the binding of muscimol and benzodiazepines to GABA$_A$ receptors (Paul and Purdy, 1992). There is considerable evidence that these neuroactive steroids are potent positive allosteric modulators of GABA$_A$ receptors because they increase the frequency and/or duration of openings of the GABA-gated chloride channel (Paul and Purdy, 1992; Lambert et al., 1995). However, very high concentrations in the micromolar range of these neuroactive steroids have been shown to exert a certain intrinsic agonistic activity in the absence of GABA (Puia et al., 1990). It has been demonstrated that pharmacological activities of benzodiazepines at GABA-gated ion channels vary with the α-subunit composition and require the presence of a γ-subunit. Such strictly defined prerequisites are not necessary for the actions of neuroactive steroids (Puia et al., 1990). Nevertheless, differences across species exist as Drosophila GABA$_A$ receptors, unlike mammalian GABA$_A$ receptors, which are almost insensitive to steroid modulation (Chen et al., 1994). Numerous studies concerning the structure–activity relationship of neuroactive steroids at GABA$_A$ receptors revealed a stereoselectivity and the presence of a 3α-hydroxy group within the A-ring of these molecules to be the critical determinant for their positive allosteric activity at GABA$_A$ receptors (Paul and Purdy, 1992; Lambert et al., 1995; Gee et al., 1988). In contrast to 3α-reduced pregnane steroids, DHEAS and PREGS display GABA-antagonistic properties (Paul and Purdy, 1992; Lambert et al., 1995; Rupprecht, 1997) (Table I). The modulatory effects of neuroactive steroids on GABA$_A$ receptors have been reviewed in more detail elsewhere (Paul and Purdy, 1992; Lambert et al., 1995).

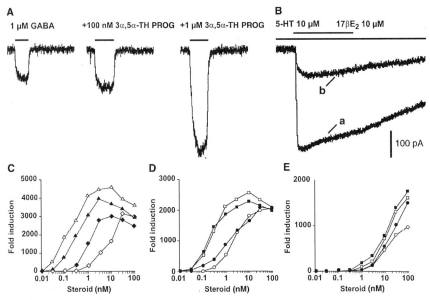

FIG. 2. (A) Positive allosteric modulation of the GABA-evoked chloride current by 3α-hydroxy-5α-pregnane steroids. Rat hypothalmic neurons were recorded in the whole-cell voltage-clamp configuration. The bar indicates the presence of 1 μM GABA. (B) Functional antagonism of 17β-estradiol on the serotonin-evoked cation current. HEK 293 cells expressing the 5-HT$_3$ receptor were recorded in the whole-cell voltage–clamp configuration. The upper bar indicates the application of 10 μM serotonin; the lower bar indicates the presence of 10 μM 17β-estradiol (17βE$_2$). Control experiments (a) without 17β-estradiol; (b) experiments with 17β-estradiol. (C–E) Progesterone-receptor mediated gene expression by steroids. Induction of the mouse mammary tumor virus (MTV) promoter after co-transfection of the chicken (cPR$_B$) or human (hPR$_B$) progesterone receptor expression vectors and incubation with steroids at the indicated concentrations in SK-N-MC cells. (C) progesterone (♦) and R 5020 (▲) after transfection of cPR$_B$; progesterone (◇) and R 5020 (△) after transfection of hPR$_B$. (D, E) 3α,5α-TH PROG (○) and 3α,5α-TH DOC (●), 5α-DH PROG (□) and 5α-DH DOC (■) after transfection of cPR$_B$ (D) and hPR$_B$ (E). The baseline activity of the MTV promoter without addition of steroid is set as 1. (Reprinted from *Trends in Neurosciences*, Vol. 22: Rupprecht, R., and Holsboer, F. Neuroactive steroids: Mechanisms of action and neuropsychopharmacological perspectives, pp. 410–416. Copyright © 1999, with permission from Elsevier Science.)

IV. Steroid Modulation of Other Neurotransmitter Receptors

The majority of studies have focused on the modulatory actions of neuroactive steroids at GABA$_A$ receptors. However, other members within this family of ligand-gated ion channels [i.e., glycine receptors, nicotinic acetylcholine receptors, and, finally, 5-HT$_3$ receptors (Fig. 2)] have

TABLE I
MODULATION OF NEUROTRANSMITTER RECEPTORS BY NEUROACTIVE STEROIDS

Receptor	Steroid	Modulation	Concentration	Reference
GABA$_A$	3α,5α-TH PROG	Positive	10^{-8}–10^{-5} M	Paul and Purdy (1992)
	3α,5β-TH PROG	Positive	10^{-8}–10^{-5} M	Lambert et al. (1995)
	3α,5α-TH DOC	Positive	10^{-8}–10^{-5} M	(for review)
	PREGS	Negative	10^{-5}–10^{-3} M	
	DHEAS	Negative	10^{-5}–10^{-3} M	
Nicotinic acetylcholine	Progesterone	Negative	10^{-6}–10^{-4} M 10^{-5}–10^{-4} M	Valera et al. (1992) Bullock et al. (1997)
	3α,5α-TH PROG	Negative	10^{-5}–10^{-4} M	Bullock et al. (1997)
Glycine	Progesterone	Negative	10^{-5}–10^{-3} M	Wu et al. (1990)
	PREGS	Negative	10^{-4} M^a	Wu et al. (1990)
Serotonin type 3 (5-HT$_3$)	Estradiol (17α and β)	Negative	10^{-6}–10^{-4} M	Wetzel et al. (1998)
	Progesterone	Negative	10^{-6}–10^{-4} M	Wetzel et al. (1998)
	Testosterone	Negative	10^{-5} M^a	Wetzel et al. (1998)
	3α,5α-TH PROG	Negative	10^{-5} M^a	Wetzel et al. (1998)
	PREGS	None	10^{-5} M^a	Wetzel et al. (1998)
NMDA	17β-Estradiol	Negative	5×10^{-5} M^a	Weaver et al. (1997a)
	PREGS	Negative	10^{-6}–10^{-3} M	Park-Chung et al. (1994)
	Pregnanolone-hemisuccinate	Negative	10^{-4} M^a	Weaver et al. (1997b)
	PREGS	Positive	10^{-5}–10^{-3} M 10^{-6}–10^{-4} M	Wu et al. (1991) Bowlby (1993)
AMPA	PREGS	Negative	10^{-4} M^a	Wu et al. (1991)
Kainate	PREGS	Negative	10^{-4} M^a	Wu et al. (1991)
	17β-Estradiol	Positive	10^{-8}–10^{-5} M	Gu and Moss (1996)
	Progesterone	Positive	10^{-5}–10^{-3} M	Wu et al. (1998)
Oxytocin	Progesterone	Negative	10^{-9}–10^{-6} M	Grazzini et al. (1998)
Sigma type 1 (σ_1)	DHEAS	Positive	10^{-7}–10^{-5} M	Monnet et al. (1995)
	PREGS	Negative	10^{-7}–10^{-5} M	Monnet et al. (1995)
	Progesterone	Antagonist	10^8–10^{-6} M	Monnet et al. (1995)

Note. GABA$_A$, γ-aminobutyric acid type A; NMDA, N-methyl-D-aspartate; AMPA, α-amino-3-hydroxy-5-methyl-4-isoxazolepropionic acid; TH PROG, tetrahydroprogesterone; TH DOC, tetrahydrodeoxycorticosterone; PREGS, pregnenolone sulfate; DHEAS, dehydroepiandrosterone sulfate.

^{a}Only one concentration tested.

meanwhile all been identified as potential targets for modulation by various neuroactive steroids (Table I). Within the glutamate receptor family, N-methyl-D-aspartate (NMDA) receptors, α-amino-3-hydroxy-5-methyl-4-isoxazolepropionic acid receptors and kainate receptors have also been demonstrated to be steroid sensitive (Table I). However, the structure–activity requirements for the modulation of ligand-gated ion channels by steroids apparently differ considerably among members of these neurotransmitter receptor families. Nevertheless, further studies are needed to address the issue of stereoselectivity of steroid modulation for these ligand-gated ion channels. Although steroid concentrations in the nanomolar range are sufficient for the activation of intracellular steroid receptors (Rupprecht, 1997; Evans, 1988) and for the positive allosteric modulation of $GABA_A$ receptors (Paul and Purdy, 1992; Lambert et al., 1995), the antagonistic modulation of various ligand-gated ion channels by neuroactive steroids usually requires concentrations in the micromolar range. Also, some doubts remain concerning the physiological relevance of the steroid modulation of neurotransmitter receptors different from $GABA_A$ receptors in view of the high concentrations needed. It cannot be excluded that the respective concentrations are reached locally in vivo considering the lipophilic properties of steroids. The modulatory actions of steroids, however, are not restricted to direct allosteric interactions with ligand-gated ion channels because G-protein-coupled mechanisms have also been shown to play a role in the steroid modulation of calcium channels (French-Mullen et al., 1994) and kainate receptors (Gu and Moss, 1996). The oxytocin receptor was recently identified as the first G-protein-coupled receptor for which steroids can be ligands (Grazzini et al., 1998). σ-Receptors, which lack detailed molecular characterization, may also bind steroids (Su et al., 1988) and are sensitive to steroid modulation (Table I).

V. A Putative Specific Steroid-Binding Site on Ligand-Gated Ion Channels

Although considerable work in the 1990s led to the hypothesis of a putative specific steroid-binding site at the $GABA_A$ receptor (Paul and Purdy, 1992; Lambert et al., 1995), as yet there is no biochemical evidence for a direct interaction of steroids with this neurotransmitter receptor, and attempts to identify such a binding site have been unsuccessful. The concept of a steroid recognition site has been supported by pharmacological studies concerning the potency, selectivity, and structure–activity relationship of the action of neuroactive steroids at this neurotransmitter receptor (Lambert et al., 1995). Studies using $GABA_A$/glycine-receptor chimeras suggest an allosteric action of neuroactive steroids at the N-terminal side of the middle of the second transmembrane domain of the GABA$_A$ receptor β_1 and/or α_2 subunits (Rick et al., 1998). However, the data available do not prove a direct binding of steroids to the receptor protein. Studies with the structurally related 5-HT$_3$ receptor could demonstrate neither a displacement of the fluorescence labeling of a cell line stably expressing the 5-HT$_3$ receptor obtained with conjugated steroids by unlabeled steroids nor a saturable steroid-binding site in binding studies (Wetzel et al., 1998). Moreover, the functional effects of steroids could be mimicked by distinct aromatic alcohols (Wetzel et al., 1998). These data were more in favor of the view that the steroids enter the membrane at the receptor–membrane interface and thereby allosterically modulate the function of this neurotransmitter receptor in a structure-specific manner (Wetzel et al., 1998). Moreover, recent studies showed that the interaction of positive and negative allosteric modulators at the GABA$_A$ receptor is not competitive (Park-Chung et al, 1999). Although research is needed to clarify definitively the issue of how steroids interact with ligand-gated ion channels at the molecular level, it appears that the allosteric modulation of neurotransmitter receptors by steroids is a highly complex phenomenon that is dependent on the molecular structure of the respective steroid, the amino acid composition of the individual receptor, and the physicochemical properties of the cell membrane (Wetzel et al., 1998). The lack of a clearly defined steroid-binding site at neurotransmitter receptors will cause difficulties in synthesizing steroid compounds with highly specific agonistic or antagonistic properties at a distinct neurotransmitter receptor.

VI. Genomic Effects of Neuroactive Steroids

In contrast to the classical genomic pathway of steroid hormones (Evans, 1988), the steroids modulating GABA$_A$ receptors were initially believed not to regulate gene expression via intracellular steroid receptors because of their inability to bind to such receptors (Gee et al., 1988; Paul and Purdy, 1992). These steroids may nevertheless affect transcription via progesterone receptors through the generation of metabolites that are ligands for steroid receptors (Rupprecht et al., 1993; Rupprecht, 1997). For example, 3α,5α-TH PROG and 3α,5α-TH DOC may induce transcription via progesterone receptors (Fig. 2) after intracellular oxidation into 5α-DH PROG and 5α-DH DOC (Fig. 1), which, in contrast to 3α,5α-TH PROG and 3α,5α-TH DOC, are ligands of progesterone receptors (Rupprecht et al., 1993). It has been shown that 3α,5α-TH PROG contributes to the development of progesterone withdrawal symptoms by enhancing the expression of the gene encoding for the α_4 subunit of the GABA$_A$ receptor (Smith *et al.*, 1998). Both 3α,5α-TH PROG and 3α,5α-TH DOC, when administered to rats, suppress the expression of vasopressin and corticotropin-releasing hormone, two neuropeptides mediating profound neuroendocrine and behavioral effects (Patchev *et al.*, 1994, 1997). These findings support the notion that neuroactive steroids regulate neuronal function through their concurrent influence on neuronal excitability and gene expression. The possibility of genomic effects of neuroactive steroids also must be taken into account when evaluating a putative side effect profile of such compounds.

VII. Neuropsychopharmacological Properties of Neuroactive Steroids

The actions of neuroactive steroids at the molecular level provide the basis for their modulation of a broad spectrum of physiological and pathological conditions (Table II). For example, the NMDA-receptor antagonism of estrogens and 3α-reduced pregnane steroids might in part account for their neuroprotective properties (Weaver, Marek *et al.*, 1997; Weaver, Park-Chung *et al.*, 1997). Moreover, in view of the reported memory-enhancing effects of pregnenolone (Flood *et al.*, 1992) and DHEA (Flood *et al.*, 1988) in animal studies, steroids that are negative allosteric modulators of GABA$_A$ receptors may become a novel therapeutic strategy for ameliorating cognitive deficits. However, although DHEA concentrations decrease with age, no consistent relationship between DHEA concentrations and cognitive function or dementia in humans has been established (Legrain *et al.*, 1995;

TABLE II
PUTATIVE NEUROPSYCHOPHARMACOLOGICAL PROPERTIES OF NEUROACTIVE STEROIDS

- Neuroprotective (17β-estradiol, 3α-reduced pregnane steroids)
- Memory enhancing (pregnenolone, DHEA)
- Sedative (progesterone, 3α-reduced pregnane steroids)
- Hypnotic (progesterone, 3α-reduced pregnane steroids)
- Anesthetic (progesterone, 3α-reduced pregnane steroids)
- Anxiolytic (progesterone, 3α-reduced pregnane steroids)
- Sleep modulating (progesterone, 3α-reduced pregnane steroids, pregnenolone, deoxyhydroepiandrosterone)
- Anticonvulsant (progesterone, 3α-reduced pregnane steroids)
- Antipsychotic (progesterone)
- Antidepressant (3α-reduced pregnane steroids)

Rupprecht, 1997). Initial studies suggest that DHEA might also be useful in the treatment of depression (Wolkowitz *et al.*, 1997). Future placebo-controlled studies need to clarify whether there is really a clinical benefit of such steroids in the treatment of dementia disorders or depression.

GABA-agonistic 3α-reduced pregnane steroids or progesterone as their precursor molecule may exert sedative (Arafat *et al.*, 1988), hypnotic (Arafat *et al.*, 1988), anesthetic (Carl *et al.*, 1990; Frye and Duncan, 1994), anxiolytic (Bitran *et al.*, 1991; Patchev *et al.*, 1997), neuroleptic-like (Rupprecht *et al.*, 1999), sleep-modulating (Lancel *et al.*, 1996, 1997), and anticonvulsant (Bäckström *et al.*, 1984; Belelli *et al.*, 1990; Herzog, 1995) effects in animals and humans. Various attempts have already been undertaken to develop such steroids for clinical use in anesthesia (Lambert *et al.*, 1995). Synthetic derivatives of 3α-reduced pregnane steroids are under clinical investigation for the treatment of epilepsy disorders (Monaghan *et al.*, 1997). However, particular attention must be drawn to the potential side effects of such steroids in view of the benzodiazepine-like effects on sleep of progesterone (Lancel *et al.*, 1996) and 3α,5α-TH PROG (Lancel *et al.*, 1997) and the increased seizure susceptibility following progesterone withdrawal (Smith *et al.*, 1998). Moreover, it must be considered that 3α-reduced pregnane steroids are not selective modulators of $GABA_A$ receptors but also affect other neurotransmitter receptors (Table I) and gene expression (Rupprecht *et al.*, 1993; Patchev *et al.*, 1994, 1997; Smith *et al.*, 1998), which may contribute to both their clinical and their side effects. The behavioral effects of neuroactive steroids are reviewed in more detail elsewhere (Rupprecht, 1997).

VIII. Modulation of Endogenous Neuroactive Steroids as a Pharmacological Principle

In addition to the putative pharmacological properties of neuroactive steroids, they play an important role as endogenous modulators of neuronal and behavioral functions. For example, increased fatigue during pregnancy has been attributed to the increased concentrations of progesterone and 3α-reduced pregnane steroids (Biedermann and Schoch, 1995), whereas rapid changes in the bioavailability of various neuroactive steroids might contribute to premenstrual syndrome or postpartum blues (Wang *et al.*, 1996; Rupprecht, 1997). These neuroactive steroids also may help to maintain homeostasis during hormonal responses to stress by counteracting the activity of the hypothalamic-pituitary-adrenal system (Purdy *et al.*, 1991; Patchev *et al.*, 1996, 1997). Thus, fluctuations in the concentrations of neuroactive steroids may in part be responsible for the discomfort and increased vulnerability to develop psychiatric disease in women in the perimenstrual phase, during pregnancy, in the postpartum period, and at the onset of menopause. However, neuroactive steroids are involved not only in physiological but also in pathological conditions. During ethanol withdrawal, rats exhibit enhanced sensitivity to the anticonvulsant actions of 3α-reduced pregnane steroids (Devaud et al., 1996), and reduced concentrations of such steroids have been reported in alcoholic patients during ethanol withdrawal (Romeo et al., 1996). Moreover, antidepressants may increase the concentrations of 3α-reduced pregnane steroids in the rat brain (Uzunov et al., 1996), and there is a disequilibrium of these steroids in patients suffering from a depressive episode that can be corrected by treatment with antidepressant drugs (Romeo et al., 1998; Uzunova et al., 1998). These examples illustrate that a disturbed homeostasis of neuroactive steroids may be a risk factor for the development of psychiatric diseases and, conversely, psychopharmacological drugs, such as antidepressants, may, at least in part, exert their clinical effects through their influence on the equilibrium of neuroactive steroids.

IX. Outlook

The term neuroactive steroids, which was initially coined for steroids that modulate neuronal excitability via specific interactions with neurotransmitter receptors, has been challenged by the identification of steroid hormones that allosterically modulate various neurotransmitter receptors and the genomic effects of neuroactive steroids modulating the $GABA_A$ receptor. These observations underscore the importance of the intracellular cross-talk between genomic and nongenomic steroid effects that provides the molecular basis for steroid action in the brain (Fig. 3). As yet, no naturally occurring steroid with a truly specific and selective action at a distinct steroid receptor or neurotransmitter receptor has been identified. Instead, steroids appear to be promiscuous molecules, which may explain their broad spectrum of biological effects. As such, it is likely that synthetic derivatives with unique pharmacological features at a distinct neurotransmitter receptor or steroid receptor may also target other receptors. Moreover, conversion of steroids into derivatives with pharmacological profiles different from their precursors may explain their broad spectrum of effects under *in vivo* conditions that must be considered when evaluating the putative clinical properties of neuroactive steroids. As an alternative to exogenous administration, drugs that interfere with steroid synthesis (e.g., enzyme inhibitors, antidepressants) may be used to influence the equilibrium of endogenous

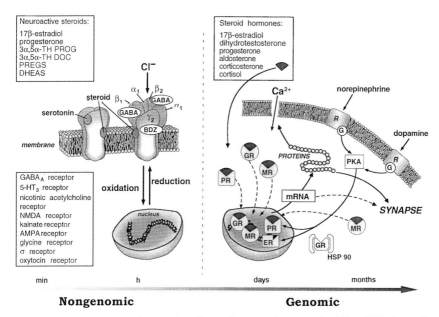

FIG. 3. Nongenomic and genomic effects of neuroactive steroids. (*Note.* BDZ, benzodiazepines; R, receptor; G, G-protein; PKA, protein kinase A; HSP 90, heat shock protein 90; GR, glucocorticoid receptor; MR, mineralocorticoid receptor; PR, progesterone receptor; ER, estrogen receptor.) (Reprinted from *Trends in Neurosciences*, Vol. 22: Rupprecht, R., and Holsboer, F. Neuroactive steroids: Mechanisms of action and neuropsychopharmacological perspectives, pp. 410–416. Copyright © 1999, with permission from Elsevier Science.)

neuroactive steroids. Such effects may help to explain the *in vivo* pharmacological profile of certain nonsteroidal compounds and open the door for novel therapeutic strategies in the treatment of neuropsychiatric diseases. In conclusion, neuroactive steroids may provide a new lead with considerable potential in neuropsychopharmacology. To fully exploit their possibilities, it will be necessary to dissect their pleiotropic actions at the molecular level. This may involve the use of drug design strategies that dissect the various actions of neuroactive steroids and apply high throughput screens that allow the identification of synthetic steroids with more targeted actions and minimized side effects. Although endogenous neuroactive steroids are of major importance for the maintenance of central nervous plasticity and homeostasis, the promiscuity of steroid molecules at the molecular level imposes considerable difficulties concerning their pharmaceutical exploitation. Whether it will be possible to identify steroids for specific molecular targets during chronic administration will determine whether such synthetic compounds may really possess advantages over already known classes of drugs. Another possibility is to screen for substances that influence the metabolism of endogenous neuroactive steroids. Which of these approaches will be more straightforward is a challenge for steroid research in neuroscience in the future.

Acknowledgments

We thank Christian Behl and Bettina Hermann for their critical reading of the manuscript. The work on neuroactive steroids at the Max Planck Institute of Psychiatry in Munich is supported by the Gerhard Heß Programm of the Deutsche Forschungsgemeinschaft to R.R.

References

Arafat, E. S., Hargrove, J. T., Maxson, W. S., Desiderio, D. M., Wentz, A. C., and Andersen, R. N. (1988). Sedative and hypnotic effects of oral administration of micronized progesterone may be mediated through its metabolites. *Am. J. Obstet. Gynecol.* **159,** 1203–1209.

Bäckström, T., Zetterlund, B., Blom, S., and Romano, M. (1984). Effects of intravenous progesterone infusions on the epileptic discharge frequency in women with partial epilepsy. *Acta Neurol. Scand.* **69,** 240–248.

Balthazart, J., and Ball, G. F. (1998). New insights into the regulation and function of brain estrogen synthase (aromatase). *Trends Neurosci.* **21,** 243–248.

Baulieu, E.-E. (1998). Neurosteroids: A novel function of the brain. *Psychoneuroendocrinology* **23,** 963–987.

Behl, C., Skutella, T., Lezoualc'h, F., Post, A., Widmann, M., Newton, C., and Holsboer, F. (1997). Neuroprotection against oxidative stress by estrogens: Structure–activity relationship. *Mol. Pharmacol.* **51,** 535–541.

Belelli, D., Lan, N. C., and Gee, K. W. (1990). Anticonvulsant steroids and the GABA/benzodiazepine receptor–chloride ionophore complex. *Neurosci. Biobehav. Rev.* **14,** 315–322.

Biedermann, K., and Schoch, P. (1995). Do neuroactive steroids cause fatigue in pregnancy? *Eur. J. Obstet. Gynecol. Reprod. Biol.* **58,** 15–18.

Bitran, D., Hilvers, R. J., and Kellogg, C. K. (1991). Anxiolytic effects of 3α-hydroxy-5α[β]pregnan-20-one: Endogenous metabolites of progesterone that are active at the $GABA_A$ receptor. *Brain Res.* **561,** 157–161.

Bowlby, M. R. (1993). Pregnenolone sulfate potentiation of N-methyl-D-aspartate receptor channels in hippocampal neurons. *Mol. Pharmacol.* **43,** 813–819.

Bullock, A. E., Clark, A. L., Grady, S. R., Robinson, S. F., Slobe, B. S., Marks, M. J., and Collins, A. C. (1997). Neurosteroids modulate nicotinic receptor function in mouse striatal and thalamic synaptosomes. *J. Neurochem.* **68,** 2412–2423.

Carl, P., Högskilde, S., Nielsen, J. W., Sörensen, M. B., Lindholm, M., Karlen, B., and Bäckström, T. (1990). Pregnanolone emulsion: A preliminary pharmacokinetic and pharmacodynamic study of a new intravenous agent. *Anaesthesia* **45,** 189–197.

Celotti, F., Melcangi, R. C., and Martini, L. (1992). The 5α-reductase in the brain: Molecular aspects and relation to brain function. *Front. Neuroendocrinol.* **13**, 163–215.

Chen, R., Belelli, D., Lambert, J. J., Peters, J. A., Reyes, A., and Lan, N. C. (1994). Cloning and functional expression of a drosophila γ-aminobutyric acid receptor. *Proc. Natl. Acad. Sci. USA* **91**, 6069–6073.

Devaud, L. L., Purdy, R. H., Finn, D. A., and Morrow, A. L. (1996). Sensitization of γ-aminobutyric acid$_A$ receptors to neuroactive steroids in rats during ethanol withdrawal. *J. Pharmacol. Exp. Ther.* **278**, 510–517.

Evans, R. M. (1988). The steroid and thyroid hormone receptor superfamily. *Science* **240**, 889–895.

French-Mullen, J. M. H., Danks, P., and Spence, K. T. (1994). Neurosteroids modulate calcium currents in hippocampal CA 1 neurons via a pertussis toxin-sensitive G-protein-coupled mechanism. *J. Neurosci.* **14**, 1963–1977.

Flood, J. F., Morley, J. E., and Roberts, E. (1992). Memory-enhancing effects in male mice of pregnenolone and of steroids metabolically derived from it. *Proc. Natl. Acad. Sci. USA* **89**, 1567–1571.

Flood, J. F., Smith, G. E., and Roberts, E. (1988). Dehydroepiandrosterone and its sulfate enhance memory retention in mice. *Brain Res.* **447**, 269–278.

Frye, C. A., and Duncan, J. E. (1994). Progesterone metatolites effective at the GABA$_A$ receptor complex modulate pain sensitivity in rats. *Brain Res.* **643**, 194–203.

Gee, K. W., Bolger, M. B., Brinton, R. E., Coirini, H., and McEwen, B. S. (1988). Steroid modulation of the chloride ionophore in rat brain: Structure–activity requirements, regional dependance and mechanism of action. *J. Pharmacol. Exp. Ther.* **246**, 803–812.

Grazzini, E., Guillon, G., Mouillac, B., and Zingg, H. H. (1998). Inhibition of oxytocin receptor function by direct binding of progesterone. *Nature* **392**, 509–512.

Gu, Q., and Moss, R. L. (1996). 17 β-Estradiol potentiates kainate-induced currents via activation of the cAMP cascade. *J. Neurosci.* **16**, 3620–3629.

Herzog, A. G. (1995). Progesterone therapy in women with complex partial and secondary generalized seizures. *Neurology* **45**, 1660–1662.

Lambert, J. J., Belelli, D., Hill-Venning, C., and Peters, J. A. (1995). Neurosteroids and GABA$_A$ receptor function. *Trends Pharmacol. Sci.* **16**, 295–303.

Lancel, M., Faulhaber, J., Holsboer, F., and Rupprecht, R. (1996). Progesterone induces changes in sleep EEG comparable to those of agonistic GABA$_A$ receptor modulators. *Am. J. Physiol.* **271**, E763–E772.

Lancel, M., Faulhaber, J., Schiffelholz, T., Romeo, E., di Michele, F., Holsboer, F., and Rupprecht, R. (1997). Allopregnanolone affects sleep in a benzodiazepine-like fashion. *J. Pharmacol. Exp. Ther.* **282**, 1213–1218.

Legrain, S., Berr, C., Frenoy, N., Gourlet, V., Debuire, B., and Baulieu, E.-E. (1995). Dehydroepiandrosterone sulfate in a long-term care aged population. *Gerontology* **41**, 343–351.

Majewska, M. D., Harrison, N. L., Schwartz, R. D., Barker, J. L., and Paul, S. M. (1986). Steroid hormone metabolites are barbiturate-like modulators of the GABA receptor. *Science* **232**, 1004–1007.

McEwen, B. S. (1991). Non-genomic and genomic effects of steroids on neural activity. *Trends Pharmacol. Sci.* **12**, 141–147.

Mellon, S. H. (1994). Neurosteroids: Biochemistry, modes of action and clinical relevance. *J. Clin. Endocrinol. Metab.* **78**, 1003–1008.

Monaghan, E. P., Navalta, L. A., Shum, L., Ashbrock, D. W., and Lee, D. A. (1997). Initial human experience with ganaxolone, a neuroactive steroid with antiepileptic activity. *Epilepsia* **38**, 1026–1031.

Monnet, F. P., Mahé, V., Robel, P., and Baulieu, E.-E. (1995). Neurosteroids, via sigma receptors, modulate the [^{3}H]norepinephrine release evoked by N-methyl-D-aspartate in the rat hippocampus. *Proc. Natl. Acad. Sci. USA* **92,** 3774–3778.

Orchinik, M., Murray, T. F., and Moore, F. L. (1991). A corticosteroid receptor in neuronal membranes. *Science* **252,** 1848–1851.

Pappas, T. D., Gametchu, B., and Watson, C. S. (1995). Membrane estrogen receptors identified by multiple antibody labeling and impeded-ligand binding. *FASEB J.* **9,** 404–410

Park-Chung, M., Wu, F. S., and Farb, D. H. (1994). 3α-Hydroxy-5β-pregnan-20-one sulfate: A negative modulator of the NMDA-induced current in cultured neurons. *Mol. Pharmacol.* **46,** 146–150.

Park-Chung, M., Malayev, A., Purdy, R. H., Gibbs, T. T., and Farb, D. H. (1999). Sulfated and unsulfated steroids modulate γ-aminobutyric acid A receptor function through distinct sites. *Brain Res.* **830,** 72–87.

Patchev, V. K., Hassan, A. H. S., Holsboer, F., and Almeida, O. F. X. (1996). The neurosteroid tetrahydroprogesterone attenuates the endocrine response to stress and exerts glucocorticoid-like effects on vasopressin gene transcription in the rat hypothalamus. *Neuropsychopharmacology* **15,** 533–541.

Patchev, V. K., Montkowski, A., Rouskova, D., Koranyi, L., Holsboer, F., and Almeida, O. F. X. (1997). Neonatal treatment of rats with the neuroactive steroid tetrahydrodeoxycorticosterone (TH DOC) abolishes the behavioral and neurondocrine consequences of adverse early life events. *J. Clin. Invest.* **99,** 962–966.

Patchev, V. K., Shoaib, M., Holsboer, F., and Almeida, O. F. X. (1994). The neurosteroid tetrahydroprogesterone counteracts corticotropin-releasing hormone-induced anxiety and alters the release and gene expression of corticotropin-releasing hormone in the rat hypothalamus. *Neuroscience* **62,** 265–271.

Paul, S. M., and Purdy, R. H. (1992). Neuroactive steroids. *FASEB J.* **6,** 2311–2322.

Puia, G., Santi, M. R., Vicini, S., Pritchett, D. B., Purdy, R. H., Paul, S. M., Seeburg, P. H., and Costa, E. (1990). Neurosteroids act on recombinant human GABA$_A$ receptors. *Neuron* **4,** 759–765.

Purdy, R. H., Morrow, A. L., Moore, P. H., and Paul, S. M. (1991). Stress-induced elevations of γ-aminobutyric acid type A receptor-active steroids in the rat brain. *Proc. Natl. Acad. Sci. USA* **8,** 4553–4557.

Ramirez, V. D., and Zheng, J. (1996). Membrane sex-steroid receptors in the brain. *Front. Neuroendocrinol.* **17,** 402–439.

Rick, C. E., Ye, Q., Finn, S. E., and Harrison, N. L. (1998). Neurosteroids act on the GABA(A) receptor at sites on the N-terminal side of the middle of TM2. *Neuroreporter* **9,** 379–383.

Romeo, E., Brancati, A., De Lorenzo, A., Fucci, P., Furnari, C., Pompili, E., Sasso, G. F., Spalletta, G., Troisi, A., and Pasini, A. (1996). Marked decrease of plasma neuroactive steroids during alcohol withdrawal. *Clin. Neuropharmacol.* **19,** 366–369.

Romeo, E., Ströhle, A., Spalletta, G., di Michele, F., Hermann, B., Holsboer, F., Pasini, A., and Rupprecht, R. (1998). Effects of antidepressant treatment on neuroactive steroids in major depression. *Am. J. Psychiatry.* **155,** 910–913.

Rupprecht, R. (1997). The neuropsychopharmacological potential of neuroactive steroids. *J. Psychiatric Res.* **31,** 297–314.

Rupprecht, R., and Holsboer, F. (1999). Neuroactive steroids: Mechanisms of action and neuropsychopharmacological perspectives. *Trends Neurosci.* **22,** 410–416.

Rupprecht, R., Koch, M., Montkowski, A., Lancel, M., Faulhaber, J., Harting, J., and Spanagel, R. (1999). Assessment of neuroleptic-like properties of progesterone. *Psychopharmacology* **143,** 29–38.

Rupprecht, R., Reul, J. M. H. M., Trapp, T., van Steensel, B., Wetzel, C., Damm, K., Zieglgänsberger, W., and Holsboer, F. (1993). Progesterone receptor-mediated effects of neuroactive steroids. *Neuron* **11,** 523–530.

Smith, S. S., Gong, Q. H., Hsu, F.-C., Markowitz, R. S., French-Mullen, J. M. H., and Li, X. (1998). $GABA_A$ receptor α_4 subunit suppression prevents withdrawal properties of an endogenous steroid. *Nature* **392,** 926–930.

Su, T.-P., London, E. D., and Jaffe, J. H. (1988). Steroid binding at σ receptors suggests a link between endocrine, nervous, and immune systems. *Science* **240,** 219–221.

Truss, M., and Beato, M. (1993). Steroid hormone receptors: Interaction with deoxyribonucleic acid and transcription factors. *Endocrine Rev.* **14,** 459–479.

Uzunov, D. P., Cooper, T. B., Costa, E., and Guidotti, A. (1996). Fluoxetine-elicited changes in brain neurosteroid content measured by negative ion mass fragmentography. *Proc. Natl. Acad. Sci. USA* **93,** 12599–12604.

Uzunova, V., Sheline, Y., Davis, J. M., Rasmusson, A., Uzunov, D. P., Costa, E., and Guidotti, A. (1998). Increase in the cerebrospinal fluid content of neurosteroids in patients with unipolar major depression who are receiving fluoxetine or fluvoxamine. *Proc. Natl. Acad. Sci. USA* **95,** 3239–3244.

Valera, S., Ballivet, M., and Bertrand, D. (1992). Progesterone modulates a neuronal nicotinic acetylcholine receptor. *Proc. Natl. Acad. Sci. USA* **89,** 9949–9953.

Wang, M., Seippel, L., Purdy, R. H., and Bäckström, T. (1996). Relationship between symptom severity and steroid variation in women with premenstrual syndrome: Study on serum pregenenolone, pregnenolone sulfate, 5α-pregnane-3,20-dione and 3α-hydroxy-5α-pregnan-20-one. *J. Clin. Endocrinol. Metab.* **81,** 1076–1082.

Weaver, C. E., Jr., Park-Chung, M., Gibbs, T. T., and Farb, D. H. (1997a). 17β-Estradiol protects against NMDA-induced excitotoxicity by direct inhibition of NMDA receptors. *Brain Res.* **761,** 338–341.

Weaver, C. E., Jr., Marek, P., Park-Chung, M., Tam, S. W., and Farb, D. H. (1997b). Neuroprotective activity of a new class of steroidal inhibitors of the N-methyl-D-aspartate receptor. *Proc. Natl. Acad. Sci. USA* **94,** 10450–10454.

Wehling, M. (1997). Specific, nongenomic actions of steroid hormones. *Ann. Rev. Physiol.* **59,** 365–393.

Weiland, N. G., and Orchinik, M. (1995). Specific subunit mRNAs of the $GABA_A$ receptor are regulated by progesterone in subfields of the hippocampus. *Mol. Brain Res.* **32,** 271–278.

Wetzel, C. H. R., Hermann, B., Behl, C., Pestel, E., Rammes, G., Zieglgänsberger, W., Holsboer, F., and Rupprecht, R. (1998). Functional antagonism of gonadal steroids at the 5-HT_3 receptor. *Mol. Endocrinol.* **12,** 1441–1451.

Wolkowitz, O. M., Reus, V. I., Roberts, E., Manfredi, F., Chan, T., Raum, W. J., Ormiston, S., Johnson, R., Canick, J., Brizendine, L., and Weingartner, H. (1997). Dehydroepiandrosterone (DHEA) treatment of depression. *Biol. Psychiatry.* **41,** 311–318.

Wu, F.-S., Gibbs, T. T., and Farb, D. H. (1990). Inverse modulation of γ-aminobutyric acid- and glycine-induced currents by progesterone. *Mol. Pharmacol.* **37,** 597–602.

Wu, F.-S., Gibbs, T. T., and Farb, D. H. (1991). Pregnenolone sulfate: A positive allosteric modulator at the N-methyl-D-aspartate receptor. *Mol. Pharmacol.* **40,** 333–336.

Wu, F.-S., Yu, H.-M., and Tsai, J.-J. (1998). Mechanism underlying potentiation by progesterone of the kainate-induced current in cultured neurons. *Brain Res.* **779,** 354–358.

Zakon, H. H. (1998). The effects of steroid hormones on electrical activity of excitable cells. *Trends Neurosci.* **21,** 202–207.

CURRENT PERSPECTIVES ON THE ROLE OF NEUROSTEROIDS IN PMS AND DEPRESSION

Lisa D. Griffin*, Susan C. Conrad[†], and Synthia H. Mellon[‡,1]

*Department of Neurology, [†]Department of Pediatrics, and [‡]Department of
Obstetrics, Gynecology, and Reproductive Sciences
University of California–San Francisco
California 94143

I. Introduction
II. Premenstrual Syndrome or Premenstrual (Late Luteal) Dysphoric Disorder
III. Allopregnanolone and Premenstrual Dysphoric Disorder
IV. Biosynthesis of Allopregnanolone
V. SSRIs in the Treatment of Premenstrual Dysphoric Disorder: Modulation of Neurosteroid Levels
VI. SSRIs, Neurosteroids, and Depression
VII. Conclusions
References

I. Introduction

The biosynthesis and mechanism of action of the $3\alpha,5\alpha$- and $3\alpha,5\beta$-reduced derivatives of progesterone have been described in the preceding chapters. The potential role that these compounds may play in *in vivo* physiology has also been discussed. A central question that remains unanswered is whether abnormal synthesis or regulation of the synthesis of these compounds is related to any human diseases. Studies have pointed to their involvement in premenstrual syndrome (PMS), postpartum depression, several types of major depression, alcoholism, and seizure disorders. What evidence supports the idea that altered synthesis of these compounds is causative of certain disorders? The evidence so far points to a role of these neurosteroids in various affective disorders, as restoration of normal levels causes relief from symptoms, but there is no evidence yet that dysregulation of neurosteroid synthesis by itself is the cause of any particular disease. Here we summarize the potential role of neurosteroids in the etiology of some affective disorders and how treatments that increase the synthesis of these compounds alleviate the symptoms of these disorders.

[1] Address to whom correspondence should be sent.

II. Premenstrual Syndrome or Premenstrual (Late Luteal) Dysphoric Disorder

PMS is a heterogeneous group of affective and somatic symptoms that occur in a phase-specific manner during the luteal phase of the menstrual cycle (reviewed in Rubinow and Schmidt, 1995; Freeman and Halbreich, 1998; Rubinow *et al.*, 1998; Endicott *et al.*, 1999; Halbreich, 1999). Premenstrual dysphoric disorder (PDD), also termed late luteal phase dysphoric disorder, is characterized by both psychological symptoms, including depression, anxiety, affective lability, and decreased interest in activities, and physical symptoms, including breast tenderness or swelling, headache, myalgia, arthralgia, and a sensation of bloating during the last week of the luteal phase of the menstrual cycle. These symptoms must have regularly occurred during the last week of the luteal phase in most menstrual cycles of the prior 12 months. The symptoms remit within the first week of the subsequent follicular phase of the next menstrual cycle and are always absent in the week following menses. At least five of the symptoms included in the fourth edition of the *Diagnostic and Statistical Manual of Mental Disorders* must be present most of the time during the last week of the luteal phase (American Psychiatric Association, 1994). This syndrome is estimated to affect 3–5% of women of reproductive age, often requiring them to miss work for several days each month. Thus, the individual and economic disability of this syndrome is clearly significant.

The etiology of PDD is unknown, but its cyclical nature, fluctuating with increased ovarian steroid hormone concentrations, suggests that it may involve one or more ovarian or hypothalamic hormones. The hypothalamic gonadotropin-releasing hormone (GnRH) is unlikely to be involved directly, as the symptoms are not seen in either spontaneous anovulatory cycles or anovulatory cycles induced pharmacologically by GnRH analogs (Hammarback and Backström, 1988; Hammarback *et al.*, 1991). Studies by several groups have indicated that there are no differences in the circulating concentrations of progesterone, estradiol, GnRH, luteinizing hormone (LH), or follicle-stimulating hormone (FSH), between women who do not have PDD and those who do. Other studies have found that in some women there was a higher number of adverse premenstrual complaints in cycles with high luteal phase serum estradiol and progesterone concentrations and that high estradiol concentrations were positively correlated with PMS symptom severity (Hammarback *et al.*, 1989).

In addition to the action of progesterone on its nuclear steroid hormone receptor, it is now known that progesterone can be metabolized by a variety of tissues, including the corpus luteum and the brain, to other more potent neuroactive compounds (neurosteroids). These progesterone derivatives are 5α- and 3α-reduced derivatives of progesterone and are given the names 5α-dihydroprogesterone (5α-pregnane-3,20-dione) and 3α,5α-tetrahydroprogesterone (3α-hydroxy-5α-pregnan-20-one or allopregnanolone). Biochemical and physiological studies have demonstrated that allopregnanolone is an agonist of the γ-aminobutyric acid type A (GABA$_A$) neurotransmitter receptor. Allopregnanolone binds to a distinct site on the GABA$_A$ receptor and affects the frequency and duration of the chloride channel opening. Hence, allopregnanolone modulates GABA ergic transmission and may affect complex behaviors such as anxiety.

III. Allopregnanolone and Premenstrual Dysphoric Disorder

Several studies have looked at the potential role of allopregnanolone in the etiology of PDD (Schmidt *et al.*, 1994; Wang *et al.*, 1996; Rapkin *et al.*, 1997; Bicikova *et al.*, 1998; Monteleone *et al.*, 2000). Interestingly, the authors of these papers come to different conclusions about the steroid concentrations found in control women and women with PDD. The results of these studies are summarized in Table I. Schmidt *et al.* (1994) did not report concentrations of steroids during the follicular phase and found no differences between controls and women with PDD during the luteal phase of the menstrual cycle. Wang *et al.* (1996) found no differences in the concentrations of progesterone or allopregnanolone between control women and women with PDD during the follicular phase of the menstrual cycle but did find differences between these groups of women during the luteal phase of the cycle in plasma concentrations of progesterone, but not in allopregnanolone. Sundstrom *et al.* (1998) also did not detect differences among women, in their concentrations of progesterone or of pregnanolone (3α,5β-tetrahydroprogesterone) during the follicular or the luteal phase (Days 15–28) of the menstrual cycle. Monteleone *et al.* (2000) reported differences between control women and women with PDD. They found significant differences in the concentrations of allopregnanolone during the luteal phase of the menstrual cycle (Days 22–26). However, unlike all of the other groups, this group also found significant differences in the concentrations of progesterone during the luteal phase. Rapkin *et al.* (1997) also found differences in the concentration of allopregnanolone between women with PDD and control women. However, they could detect this difference only on Day 26 of the cycle, but not on day 19. The concentrations of allopregnanolone reported by Rapkin *et al.* (1997) are greater than 10 times those reported in any of the other studies and may be an error in calculation or may be an effect of the method of allopregnanolone purification and

TABLE I
Comparison of Blood Steroid Concentrations during the Menstrual Cycle in Controls and Women with PMS

Study	Subjects PMS	Subjects Control	Timing	Progesterone PMS	Progesterone Control	P	Allopregnanolone PMS	Allopregnanolone Control	P	Pregnanolone PMS	Pregnanolone Control	P
Schmidt et al. (1994)	15	12	Follicular (not reported)									
			Luteal (4–8 DBM[a])	30	24.2	NS	4.3	3.5	NS	3.5	3.1	NS
Wang et al. (1996)	12	8	Follicular (d 1–14)	3.2	5	NS	1.9	1.8	NS			
			Luteal (d 15–28)	30.4	34.7	<0.02	3.6	3.6	NS			
Rapkin et al. (1997)	35	36	Follicular (not reported)									
			Luteal (d 19)	33.9	28.2	NS	19.8	24.2	NS			
			Luteal (d 26)	24.3	20.3	NS	11.3	23.6	<0.01			
Sundstrom et al. (1998)	12	12	Follicular (d 6–12)	1.4	1.2	NS				0.64	0.56	NS
			Luteal (1–7 DBM[a])	24	16.2	NS				0.79	1.12	NS
Bicikova et al. (1998)	23	17	Follicular (d 4–9)	0.78	2.3	<0.05	0.09	0.45	<0.05			
			Luteal (1–6 DBM[a])	12.2	13.3	NS	1.56	1.2	NS			
Monteleone et al. (2000)	28	28	Follicular (d 5–8)	2.4–2.5	5.2–5.5	<0.0001	0.85–0.88	0.77–0.81	<0.0001			
			Luteal (d 22–26)	6.6–6.8	34.9–36	<0.0001	1.5–1.7	3.7–3.9	<0.0001			

Note: Data expressed in nanomoles per liter; NS, no significant difference.
[a]Days before menses.

analysis by radioimmunoassay. Finally, the data from Bicikova *et al.* (1998) unlike those from all of the other studies, reported differences between women with PDD and controls in both progesterone and allopregnanolone during the *follicular* phase of the menstrual cycle but not during the luteal phase of the cycle. These results would be inconsistent with the behavioral data and diagnosis of PDD, which occurs during the late luteal phase of the cycle and remits within the first week of the follicular phase of the cycle.

There may be several reasons for the discrepancies among the published clinical studies:

1. The diagnosis of PDD may differ among the groups. Hence, groups that found differences in hormone concentrations may have excluded women with less severe symptoms.
2. The time of the hormone measurement in blood may have differed among the studies. Although both follicular and luteal measurements were made, the exact day within the menstrual cycle when the blood samples were taken may have differed. This is borne out by the study by Rapkin *et al.* (1997) who did not see differences in allopregnanolone concentrations at Day 19 but did see differences at Day 26 of the menstrual cycle.
3. The differences in the assay techniques may also cause differences in the results. Some investigators purified the steroids by high-performance liquid chromatography (HPLC) before radioimmunoassay; others used different purification techniques. Even purification by HPLC may be different, as some systems may separate allopregnanolone (the $3\alpha,5\alpha$-stereoisomer) from all of the other isomers (e.g., the neuroinactive $3\beta,5\alpha$- or $3\beta,5\beta$-stereoisomers) or from 5α-dihydroprogesterone. Also, the antibody used for the allopregnanolone assay may be different, with a different amount of cross-reactivity with other progesterone derivatives.

IV. Biosynthesis of Allopregnanolone

Allopregnanolone can be synthesized from progesterone by the sequential activities of two steroidogenic enzymes, 5α-reductase and 3α-hydroxysteroid dehydrogenase (3α-HSD) (Fig. 1). These enzymes are found in specific regions of the brain (reviewed in Chapter 1 and in Compagnone and Mellon, 2000). These enzymes are also found in other tissues, including the kidney, thymus, corpus luteum, uterus, and lymphocytes. Hence,

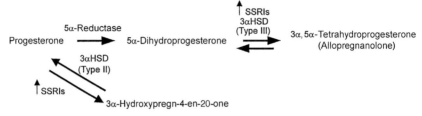

FIG. 1. Biochemical pathways leading to the synthesis of neuroactive steroids. The synthesis of the neuroactive compounds 5α-dihydroprogesterone, allopregnanolone, and 3α-hydroxypreg-4-en-20-one from progesterone substrate are shown. The enzymes mediating these conversions are 5α-reductase type I and 3αHSD. The synthesis of allopregnanolone in humans is mediated by the type III isoform, while the synthesis of 3α-hydroxypregn-4-en-20-one in humans is mediated by the type II isoform. The reactions enhanced by SSRIs (i.e., 3αHSD activities) are indicated. The synthesis of other neuroactive compounds, such as pregnanolone, is not shown; the substrate for this reaction, and enzymes mediating this synthesis, are not yet established.

a large variety of tissues could conceivably contribute to the synthesis of allopregnanolone from circulating progesterone.

Another neuroactive compound has also been identified in rat pituitaries and in rat Sertoli cells (Wiebe and Wood, 1987; Wood and Wiebe, 1989; Dhanvantari and Wiebe, 1994; Wiebe et al., 1994; Wiebe et al., 1997). This compound was identified as 3α-dihydroprogesterone [3α-hydroxypregn-4-en-20-one (3α-DH PROG)], which was found to be as active as a potentiator of the GABA$_A$ receptor as allopregnanolone (Morrow et al., 1990). It could be synthesized from progesterone using rat pituitary extracts or rat Sertoli cells. In addition to working via a GABA$_A$ receptor as a "neurosteroid," this compound was found to suppress FSH, but not LH, release in cultures of anterior pituitary cells. 3α-DH PROG was shown to inhibit the GnRH-stimulated FSH release from pituitaries and may be interacting with the Ca^{2+} channel component of the GnRH signal transduction mechanism. In addition, 3α-DH PROG may also suppress FSH release through direct action at the level of the gonadotrope membrane.

We have performed a series of experiments to determine the enzyme(s) involved in the synthesis of 3α-DH PROG (Griffin and Mellon, 2001). We used cloned rat 3αHSD in an *in vitro* assay to determine directly if that enzyme could convert progesterone to its 3α-reduced derivative without conversion to the 5α-reduced intermediate 5α-dihydroprogesterone. Our results indicated that, indeed, rat 3αHSD could convert progesterone to 3α-DH PROG. However, this reaction converts only 4.5% of the progesterone substrate, in comparison to the nearly 100% conversion of 5α-DH PROG to 3α,5α-TH PROG (allopregnanolone). In addition, the synthesis of 3α-DH PROG by pituitary extracts was inhibited by indomethacin, a known

TABLE II
Enzymatic Activity of the Different Isoforms of 3αHSD and 3αHSD from Rats and Humans

Enzyme	3αHSD activity substrate			20αHSD activity substrate	17βHSD activity substrate
	Progesterone	5α-Dihydroprogesterone	Dihydrotestosterone	Progesterone	Androstanediol/DHT
Rat 3α HSD	+	+	+	−	−
Human type II	+	−	+	+	+
Human type III	−	+	+	+		

inhibitor of 3αHSD activity. Thus, the rat 3αHSD could be responsible for the synthesis of 3α-DH PROG *in vivo*.

Humans have at least three forms of 3αHSD, all with different activities, summarized in Table II. 3αHSD types II (brain) and III are found in the brain (Griffin and Mellon, 1999), whereas 3αHSD type I is found exclusively in the liver. We tested both of these enzymes *in vitro* for their ability to synthesize 3α-DH PROG from progesterone. We found that the type III enzyme, which uses progestins (5α-DH PROG) as substrates, did not convert progesterone to 3α-DH PROG, while the type II$_{brain}$ enzyme did mediate this conversion. We previously showed that the type II$_{brain}$ 3αHSD had 20αHSD activity, could convert progesterone to 3α-DH PROG, could not convert 5α-DH PROG to 3α,5α-TH PROG but could convert 5α-dihydrotestosterone to 5α-androstane-3α,17β-diol. Our data now indicate that if 3α-DH PROG is indeed synthesized in the human brain or pituitary gland, it is likely that 3αHSD Type II$_{brain}$ is the enzyme mediating this conversion.

What is the potential importance of this finding? First these data indicate that this progesterone derivative could be synthesized *in vivo* in humans. Its local synthesis in the pituitary gland may be important for the regulation of GnRH action at the pituitary, and thus this compound may be an important regulator of reproductive function. Second, others had found that inhibition of 5α-reductase activity *in vivo* still resulted in the production of a compound that was neuroactive at GABA$_A$ receptors. 3α-DH PROG may indeed be that compound, and it may substitute for allopregnanolone modulation of GABA$_A$ receptors.

V. SSRIs in the Treatment of Premenstrual Dysphoric Disorder: Modulation of Neurosteroid Levels

PDD shares many features of depression and anxiety states (Pearlstein *et al.*, 1990; Endicott, 1993; Wurtman, 1993) that are associated with the dysregulation of the serotonergic system (Rapkin *et al.*, 1987; Rojansky *et al.*,

1991; Rapkin, 1992; Steiner, 1992; Yatham, 1993). Hence, treatment of PDD with drugs that directly affect the serotonergic system [e.g., selective serotonin reuptake inhibitors (SSRIs)] were evaluated by several different groups during the 1990s (Stone *et al.*, 1990; Stone *et al.*, 1991; Menkes *et al.*, 1992; Wood *et al.*, 1992; Menkes *et al.*, 1993; Steiner *et al.*, 1995; Steiner, 1997; Su *et al.*, 1997; Endicott *et al.*, 1999; Eriksson, 1999; Masand and Gupta, 1999; Kessel, 2000; Steiner and Pearlstein, 2000). In 2000, Dimmock *et al.* evaluated by meta-analysis 29 published studies of the use of SSRIs in PDD to determine the efficacy of SSRIs in the treatment of PDD. The results from the individual studies and from the meta-analysis indicate that SSRIs were effective in treating both physical and behavioral symptoms of PDD. Furthermore, the results from these studies indicated that intermittent dosing of SSRIs during only part of the menstrual cycle was equally effective in treating the symptoms of PDD.

The mechanism by which SSRIs are effective in treating the symptoms of PDD is unknown but has been thought to be solely the result of modulation of the serotonergic system. However, the cyclic nature of the disease suggested to others that perhaps ovarian steroids, or neurosteroids that are metabolites of the ovarian steroid progesterone, may be involved in the etiology of PDD. Furthermore, because SSRIs were effective at treating the symptoms of PDD, perhaps SSRIs have additional functions (i.e., altering neurosteroid synthesis) (Uzunov *et al.*, 1996).

The study by Uzunov *et al.* (1996) was performed in laboratory rats. They showed that, in adrenalectomized rats, fluoxetine could increase the abundance of the neurosteroid allopregnanolone, and not the precursor 5α-dihydroprogesterone (DHP) in the rat brain. This increase in allopregnanolone abundance occurred rapidly after SSRI treatment (within 15 min), in a region-specific fashion.

Our laboratory built on these findings and sought the biochemical mechanism by which SSRIs may increase allopregnanolone production (Griffin and Mellon, 1999). Using recombinant proteins for 5α-reductase type I, for rat 3αHSD, and for types II and III human 3αHSD, we tested the kinetics of the formation of DHP (5α-reductase type I activity) and of the formation of allopregnanolone from DHP (3αHSD reductive activity) or of the formation of 5α-DH PROG from $3\alpha,5\alpha$-TH PROG (3αHSD, oxidative activity). Our results indicated that SSRIs had no effect on the enzymatic activity of 5α-reductase type I and hence did not increase the production of 5α-DH PROG from progesterone. However, SSRIs could increase the reductive enzymatic activity of 3αHSD, thereby increasing the synthesis of $3\alpha,5\alpha$-TH PROG from 5α-DH PROG. As described in Chapter 2, human 3αHSD has multiple enzymatic activities in addition to its 3αHSD activity (also shown in Table II). 3αHSD is also found as multiple isoforms in humans, and each of these isoforms has specificity with respect to substrate and localization within the body and within the brain. Thus, only 3αHSD type III—but not type II$_{brain}$—uses 5α-DH PROG as substrate and hence is involved in allopregnanolone production. This enzyme activity, using 5α-DH PROG as substrate, is increased by several SSRIs tested but is not increased by another antidepressant, imipramine. 3αHSD type II$_{brain}$ activity could also be increased by SSRIs using androgens as substrates. SSRIs increased the production of 5α-androstane-3α,17β-diol from 5α-dihydrotestosterone. In all cases, SSRIs affected the apparent K_m of the enzyme.

In another study, it has been found that SSRIs could also potentiate the antidepressant activity of the neurosteroid allopregnanolone *in vivo* in rats (Khisti and Chopde, 2000). Several lines of evidence also suggested that the SSRIs worked through increased allopregnanolone production, although allopregnanolone was not measured. Indomethacin, a known inhibitor of 3αHSD activities, also inhibited the potentiation by fluoxetine of the antidepressant-like effect of allopregnanolone, again suggesting that fluoxetine works in part through modulation of 3αHSD activity. Although the authors found that activation of 5-HT receptors potentiated the antidepressant-like effect of allopregnanolone in a forced swim test, they also found that pretreatment with (I)-p-chlorophenylalanine (p-CPA), a compound that depletes neuronal stores of serotonin, did not block the potentiation effect of SSRIs. These results suggest again that SSRIs were working through a mechanism unrelated to serotonin release.

Thus, SSRIs have additional effects on neurosteroidogenesis and on the efficacy of neurosteroid action *in vivo* at the GABA$_A$ receptor that may play a role in their effective treatment of PDD. Furthermore, the data also suggest that alterations in normal neurosteroidogenesis and/or in GABA$_A$-receptor sensitivity may be involved in the etiology of PDD.

VI. SSRIs, Neurosteroids, and Depression

In addition to a potential role of neurosteroids in hormonally related depression, such as PDD, neurosteroids may play a role in other forms of depression. Several studies have indicated that the concentrations of certain neurosteroids may be lower in depressed patients in both men and women, than in people who are not clinically diagnosed as depressed (Romeo *et al.*, 1998; Uzunova *et al.*, 1998). In a study by Uzunova *et al.* (1998) the cerebral spinal fluid (CSF) concentrations of neurosteroids in both men and women were assessed by gas chromatography–mass spectrometry. The authors found that before treatment, the concentrations of allopregnanolone were lower in patients diagnosed with unipolar depression (mean = 15 ± 1.5 pM; range: 7–30 pM) than in normal patients (mean = 39 ± 4.9 pM) who did not have this diagnosis. Steroids were measured in four different cisternal–lumbar CSF fractions in each patient, with no differences found in the different fractions. Following treatment for 8 weeks with fluvoxamine or fluoxetine, titrated to obtain an optimal clinical response with minimal side effects, CSF levels of the same neurosteroids were assessed. All patients were rated by the Hamilton Rating Scale for Depression (HAM-D) before and after treatment. In patients in whom there was a response to the SSRIs, assessed by a decrease in their HAM-D scores, there was a concomitant increase in the concentrations of allopregnanolone, and the 5β-stereoisomer pregnanolone, in their CSF. Allopregnanolone content doubled in the CSF, giving an average increase of 16 pM, to a final average of 31 ± 4.7 pM, similar to those concentrations found in patients who were not depressed (39 pM). This concentration of allopregnanolone was low (about 30 pM), while the concentration of allopregnanolone needed to modulate $GABA_A$ receptors has been reported to be in the low nanomolar range. Extrapolating from animal studies, the CSF allopregnanolone concentration is only about 10% of the brain content, and hence the actual brain concentration of allopregnanolone may have reached the low nanomolar concentration. The largest increase in the CSF content of allopregnanolone occurred in patients who had the greatest improvement in HAM-D scores, while patients who did not improve also did not show a significant increase in CSF allopregnanolone content. Neither pregnenolone nor progesterone concentrations fluctuated in a manner similar to that found for allopregnanolone, indicating that the effect of the SSRIs was specific. Thus, the authors concluded that depressed patients have lower than normal concentrations of allopregnanolone that can be increased by SSRI treatment.

In another study (Romeo et al., 1998), 8 drug-naïve outpatients with major depression were studied before and during treatment with fluoxetine, and 11 inpatients with major depression were studied during a severe depressive episode and also after recovery following treatment with various antidepressants. In this study, *plasma* concentrations of neurosteroids were assayed by gas chromatography–mass spectrometry. The authors found that patients diagnosed with unipolar depression had significantly lower concentrations of the neuroactive steroids allopregnanolone and pregnanolone than in normal control subjects. They also found an increase in the neuroinactive compound epiallopregnanolone ($3\beta,5\alpha$-TH PROG). Actual ranges of the concentrations of these steroids were not given. Treatment with fluoxetine for ~7 weeks (50 days) caused a significant increase in allopregnanolone and pregnanolone and a decrease in epiallopregnanolone concentrations ($P < 0.05$). Similar results, but not of the same magnitude, were seen in a group of 11 patients given a variety of psychopharmacologic treatments (mean 55 days), including amitriptyline, clomipramine, nortriptyline, viloxazine, and lithium. These data suggest that changes in plasma concentrations of neuroactive steroids during antidepressant treatment may not be unique to SSRI treatment but may be a common feature of treatment with antidepressants. Unlike the Uzunov *et al.* (1996) study, this study was unable to correlate changes in neurosteroid concentrations with changes in mood assessment (based on HAM-D scores). These results indicate that increasing the low concentrations of neurosteroids in depressed patients may play a role in successful antidepressant therapy and further suggest that dysregulation of neurosteroid synthesis may play a role in major depression.

VII. Conclusions

The studies presented in this chapter are the initial studies that link neurosteroid concentrations and changes in their concentrations to human affective disorders. The number of studies is small, demonstrating the need for more and larger studies to demonstrate definitively a role for these neuroactive compounds in various diseases. The results are certainly enticing and, although not yet conclusive, point to a role for neurosteroids in hormonally regulated depression, such as PDD, as well as in other forms of unipolar depression. The hypothesis that changes in neurosteroid concentrations may underlie effective antidepressant therapy certainly warrants further study.

Acknowledgments

This work was supported by grants from the NIH to SHM (HD 27970) and to LDG (NS 01979), and by grants from The March of Dimes and Ara Parseghian Medical Research Foundation (to SHM) by an NIH training grant (DK 07161).

References

American Psychiatric Association. (1994). "Diagnostic and statistical manual of mental disorders" (4th edition). Washington, DC.
Bicikoya, M., Dibbelt, L., Hill, M., Hampl, R., and Stárka, L. (1998). Allopregnanolone in women with premenstrual syndrome. *Horm. Metab. Res.* **30,** 227–230.
Compagnone, N. A., and Mellon, S. H. (2000). Neurosteroids: Biosynthesis and function of these novel neuromodulators. *Front. Neuroendocrinol.* **21,** 1–56.

Dhanvantari, S., and Wiebe, J. P. (1994). Suppression of follicle-stimulating hormone by the gonadal- and neurosteroid 3 alpha-hydroxy-4-pregnen-20-one involves actions at the level of the gonadotrope membrane/calcium channel. *Endocrinology* **134,** 371–376.

Dimmock, P. W., Wyatt, K. M., Jones, P. W., and O'Brien, P. M. (2000). Efficacy of selective serotonin-reuptake inhibitors in premenstrual syndrome: A systematic review. *Lancet* **356,** 1131–1136.

Endicott, J. (1993). The menstrual cycle and mood disorders. *J. Affect. Disord.* **29,** 193–200.

Endicott, J., Amsterdam, J., Eriksson, E., Frank, E., Freeman, E., Hirschfeld, R., Ling, F., Parry, B., Pearlstein, T., Rosenbaum, J., Rubinow, D., Schmidt, P., Severino, S., Steiner, M., Stewart, D. E., and Thys-Jacobs, S. (1999). Is premenstrual dysphoric disorder a distinct clinical entity? *J. Womens Health Gend. Based Med.* **8,** 663–679.

Eriksson, E. (1999). Serotonin reuptake inhibitors for the treatment of premenstrual dysphoria. *Int. Clin. Psychopharmacol.* **14 Suppl 2,** S27–S33.

Freeman, E. W., and Halbreich, U. (1998). Premenstrual syndromes. *Psychopharmacol. Bull.* **34,** 291–295.

Griffin, L. D., and Mellon, S. H. (1999). Selective serotonin reuptake inhibitors directly alter activity of neurosteroidogenic enzymes. *Proc. Natl. Acad. Sci. USA* **96,** 13512–13517.

Griffin, L. D., and Mellon, S. H. (2001). Biosynthesis of the neurosteroid 3α-hydroxy-4prenene-20-one (3α-HP), a specific inhibitor of FSH release. *Endocronology* (In Press).

Halbreich, U. (1999). Premenstrual syndromes: closing the 20th century chapters. *Curr. Opin. Obstet. Gynecol.* **11,** 265–270.

Hammarback, S., and Backström, T. (1988). Induced anovulation as treatment of premenstrual tension syndrome. A double-blind cross-over study with GnRH-agonist versus placebo. *Acta Obstet. Gynecol. Scand.* **67,** 159–166.

Hammarback, S., Damber, J. E., and Backström, T. (1989). Relationship between symptom severity and hormone changes in women with premenstrual syndrome. *J. Clin. Endocrinol. Metab.* **68,** 125–130.

Hammarback, S., Ekholm, U. B., and Backström, T. (1991). Spontaneous anovulation causing disappearance of cyclical symptoms in women with the premenstrual syndrome. *Acta Endocrinol. (Copenhagen)* **125,** 132–137.

Kessel, B. (2000). Premenstrual syndrome. Advances in diagnosis and treatment. *Obstet. Gynecol. Clin. North Am.* **27,** 625–639.

Khisti, R. T., and Chopde, C. T. (2000). Serotonergic agents modulate antidepressant-like effect of the neurosteroid 3alpha-hydroxy-5alpha-pregnan-20-one in mice. *Brain Res.* **865,** 291–300.

Masand, P. S., and Gupta, S. (1999). Selective serotonin-reuptake inhibitors: An update. *Harv. Rev. Psychiatry* **7,** 69–84.

Menkes, D. B., Taghavi, E., Mason, P. A., and Howard, R. C. (1993). Fluoxetine's spectrum of action in premenstrual syndrome. *Int. Clin. Psychopharmacol.* **8,** 95–102.

Menkes, D. B., Taghavi, E., Mason, P. A., Spears, G. F., and Howard, R. C. (1992). Fluoxetine treatment of severe premenstrual syndrome. *Br. Med. J.* **305,** 346–347.

Monteleone, P., Luisi, S., Tonetti, A., Bernardi, F., Genazzani, A. D., Luisi, M., Petraglia, F., and Genazzani, A. R. (2000). Allopregnanolone concentrations and premenstrual syndrome. *Eur. J. Endocrinol.* **142,** 269–273.

Morrow, A. L., Pace, J. R., Purdy, R. H., and Paul, S. M. (1990). Characterization of steroid interactions with gamma-aminobutyric acid receptor-gated chloride ion channels: Evidence for multiple steroid recognition sites. *Mol. Pharmacol.* **37,** 263–270.

Pearlstein, T. B., Frank, E., Rivera-Tovar, A., Thoft, J. S., Jacobs, E., and Mieczkowski, T. A. (1990). Prevalence of axis I and axis II disorders in women with late luteal phase dysphoric disorder. *J. Affect. Disord.* **20,** 129–134.

Rapkin, A. J. (1992). The role of serotonin in premenstrual syndrome. *Clin. Obstet. Gynecol.* **35**, 629–636.

Rapkin, A. J., Edelmuth, E., Chang, L. C., Reading, A. E., McGuire, M. T., and Su, T. P. (1987). Whole-blood serotonin in premenstrual syndrome. *Obstet. Gynecol.* **70**, 533–537.

Rapkin, A. J., Morgan, M., Goldman, L., Brann, D. W., Simone, D., and Mahesh, V. B. (1997). Progesterone metabolite allopregnanolone in women with premenstrual syndrome. *Obstet. Gynecol.* **90**, 709–714.

Rojansky, N., Halbreich, U., Zander, K., Barkai, A., and Goldstein, S. (1991). Imipramine receptor binding and serotonin uptake in platelets of women with premenstrual changes. *Gynecol. Obstet. Invest.* **31**, 146–152.

Romeo, E., Strohle, A., Spalletta, G., di Michele, F., Hermann, B., Holsboer, F., Pasini, A., and Rupprecht, R. (1998). Effects of antidepressant treatment on neuroactive steroids in major depression. *Am. J. Psychiatry* **155**, 910–913.

Rubinow, D. R., and Schmidt, P. J. (1995). The neuroendocrinology of menstrual cycle mood disorders. *Ann. NY Acad. Sci.* **771**, 648–659.

Rubinow, D. R., Schmidt, P. J., and Roca, C. A. (1998). Hormone measures in reproductive endocrine-related mood disorders: diagnostic issues. *Psychopharmacol. Bull.* **34**, 289–290.

Schmidt, P. J., Purdy, R. H., Moore, P. H., Paul, S. M., and Rubinow, D. R. (1994). Circulating levels of anxiolytic steroids in the luteal phase in women with premenstrual syndrome and in control subjects. *J. Clin. Endocrinol. Metab.* **79**, 1256–1260.

Steiner, M. (1992). Female-specific mood disorders. *Clin. Obstet. Gynecol.* **35**, 599–611.

Steiner, M. (1997). Premenstrual syndromes. *Annu. Rev. Med.* **48**, 447–455.

Steiner, M., and Pearlstein, T. (2000). Premenstrual dysphoria and the serotonin system: Pathophysiology and treatment. *J. Clin. Psychiatry.* **61**, 17–21.

Steiner, M., Steinberg, S., Stewart, D., Carter, D., Berger, C., Reid, R., Grover, D., and Streiner, D. (1995). Fluoxetine in the treatment of premenstrual dysphoria. Canadian Fluoxetine/Premenstrual Dysphoria Collaborative Study Group. *N. Engl. J. Med.* **332**, 1529–1534.

Stone, A. B., Pearlstein, T. B., and Brown, W. A. (1990). Fluoxetine in the treatment of premenstrual syndrome. *Psychopharmacol. Bull.* **26**, 331–335.

Stone, A. B., Pearlstein, T. B., and Brown, W. A. (1991). Fluoxetine in the treatment of late luteal phase dysphoric disorder. *J. Clin. Psychiatry* **52**, 290–293.

Su, T. P., Schmidt, P. J., Danaceau, M. A., Tobin, M. B., Rosenstein, D. L., Murphy, D. L., and Rubinow, D. R. (1997). Fluoxetine in the treatment of premenstrual dysphoria. *Neuropsychopharmacology* **16**, 346–356.

Sundström, I., Andersson, A., Nyberg, S., Ashbrook, D., Purdy, R. H., and Bäckström, T. (1988). Patients with premenstrual syndrome have a different sensitivity to a neuroactive steroid during the menstrual cycle compared to control subjects. *Neuroendocrinology* **67**, 126–138.

Uzunov, D. P., Cooper, T. B., Costa, E., and Guidotti, A. (1996). Fluoxetine-elicited changes in brain neurosteroid content measured by negative ion mass fragmentography. *Proc. Natl. Acad. Sci. USA* **93**, 12599–12604.

Uzunova, V., Sheline, Y., Davis, J. M., Rasmusson, A., Uzunov, D. P., Costa, E., and Guidotti, A. (1998). Increase in the cerebrospinal fluid content of neurosteroids in patients with unipolar major depression who are receiving fluoxetine or fluvoxamine. *Proc. Natl. Acad. Sci. USA* **95**, 3239–3244.

Wang, M., Seippel, L., Purdy, R. H., and backstrom, T. (1996). Relationship between symptom severity and steroid variation in women with premenstrual syndrome: Study on serum pregnenolone, pregnenolone sulfate, 5 alpha-pregnane- 3,20-dione and 3 alpha-hydroxy-5 alpha-pregnan-20-one. *J. Clin. Endocrinol. Metab.* **81**, 1076–1082.

Wiebe, J. P., Boushy, D., and Wolfe, M. (1997). Synthesis, metabolism and levels of the neuroactive steroid, 3alpha-hydroxy-4-pregnen-20-one (3alphaHP), in rat pituitaries. *Brain Res.* **764,** 158–166.

Wiebe, J. P., Dhanvantari, S., Watson, P. H., and Huang, Y. (1994). Suppression in gonadotropes of gonadotropin-releasing hormone-stimulated follicle-stimulating hormone release by the gonadal- and neurosteroid 3 alpha-hydroxy-4-pregnen-20-one involves cytosolic calcium. *Endocrinology* **134,** 377–382.

Wiebe, J. P., and Wood, P. H. (1987). Selective suppression of follicle-stimulating hormone by 3 alpha-hydroxy-4-pregnen-20-one, a steroid found in Sertoli cells. *Endocrinology* **120,** 2259–2264.

Wood, P. H., and Wiebe, J. P. (1989). Selective suppression of follicle-stimulating hormone secretion in anterior pituitary cells by the gonadal steroid 3 alpha-hydroxy-4-pregnen-20-one. *Endocrinology* **125,** 41–48.

Wood, S. H., Mortola, J. F., Chan, Y. F., Moossazadeh, F., and Yen, S. S. (1992). Treatment of premenstrual syndrome with fluoxetine: A double-blind, placebo-controlled, crossover study. *Obstet. Gynecol.* **80,** 339–344.

Wurtman, J. J. (1993). Depression and weight gain: The serotonin connection. *J. Affect. Disord.* **29,** 183–192.

Yatham, L. N. (1993). Is 5HT1A receptor subsensitivity a trait marker for late luteal phase dysphoric disorder? A pilot study. *Can. J. Psychiatry* **38,** 662–664.

INDEX

A

ACTH, *see* Adrenocorticotropic hormone
AD, *see* Alzheimer's disease
Adrenocorticotropic hormone (ACTH)
  chronic stress release, 255–257
  induction of steroid synthesis, 128–129
Adrenodoxin reductase
  brain expression, 51
  function, 39
  genes, 39
Aggression
  animal models, 333
  ethanol effects, 334
  5-HT$_{1B}$ role, 333–334
  neurosteroid effects on behavior, 334–335
Aging, *see also* Menopause
  dehydroepiandrosterone and sulfate in human aging cognition
    decline of levels, 303–304
    overview of cognitive effects, 309
    plasma levels and cognitive function
      Alzheimer's disease patients, 307
      healthy elderly population, 305
      residential care population, 305, 307
      study design, 306–307
    stable analog studies, 311
    sulfate effects in cognition, 310–311
    treatment effects on cognitive performance, 307–308
  learning and memory animal model studies
    excitatory neurosteroid effects, 286–287
    pregnenolone sulfate levels and aging rat performance, 301–303
  myelin protein expression, 162
  neurosteroid level effects
    dehydroepiandrosterone, 303–304, 411
    3α-hydroxy-5α-pregnan-20-one, 410–411
  steroidogenesis enzyme expression effects, 161

Alcohol, *see* Ethanol
Allopregnanolone, *see* 3α-Hydroxy-5α-pregnan-20-one
Allotetrahydrodeoxycorticosterone, *see* 3α,5α-Tetrahydrodeoxycorticosterone
Alzheimer's disease (AD)
  cholinergic system degradation, 293, 389
  cortisol levels, 412–413
  dehydroepiandrosterone levels and cognitive function, 307, 389, 411–413
  dehydroepiandrosterone sulfate neuroprotection, 389–390
  estrogen replacement therapy effects, 429–430
  3α-hydroxy-5α-pregnan-20-one levels, 412–413
  prevalence, 429
γ-Aminobutyric acid type A receptor (GABA$_A$)
  benzodiazepine action regulation, 118–119, 215–216, 244, 338–339
  binding sites for steroids, 196–197, 468–469
  brain localization and steroid effects, 381–382
  central depressive effects of steroids, 178
  dehydroepiandrosterone sulfate neuroprotection mediation, 386–387
  discriminative stimulus modulation, 338–339, 341
  ethanol effects
    behavioral effects, 351
    brain receptor response by region, 352–353
    cell-type responsiveness, 353–354
    chloride flux enhancement, 352
    3α-hydroxy-5α-pregnan-20-one modulation, *see* 3α-Hydroxy-5α-pregnan-20-one
    overview, 350
    phosphorylation effects, 353

γ-Aminobutyric acid type A receptor (GABA$_A$) (*continued*)
  prospects for study, 369
  tolerance role, 363–364
  3β-hydroxysteroid dehydrogenase activity regulation, 63–64
  inhibitory synaptic transmission modulation by neurosteroids, 190–192
  ion channel gating effects of neurosteroids, 188–189, 208, 214–215
  learning and memory role
    neurosteroid interactions, 295
    overview, 293
  ligand diversity, 178
  nerve transection effects on levels, 164
  nervous system distribution, 146
  oral contraceptives and plasticity effects, 231–234
  pharmacological profile, 120, 246–247
  phosphorylation effects on synaptic transmission modulation by neurosteroids, 192–194
  pregnancy effects
    brain distribution of expression
      cerebral cortex, 225–228
      hippocampus, 225–228
      hypothalamic magnocellular neurons, 228–229, 231
      subunit expression, 226–229, 231
    chloride channel, 222
    expression levels, 220–221
    finasteride effects, 224
  premenstrual syndrome expression levels, 404
  progesterone effects
    chloride current, 214–216, 219, 225
    expression response, 211–212, 214, 224–225, 235
    mechanism of plasticity effects, 234–235
    withdrawal effects
      behavioral response, 220
      ligand specificity of receptor, 217–219
      pseudo-pregnancy studies in rat, 218–220
      subunit expression, 216–217, 219–220
  steroids
    modulation overview, 18–19, 170, 179, 198, 246–247, 381–384, 424, 465
    Schwann cell receptor interactions, 166–167
  uncoupling of recognition sites, 209–210
  stress response
    anxiety behavior mediation, 245, 325–328
    anxiolytic neurosteroid action, 325–326
    downregulation response, 244–246, 257, 259–261, 264
    hypothalamic–pituitary–adrenal axis activation role, 250–253
    transmission studies in acute stress, 248, 250
  structure–activity relationships for steroid ligands
    modifaction of behavioral response, 198–199
    oral bioavailability modification effects, 195–196
    overview, 194, 462
    water-soluble steroids, 196
  subunit composition
    effects on behavioral responses, 178–179
    effects on neurosteroid action
      α subunits, 185–186, 382
      β subunits, 186
      γ subunits, 187, 233–234, 382
      δ subunits, 187–188
      ε subunits, 188
    oral contraceptive effects on expression, 232–233
    pregnancy effects, 226–228
    progesterone withdrawal effects, 216–217, 219–220
α-Amino-3-hydroxy-5-methyl-4-isopropionic acid (AMPA) receptor, steroid binding selectivity, 184–185, 468
AMPA receptor, *see* α-Amino-3-hydroxy-5-methyl-4-isopropionic acid receptor
Anorexia nervosa, allopregnanolone levels, 405

B

Brain transport, neurosteroids, 37–38

C

CAH, *see* Congenital adrenal hyperplasia
Cancer, peripheral-type benzodiazepine receptor role, 134–135

INDEX

Cholesterol
  peripheral-type benzodiazepine
    receptor-mediated transport,
    126–127, 137
  synthesis, 7
  transport
    overview, 118
    steroidogenic acute regulatory protein
      role, 136
Cognition, see Learning and memory
Congenital adrenal hyperplasia (CAH),
  steroidogenic acute regulatory protein
  mutations, 136
Corticosteroid
  glial cell effects, 157
  synthesis in nervous system, 10–11
Cortisol, Alzheimer's disease levels, 412–413
Cytochrome b_5
  adrenal distribution, 42
  brain expression, 53
  function, 41–42

D

DBI, see Diazepam-binding inhibitor
Dehydroepiandrosterone (DHEA)
  aging effects on levels, 411
  Alzheimer's disease levels, 411–413
  behavioral effects in mice, 12
  brain distribution, 6–7
  cognition and human aging studies
    decline of levels, 303–304, 384
    overview of cognitive effects, 309
    plasma levels and cognitive function
      Alzheimer's disease patients, 307
      healthy elderly population, 305
      residential care population, 305, 307
      study design, 306–307
    stable analog studies, 311
    sulfate effects in cognition, 310–311
    treatment effects on cognitive
      performance, 307–308
  depression treatment, 471
  fetal synthesis, 407
  7-hydroxylation
    anti-glucocorticoid effects
      dexamethasone-induced apoptosis
        prevention, 84–85
      glucocorticoid receptor binding,
        84–85
    hydroxylation specificity of effects,
      86, 88
    7α-hydroxylation, see P4507B1
    7β-hydroxylation in brain, 83–84
    serum levels, 83–84
    tissue distribution of metabolism, 88, 90
  learning and memory effects
    aging studies, 286–287
    developmental effects, 281
    dose dependency, 289
    learning session administration effects,
      285–286
    post-training administration effects,
      282–284
    post-training versus preretention
      administration effects, 285
    pretraining administration effects, 282
    pretraining versus post-training
      effects, 284
    pretraining versus post-training versus
      preretention administration
      effects, 285
    summary of studies, 289–292
    time-dependent effects, 289
  menopause supplementation studies,
    407–408
  metabolism in brain, 10
  neuroprotection mechanisms, 412
  premenstrual syndrome levels, 404
  puberty levels, 402
  supplementation effects in elderly, 411
  synaptic plasticity role, 384
  synthesis, 9, 80
Dehydroepiandrosterone sulfate (DHEAS)
  γ-aminobutyric acid receptor
    interactions, 19
  brain distribution, 3, 6–7
  conjugation in brain, 9–10, 12
  learning and memory effects
    aging studies, 286–287
    developmental effects, 281
    dose dependency, 289
    learning session administration effects,
      285–286
    long-term potentiation role, 300
    neurotransmitter system interactions,
      294–295, 298–299
    overview, 329–333
    post-training administration effects,
      282–284

Dehydroepiandrosterone sulfate
(DHEAS) (continued)
post-training versus preretention
administration effects, 285
pretraining administration effects, 282
pretraining versus post-training
effects, 284
pretraining versus post-training versus
preretention administration
effects, 285
summary of studies, 289–292
time-dependent effects, 289
menopause supplementation
studies, 409
neuroprotection
Alzheimer's disease, 389–390
decline with age, 384–385, 391
stroke and ischemia studies, 385–387, 391
pregnancy levels, 407
synaptic plasticity role, 384
Depression
dehydroepiandrosterone sulfate
treatment, 471
neurosteroid concentrations, 209, 487–489
DHEA, see Dehydroepiandrosterone
DHEAS, see Dehydroepiandrosterone sulfate
Diazepam-binding inhibitor (DBI)
anxiolytic neurosteroid interactions, 326–327
brain expression, 131
peripheral-type benzodiazepine receptor
ligand, 120
stress response, 262
3α-Dihydroprogesterone
comparison with, 3α-hydroxy-5α-
pregnan-20-one potency, 484
selective serotonin reuptake inhibitor
interactions, 486–487
synthesis, 484–485
Discriminative stimulus
drug discrimination, 338–339
features, 337–338
neurosteroid effectss, 339–341

E

β-Endorphin, marker for hormone
replacement therapy, 408–409

Epilepsy
cortocosteroid effects, 442
medroxyprogesterone acetate effects, 444–446
menstrual dysfunction, 443–445
premenstrual onset, 441–442, 445
seizure threshold effects
estradiol, 442
progesterone, 442, 445, 447
testosterone, 446
Estrogen
epilepsy seizure threshold effects, 442
hormone replacement therapy
Alzheimer's disease, 429–430
β-endorphin as marker, 408–409
hot flush effects in menopause, 437–438
mood change effects in menopause, 438–439
sensory-motor and cognitive function
effects, 427–429
side effects, 439–441
N-methyl-D-aspartate receptor modulation
of action, 423–424, 470
overview of central nervous system
function effects, 421–422
premenstrual syndrome effects, 432–433
receptors, 422–423
Schwann cell proliferation effects, 169–170
synaptic plasticity role, 423
Ethanol
aggression effects, 334
cortisol response in sons of alcoholics, 363
$GABA_A$ effects
behavioral effects, 351
brain receptor response by region, 352–353
cell-type responsiveness, 353–354
chloride flux enhancement, 352
overview, 350
phosphorylation effects, 353
prospects for study, 369
tolerance role, 363–364
3α-hydroxy-5α-pregnan-20-one role
in action
behavioral effects, 358–359
brain concentrations in response
to ethanol
acute administration, 354–355, 357
chronic administration, 361–362
dependence role

GABA$_A$ sensitivity and function
  relationship with altered steroid
  levels, 367–368
 sensitization in ethanol-dependent
  rats, 364–365
 withdrawal seizure sensitivity
  associated with altered levels
  in mice, 365–367
 firing rates of medial septum/diagonal
  band of Broca neurons, 359
 indomethacin effects on ethanol
  induction, 357–358
 self-admistration of ethanol alterations,
  359–361
 tolerance role, 361–364
hypothalamic–pituitary–adrenal axis
 activation, 350
withdrawal symptoms, 361

F

Feeding, neurosteroid regulation, 336–337, 382
Finasteride
 effects on ethanol induction of
  3α-hydroxy-5α-pregnan-20-one,
  358–359
 inhibition of hormone synthesis in
  pregnancy, 224
Flumazenil, anxiolytic neurosteroid
 inhibition, 326
Flunitrazepam, inhibition of hormone
 synthesis, 125–126

G

GABA$_A$, see γ-Aminobutyric acid
 type A receptor
Gas chromatography–mass spectrometry
 (GC–MS)
 columns for gas chromatography, 106–107
 data acquisition modes, 107–108
 derivatization, 103, 106
 high-performance liquid chromatography
  preparation, 99–100
 isotopic dilution for quantitative
  analysis, 107
 sulfated steroids
  atmospheric pressure chemical
   ionization, 111

derivatization, 108–109
 electron impact ionization, 109
 electrospray ionization, 109–112
 fast atom bombardment, 109
 Fourier transform ion cyclotron
  resonance detection, 111–112
 tandem mass spectrometry, 108
 unconjugated neurosteroid analysis,
  100–101
GC–MS, see Gas chromatography–mass
 spectrometry
GFAP, see Glial fibrillary acidic protein
Ginkgo biloba extract, effects on
 peripheral-type benzodiazepine
 receptor and glucocorticoid expression,
 129–130
Glial cell
 neuron interactions
  neurosteroid synthesis effects, 154–155
  overview of effects, 153–154
  systems for study, 154
 steroid effects in central nervous system
  glial fibrillary acidic protein expression
   corticosteroids, 157
   sex steroids, 156–157
  oligodendrocytes, 160–161
Glial fibrillary acidic protein (GFAP), effects
 on expression
 corticosteroids, 157
 sex steroids, 156–157
Glucocorticoid receptor,
 dehydroepiandrosterone
 7-hydroxylated derivative effects
 binding, 84–85
 dexamethasone-induced apoptosis
  prevention, 84–85
 hydroxylation specificity of effects, 86, 88
Glucocorticoid response element, astrocyte
 genes, 157, 160
Glycine receptor
 steroid binding selectivity, 181, 467
 subunits, 180–181

H

hCG, see Human chorionic gonadotropin
High-performance liquid chromatography
 (HPLC), neurosteroid preparation
 prior to gas chromatography–mass
 spectrometry, 99–100, 400

Hormone replacement therapy (HRT), see Estrogen; Progesterone
HPLC, see High-performance liquid chromatography
HRT, see Hormone replacement therapy
HST, see Sulfotransferase
5-HT receptor, see Serotonin receptor
Human chorionic gonadotropin (hCG), regulation of peripheral-type benzodiazepine receptor, 126
7α-Hydroxylase
  function, 48–49
  tissue distribution, 49
24-Hydroxylase
  brain distribution, 48
  function, 47–48
26α-Hydroxylase, tissue distribution and expression regulation, 48
3α-Hydroxy-5α-pregnan-20-one
  adrenal synthesis, 401–402
  aging effects on levels, 410–411
  Alzheimer's disease levels, 412–413
  anesthesia effects on levels, 427
  anorexia nervosa levels, 405
  antidepressant effects on brain levels, 472
  anxiolytic properties, 324–326
  brain
    distribution and levels of steroid, 425
    synthesis, 400, 483–484
  cerebrospinal fluid content, 401
  depression concentrations, 209, 487–489
  discriminative stimulus effects, 339–341
  ethanol action role
    behavioral effects, 358–359
    brain concentrations in response to ethanol
      acute administration, 354–355, 357
      chronic administration, 361–362
    dependence role
      GABA$_A$ sensitivity and function relationship with altered steroid levels, 367–368
      sensitization in ethanol-dependent rats, 364–365
      withdrawal seizure sensitivity associated with altered levels in mice, 365–367
    firing rates of medial septum/diagonal band of Broca neurons, 359

  indomethacin effects on ethanol induction, 357–358
  self-admistration of ethanol alterations, 359–361
  tolerance role, 361–364
feeding effects, 336–337
GABA$_A$ receptor interactions
  expression response, 211–212, 214, 234
  inhibitory synaptic transmission modulation, 190–192
  ion channel modulation, 188–189, 208, 210, 351
  phosphorylation effects on modulation, 192–194
  structure–activity relationships, 194–196
  subunit composition effects, 185–188
  uncoupling of recognition sites, 209–210
learning and memory effects, 287–288, 292
menstrual cycle changes, 209
ovary synthesis, 401–402
pregnancy
  hypertension control, 406–407
  levels, 220–221, 231, 405–406
premenstrual syndrome
  effects, 433–434
  levels, 403–404, 481, 483
radioimmunoassay, 399–400
receptor modulation of transcription, 469–470
reinforcement regulation, 337
selective serotonin reuptake inhibitor interactions, 486–487
serotonin receptor interactions, 183
sleep effects, 335–336
stress effects
  acute stress
    brain concentration, 247–248, 322–323, 426
    GABAergic transmission studies, 248, 250
    hypothalamic–pituitary–adrenal axis activation, 250–253, 261
    physiological role of response, 250
    plasma concentration, 247
  chronic stress
    adrenocorticotropic hormone response, 255–257
    brain concentration, 254–255, 260

GABA$_A$ modulation, 260–261
plasma concentration, 255, 260
rodent models, 253
social isolation effects, 253–255, 260
mechanisms, 261–264
therapeutic prospects, 471–474
3α-Hydroxysteroid dehydrogenase
aging effects on expression, 161
brain expression, 58–59, 149
function, 45–46, 147–148, 485
genes, 46, 148
isoforms, 150–151, 485
peripheral nervous system expresson, 161–162
progesterone response in cultured cells, 211
regulation in central nervous system, 151–153
substrate specificity, 147, 151
3β-Hydroxysteroid dehydrogenase
activities, 40
genes, 40
nervous system expression, 53–54
regulation of expression, 63–65
11β-Hydroxysteroid dehydrogenase
brain expression, 56
function, 44
isoforms, 44
17β-Hydroxysteroid dehydrogenase
brain expression, 54–55
types and function, 43
Hypothalamic amenorrhea
anorexia nervosa as cause, 405
hormone levels, 405

K

Kainate receptor, steroid binding selectivity, 184–185, 468
17-Ketosteroid reductase, types, 43, 43

L

LC–MS, see Liquid chromatography–mass spectrometry
Learning and memory
animal models
complex learning tasks, 279–280
conditioning learning tasks, 278–279
maze tests, 328–329

overview, 277, 328
simple learning tasks, 277–278
definitions, 275
dehydroepiandrosterone and sulfate in human aging
decline of levels, 303–304
overview of cognitive effects, 309
plasma levels and cognitive function
Alzheimer's disease patients, 307
healthy elderly population, 305
residential care population, 305, 307
study design, 306–307
stable analog studies, 311
sulfate effects in cognition, 310–311
treatment effects on cognitive performance, 307–308
estrogen replacement therapy effects, 427–429
excitatory neurosteroid effects
aging studies, 286–287
developmental effects, 281
dose dependency, 289, 332
learning session administration effects, 285–286
overview, 329–333
post-training administration effects, 282–284
post-training versus preretention administration effects, 285
pretraining administration effects, 282
pretraining versus post-training effects, 284
pretraining versus post-training versus preretention administration effects, 285
summary of studies, 289–292
time-dependent effects, 289
inhibitory neurosteroid effects
developmental effects, 287
3α-hydroxy-5α-pregnan-20-one effects, 287–288, 292
progesterone effects, 288–289
learning types, 276
long-term potentiation and neurosteroid effects, 299–300, 332–333
memory types
declarative, 276–277
episodic, 277
nondeclarative, 277
semantic, 277

Learning and memory (*continued*)
  neurosteroid role
    interconversion of neurosteroids, 309–310
    overview, 274–275
    quantitative analysis, 311
  neurotransmitter systems in function
    cholinergic systems
      dehydroepiandrosterone sulfate interactions, 294–295
      overview, 292–293
      pregnenolone sulfate interactions, 294–295
    GABAergic systems
      neurosteroid interactions, 295, 332–333
      overview, 293
    N-methyl-D-aspartate receptor
      neurosteroid interactions, 295, 298, 330
      overview, 293–294
    sigma receptors
      neurosteroid interactions, 298–299, 331
      overview, 294
  pregnenolone sulfate
    rat hippocampus, 20, 22–23, 25, 300
    levels and aging rat performance, 301–303
Liquid chromatography–mass spectrometry (LC–MS)
  sulfated steroid analysis, 110–111
  tandem mass spectrometry, 108
  thermospray liquid chromatography–mass spectrometry, 111
Long-term potentiation (LTP), neurosteroid effects, 299–300, 332, 423
LTP, *see* Long-term potentiation

M

MAP2, *see* Microtubule-associated protein 2
Mass spectrometry, *see* Gas chromatography–mass spectrometry; Liquid chromatography–mass spectrometry
Medroxyprogesterone acetate
  premenstrual syndrome management, 441
  seizure inhibition, 444–446

Memory, *see* Learning and memory
Menopause
  age of onset, 437
  allopregnanolone variations, 424
  dehydroepiandrosterone
    sulfate supplementation, 409
    supplementation studies, 407–408
  β-endorphin as marker for hormone replacement therapy, 408–409
  gonadotropin pulses, 436
  hot flushes, 437–438
  mood changes, 438–439
  progesterone levels, 409
  steroid hormone synthesis and metabolism, 436–437
Menstrual cycle, *see also* Premenstrual syndrome
  3α-hydroxy-5α-pregnan-20-one changes, 209
  premenstrual epilepsy, *see* Epilepsy
Metabolism, neurosteroids
  7-hydroxylation products, *see* P4507B1
  pathway elucidation, 4–5
N-Methyl-D-aspartate (NMDA) receptor
  behavior mediation of neurosteroids, 383–384
  estrogen modulation, 423–424
  learning and memory role
    neurosteroid interactions, 295, 298, 330
    overview, 293–294
  steroid binding selectivity, 184–185, 383
  subunits, 383
Microtubule-associated protein 2 (MAP2), neurosteroid binding, 23, 25
Myelin
  aging effects on protein expression, 162
  castration effects on protein expression, 165
  progesterone effects in rats
    adult rats, 163
    aged rats, 162–163
    γ-aminobutyric acid receptor modulation, 166–167
    nerve repair, 163–165
    Schwann cell culture, 165–166

N

Neurosteroid, definition, 34–35, 37, 246, 274, 422

Nicotinic acetylcholine receptors, steroid binding selectivity, 182
NMDA receptor, see N-Methyl-D-aspartate receptor

O

Oral contraceptives
  $GABA_A$ plasticity effects, 231–234
  mood side effects, 434–435
  neuroactive steroid response, 232
  premenstrual syndrome management, 434–436
Oxytocin receptor, steroid binding selectivity, 468

P

P450 aro
  brain expression, 55–56
  function, 44
  gene, 44
P4507B1
  brain distribution, 81
  dehydroepiandrosterone as substrate in brain, 81
  inhibition studies, 82
  mouse strain differences, 81–82
  pregnenolone as substrate in brain, 81
P450c11
  functions, 43
  genes, 42–43
  nervous system expression, 54
P450c17
  activities, 40–41
  adrenal distribution, 41
  developmental regulation of expression, 52–53
  regulation of expression, 62–63
P450c21, function, 42
P450 reductase
  brain expression, 53
  function, 41
P450scc
  brain distribution, 7–8
  brain expression, 49–50
  developmental regulation, 50–51
  genes, 39
  glial cells, 132
  reaction rate, 38, 118
  regulation of expression, 8–9, 61–62
PBR, see Peripheral-type benzodiazepine receptor
Peripheral-type benzodiazepine receptor (PBR)
  brain expression, 131
  cholesterol synthesis role, 7
  developmental expression, 128
  discovery, 119, 121
  disease roles
    cancer, 134–135
    neurodegenerative diseases, 135
  gene cloning, 120–121
  pharmacological profile, 120
  pregnancy levels, 130
  steroidogenesis role
    adrenocorticotropic hormone induction of steroid synthesis, 128–129
    cholesterol transport and mutagenesis effects, 126–127, 137
    evidence, 124–125
    flunitrazepam inhibition of hormone synthesis, 125–126
    Ginkgo biloba extract effects on receptor expression, 129–130
    glial cell steroidogenesis role
      human, 133–134
      rat, 131–133
    hormonal regulation of receptor, 126
    3β-hydroxysteroid dehydrogenase activity regulation, 64–65
    targeted disruption effects, 127–128
  stress response, 130
  structure of complex
    adenine nucleotide carrier, 121
    atomic force microscopy, 123
    diazepam-binding inhibitor, 120
    modeling, 123–124
    PAP7, 122
    topography in mitochondrial membrane, 123
    transmembrane domains of core protein, 121
    two-hybrid screening, 121–122
    voltage-dependent anion channel, 121
PMS, see Premenstrual syndrome

Pregnancy
  dehydroepiandrosterone sulfate levels, 407
  GABA$_A$ response
    brain distribution of expression
      cerebral cortex, 225–228
      hippocampus, 225–228
      hypothalamic magnocellular neurons, 228–229, 231
      subunit expression, 226–229, 231
    chloride channel, 222
    expression levels, 220–221
    finasteride effects, 224
    3α-hydroxy-5α-pregnan-20-one levels, 220–221, 231, 405–407
  hypertension control by neurosteroids, 406–407
  peripheral-type benzodiazepine receptor levels, 130
  progesterone levels, 220–221, 405
  3α,5α-tetrahydrodeoxycorticosterone levels, 220–221
Pregnenolone
  anesthesia effects on levels, 427
  brain
    distribution, 4, 6–7
    levels, 426
  7-hydroxylation
    7α-hydroxylation, see P4507B1
    7β-hydroxylation in brain, 83–84
  tissue distribution of metabolism, 88, 90
  learning and memory effects
    aging studies, 286–287
    developmental effects, 281
    dose dependency, 289
    learning session administration effects, 285–286
    post-training administration effects, 282–284
    post-training versus preretention administration effects, 285
    pretraining administration effects, 282
    pretraining versus post-training effects, 284
    pretraining versus post-training versus preretention administration effects, 285
    summary of studies, 289–292
    time-dependent effects, 289
  metabolism in brain, 10
  nerve lesion induction, 14, 16

  neuroprotection with hemisoccinate, 387, 389
  peripheral nervous system distribution, 14
  premenstrual syndrome effects, 433–434
  synthesis, 7–8, 12, 11, 79–80
Pregnenolone sulfate
  γ-aminobutyric acid receptor interactions, 19
  behavioral effects in mice, 12
  brain levels, 426
  cognition studies in rat, 20, 22–23, 25
  conjugation in brain, 9–10
  glutamate receptor interactions, 19–20
  learning and memory effects
    aging rat performance and levels, 301–303
    aging studies, 286–287
    developmental effects, 281
    dose dependency, 289
    learning session administration effects, 285–286
    neurotransmitter system interactions, 294–295, 298–299
    overview, 329–333
    post-training administration effects, 282–284
    post-training versus preretention administration effects, 285
    pretraining administration effects, 282
    pretraining versus post-training effects, 284
    pretraining versus post-training versus preretention administration effects, 285
    summary of studies, 289–292
    time-dependent effects, 289
Premenstrual syndrome (PMS)
  alopregnanolone levels, 481, 483
  definition, 430–431, 480
  endocrine profile, 431–432, 480
  hormone replacement therapy effects, 432–433
  medroxyprogesterone acetate management, 441
  neurosteroid levels and symptom severity, 433–434
  oral contraceptive management, 434–436
  premenstrual dysphoric disorder comparison, 430, 480
  progesterone levels, 403–404, 481, 483

selective serotonin reuptake inhibitor
 treatment, 485–487
terminology, 430, 480
Progesterone, *see also* Oral contraceptives;
 Pregnancy
 anxiolytic properties, 325
 behavioral effects in mice, 12
 brain
  distribution, 4, 6–7
  levels, 424–426
  metabolism, 10, 90
  transport, 37–38
 epilepsy seizure threshold effects, 442,
  445, 447
 GABA$_A$ effects
  chloride current, 214–216, 219, 225
  expression response, 211–212, 214,
   224–225, 235
  mechanism of plasticity effects, 234–235
  overview, 19
  withdrawal effects
   behavioral response, 220
   ligand specificity of receptor, 217–219
   pseudo-pregnancy studies in rat,
    218–220
   subunit expression, 216–217, 219–220
 glial fibrillary acidic protein expression
  effects, 156–159
 hormone replacement therapy
  Alzheimer's disease, 429–430
  hot flush effects in menopause, 437–438
  mood change effects in menopause,
   438–439
  seizure prevention, 447
  sensory-motor and cognitive function
   effects, 427–429
  side effects, 439–441
 learning and memory effects, 288–289
 menopause levels, 409–410
 myelin effects in rats
  adult rats, 163
  aged rats, 162–163
  γ-aminobutyric acid receptor
   modulation, 166–167
  nerve repair, 163–165
  Schwann cell
   culture studies, 165–166
   proliferation effects, 169–170
 myelination promotion, 16–18, 25
 nerve lesion induction, 14, 16

neuronal effects, 149–150
overview of central nervous system
 function effects, 421–422
peripheral nervous system
 distribution, 14
pregnancy levels, 220–221, 405
premenstrual syndrome levels and effects,
 432–433, 481, 483
receptor
 brain distribution, 13, 25
 estradiol induction, 13
 modulation of transcription, 469–470
 tissue distribution, 2
stress effects
 acute stress
  brain concentration, 247–248
  GABAergic transmission studies,
   248, 250
  hypothalamic–pituitary–adrenal axis
   activation, 250–253, 261
  physiological role of response, 250
  plasma concentration, 247
 chronic stress
  adrenocorticotropic hormone
   response, 255–257
  brain concentration, 254–255, 260
  GABA$_A$ modulation, 260–261
  plasma concentration, 255, 260
  rodent models, 253
  social isolation effects, 253–255,
   260
 mechanisms, 261–264
synthesis in Schwann cells, 13
Progestogens, *see* Oral contraceptives
Puberty, neurosteroid synthesis, 402–403

R

Radioimmunoassay (RIA), neurosteroids,
 99, 399–400
5α-Reductase
 aging effects on expression, 161
 brain expression, 56–58, 148–150
 function, 44–45
 genes, 45, 147
 isoforms, 147, 150
 peripheral nervous system expresson,
  161–162
 progesterone response in cultured
  cells, 211

5α-Reductase (*continued*)
  regulation in central nervous system, 151–153
  substrate specificity, 147
Reinforcement
  ethanol self-admistration, 3α-hydroxy-5α-pregnan-20-one alterations, 359–361
  neurosteroid regulation, 337
RIA, *see* Radioimmunoassay

S

Schwann cell
  androgen effects, 167–169
  estrogen effects on proliferation, 169–170
  progesterone effects
    γ-aminobutyric acid receptor modulation, 166–167
    cell culture studies, 165–166
    proliferation, 169–170
Selective serotonin reuptake inhibitors (SSRIs)
  allopregnanolone interactions, 486–487
  premenstrual syndrome treatment, 485–487
Serotonin (5-HT) receptor
  5-HT$_{1B}$ role in aggression, 333–334
  5-HT$_3$
    function, 183
    steroid binding selectivity
      5-HT$_{3A}$, 183–184
      5-HT$_{3B}$, 184
    subunits, 182
Sigma receptor
  learning and memory role
    neurosteroid interactions, 298–299, 331
    overview, 294
  N-methyl-D-aspartate receptor interactions, 383
  steroid binding selectivity, 468
  subtypes, 383
Sleep, neurosteroid regulation, 335–336
SSRIs, *see* Selective serotonin reuptake inhibitors
StAR, *see* Steroidogenic acute regulatory protein
Steroidogenic acute regulatory protein (StAR)
  brain expression, 51–52, 136–137
  cholesterol transport role, 136
  congenital adrenal hyperplasia mutations, 136
  discovery, 136
  knockout mouse phenotype, 136
  neurosteroidogenesis regulation, 39–40
Stress
  anxiolytic properties of neurosteroids, 324–325
  definition, 322
  diazepam-binding inhibitor response, 262
  GABA$_A$
    anxiety behavior mediation, 245
    downregulation response, 244–246, 257, 259–261, 264
    hypothalamic–pituitary–adrenal axis activation role, 250–253
    transmission studies in acute stress, 248, 250
  peripheral-type benzodiazepine receptor response, 130
  stress effects on neuroactive steroids
    acute stress
      brain concentration, 247–248, 322–323
      GABAergic transmission studies, 248, 250
      hypothalamic–pituitary–adrenal axis activation, 250–253, 261, 323
      physiological role of response, 250
      plasma concentration, 247
    chronic stress
      adrenocorticotropic hormone response, 255–257
      brain concentration, 254–255, 260
      GABA$_A$ modulation, 260–261
      plasma concentration, 255, 260
      rodent models, 253
      social isolation effects, 253–255, 260
    mechanisms, 261–264
Stroke, neurosteroid neuroprotection
  dehydroepiandrosterone sulfate, 385–387, 391
  pregnenolone hemisoccinate, 387, 389
STS, *see* Sulfatase
Sulfatase (STS)
  brain expression, 60–61
  function, 47
  gene, 47

Sulfotransferase (HST)
  brain expression, 59–60
  function, 46–47
Synthesis, neurosteroids
  adrenodoxin reductase
    brain expression, 51
    function, 39
    genes, 39
  brain, 37–38, 380, 463–465
  cytochrome b$_5$
    adrenal distribution, 42
    brain expression, 53
    function, 41–42
  enzyme classification, 35, 37
  7α-hydroxylase
    function, 48–49
    tissue distribution, 49
  24-hydroxylase
    brain distribution, 48
    function, 47–48
  26α-hydroxylase, tissue distribution and expression regulation, 48
  3α-hydroxysteroid dehydrogenase
    brain expression, 58–59
    function, 45–46
    genes, 46
  3β-hydroxysteroid dehydrogenase
    activities, 40
    genes, 40
    nervous system expression, 53–54
    regulation of expression, 63–65
  11β-hydroxysteroid dehydrogenase
    brain expression, 56
    function, 44
    isoforms, 44
  17β-hydroxysteroid dehydrogenase
    brain expression, 54–55
    types and function, 43
  17-ketosteroid reductase, 43, 43
  P450 aro
    brain expression, 55–56
    function, 44
    gene, 44
  P450c11
    functions, 43
    genes, 42–43
    nervous system expression, 54
  P450c17
    activities, 40–41
    adrenal distribution, 41

  developmental regulation of expression, 52–53
  regulation of expression, 62–63
  P450c21, 42
  P450 reductase
    brain expression, 53
    function, 41
  P450scc
    brain expression, 49–50
    developmental regulation, 50–51
    genes, 39
    reaction rate, 38
    regulation of expression, 61–62
  pathway elucidation, 4, 35–37, 380–381
  prospects for study, 66
  5α-reductase
    brain expression, 56–58
    function, 44–45
    genes, 45
  steroidogenic acute regulatory protein
    brain expression, 51–52
    neurosteroidogenesis regulation, 39–40
  sulfatase
    brain expression, 60–61
    function, 47
    gene, 47
  sulfotransferase
    brain expression, 59–60
    function, 46–47

T

Testosterone
  epilepsy seizure threshold effects, 446
  metabolism in brain, 90
  5α-reductase activity effects, 153
  Schwann cell effects, 167–169
3α,5α-Tetrahydrodeoxycorticosterone
  antidepressant effects on brain levels, 472
  anxiolytic properties, 325
  feeding effects, 336–337
  GABA$_A$ receptor expression effects
    cortex, 221–222, 224–225, 227
    hippocampus, 221–222, 224–225, 227
    mechanisms, 234–235
  oral contraceptive response, 231–233
  pregnancy levels, 220–221
  receptor modulation of transcription, 469–470

3α,5α-Tetrahydrodeoxycorticosterone (*continued*)
  sleep effects, 335–336
  stress effects
    acute stress
      brain concentration, 247–248
      GABAergic transmission studies, 248, 250
      hypothalamic–pituitary–adrenal axis activation, 250–253, 261
      physiological role of response, 250
      plasma concentration, 247
    chronic stress
      adrenocorticotropic hormone response, 255–257
      brain concentration, 254–255, 260
      $GABA_A$ modulation, 260–261
      plasma concentration, 255, 260
      rodent models, 253
      social isolation effects, 253–255, 260
    mechanisms, 261–264
    therapeutic prospects, 471–474
TGF-β, *see* Transforming growth factor-β
Transforming growth factor-β (TGF-β), 5α-reductase inhibition, 152–153

V

Vascular dementia, cortisol levels, 412

CONTENTS OF RECENT VOLUMES

Volume 33

Olfaction
S. G. Shirley

Neuropharmacologic and Behavioral Actions of Clonidine: Interactions with Central Neurotransmitters
Jerry J. Buccafusco

Development of the Leech Nervous System
Gunther S. Stent, William B. Kristan, Jr., Steven A. Torrence, Kathleen A. French, and David A. Weisblat

$GABA_A$ Receptors Control the Excitability of Neuronal Populations
Armin Stelzer

Cellular and Molecular Physiology of Alcohol Actions in the Nervous System
Forrest F. Weight

INDEX

Volume 34

Neurotransmitters as Neurotrophic Factors: A New Set of Functions
Joan P. Schwartz

Heterogeneity and Regulation of Nicotinic Acetylcholine Receptors
Ronald J. Lukas and Merouane Bencherif

Activity-Dependent Development of the Vertebrate Nervous System
R. Douglas Fields and Phillip G. Nelson

A Role for Glial Cells in Activity-Dependent Central Nervous Plasticity? Review and Hypothesis
Christian M. Müller

Acetylcholine at Motor Nerves: Storage, Release, and Presynaptic Modulation by Autoreceptors and Adrenoceptors
Ignaz Wessler

INDEX

Volume 35

Biochemical Correlates of Long-Term Potentiation in Hippocampal Synapses
Satoru Otani and Yehezkel Ben-Ari

Molecular Aspects of Photoreceptor Adaptation in Vertebrate Retina
Satoru Kawamura

The Neurobiology and Genetics of Infantile Autism
Linda J. Lotspeich and Roland D. Ciaranello

Humoral Regulation of Sleep
Levente Kapás, Ferenc Obál, Jr., and James M. Krueger

Striatal Dopamine in Reward and Attention: A System for Understanding the Symptomatology of Acute Schizophrenia and Mania
Robert Miller

Acetylcholine Transport, Storage, and Release
Stanley M. Parsons, Chris Prior, and Ian G. Marshall

Molecular Neurobiology of Dopaminergic Receptors
David R. Sibley, Frederick J. Monsma, Jr., and Yong Shen

INDEX

Volume 36

Ca^{2+}, N-Methyl-D-aspartate Receptors, and AIDS-Related Neuronal Injury
  Stuart A. Lipton

Processing of Alzheimer Aβ-Amyloid Precursor Protein: Cell Biology, Regulation, and Role in Alzheimer Disease
  Sam Gandy and Paul Greengard

Molecular Neurobiology of the GABA$_A$ Receptor
  Susan M. J. Dunn, Alan N. Bateson, and Ian L. Martin

The Pharmacology and Function of Central GABA$_B$ Receptors
  David D. Mott and Darrell V. Lewis

The Role of the Amygdala in Emotional Learning
  Michael Davis

Excitotoxicity and Neurological Disorders: Involvement of Membrane Phospholipids
  Akhlaq A. Farooqui and Lloyd A. Horrocks

Injury-Related Behavior and Neuronal Plasticity: An Evolutionary Perspective on Sensitization, Hyperalgesia, and Analgesia
  Edgar T. Walters

INDEX

Volume 37

Section I: Selectionist Ideas and Neurobiology

Selectionist and Instructionist Ideas in Neuroscience
  Olaf Sporns

Population Thinking and Neuronal Selection: Metaphors or Concepts?
  Ernst Mayr

Selection and the Origin of Information
  Manfred Eigen

Section II: Development and Neuronal Populations

Morphoregulatory Molecules and Selectional Dynamics during Development
  Kathryn L. Crossin

Exploration and Selection in the Early Acquisition of Skill
  Esther Thelen and Daniela Corbetta

Population Activity in the Control of Movement
  Apostolos P. Georgopoulos

Section III: Functional Segregation and Integration in the Brain

Reentry and the Problem of Cortical Integration
  Giulio Tononi

Coherence as an Organizing Principle of Cortical Functions
  Wolf Singer

Temporal Mechanisms in Perception
  Ernst Pöppel

Section IV: Memory and Models

Selection versus Instruction: Use of Computer Models to Compare Brain Theories
  George N. Reeke, Jr.

Memory and Forgetting: Long-Term and Gradual Changes in Memory Storage
  Larry R. Squire

Implicit Knowledge: New Perspectives on Unconscious Processes
  Daniel L. Schacter

Section V: Psychophysics, Psychoanalysis, and Neuropsychology

Phantom Limbs, Neglect Syndromes, Repressed Memories, and Freudian Psychology
  V. S. Ramachandran

Neural Darwinism and a Conceptual Crisis in Psychoanalysis
  Arnold H. Modell

A New Vision of the Mind
  Oliver Sacks

INDEX

Volume 38

Regulation of GABA$_A$ Receptor Function and Gene Expression in the Central Nervous System
  A. Leslie Morrow

Genetics and the Organization of the Basal Ganglia
Robert Hitzemann, Yeang Olan, Stephen Kanes, Katherine Dains, and Barbara Hitzemann

Structure and Pharmacology of Vertebrate GABA$_A$ Receptor Subtypes
Paul J. Whiting, Ruth M. McKernan, and Keith A. Wafford

Neurotransmitter Transporters: Molecular Biology, Function, and Regulation
Beth Borowsky and Beth J. Hoffman

Presynaptic Excitability
Meyer B. Jackson

Monoamine Neurotransmitters in Invertebrates and Vertebrates: An Examination of the Diverse Enzymatic Pathways Utilized to Synthesize and Inactivate Biogenic Amines
B. D. Sloley and A. V. Juorio

Neurotransmitter Systems in Schizophrenia
Gavin P. Reynolds

Physiology of Bergmann Glial Cells
Thomas Müller and Helmut Kettenmann

INDEX

Volume 39

Modulation of Amino Acid-Gated Ion Channels by Protein Phosphorylation
Stephen J. Moss and Trevor G. Smart

Use-Dependent Regulation of GABA$_A$ Receptors
Eugene M. Barnes, Jr.

Synaptic Transmission and Modulation in the Neostriatum
David M. Lovinger and Elizabeth Tyler

The Cytoskeleton and Neurotransmitter Receptors
Valerie J. Whatley and R. Adron Harris

Endogenous Opioid Regulation of Hippocampal Function
Michele L. Simmons and Charles Chavkin

Molecular Neurobiology of the Cannabinoid Receptor
Mary E. Abood and Billy R. Martin

Genetic Models in the Study of Anesthetic Drug Action
Victoria J. Simpson and Thomas E. Johnson

Neurochemical Bases of Locomotion and Ethanol Stimulant Effects
Tamara J. Phillips and Elaine H. Shen

Effects of Ethanol on Ion Channels
Fulton T. Crews, A. Leslie Morrow, Hugh Criswell, and George Breese

INDEX

Volume 40

Mechanisms of Nerve Cell Death: Apoptosis or Necrosis after Cerebral Ischemia
R. M. E. Chalmers-Redman, A. D. Fraser, W. Y. H. Ju, J. Wadia, N. A. Tatton, and W. G. Tatton

Changes in Ionic Fluxes during Cerebral Ischemia
Tibor Kristian and Bo K. Siesjo

Techniques for Examining Neuroprotective Drugs *in Vivo*
A. Richard Green and Alan J. Cross

Techniques for Examining Neuroprotective Drugs *in Vitro*
Mark P. Goldberg, Uta Strasser, and Laura L. Dugan

Calcium Antagonists: Their Role in Neuroprotection
A. Jacqueline Hunter

Sodium and Potassium Channel Modulators: Their Role in Neuroprotection
Tihomir P. Obrenovich

NMDA Antagonists: Their Role in Neuroprotection
Danial L. Small

Development of the NMDA Ion-Channel Blocker, Aptiganel Hydrochloride, as a Neuroprotective Agent for Acute CNS Injury
Robert N. McBurney

The Pharmacology of AMPA Antagonists and Their Role in Neuroprotection
Rammy Gill and David Lodge

GABA and Neuroprotection
  Patrick D. Lyden

Adenosine and Neuroprotection
  Bertil B. Fredholm

Interleukins and Cerebral Ischemia
  Nancy J. Rothwell, Sarah A. Loddick,
  and Paul Stroemer

Nitrone-Based Free Radical Traps as Neuroprotective Agents in Cerebral Ischemia and Other Pathologies
  Kenneth Hensley, John M. Carney,
  Charles A. Stewart, Tahera Tabatabaie,
  Quentin Pye, and Robert A. Floyd

Neurotoxic and Neuroprotective Roles of Nitric Oxide in Cerebral Ischemia
  Turgay Dalkara and Michael A. Moskowitz

A Review of Earlier Clinical Studies on Neuroprotective Agents and Current Approaches
  Nils-Gunnar Wahlgren

INDEX

Volume 41

Section I: Historical Overview

Rediscovery of an Early Concept
  Jeremy D. Schmahmann

Section II: Anatomic Substrates

The Cerebrocerebellar System
  Jeremy D. Schmahmann and Deepak N. Pandya

Cerebellar Output Channels
  Frank A. Middleton and Peter L. Strick

Cerebellar-Hypothalamic Axis: Basic Circuits and Clinical Observations
  Duane E. Haines, Espen Dietrichs,
  Gregory A. Mihailoff, and E. Frank McDonald

Section III. Physiological Observations

Amelioration of Aggression: Response to Selective Cerebellar Lesions in the Rhesus Monkey
  Aaron J. Berman

Autonomic and Vasomotor Regulation
  Donald J. Reis and Eugene V. Golanov

Associative Learning
  Richard F. Thompson, Shaowen Bao, Lu Chen,
  Benjamin D. Cipriano, Jeffrey S. Grethe,
  Jeansok J. Kim, Judith K. Thompson,
  Jo Anne Tracy, Martha S. Weninger,
  and David J. Krupa

Visuospatial Abilities
  Robert Lalonde

Spatial Event Processing
  Marco Molinari, Laura Petrosini,
  and Liliana G. Grammaldo

Section IV: Functional Neuroimaging Studies

Linguistic Processing
  Julie A. Fiez and Marcus E. Raichle

Sensory and Cognitive Functions
  Lawrence M. Parsons and Peter T. Fox

Skill Learning
  Julien Doyon

Section V: Clinical and Neuropsychological Observations

Executive Function and Motor Skill Learning
  Mark Hallett and Jordon Grafman

Verbal Fluency and Agrammatism
  Marco Molinari, Maria G. Leggio, and
  Maria C. Silveri

Classical Conditioning
  Diana S. Woodruff-Pak

Early Infantile Autism
  Margaret L. Bauman, Pauline A. Filipek,
  and Thomas L. Kemper

Olivopontocerebellar Atrophy and Friedreich's Ataxia: Neuropsychological Consequences of Bilateral versus Unilateral Cerebellar Lesions
  Thérèse Botez-Marquard and Mihai I. Botez

Posterior Fossa Syndrome
  Ian F. Pollack

Cerebellar Cognitive Affective Syndrome
  Jeremy D. Schmahmann and Janet C. Sherman

Inherited Cerebellar Diseases
  Claus W. Wallesch and Claudius Bartels

Neuropsychological Abnormalities in Cerebellar Syndromes—Fact or Fiction?
  Irene Daum and Hermann Ackermann

Section VI: Theoretical Considerations

Cerebellar Microcomplexes
Masao Ito

Control of Sensory Data Acquisition
James M. Bower

Neural Representations of Moving Systems
Michael Paulin

How Fibers Subserve Computing Capabilities: Similarities between Brains and Machines
Henrietta C. Leiner and Alan L. Leiner

Cerebellar Timing Systems
Richard Ivry

Attention Coordination and Anticipatory Control
Natacha A. Akshoomoff, Eric Courchesne, and Jeanne Townsend

Context-Response Linkage
W. Thomas Thach

Duality of Cerebellar Motor and Cognitive Functions
James R. Bloedel and Vlastislav Bracha

Section VII: Future Directions

Therapeutic and Research Implications
Jeremy D. Schmahmann

Volume 42

Alzheimer Disease
Mark A. Smith

Neurobiology of Stroke
W. Dalton Dietrich

Free Radicals, Calcium, and the Synaptic Plasticity–Cell Death Continuum: Emerging Roles of the Trascription Factor NFκB
Mark P. Mattson

AP-I Transcription Factors: Short- and Long-Term Modulators of Gene Expression in the Brain
Keith Pennypacker

Ion Channels in Epilepsy
Istvan Mody

Posttranslational Regulation of Ionotropic Glutamate Receptors and Synaptic Plasticity
Xiaoning Bi, Steve Standley, and Michel Baudry

Heritable Mutations in the Glycine, $GABA_A$, and Nicotinic Acetylcholine Receptors Provide New Insights into the Ligand-Gated Ion Channel Receptor Superfamily
Behnaz Vafa and Peter R. Schofield

INDEX

Volume 43

Early Development of the *Drosophila* Neuromuscular Junction: A Model for Studying Neuronal Networks in Development
Akira Chiba

Development of Larval Body Wall Muscles
Michael Bate, Matthias Landgraf, and Mar Ruiz Gómez Bate

Development of Electrical Properties and Synaptic Transmission at the Embryonic Neuromuscular Junction
Kendal S. Broadie

Ultrastructural Correlates of Neuromuscular Junction Development
Mary B. Rheuben, Motojiro Yoshihara, and Yoshiaki Kidokoro

Assembly and Maturation of the *Drosophila* Larval Neuromuscular Junction
L. Sian Gramates and Vivian Budnik

Second Messenger Systems Underlying Plasticity at the Neuromuscular Junction
Frances Hannan and Yi Zhong

Mechanisms of Neurotransmitter Release
J. Troy Littleton, Leo Pallanck, and Barry Ganetzky

Vesicle Recycling at the *Drosophila* Neuromuscular Junction
Daniel T. Stimson and Mani Ramaswami

Ionic Currents in Larval Muscles of *Drosophila*
Satpal Singh and Chun-Fang Wu

Development of the Adult Neuromuscular System
Joyce J. Fernandes and Haig Keshishian

Controlling the Motor Neuron
*James R. Trimarchi, Ping Jin,
and Rodney K. Murphey*

What Neurological Patients Tell Us about the Use of Optic Flow
L. M. Vaina and S. K. Rushton

INDEX

Volume 44

Human Ego-Motion Perception
A. V. van den Berg

Optic Flow and Eye Movements
M. Lappe and K.-P. Hoffman

The Role of MST Neurons during Ocular Tracking in 3D Space
K. Kawano, U. Inoue, A. Takemura, Y. Kodaka, and F. A. Miles

Visual Navigation in Flying Insects
M. V. Srinivasan and S.-W. Zhang

Neuronal Matched Filters for Optic Flow Processing in Flying Insects
H. G. Krapp

A Common Frame of Reference for the Analysis of Optic Flow and Vestibular Information
B. J. Frost and D. R. W. Wylie

Optic Flow and the Visual Guidance of Locomotion in the Cat
H. Sherk and G. A. Fowler

Stages of Self-Motion Processing in Primate Posterior Parietal Cortex
F. Bremmer, J.-R. Duhamel, S. B. Hamed, and W. Graf

Optic Flow Analysis for Self-Movement Perception
C. J. Duffy

Neural Mechanisms for Self-Motion Perception in Area MST
R. A. Andersen, K. V. Shenoy, J. A. Crowell, and D. C. Bradley

Computational Mechanisms for Optic Flow Analysis in Primate Cortex
M. Lappe

Human Cortical Areas Underlying the Perception of Optic Flow: Brain Imaging Studies
M. W. Greenlee

Volume 45

Mechanisms of Brain Plasticity: From Normal Brain Function to Pathology
Philip. A. Schwartzkroin

Brain Development and Generation of Brain Pathologies
Gregory L. Holmes and Bridget McCabe

Maturation of Channels and Receptors: Consequences for Excitability
David F. Owens and Arnold R. Kriegstein

Neuronal Activity and the Establishment of Normal and Epileptic Circuits during Brain Development
John W. Swann, Karen L. Smith, and Chong L. Lee

The Effects of Seizures of the Hippocampus of the Immature Brain
Ellen F. Sperber and Solomon L. Moshe

Abnormal Development and Catastrophic Epilepsies: The Clinical Picture and Relation to Neuroimaging
Harry T. Chugani and Diane C. Chugani

Cortical Reorganization and Seizure Generation in Dysplastic Cortex
G. Avanzini, R. Preafico, S. Franceschetti, G. Sancini, G. Battaglia, and V. Scaioli

Rasmussen's Syndrome with Particular Reference to Cerebral Plasticity: A Tribute to Frank Morrell
Fredrick Andermann and Yvonne Hart

Structural Reorganization of Hippocampal Networks Caused by Seizure Activity
Daniel H. Lowenstein

Epilepsy-Associated Plasticity in gamma-Amniobutyric Acid Receptor Expression, Function and Inhibitory Synaptic Properties
Douglas A. Coulter

Synaptic Plasticity and Secondary Epileptogenesis
Timothy J. Teyler, Steven L. Morgan, Rebecca N. Russell, and Brian L. Woodside

Synaptic Plasticity in Epileptogenesis: Cellular Mechanisms Underlying Long-Lasting Synaptic Modifications that Require New Gene Expression
Oswald Steward, Christopher S. Wallace, and Paul F. Worley

Cellular Correlates of Behavior
Emma R. Wood, Paul A. Dudchenko, and Howard Eichenbaum

Mechanisms of Neuronal Conditioning
David A. T. King, David J. Krupa, Michael R. Foy, and Richard F. Thompson

Plasticity in the Aging Central Nervous System
C. A. Barnes

Secondary Epileptogenesis, Kindling, and Intractable Epilepsy: A Reappraisal from the Perspective of Neuronal Plasticity
Thomas P. Sutula

Kindling and the Mirror Focus
Dan C. McIntyre and Michael O. Poulter

Partial Kindling and Behavioral Pathologies
Robert E. Adamec

The Mirror Focus and Secondary Epileptogenesis
B. J. Wilder

Hippocampal Lesions in Epilepsy: A Historical Review
Robert Naquet

Clinical Evidence for Secondary Epileptogensis
Hans O. Luders

Epilepsy as a Progressive (or Nonprogressive "Benign") Disorder
John A. Wada

Pathophysiological Aspects of Landau-Kleffner Syndrome: From the Active Epileptic Phase to Recovery
Marie-Noelle Metz-Lutz, Pierre Maquet, Annd De Saint Martin, Gabrielle Rudolf, Norma Wioland, Edouard Hirsch and Chriatian Marescaux

Local Pathways of Seizure Propagation in Neocortex
Barry W. Connors, David J. Pinto, and Albert E. Telefeian

Multiple Subpial Transection: A Clinical Assessment
C. E. Polkey

The Legacy of Frank Morrell
Jerome Engel, Jr.